# RNAi

## A Guide to Gene Silencing

# RNAi

## A Guide to Gene Silencing

EDITED BY

## Gregory J. Hannon

*Cold Spring Harbor Laboratory*

COLD SPRING HARBOR LABORATORY PRESS
Cold Spring Harbor, New York

# RNAi
A Guide to Gene Silencing

| | |
|---|---|
| **Publisher** | John Inglis |
| **Commissioning Editors** | John Inglis and David Crotty |
| **Project Manager and** | |
| **  Developmental Editor** | Judy Cuddihy |
| **Developmental Editor** | Michael Zierler |
| **Project Coordinator** | Mary Cozza |
| **Permissions Coordinator** | Nora Rice |
| **Production Manager** | Denise Weiss |
| **Production Editor** | Dorothy Brown |
| **Desktop Editor** | Susan Schaefer |
| **Cover Designer** | Ed Atkeson |

**Front Cover:** *(Background)* Petunia flower resulting from attempted high-level overexpression of chalcone synthase, a key enzyme in anthocyanin pigment biosynthesis. This particular pigment pattern requires overexpression of a sense transgene, so it is characteristic of the sense cosuppression mode of RNAi. (Courtesy of Richard A. Jorgensen, University of Arizona.) *(Inset: Left panel)* C. elegans ubiquitously expressing a green fluorescent protein (GFP) reporter construct. *(Right panel)* C. elegans carrying the same construct shown after feeding with bacteria that express dsRNA directed against GFP. The GFP signal is lost throughout the animal (some neurons in the head region are resistant to this effect). (Courtesy of Ketting et al., Chapter 3, this volume.)

Library of Congress Cataloging-in-Publication Data

RNAi : a guide to gene silencing / edited by Gregory J. Hannon.
   p. cm.
  Includes bibliographical references and index.
  ISBN 0-87969-641-9 (hardcover: alk. paper) -- ISBN 0-87969-704-0 (paperback: alk. paper)
  1. Gene silencing. 2. RNA. I. Hannon, Gregory J., 1964–

QH450.R628 2003
572.8′65--dc21                                    200305087

10  9  8  7  6  5  4  3  2  1

All Cold Spring Harbor Laboratory Press publications may be ordered directly from Cold Spring Harbor Laboratory Press, 500 Sunnyside Blvd., Woodbury, N.Y. 11797-2924. Phone: 1-800-843-4388 in Continental U.S. and Canada. All other locations: (516) 422-4100. FAX: (516) 422-4097. E-mail: cshpress@cshl.edu. For a complete catalog of all Cold Spring Harbor Laboratory Press publications, visit our World Wide Web Site http://www.cshlpress.com.

# Contents

# Preface

MY FIRST EXPOSURE TO RNA INTERFERENCE (RNAi) came at a Pew Scholars meeting in Puerta Vallarta, Mexico in 1998. Craig Mello gave a short, informal talk about a remarkable observation—exposing *Caenorhabditis elegans* to double-stranded RNA (dsRNA) caused potent and sequence-specific silencing of homologous genes. Beyond this, the response to dsRNA had a number of nearly unbelievable properties. Silencing occurred not only in cells directly exposed to the dsRNA, but also spread throughout the organism and into its progeny. We were all confounded and amazed by this seemingly inexplicable set of observations.

I am fairly certain that, at the time, Craig had not yet realized the extent to which his observations would take hold of biology to create an exciting new field, which at the time of the writing of this book, seems to be expanding at an exponential rate. Shortly after the original report of RNAi, Rich Carthew, another Pew Scholar, extended many of Andrew Fire's and Mello's basic observations about the ability of dsRNA to suppress gene expression into *Drosophila* embryos. Biochemical and genetic studies led to the realization of a common silencing response that extended throughout eukaryotes from plants (where the phenomena were really first noted) to fungi to animals.

Since its inception, the study of RNAi has proceeded along two parallel tracks. The first is an effort to understand the biology of this response—its mechanism and its role in the organism. The second has been the desire to develop RNAi as an experimental and possibly therapeutic tool. This duality has made the planning and construction of this book a challenge. In the end, it was decided that the volume should provide both an introduction to the biology of RNAi and practical advice, including detailed protocols, on its application in numerous systems.

The work described herein represents a field that is moving at an astonishing pace. Therefore, I am all the more grateful to the authors, who took time from their very busy schedules to contribute to this volume. This book would also not have been possible without the constant support (read nagging and prodding) of the people at Cold Spring Harbor Laboratory Press, namely, Mary Cozza, Judy Cuddihy, David Crotty, and John Inglis. I am also indebted to those colleagues in the field who have provided encouragement and support of our work over the past several years. I am especially thankful for the talented group of graduate students and postdoctoral fellows including Scott Hammon, Emily Bernstein, Yvette Seger, Amy Caudy, Ahmert Denli, Jian Du, Izabela Sujka, Jose Silva, Patrick Paddison, Doug Conklin, and Michelle Carmell. I am grateful to my family, Gretchen, Will, and Claire, for their support and tolerance of the time that this endeavor has consumed.

**Greg Hannon**

# Introduction

## Gregory J. Hannon

*Watson School of Biological Sciences, Cold Spring Harbor Laboratory, Cold Spring Harbor, New York 11724*

THE FIELD OF INQUIRY THAT FOCUSES ON RNA INTERFERENCE (RNAi) was not born out of the desire to answer a specific biological question. Instead, it coalesced around an observation by Craig Mello and Andrew Fire that double-stranded RNA (dsRNA) could induce a sequence-specific silencing response in *Caenorhabditis elegans*. This observation led immediately to the hypothesis that a set of homology-dependent silencing mechanisms might be related as responses to a common silencing trigger, namely, dsRNA. This hypothesis could certainly explain virus-induced gene silencing and some types of transgene cosuppression, particularly those caused by the presence of complex transgene arrays. However, it was (and remains) less clear how other types of homology-dependent silencing responses, for example, copy-number-dependent silencing of unlinked transgenes, might trigger a generic RNAi pathway.

As the field has moved forward, other types of homology-dependent regulatory mechanisms have been drawn into the fold. For example, small temporal RNAs, discovered approximately 10 years ago, are now known to regulate their protein-coding targets via the RNAi machinery. The small temporal RNAs are now recognized as the founding members of a large class of similarly structured noncoding RNAs, most with unknown cellular roles. Maintenance of facultative heterochromatin at the fission yeast centromere requires RNAi, drawing a parallel between dsRNA-mediated phenomena in fungi and the previously demonstrated ability of dsRNA to induce heritable silencing in plants by triggering alterations in chromatin structure. The notion that dsRNA could induce epigenetic change via the RNAi machinery by acting on the genome was cited as the 2002 "Breakthrough of the Year" by *Science*. However, it is not yet clear to what extent the ever-widening net of RNAi will also be cast over other epigenetic phenomena to include, for example, imprinting and X-chromosome inactivation and epigenetic memory during development.

During the past 5 years, tremendous progress has been made in lifting the veil that hid the underlying mechanism of RNAi and related silencing responses. The molecular underpinnings of the basic aspects of the silencing process have emerged from a combination of biochemical and genetic studies in several model systems. However, as is reflected in the content of this volume, we have really only reached the first level in our understanding of how dsRNA directs the suppression of homologous genes. Specifically, we have identified a number of the participants in the silencing process. However, the precise mechanisms by which they function remain unknown. For example, it is now clear that short interfering RNAs (siRNAs) direct the silencing machinery to its targets, but we have little understanding of how the transcriptome—or for that matter perhaps the genome—is effectively searched. We do not even know precisely which proteins contact the siRNAs within the RNA-induced silencing complex (RISC) or which components of this complex represent its catalytic and functional core.

We do have a basic understanding of how siRNAs are produced, as this seems to occur through the action of a single enzyme, Dicer. However, we do not yet know how Dicer might be regulated, how it recognizes dsRNA substrates with different structures, nor how it might facilitate the incorporation of siRNAs into effector complexes.

The identification of components of the RNAi machinery has led to a number of significant advances. Through the use of genetics, a number of remarkable biological functions for this gene-silencing pathway have been uncovered. One of the key functions of the RNAi machinery in plants was foreshadowed by the observation that infection of a plant with an RNA virus that included a fragment of an endogenous gene caused silencing of that gene. It is now clear that RNAi viruses are combated to a large degree in plants by posttranscriptional gene silencing (PTGS; the plant nomenclature for the RNAi response). Notably, many plant viruses encode proteins that antagonize PTGS, and these proteins are essential for viral replication. Such virulence determinants become dispensable in plants harboring mutations that compromise PTGS, creating a beautiful interlock of genetic experiments that argue strongly for the importance of PTGS as the major antiviral defense in plants.

Genetics studies of RNAi in *C. elegans* have also revealed a role for RNAi in the control of endogenous parasitic nucleic acids. A distinct subset of RNAi-resistant mutants release transposons from repression, creating a mutator phenotype. Perhaps not unrelated are roles in managing repetitive elements at the *Schizosaccharomyces pombe* centromere, where an intact RNAi machinery is essential for formation of heterochromatin, which is in turn required for proper centromere function. In *Tetrahymena*, preliminary evidence, namely, the presence of small RNAs and the requirement for an Argonaute family member, suggests that this organism's strategy for managing repetitive elements also depends on an RNAi-like mechanism. Transposon-like sequences are eliminated from the transcriptionally active macronucleus of *Tetrahymena* in a process that seems to be determined by small RNAs that guide histone methylation.

Considered together, evidence that has emerged over the past few years has perhaps suggested an ancestral function for RNAi in controlling parasitic and pathogenic nucleic acids. This function may have evolved into a more general mechanism for managing repetitive elements. However, it has also become clear that the RNAi machinery is intimately involved in the regulation of endogenous, protein-coding genes, particularly those that control development and possibly also stem cell maintenance. This conclusion grew largely out of the observed phenotypes of Dicer mutants, which connected RNAi to previously characterized small RNAs, the stRNAs, that regulate developmental timing. Since that time, many similarly structured, endogenous nonconding RNAs have joined this family, which are now known collectively as microRNAs or miRNAs. Presently, the functions of almost all of these are unknown.

In addition to a wealth of biological insight, mechanistic studies of RNAi have revealed how to exploit the dsRNA response as a tool for experimental biology with ever-increasing effectiveness. Tools based on RNAi have in some cases begun to revolutionize experimental biology. Ironically, in many ways, the discovery of RNAi itself grew out of attempts to design an approach to performing reverse genetics in *C. elegans*, specifically the use of antisense RNA. Now the phenomenon has come full circle, having taken experimentalists working in numerous model systems by storm and with RNAi largely supplanting the use of antisense technologies by experimental biologists. Of course, the use of RNAi has been *de rigueur* in *C. elegans* for several years, but more recently, the use of dsRNA as a silencing tool has been extended into traditional genetic models such as *Drosophila*, into nontraditional models such as trypanosomes, and into mammalian systems.

With *C. elegans*, the development of genome-wide libraries of RNAi inducers is now permitting investigators to identify nearly every gene required for a specific biological phenotype in a rapid and efficient fashion. Similar libraries are on the horizon for *Drosophila*, trypanosomes, and mammals. Although these libraries will never usurp the position of traditional genetics, they offer a revolutionary capability in somewhat genetically intractable models such as mammalian cells. Furthermore, in many experimental systems, including *C. elegans* and mice, RNAi complements traditional genetic technologies by offering a much more rapid and cost-effective way to mimic the effects of hypomorphic mutations in living animals.

Given the rapid pace of advance in this field, it is difficult to predict a future that will still be the future by the time this volume is published. However, given recent advances in the use of RNAi to engineer heritable silencing in mammals, to alter stem cells for organ reconstitution, and to alter the course of disease in model systems, the transition of RNAi from a tool that permits the analysis and modeling of disease to one that provides a therapy for disease must be anticipated. Although significant hurdles remain to be overcome, the course to therapeutic application of RNAi has already been mapped out by proof-of-principle studies in the literature. These include notable successes in blocking viral replication in cultured cells, as well as studies with HIV and with poliovirus. In the former case, a combination of in vitro studies with the demonstrated ability to stably engineer hematopoietic stem cells may provide the means to immunize cells against HIV infection via autologous transplantation. In vivo, RNAi has been shown to antagonize the effects of hepatitis C virus in mice.

In addition to its use as a therapeutic tool, the RNAi machinery itself may be altered in human disease. Recently, mechanistic studies have revealed potential connections between the Fragile X syndrome and the RNAi machinery. More broadly, genetic and biochemical analyses in *Drosophila* and more recently zebrafish have suggested potential roles for the RNAi machinery in the localization and storage of mRNAs, which can be released upon some stimulus (e.g., neuronal stimulation or fertilization of an egg) to permit regulated protein synthesis. Finally, a consideration of genetic data from plants and *Drosophila* and expression data from mammals suggests a central role for the RNAi machinery in maintenance of stem cell identity. Given all of these suggestions, it seems likely that not only will RNAi be harnessed as a therapeutic tool, but also the RNAi machinery itself may become a therapeutic target.

The extent to which studies of the RNAi machinery have infiltrated biology as a whole is reflected in the diversity of chapters in this volume. This book contains many chapters that are devoted to the application of RNAi as a tool in numerous biological systems. These have been designed as practical guides, discussing both the strengths and weaknesses of using RNAi in a given setting and providing detailed protocols. A number of chapters that illustrate the diversity of silencing processes that are triggered by dsRNA and introduce their biological relevance have also been included. Additional general background is provided in key areas that are essential to understanding RNAi, such as epigenetics and the enzymology of the ribonuclease III family. Given this balanced organization, it is hoped that this volume can provide a starting point to those interested in using RNAi as a tool while also providing a strong background in the biology of this gene silencing response.

# CHAPTER 1

# Sense Cosuppression in Plants: Past, Present, and Future

Richard A. Jorgensen

*Department of Plant Sciences and Interdisciplinary Program in Genetics, University of Arizona, Tucson 85721-0036*

COSUPPRESSION IS DEFINED OPERATIONALLY as the simultaneous reduction in expression of a transgene and homologous, endogenous genes, in which the cosuppressed state is identified by comparison with an opposite state of coexpression of both transgene and homologous endogenes (Napoli et al. 1990; van der Krol et al. 1990). Typically, the coexpression state is observed either in a different physiological or developmental context from the cosuppressed state or as an epigenetic revertant of the cosuppressed state. Because the definition of cosuppression refers to an outcome (*not* to a mechanism), it may be applied correctly to both transcriptional and posttranscriptional silencing (although it has been most commonly used to refer to posttranscriptional silencing).

It was recognized from the outset that cosuppression is phenomenologically distinct from paramutation (and paramutation-like examples of transgene silencing), but that the mechanisms underlying these two phenomena might nevertheless be related (Napoli et al. 1990). Paramutation is the imposition of an epigenetic silencing state on a gene by another silent allele (or homologous gene) that persists after segregation of the causative allele (or gene). Because the paramutagenic (i.e., causative) gene is silent prior to the interaction, the interacting alleles or genes are never observed to be coexpressed, and so cannot properly be referred to as cosuppressed in the silent state. Generally, the cosuppression state is completely reversible for all interacting genes after segregation or epigenetic reversion, whereas paramutation, by definition, can be maintained without the continuing presence of the causative allele in sexual progeny, although perhaps only in some progeny or only for short times.

Over time, the term "cosuppression" came to be used inappropriately to refer to almost any example of transgene silencing in plants, including many examples in which no corresponding coexpression state was demonstrated, and so, through misuse, it fell into disuse. It came to be replaced in the plant literature by more mechanistic terms, such as transcriptional gene silencing (TGS) and posttranscriptional gene silencing (PTGS), both of which could encompass interactions between homologous genes that result in either cosuppression or paramutation, as well as other examples of silencing not even dependent on gene duplication. PTGS now seems to be giving way to the term "RNA silencing," which is more accurate, because the primary target for silencing is a gene's transcript rather than the gene itself (i.e., the gene remains fully active at the transcriptional level). The original meaning of the term cosuppression was revived, and applied in animals, in an influential review by Birchler et al. (1999), and as a result, cosuppression is now widely used to refer to RNA silencing of transposable elements in animals.

Many recent reviews address RNA silencing in plants (Hammond et al. 2001; Matzke et al. 2001; Vance and Vaucheret 2001; Vaucheret et al. 2001; Voinnet 2001; Waterhouse et al. 2001). To not merely reiterate the content of these excellent reviews in this chapter, the principal goals here are to (1) address the nature, mechanism, and implications of "sense cosuppression," a type of cosuppression that is caused by a sense transgene engineered for overexpression of an endogenous protein; (2) provide an historical context for the data and concepts that are relevant to understanding sense cosuppression; and (3) consider how sense cosuppression might have a role in normal posttranscriptional control of supracellular patterns of gene expression and information processing in plants. The important distinction between sense cosuppression and RNA interference (RNAi) for the purposes of this chapter is that the latter is triggered by double-stranded RNA (dsRNA) molecules injected into a cell or produced by transcription of inversely repeated DNA sequences in the genome, whereas the former is initiated by a cellular RNA-directed RNA polymerase (RdRP) that uses sense transcripts as templates to produce the dsRNA molecules that trigger RNA silencing.

## EARLY CONCEPTS FOR MECHANISMS OF HOMOLOGY-BASED GENE SILENCING IN PLANTS

At least three broad classes of hypotheses were proposed to explain cosuppression in the first years after its discovery:

- A sensitive response to exceeding some threshold of transcript production (van der Krol et al. 1990; Lindbo et al. 1993; Meins and Kunz 1994).

- Unintended production of antisense RNA, e.g., by readthrough transcription from neighboring promoters (Grierson et al. 1991; Mol et al. 1991).

- Homology-based ectopic pairing causing a perturbation in chromatin state or nuclear localization and resulting in altered transcript fate or transcription rate (Jorgensen 1990; Napoli et al. 1990).

For a detailed review of the early data and concepts, see Jorgensen (1992). Because no single hypothesis seemed able to account for all the data, many participants in the field believed that multiple mechanisms were likely to exist. In time, it came to be believed that promoter homology-based silencing, which is transcriptional, was most likely based on an ectopic pairing (DNA:DNA interaction) mechanism, whereas transcript homology-based silencing, which is posttranscriptional, involved either aberrant transcripts or a threshold response to excessive accumulation of transcripts (for review, see Matzke and Matzke 1995). However, arguments persisted for a DNA pairing-initiated process resulting in a cytoplasmic posttranscriptional state (Jorgensen 1992; Flavell 1994; Baulcombe and English 1996; Que et al. 1997; Stam et al. 1998).

The early data on homology-based gene silencing in plants were based on a variety of types of transgene constructs introduced into different plant species. These diverse experimental systems often produced apparently conflicting results, making it very difficult to draw general conclusions. Thus, it became important to focus on model systems. Three plant species led to significant progress in understanding the complexity of RNA-silencing phenomenology and mechanisms in plants. Due to its long history as an excellent model system in both plant virology and transgenesis, tobacco (and some of its wild relatives) has been particularly useful for RNA-silencing experiments involving viruses, as well as for investigating systemic movement of RNA-silencing signals. *Arabidopsis* has been especially useful for genetic analysis, in particular, the identification of genes necessary for RNA silencing, and will be the key model system for most future efforts.

Petunia had a significant early role because it offered a convenient visible phenotype, flower pigmentation, which was useful for defining different modes of silencing and detecting and monitoring infrequent epigenetic events. Results from each system are described here to the extent that they relate to sense cosuppression.

## COPY RNA AND THE INVOLVEMENT OF AN RNA-DEPENDENT RNA POLYMERASE

A major conceptual breakthrough came from the recognition by plant virologists that sense transgenes expressing RNA homologous to a viral genome conferred resistance to that virus, even if the transcript produced no protein product, i.e., the transgene-induced resistance was RNA-mediated (van der Vlugt et al. 1992; Lindbo et al. 1993). Measurement of the transcription rates of different transgenes showed that the highest rates occurred in plants that were resistant to homologous viruses, leading to the conclusion that RNA-mediated resistance might be a threshold phenomenon (Lindbo et al. 1993; Smith et al. 1994; Mueller et al. 1995; English et al. 1996; Goodwin et al. 1996).

To explain threshold-dependent, RNA-mediated silencing, Dougherty and colleagues proposed involvement of a plant-encoded RdRP that would copy a small segment or segments of an RNA that had accumulated to unacceptably elevated levels. These small RNAs would then hybridize with the target RNA, rendering the RNA nonfunctional, and RNases would target the partially double-stranded messenger or viral RNA complex for degradation. A system mediated by RNA has appeal in its relative simplicity, specificity, and the limited amount of genetic information required (Lindbo et al. 1993).

Dougherty and Parks (1995) noted that gene specificity could be provided by RNA molecules as short as 18 bp or so, a suggestion that hit remarkably close to the mark: Guide RNAs that direct RNA silencing via the RNA-induced silencing complex (RISC) are now known to be approximately 21–22 nucleotides long. The proposed involvement of an RdRP enzyme has received strong support from the subsequent discovery that genes homologous to a tomato RdRP are required for RNA silencing in plants, animals, and fungi, although it still remains to be shown whether the proteins encoded by these genes have RdRP activity. In addition, the involvement of a dsRNase, known as Dicer, has now been demonstrated. The evidence for guide RNAs, RdRP homologs, and dsRNases has been reviewed by Hannon (2002) and elsewhere in this volume.

From an historical perspective, it is interesting to note that the first suggestion that a dsRNase might be involved in homology-based gene silencing was actually made by Cameron and Jennings (1991), after observing silencing of one transgene by a homologous transgene in cultured animal cells. These authors suggested that at high concentrations, short complementary regions of sense RNA molecules that normally pair intramolecularly might participate in intermolecular interactions, creating aggregates that might be subject to recognition by a dsRNase.

## TWO MODES OF INDUCTION OF RNA SILENCING BY SENSE OVEREXPRESSION TRANSGENES

Recognition that sense transgenes can cause two distinct modes of RNA silencing came from analyses of the frequencies and patterns of chalcone synthase silencing triggered by different types of transgene constructs in petunia flowers (Que et al. 1997). First, silencing of chalcone synthase genes occurred at strikingly different efficiencies, depending on the nature of the transgene construct: A high frequency of silencing (80% of trans-

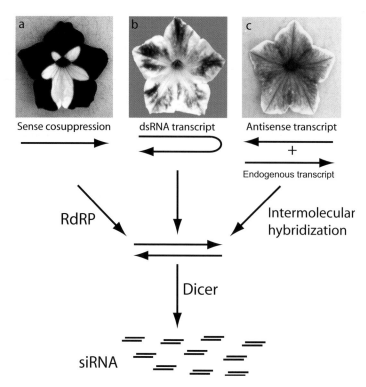

**FIGURE 1.1.** Modes of dsRNA-mediated silencing of chalcone synthase in petunia induced by (*a*) sense overexpression transgene, (*b*) inverted repeat transgene, and (*c*) antisense transgene. (*a* and *c* reprinted, with permission, from Que et al. 1998 © Blackwell Science.)

genotes) was obtained with constructs engineered for high-level overexpression of chalcone synthase protein (Napoli et al. 1990; Jorgensen et al. 1996), whereas a low frequency of silencing (5–15%) was observed with sense constructs not designed for protein overexpression (van der Krol et al. 1990; van Blokland et al. 1994). The former type of construct also resulted in a greater diversity of phenotypes, ranging from highly ordered, morphology-based, flower color patterns to disordered complex patterns. Single-copy or dispersed transgenes were associated with ordered patterns, whereas inverted repeat transgenes (a very common arrangement of *Agrobacterium*-transferred DNA [T-DNA] molecules) were associated with complex, disordered patterns (Jorgensen et al. 1996; Que et al. 1997). Examples of the patterns produced by these two types of sense transgenes are shown in Figure 1.1 (and are also contrasted with the pattern of silencing that is typical of an antisense construct). Importantly, constructs that were not engineered for protein overexpression caused RNA silencing *only* when the transgene was integrated into the genome as an inverted repeat (Que et al. 1997; Stam et al. 1997).

Two modes of induction of RNA silencing by sense transgenes were thus proposed: one initiated only by sense overexpression transgenes (sense cosuppression) and the other initiated by inverted repeat integrants, regardless of how the transgene was engineered (Que et al. 1997). Further experimental support for the existence of these two modes has been provided by Dalmay et al. (2000b) and Beclin et al. (2002), as discussed in a review by Vance and Vaucheret (2001). Distinct features of the two modes of silencing that can be induced by sense overexpression transgenes are summarized in Table 1.1 and are discussed below.

---

**TABLE 1.1.** Two modes of RNA silencing induced by sense expression transgenes in plants

1. dsRNA Transcript-mediated Silencing
   - Requires inverted repeat integrant.
   - Does not require strong promoter.
   - Premature nonsense codons do not block silencing.
   - Does not require RNA-dependent RNA polymerase homolog (in *Arabidopsis*).

2. Sense Cosuppression
   - Requires strong promoter driving sense transcript.
   - Single-copy transgene is sufficient.
   - Premature nonsense codons reduce silencing dramatically.
   - Requires RNA-dependent RNA polymerase homolog (in *Arabidopsis*).

---

## dsRNA Transcripts as Triggers of RNA Silencing

The first indication that dsRNA might be a trigger of RNA silencing in plants came from experiments by Waterhouse et al. (1998), who reported that plants carrying both a sense and an antisense transgene homologous to a viral genome were much more resistant to that virus than were plants carrying either a sense or an antisense transgene alone. Hypothesizing that this might be a reflection of involvement of dsRNA in triggering RNA silencing, Waterhouse et al. engineered transgenes with transcribed inverted repeats homologous to the target RNA to produce dsRNA by intramolecular base pairing and found that such constructs were much more efficient inducers of RNA silencing than were sense or antisense transgenes; such constructs producing self-complementary transcripts are now widely used for functional genomics in plants. These experiments immediately suggested that plant transgenes that had integrated as inverted repeats and that were exhibiting RNA silencing were likely to be doing so by means of dsRNA produced by readthrough transcription from one T-DNA copy into another.

The independent parallel discoveries of dsRNA as a trigger of RNA silencing in plants and dsRNA as the trigger of RNAi in worms (Fire et al. 1998) suggested the possibility of a common underlying mechanism. Within a few years, it was demonstrated that plants, animals, and fungi all required several similar genes for dsRNA-mediated silencing (including an RdRP homolog), proving the existence of a common underlying mechanism (for review, see Hammond et al. 2001). Elucidation of the molecular basis of dsRNA-mediated degradation of homologous transcripts in animals, plants, and fungi is progressing rapidly, as can be seen throughout this volume, and it is broadly accepted that dsRNA is the ultimate trigger for this process, even if not the initial trigger in cases such as sense cosuppression.

## How Do Single-copy Sense Transgenes Act as Triggers of RNA Silencing?

Key features of a transgene construct that determine whether a single-copy sense transgene can produce RNA silencing in plants are (1) a strong promoter and (2) the absence of a premature nonsense codon in the protein-coding sequence (Que et al. 1997). This suggests that a high concentration of translatable sense transcript must be produced for a single-copy transgene to trigger cosuppression.

It was proposed that single-copy sense transgenes only cause RNA silencing when they are integrated in such a way that antisense transcripts are produced by readthrough from adjacent promoters (Grierson et al. 1991; Mol et al. 1991; Fire and Montgomery 1998). This hypothesis seems inconsistent with the fact that nearly all single-copy transgene integrants are able to trigger RNA silencing if the transgene construct was engineered for overexpression (Jorgensen et al. 1996). In addition, introduction of an early nonsense codon to the coding sequence drastically reduces the ability of single-copy transgenes to trigger silencing (Que et al. 1997). It seems unlikely that a change in several base pairs at a single location in a 1.1-kb transcript (to introduce a nonsense codon) would affect the ability of an antisense readthrough transcript to form dsRNA and trigger silencing, unless one accepts that a slightly lower accumulation of an early nonsense transcript (i.e., the "threshold" hypothesis) could be responsible for its failure to cause silencing. Additional observations inconsistent with a frequent role for antisense readthrough transcripts in sense cosuppression come from a series of constructs expected to be subject to readthrough antisense transcription into chalcone synthase from an adjacent selective marker (Que and Jorgensen 1998). No increase in silencing was observed, contrary to expectation in the antisense readthrough hypothesis.

More direct support for the conclusion that single-copy sense transgenes produce a distinct mode of silencing comes from the demonstration that a single-copy antisense transgene (which carries multiple nonsense codons in its transcript) does not produce the ordered pattern of silencing produced by an allelic sense transgene created by Cre/lox-mediated inversion of the coding sequence (Que et al. 1998). Instead, single-copy antisense transgenes reduce pigmentation quantitatively throughout the corolla of the flower (Figure 1.1c). Presumably, the antisense transcript acts by pairing with the endogenous sense transcript to interfere with its translation and/or to produce dsRNA leading to RNA degradation. Finally, removal of nonsense codons from an antisense chalcone synthase transgene produces patterns characteristic of sense transgenes (superimposed on the typical antisense pattern), indicating that an antisense transcript has the capacity to trigger sense cosuppression if translation is not terminated by early nonsense codons (N. Doetsch and R. Jorgensen, unpubl.).

Once a role for dsRNA was suggested by the experiments of Waterhouse et al. (1998), the RdRP that had been proposed by Dougherty and colleagues to copy sense transcripts into cRNA molecules came to be viewed as the source of dsRNA needed for RNA silencing. From this perspective, recognition of a sense transcript to serve as a template for the RdRP enzyme would be the initial event in the sense cosuppression mode of silencing. An RdRP homolog in *Arabidopsis* (SDE1/SGS2) was shown to be required for (certain examples of) transgene-induced RNA silencing, but not for RNA virus-induced RNA silencing (Dalmay et al. 2000b; Mourrain et al. 2000). The likely interpretation of these observations is that because viruses replicate via a dsRNA intermediate produced by a virus-encoded RdRP enzyme, a plant-encoded RdRP is not needed to trigger RNA silencing of replicating RNA viruses, whereas in the case of silencing that is initiated by a sense transgene, it is.

Two important features of RNA silencing in plants are (1) the ability to maintain silencing after removal of the source of the initiator RNA (Lindbo et al. 1993; Palauqui and Vaucheret 1998; Ruiz et al. 1998; Voinnet et al. 1998) and (2) spreading of RNA targeting from the region homologous to the initiator RNA to adjacent regions, both 5′ and 3′ to the region of homology (Voinnet et al. 1998). Investigation of the relationship between maintenance and spreading in *Arabidopsis* showed both processes to be dependent on transcription of the target gene, as well as on the presence of functional SDE1/SGS2, suggesting that maintenance involves production of dsRNA by SDE1/SGS2 using the entire length of the target RNA as template (Dalmay et al. 2000b; Vaistij et al. 2002). Thus, the role of RdRP in sense cosuppression appears to be twofold: (1) to initiate silencing by converting sense transcripts to dsRNA substrates for the DICER nuclease and (2) to produce cRNA molecules for incorporation into the RISC complex for targeting homologous transcripts.

## Complications for Interpretation of the Plant Literature

Given the existence of two modes of RNA silencing in transgenic plants, it is clear that complications may arise when a transgene is engineered for overexpression and is then integrated in the plant genome as an inverted repeat. Thus, inverted repeat transgene complexes can trigger both modes of RNA silencing simultaneously, as well as alternately during the development of a plant: For example, some branches produce flower color patterns typical of single-copy chalcone synthase overexpression transgenes, whereas other branches of the plant produce flowers with no silencing or with complex patterns typical of those produced by inverted repeat transgenes driven by weak promoters (Que et al. 1997). In addition, inverted repeats are subject to DNA methylation in plants, and this can result in partial or complete transcriptional silencing of the sense overexpression transgene and/or other promoters in the T-DNA that may be responsible for dsRNA production by readthrough expression. Epigenetic changes in patterns of DNA methylation that occur in inverted repeat integrants can lead to alternate transgene states: For instance, one in which the promoter is methylated, and another in which the transcribed sequences are methylated, but not the promoter (Stam et al. 1998).

Another example of alternate epigenetic states would be a plant in which one state expresses the selective marker transgene while the adjacent sense overexpression transgene has been subjected to transcriptional silencing, and the reciprocal state exhibits transcriptional silencing of the selective marker transgene concomitant with high-level transcription of the sense overexpression transgene. A possible consequence of such alternation of epigenetic states in an inverted repeat integrant is (1) transcriptional dsRNA production via readthrough transcription from the marker transgene into the overexpression transgene in one epigenetic state of the inverted repeat transgene complex and (2) high-level production of sense transcripts from the overexpression transgene in the other state potentially triggering sense cosuppression. In the case of chalcone synthase in petunia flowers, these two distinct modes of silencing can be monitored by observing the patterns of silencing which distinguish the two modes (shown in Figure 1.1). Frequently, however, the two phenotypes are superimposed in the same flower (Jorgensen et al. 1996), indicating simultaneous or coincident occurrence of both modes not only in the same plant, but even in the same organ.

Clearly then, if two distinct modes of initiating RNA silencing can occur in the same plant alternately or simultaneously, interpretation of the results is likely to be compromised. Before this was understood, nearly the entire literature on RNA silencing in plants was based on transgenes engineered in such a way that single-copy integrants were unable to cause RNA silencing, whereas inverted repeat integrants could trigger silencing, whether by sense cosuppression or by dsRNA production via readthrough transcription. In retrospect, analyses of inverted repeat integrants carrying sense transgenes were likely a prime cause of the difficulty of fitting all observations to a single model for the mechanism of RNA silencing. Fortunately, the solution to this problem is relatively simple, although infrequently adopted: Either one should analyze single-copy integrants of sense transgenes designed for protein overexpression (taking care to avoid unnecessary readthrough from neighboring genes, such as selective marker transgenes; see, e.g., Que et al. 1997; Que and Jorgensen 1998), or one should analyze transgenes designed to produce dsRNA transcriptionally via self-complementary transcripts in the manner of Waterhouse et al. (1998).

Keeping in mind the caveat that much of the plant literature is based on inverted repeat integrants of sense transgenes, there are nonetheless interesting observations in that literature relevant to the phenomenon of sense cosuppression, such as a possible role for translation in the initiation of silencing and for DNA methylation in maintaining or propagating silencing states.

## A ROLE FOR TRANSLATION IN SENSE COSUPPRESSION?

An indication that translation might be involved in sense cosuppression first came from experiments intended to both overexpress and silence ACC (1-aminocyclopropane-1-carboxylic acid) synthase in transgenic tomato plants in order to investigate the control of fruit ripening (Lee et al. 1997). ACC synthase is responsible for biosynthesis of the ripening hormone ethylene. Lee et al. overexpressed ACC synthase by using the same expression cassette that had been used by Napoli et al. (1990) to silence chalcone synthase in petunia by sense cosuppression. Two classes of tomato transformants were observed: those that overexpressed the enzyme and produced the expected phenotype due to ethylene overproduction (known as epinasty) and those that showed cosuppression of both endogenous and introduced ACC synthase transcripts. What was particularly interesting was that although full-length transgene transcripts were drastically reduced in leaves of silenced plants, the level of abundance of the 5´-most sequences of the transgene transcript was not reduced, whereas abundance of the 3´-most sequences was reduced. In addition, these truncated transcripts were found to be associated with polyribosomes. The implication was that transgene transcript degradation occurs on polysomes and is targeted predominantly to the 3´-most sequences. In ripening fruit (where the endogenous gene is expressed to high levels), the transgene transcript was affected in same manner as in leaves (where the endogenous gene is expressed at low levels or not at all). In constrast, the endogenous gene transcript was reduced in abundance throughout its length. In similar experiments with tobacco plants expressing viral transcripts, RNAs truncated at their 3´ ends were also found to be associated with polysomes and were observed only in silenced plants (Tanzer et al. 1997). Inhibitor studies showed that the degradation process itself appeared to be translation-independent (Jacobs et al. 1997; Tanzer et al. 1997), suggesting that the initiation of RNA silencing, rather than degradation, is associated with the translation process.

A reasonable explanation for these observations is that the transgene transcript triggers RNA silencing by being recognized as aberrant in some way (e.g., high concentration or abnormal secondary structure), perhaps while being translated, and thereby becomes a substrate for the RdRP enzyme, which then produces cRNA using the more accessible 3´ end of the transcript as template. Due perhaps to an unusual structure of the ACC synthase transgene transcript (which possesses a 5´ leader sequence of viral origin thought to promote translation), the 5´ end of the transcript is somehow protected from RdRP transcription by the presence of active ribosomes. Targeting of the entire endogenous transcript for degradation presumably occurs because this transcript is susceptible to copying by the RdRP, perhaps because of a lower translation rate and/or a distinct transcript structure, allowing RNA silencing to "spread" along the target RNA.

In view of this, it is perhaps interesting that RNAi in *Caenorhabditis elegans* has been linked to nonsense-mediated decay, a putatively translation-associated process (Domeier et al. 2000). Interestingly, although mutations in three genes (*smg-2, smg-5,* and *smg-6*) that are necessary for nonsense-mediated decay did not block silencing initially, they did allow target gene expression to recover. Thus, Domeier et al. proposed that these genes participate in the amplification of a signal required for persistence of silencing, possibly by unwinding dsRNA to provide a template for RdRP. An SMG-2 homolog in *Arabidopsis*, SDE3, was identified in a screen for mutants required for RNA silencing by Dalmay et al. (2001). SDE3 is not an ortholog of SMG-2, and so presumably has a role distinct from, although possibly overlapping with, an SMG-2 ortholog. SDE3 is required for transgene-induced RNA silencing (probably for sense cosuppression), but not for virus-induced RNA silencing (Dalmay et al. 2001), leading to speculation that SDE3 participates with

SDE1/SGS2 (the RdRP homolog) in the conversion of single-stranded sense transcripts to dsRNA. SDE3 and SMG-2 share a highly similar RNA helicase domain; possibly then, SDE3 has a role in recognition or unwinding of intramolecular secondary structure in the target RNA to identify as aberrant or make accessible the RNA as a template for RdRP. Alternatively, or in addition, it might unwind dsRNA to permit new rounds of copying by RdRP, as suggested by Domeier et al. (2000).

Extension of the suggestion of Domeier et al. (2000) to both roles of the RdRP in sense cosuppression leads to the idea that a sense transcript can trigger silencing if it is recognized by an RdRP or an RNA helicase while being translated on polysomes, perhaps in a manner that is analogous to the recognition of early nonsense codons by SMG proteins. This hypothesis fits with the requirement for SMG-2 homolog SDE3 in sense cosuppression and could explain the observation that introduction of premature nonsense codons to a sense transgene drastically reduces its ability to trigger sense cosuppression (Que et al. 1997), if one assumes that the nonsense-mediated decay process interferes or competes with the initiation of RNA silencing. A third gene involved in sense cosuppression in *Arabidopsis* is *AGO1* (Beclin et al. 2002). AGO1, which is related to RDE-1 in *C. elegans*, is a PAZ domain protein that has homology with the rabbit translation initiation factor eIF2C, an intriguing fact in light of the suggestion that sense cosuppression is triggered during translation. Whether AGO1 interacts with polyribosome complexes, however, remains to be determined.

Investigations are needed that distinguish the multiple roles of RdRP and determine the subcellular location and molecular context in which the RdRP (SDE1/SGS2), RNA helicase (SDE3), and/or AGO1 recognizes particular sense RNA molecules that can act as primary triggers of RNA silencing. The special features of a sense transcript that are required for it to be recognized and act as a trigger also remain to be determined. Is high concentration sufficient to label a sense transcript as a template for RdRP? Or does the transcript have to possess some aberrant feature, such as unusual secondary structure? The idea that some aberrant feature of transgene transcripts acts as the initial trigger has been discussed for many years, first by Lindbo et al. (1993), but few specifics have been proposed. One suggestion is that a trigger RNA molecule possesses a 3´ end that is able to fold back to act as a primer for the RdRP. Presumably, this would require depolyadenylation and 3´→5´ degradation (as in the molecules observed by Lee et al. [1997], Tanzer et al. [1997], and Metzlaff et al. [2000]), to allow the 3´ end to anneal and prime cRNA synthesis. However, RdRP is not primer-dependent (Schiebel et al. 1998), and so other possibilities remain, such as that aberrant secondary structure is recognized by the putative RNA helicase, SDE3, a possibility that echoes an early suggestion by Cameron and Jennings (1991), or that high local concentration of homologous transcripts are sufficient to trigger recognition, perhaps by intermolecular hybridization, also suggested by Cameron and Jennings (1991).

Another interesting consideration is that the primary block in RNA silencing might not be RNA turnover, but inhibition of translation. Translational inhibition would be sufficient to block gene function, and it is thus at least conceivable that the primary role of RNA degradation in RNA silencing could simply be to produce enough short interfering RNA (siRNA) molecules to interfere with translation of homologous transcripts. Translational inhibition by 21-nucleotide RNAs is well known in *C. elegans; let-7* and *lin-4* produce what are known as small temporal RNAs (stRNAs) that do not degrade mRNA but interfere with translation (for review, see Ruvkun 2001). Thus, even if turnover of the target RNA were minimal, enough siRNA molecules might be produced for complete suppression of phenotype via translation arrest. Experimental investigations to determine whether inhibition of translation has a primary role in RNA silencing would seem to be a worthwhile endeavor.

## AN ASSOCIATION BETWEEN DNA METHYLATION AND RNA SILENCING

Many examples of RNA silencing in plants have been found to be associated with DNA methylation. An excellent analysis of this subject was published by Bender (2001), and so it will be addressed only briefly here, with emphasis on observations most relevant to sense cosuppression. Nearly all early examples of RNA silencing were based on plants carrying multiple transgene copies, usually organized as inverted repeats. Because repeated sequences are commonly methylated, it was not clear whether DNA methylation was related to RNA silencing or merely associated with the repetitiveness of transgenes. An important conceptual breakthrough came from the recognition that RNA could direct methylation of homologous DNA sequences in the plant nuclear genome (Wassenegger et al. 1994). Interestingly, this DNA methylation was noncanonical, i.e., it occurred at virtually all cytosines, not just those found in the CpG and CpXpG sequences that are normally targets of cytosine methylation in plants. Association of posttranscriptional silencing and noncanonical methylation of an active transgene locus by a homologous silencing locus was first demonstrated by Ingelbrecht et al. (1994). Both silencing and DNA methylation were reversible following segregation of the causative locus, suggesting that a common RNA-directed mechanism might cause both outcomes, one in the cytoplasm and one in the nucleus.

A stronger association between silencing and methylation was provided by Jones et al. (1998) with the observation that a cytoplasmically replicating RNA virus could trigger not only RNA silencing, but also methylation of the target transgene. Strikingly, induction of systemic RNA silencing by localized introduction of a transgene via Agroinfiltration also resulted in systemic methylation of the target transgene, as well as spreading of DNA methylation to sequences adjacent to the trigger, concomitant with spreading of RNA targeting along the target RNA (Jones et al. 1999). The association of silencing and methylation suggested that RNA degradation products (e.g., 21–23-nucleotide siRNAs) might be the cause of DNA methylation. However, blocking RNA silencing by expression of the viral inhibitor HC-Pro does not block DNA methylation or the production of a systemic RNA-silencing signal (Mallory et al. 2001). One interpretation of these results is that much lower levels of siRNAs are sufficient to cause methylation and to act as the systemic signal than are needed to trigger RNA silencing, and another is that molecules other than siRNAs are responsible for DNA methylation and/or systemic transmission of RNA silencing. Hamilton et al. (2002) observed a longer (24–26 nucleotides) class of siRNAs to be correlated with systemic silencing, whereas the shorter (21–22 nucleotides) class was not.

DNA methylation may be more than just a consequence of RNA silencing; it may also be involved in maintenance or amplification of RNA-silencing states. Following elimination of viral RNA that had initiated RNA silencing, silencing of a homologous transgene was maintained and the transgene was methylated, leading to the suggestion that transgene methylation might be the basis for maintenance (Jones et al. 1999; Dalmay et al. 2000a). Experimental support for this proposal comes from the finding that two *Arabidopsis* mutations, *ddm1* and *met1*, which are deficient in DNA methylation, can cause partial impairment of RNA silencing (Morel et al. 2000).

## POTENTIAL ROLES FOR SENSE COSUPPRESSION IN PLANT PHYSIOLOGY AND DEVELOPMENT

### RNA-silencing Signals and Functional Endogenous mRNAs Traffic Cell to Cell via Plasmodesmata and Are Phloem Mobile

One of the most remarkable features of RNA silencing in plants is the systemic spread of the silencing state via the plant's macromolecular trafficking system, which is composed of cell-to-cell transport via plasmodesmata and long-distance transport via the phloem translocation system (Palauqui et al. 1997; Voinnet et al. 1998). The primary function of systemic silencing is thought to be to transmit virus resistance throughout the plant to thwart an advancing viral infection. Parallel to this systemic silencing capability is the equally remarkable fact that a select subset of endogenous transcripts are able to move to distant sites in the plant via the phloem (Ruiz-Medrano et al. 1999). Kim et al. (2001) have argued that these transported transcripts can be functional at the sites to which they have been transported on the basis of their observation that the dominant *Mouse ears* mutant of tomato, which determines altered leaf morphology, is due to transcriptional fusion of a homeobox gene controlling leaf morphology with an unrelated gene whose transcript carries signals for phloem transmission. The *Mouse ears* fusion transcript was found to be able to move through the phloem from a rootstock into a wild-type scion (i.e., a shoot that had been grafted onto the rootstock). Leaves of the grafted scion were observed to express the *Mouse ears* morphology. Thus, plants possess a capacity to transmit information over long distances and to control gene expression in a supracellular, organismic manner by trafficking of functional transcripts throughout the plant (Jorgensen et al. 1998; Lucas et al. 2001). Whether endogenous RNA-silencing signals are also used by plants to traffic information and control gene expression at a distance remains to be determined, but it is expected to be an important area for investigation. An interesting possibility to consider in relation to sense cosuppression is that translocation of sense transcripts to distant sites in the plant might act not only positively as in *Mouse ears*, but also negatively, to down-regulate homologous gene expression in recipient organs, even while maintaining positive expression in the source organ. In addition, it is now clear that microRNAs (miRNAs) are produced by plant genomes in a manner similar to the production of miRNAs in animals (Llave et al. 2002; Park et al. 2002; Reinhart et al. 2002). Whether miRNAs are also subject to translocation between cells and to distant sites in the plants will be interesting to determine, as will the relative importance of miRNAs as compared to sense cosuppression in the control of endogenous plant gene expression.

## Epigenetic Changes in Sense Cosuppression Phenotypes

Epigenetic changes occur frequently in plants exhibiting cosuppression, most commonly in multicopy transgenic plants. In fact, the original demonstration of the cosuppression state was based on the availability of epigenetic variants (i.e., revertants to wild-type phenotype) that exhibited coexpression of transgenes and endogenes (Napoli et al. 1990; Jorgensen 1995). The diversity of epigenetic states that can be produced by the same transgene locus was illustrated most strikingly in a petunia transformant carrying two directly repeated copies of a chalcone synthase (*Chs*) transgene (Jorgensen 1995; Jorgensen et al. 1996; Jorgensen and Napoli 1996). In addition to the completely cosuppressed (white) and completely coexpressed (fully pigmented) epigenetic states, four distinct types of flower color patterns could be distinguished that followed basic determinants of petal morphology. Each state was found to be sexually transmissible, reversible, and changeable into qualitatively different states.

In contrast, petunias carrying a single copy of a *Chs* transgene are epigenetically stable and only produce patterns based on the fusion zone (i.e., junction) between adjacent petals (Figures 1.1a and 1.2), referred to as a "junction" pattern. Dispersed repeat *Chs* transformants also produce junction patterns, but are epigenetically unstable, giving rise to petal vein-based patterns (Figure 1.2). Intercrosses between transgene loci increase the frequency of these epigenetic events (Que and Jorgensen 1998), as do homozygosity and intercrossing epigenetic variants of the same locus (Jorgensen 1995).

Originally, it was proposed that these epigenetic changes might reflect alterations in the transgene promoter that modify the spatial transcriptional pattern of the transgene in

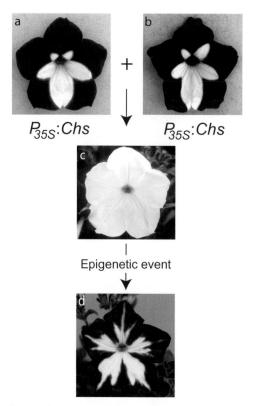

$P_{35S}$:Chs + $P_{35S}$:Chs

Epigenetic event

**FIGURE 1.2.** Duplication of *35S:Chs* transgene (*a,b*) leads to complete cosuppression (*c*), followed by an epigenetic event producing vein-based pattern (*d*). (*b* reprinted from Que et al. 1998 ©Blackwell Science. *d* reprinted, with permission, from Jorgensen 1993.)

the flower (Que et al. 1997). Because the patterns were based on the vasculature, however, the discovery of systemic RNA silencing suggested a different hypothesis, i.e., that epigenetic events induced or were responsive to systemic silencing signals. The transcriptional hypothesis would predict that promoter duplication would lead to vein-based patterns, whereas the systemic silencing hypothesis would predict that transcribed sequence duplications would do so. To distinguish these possibilities, constructs carrying the cauliflower mosaic virus (CaMV) 35S promoter or the *Chs*-coding sequence supertransformed into a host carrying a single-copy *35S:Chs* sense transgene and producing the typical junction pattern of cosuppression. The result was that the coding sequence duplication, but not the promoter duplication, produced transformants with vein-based patterns of cosuppression (R. Jorgensen et al. unpubl.).

Progeny of plants with the promoter duplication, however, did produce vein-based patterns, suggesting a delayed spreading of an epigenetic mark from the promoter to the coding sequence. Single-copy coding sequence integrants were able to induce vein-based patterns in the single-copy reporter transgene with an efficiency similar to that of inverted repeat integrants, whereas only inverted repeat integrants of the promoter construct could cause vein patterns in progeny. These inverted repeats were also associated with loss of cosuppression in the primary transformants, consistent with dsRNA-mediated transcriptional silencing of the 35S promoter (Mette et al. 1999; Jones et al. 2001). The ability of single-copy coding sequence integrants to alter the epigenetic state of the single-copy *35S:Chs* transgene and to produce vein-based patterns mirrors the occurrence of these patterns as a result of epigenetic events in dispersed integrants carrying multiple copies of the *35S:Chs* transgene. These results support the systemic silencing hypothesis for the epigenetic changes that lead to vein-based patterns.

Significantly, sexual progeny inherited the vein-based pattern even after the causative locus had segregated away, i.e., the heritable epigenetic alteration that is induced by a single-copy *Chs*-coding sequence is not only an epimutation (i.e., a change in DNA methylation or chromatin state), but also a paramutation (i.e., an epigenetic imprint transferred between alleles or homologous genes). How might paramutations lead to vein-based patterns of sense cosuppression? One possibility is that the induced epigenetic state of the transgene is responsive to a systemic silencing signal, whereas the normal state of a single-copy transgene is not responsive. Another is that the altered epigenetic state of the transgene is responsible for producing the systemic signal. Grafting experiments will be necessary to distinguish these possibilities. Considering that maintenance of RNA silencing seems to involve DNA methylation, as was explained above, it seems likely that the maintenance process is relevant to these sense transgene paramutants.

Finally, it is important to note that multiple types of epigenetic changes occur in dispersed integrants with *35S:Chs* sense transgenes. Three vein-based pattern types are morphologically distinct in phenotype and each of these three epigenetic states is heritable, both somatically and germinally (Jorgensen 1995; Jorgensen and Napoli 1996). Thus, the transgene can be imprinted in several different ways, and these distinct epigenetic states determine the phenotype of sense cosuppression. This is indicative of either multiple types of responses to a silencing signal or the existence of multiple types of signal molecules, possibilities that remain to be distinguished.

## A Speculative Hypothesis for Supracellular Information Processing and Storage in Plants

The possibility that epigenetic imprints on sense overexpression transgenes influence their participation in systemic RNA silencing could have significant implications for plant biology. If endogenous genes can also be imprinted with information that influences production of silencing signals and/or responses of genes to such signals, plants might possess a large capacity for information storage and processing, considering the large number of loci that could potentially be reservoirs of such information. Similarly, the ability to imprint information on promoters could have important consequences for transcriptional control.

There would seem to be little doubt that plants could benefit from a system for storing and integrating information accumulated during growth and development and in response to the environment. Large amounts of such information might be stored in the genome as chromatin states and/or DNA methylation patterns, distributed among hundreds, or even thousands, of loci throughout the genome.

The capacity of plants to traffic RNA, an informational macromolecule, throughout the plant via the phloem and plasmodesmata, taken together with the ability to imprint RNA-mediated information on the genome in a cell-, tissue-, or organ-specific manner, could potentially constitute a high-capacity device for long-term storage of information reflecting the developmental and physiological history of the plant. These "memory states" could conceivably be modified, reprogrammed, or reset by new information, arriving via signal transduction pathways, newly synthesized transcription factors and transcripts, and transcript turnover products. Obviously, such a mechanism would be regulated. In this regard, it is important to note that in transgenic experiments, endogenous plant genes are generally resistant to DNA methylation induced by dsRNA and unable to support either maintenance or spreading of RNAi (Jones et al. 1999; Vaistij et al. 2002), as might be expected if endogenous genes were only susceptible to these mechanisms under specific conditions.

In principle, regulated systemic transport of RNA signal molecules (e.g., siRNA and miRNA molecules, as well as longer molecules) creates the potential for long-distance transmission of epigenetic states that can be imprinted on the genomes of select target cells for the establishment and maintenance of supracellular patterns of gene expression. If this potential has been realized during the evolution of higher plants, then conceivably they could possess a robust system for information processing, storage, and transmission that integrates and assesses large amounts of information, potentially allowing plants to make "informed decisions" in growth, development, and physiology. This hypothesis is an updated and expanded version of a proposal made by the discoverer of paramutation, R. Alexander Brink (1960), for a parachromatin-centered information processing capability in plants.

## ACKNOWLEDGMENTS

I thank my colleagues, postdocs, and students, past and present (especially Carolyn Napoli and Qiudeng Que) for the many interesting and fruitful discussions over the years that contributed to the development of the perspectives presented here. The responsibility for any errors or oversights is mine alone. I apologize to the many authors whose important work in this field was not cited herein due to the emphasis of this discussion on special aspects of the phenomenon of RNA silencing in plants that are of particular interest to the author. My laboratory's research on RNA silencing mechanisms is currently funded by the Energy Biosciences Program of the U.S. Department of Energy.

## REFERENCES

Baulcombe D.C. and English J.J. 1996. Ectopic pairing of homologous DNA and post-transcriptional gene silencing in transgenic plants. *Curr. Opin. Biotechnol.* **7:** 173–180.

Beclin C., Boutet S., Waterhouse P., and Vaucheret H. 2002. A branched pathway for transgene-induced RNA silencing in plants. *Curr. Biol.* **12:** 684–688.

Bender J. 2001. A vicious cycle: RNA silencing and DNA methylation in plants. *Cell* **106:** 129–132.

Birchler J.A., Pal-Bhadra M., and Bhadra U. 1999. Less from more: Cosuppression of transposable elements. *Nat. Genet.* **21:** 148–149.

Brink R.A. 1960. Paramutation and chromosome organization. *Q. Rev. Biol.* **35:** 120–137.

Cameron F.H. and Jennings P.A. 1991. Inhibition of gene expression with a short sense fragment. *Nucleic Acids Res.* **19:** 469–474.

Dalmay T. Hamilton A., Mueller E., and Baulcombe D.C. 2000a. Potato virus X amplicons in *Arabidopsis* mediate genetic and epigenetic gene silencing. *Plant Cell* **12:** 369–379.

Dalmay T., Horsefield R., Braunstein T.H., and Baulcombe D.C. 2001. *SDE3* encodes an RNA helicase required for post-transcriptional gene silencing in *Arabidopsis*. *EMBO J.* **20:** 2069–2077.

Dalmay T., Hamilton A., Rudd S., Angell S., and Baulcombe D.C. 2000b. An RNA-dependent RNA polymerase gene in *Arabidopsis* is required for posttranscriptional gene silencing mediated by a transgene but not a virus. *Cell* **101:** 543–553.

Domeier M.E., Morse D.P., Knight S.W., Portereiko M., Bass B.L., and Mango S.E. 2000. A link between RNA interference and nonsense-mediated decay in *Caenorhabditis elegans*. *Science* **289:** 1928–1930.

Dougherty W.G. and Parks T.D. 1995. Transgenes and gene suppression: Telling us something new? *Curr. Opin. Cell. Biol.* **7:** 399–405.

English J.J., Mueller E., and Baulcombe D. 1996. Suppression of virus accumulation in transgenic plants exhibiting silencing of nuclear genes. *Plant Cell* **8:** 179–188.

Fire A. and Montgomery M. 1998. Double-stranded RNA as a mediator in sequence-specific genetic silencing and cosuppression. *Trends Genet.* **14:** 255–258.

Fire A., Xu S., Montgomery M.K., Kostas S.A., Driver S.E., and Mello C.C. 1998. Potent and specific genetic interference by double-stranded RNA in *Caenorhabditis elegans. Nature* **391:** 806–811.

Flavell R.B. 1994. Inactivation of gene expression in plants as a consequence of novel sequence duplication. *Proc. Natl. Acad. Sci.* **91:** 3490–3496.

Goodwin J., Chapman K., Swaney S., Parks T.D., Wernsman E.A., and Dougherty W.G. 1996. Genetic and biochemical dissection of transgenic RNA-mediated virus resistance. *Plant Cell* **8:** 95–105.

Grierson D., Fray R.G., Hamilton A.J., Smith C.J.S., and Watson C.F. 1991. Does co-suppression of sense genes in transgenic plants involve antisense RNA? *Trends Bio/Technology* **9:** 122–123.

Hamilton A., Voinnet O., Chappell L., and Baulcombe D. 2002. Two classes of short interfering RNA in RNA silencing. *EMBO J.* **21:** 4671–4679.

Hammond S.M., Caudy A.A., and Hannon G.J. 2001. Post-transcriptional gene silencing by double-stranded RNA. *Nat. Rev. Genet.* **2:** 110–119.

Hannon G.J. 2002. RNA interference. *Nature* **418:** 244–251.

Ingelbrecht I., Van Houdt H., Van Montagu M., and Depicker A. 1994. Posttranscriptional silencing of reporter transgenes in tobacco correlates with DNA methylation. *Proc. Natl. Acad. Sci.* **91:** 10502–10506.

Jacobs J.J.M.R., Litiere K., van Dijk V., van Eldik G.J., Van Montagu M., and Cornelissen M. 1997. Post-transcriptional β-1,3-glucanase gene silencing involves increased transcript turnover that is translation-independent. *Plant J.* **12:** 885–893.

Jones A.L., Thomas C.L., and Maule A.J. 1998. De novo methylation and co-suppression induced by a cytoplasmically replicating plant RNA virus. *EMBO J.* **17:** 6385–6393.

Jones L., Ratcliff F., and Baulcombe D.C. 2001. RNA-directed transcriptional gene silencing in plants can be inherited independently of the RNA trigger and requires Met1 for maintenance. *Curr Biol.* **11:** 747–757.

Jones L., Hamilton A.J., Voinnet O., Thomas C.L., Maule A.J., and Baulcombe D.C. 1999. RNA-DNA interactions are DNA methylation in post-transcriptional gene silencing. *Plant Cell* **11:** 2291–2301.

Jorgensen R. 1990. Altered gene expression in plants due to *trans* interactions between homologous genes. *Trends Biotechnol.* **8:** 340–344.

——. 1992. Silencing of plant genes by homologous transgenes. *Agbiotech News Info.* **4:** 265N–273N.

——. 1993. The germinal inheritance of epigenetic information in plants. *Philos. Trans. R Soc. Lond. B* **339:** 173–181.

——. 1995. Cosuppression, flower color patterns, and metastable gene expression states. *Science* **268:** 686–691.

Jorgensen R.A. and Napoli C.A. 1996. A responsive regulatory system is revealed by sense suppression of pigment genes in *Petunia* flowers. In *Genomes: Proceedings of the 22nd Stadler Genetics Symposium* (ed. J.P. Gustafson and R.B. Flavell), pp. 159–176. Plenum Press, New York.

Jorgensen R.A., Atkinson R.G., Forster R.L.S., and Lucas W.J. 1998. An RNA-based information superhighway in plants. *Science* **279:** 1486–1487.

Jorgensen R.A., Cluster P.D., English J., Que Q., and Napoli C.A. 1996. Chalcone synthase cosuppression phenotypes in petunia flowers: Comparison of sense vs. antisense constructs and single-copy vs. complex T-DNA sequences. *Plant Mol. Biol.* **31:** 957–973.

Kim M., Canio W., Kessler S., and Sinha N. 2001. Developmental changes due to long-distance movement of a homeobox fusion transcript in tomato. *Science* **293:** 287–289.

Lee K.Y., Baden C., Howie W.J., Bedbrook J., and Dunsmuir P. 1997. Post-transcriptional gene silencing of ACC synthase in tomato results from cytoplasmic RNA degradation. *Plant J.* **12:** 1127–1137.

Lindbo J.A., Silva-Rosales L., Proebsting W.M., and Dougherty W.G. 1993. Induction of a highly specific antiviral state in transgenic plants: Implications for gene regulation and virus resistance. *Plant Cell* **5:** 1749–1759.

Llave C., Kasschau K.D., Rector M.A., and Carrington J.C. 2002. Endogenous and silencing-associated small RNAs in plants. *Plant Cell* **14:** 1605–1619.

Lucas W.J., Yoo B.C., and Kragler F. 2001. RNA as a long-distance information macromolecule in plants. *Nat. Rev. Mol. Cell. Biol.* **11:** 849–857.

Mallory A.C., Ely L., Smith T.H., Marathe R., Anandalakshmi R., Fagard M., Vaucheret H., Pruss G.,

Bowman L., and Vance V.B. 2001. HC-Pro suppression of transgene silencing eliminates the small RNAs but not transgene methylation or the mobile signal. *Plant Cell* **13:** 571–583.

Matzke M.A. and Matzke A.J.M. 1995. How and why do plants inactivate homologous (trans)genes? *Plant Physiol.* **107:** 679–685.

Matzke M., Matzke A.J., and Kooter J.M. 2001. RNA: Guiding gene silencing. *Science* **293:** 1080–1083.

Meins Jr., F. and Kunz C. 1994. Silencing of chitinase expression in transgenic plants: An autoregulatory model. In *Homologous recombination and gene silencing in plants* (ed. J. Paszkowski), pp. 335–348. Kluwer Academic, Dordrecht.

Mette M.F., van der Winden J., Matzke M.A., and Matzke A.J.M. 1999. Production of aberrant promoter transcripts contributes to methylation and silencing of unlinked homologous promoters in trans. *EMBO J.* **18:** 241–248.

Metzlaff M., O'Dell M., Hellens R., and Flavell R.B. 2000. Developmentally and transgene regulated nuclear processing of primary transcripts of chalcone synthase A in petunia. *Plant J.* **23:** 63–72.

Mol J., van Blokland R., and Kooter J. 1991. More about cosuppression. *Trends Biotechnol.* **9:** 182–183.

Morel J.-B., Mourrain P., Beclin C., and Vaucheret H. 2000. DNA methylation and chromatin structure affect post-transcriptional transgene silencing in *Arabidopsis. Curr. Biol.* **10:** 1591–1594.

Mourrain P., Beclin C., Elmayan T., Feuerbach F., Godon C., Morel J.-B., Jouette D., Lacombe A.-M., Nikic S., Picault N., Remoue K., Sanial M., Vo T.-A., and Vaucheret H. 2000. *Arabidopsis SGS2* and *SGS3* genes are required for posttranscriptional gene silencing and natural virus resistance. *Cell* **101:** 533–542.

Mueller E., Gilbert J.E., Davenport G., Brigneti G., and Baulcombe D.C. 1995. Homology-dependent resistance: Transgenic virus resistance in plants related to homology-dependent gene silencing. *Plant J.* **7:** 1001–1013.

Napoli C., Lemieux C., and Jorgensen R. 1990. Introduction of a chimeric chalcone synthase gene into petunia results in reversible co-suppression of homologous genes *in trans. Plant Cell* **2:** 279–289.

Palauqui J.C. and Vaucheret H. 1998. Transgenes are dispensable for the RNA degradation step of cosuppression. *Proc. Natl. Acad. Sci.* **95:** 9675–9680.

Palauqui J.C., Elmayan T., Pollien J.-M., and Vaucheret H. 1997. Systemic acquired silencing: Transgene-specific post-transcriptional silencing is transmitted by grafting from silenced stocks to non-silenced scions. *EMBO J.* **16:** 4738–4745.

Park W., Li J., Song R., Messing J., and Chen X. 2002. CARPEL FACTORY, a Dicer homolog, and HEN1, a novel protein, act in microRNA metabolism in *Arabidopsis thaliana. Curr. Biol.* **12:** 1484–1495.

Que Q. and Jorgensen R.A. 1998. Homology-based control of gene expression patterns in transgenic petunia flowers. *Dev. Genet.* **22:** 100–109.

Que Q., Wang H.-Y., and Jorgensen R.A. 1998. Distinct patterns of pigment suppression are produced by allelic sense and antisense chalcone synthase transgenes in petunia flowers. *Plant J.* **13:** 401–409.

Que Q., Wang H.-Y., English J., and Jorgensen R.A. 1997. The frequency and degree of cosuppression by sense chalcone synthase transgenes are dependent on transgene promoter strength and are reduced by premature nonsense codons in the transgene coding sequence. *Plant Cell* **9:** 1357–1368.

Reinhart B.J., Weinstein E.G., Rhoades M.W., Bartel B., and Bartel D.P. 2002. MicroRNAs in plants. *Genes Dev.* **16:** 1616–1626.

Ruiz M.T., Voinnet O., and Baulcombe D.C. 1998. Initiation and maintenance of virus-induced gene silencing. *Plant Cell* **10:** 937–946.

Ruiz-Medrano R., Xoconostle-Cazares B., and Lucas W.J. 1999. Phloem long-distance transport of CmNACP mRNA: Implications for supracellular regulation in plants. *Development* **126:** 4405–4419.

Ruvkun G. 2001. Glimpses of a tiny RNA world. *Science* **294:** 797–799.

Schiebel W., Pelissier T., Riedel L., Thalmeir S., Schiebel R., Kempe D., Lottspeich F., Sanger H.L., and Wassenegger M. 1998. Isolation of an RNA-directed RNA polymerase-specific cDNA clone from tomato. *Plant Cell* **10:** 2087–2101.

Smith H.A., Swaney S.L., Parks T.D., Wernsman E.A., and Dougherty W.G. 1994. Transgenic plant virus resistance mediated by untranslatable sense RNAs: Expression, regulation, and fate of nonessential RNAs. *Plant Cell* **6:** 1441–1453.

Stam M., Viterbo A., Mol J.N.M., and Kooter J.M. 1998. Position-dependent methylation and transcriptional silencing of transgenes in inverted T-DNA repeats: Implications for posttranscriptional silencing of homologous host genes in plants. *Mol. Cell. Biol.* **18:** 6165–6177.

Stam M., De Bruin R., Kenter S., Van der Hoorn R.A.L., Van Blokland R., Mol J.N.M., and Kooter J.M. 1997. Post-transcriptional silencing of chalcone synthase in Petunia by inverted transgene repeats. *Plant J.* **12:** 63–82.

Tanzer M.M., Thompson W.F., Law M.D., Wernsman E.A., and S. Uknes. 1997. Characterization of post-transcriptionally suppressed transgene expression that confers resistance to tobacco etch virus infection in tobacco. *Plant Cell* **9:** 1411–1423.

van Blokland N., van der Geest N., Mol J.N.M., and Kooter J.M. 1994. Transgene-mediated suppression of chalcone synthase expression in Petunia hybrida results from an increase in RNA turnover. *Plant J.* **6:** 861–877.

van der Krol A.R., Mur L.A., Beld M., Mol J.N.M., and Stuitje A.R. 1990. Flavonoid genes in petunia: Addition of a limited number of gene copies may lead to a suppression of gene expression. *Plant Cell* **2:** 291–299.

van der Vlugt R.A.A., Ruiter R.K., and Goldbach R. 1992. Evidence for sense RNA-mediated protection of PVY[N] in tobacco plants transformed with the viral coat protein cistron. *Plant Mol. Biol.* **20:** 631–639.

Vance V. and Vaucheret H. 2001. RNA silencing in plants—Defense and counterdefense. *Science* **292:** 2277–2280.

Vastaij F.E., Jones L., and Baulcombe D.C. 2002. Spreading of RNA targeting and DNA methylation in RNA silencing requires transcription of the target gene and a putative RNA-dependent RNA polymerase. *Plant Cell* **14:** 857–867.

Vaucheret H., Beclin C., and Fagard M. 2001. Post-transcriptional gene silencing in plants. *J. Cell. Sci.* **114:** 3083–3091.

Voinnet O. 2001. RNA silencing as a plant immune system against viruses. *Trends Genet.* **17:** 449–459.

Voinnet O., Vain P., Angell S., and Baulcombe D.C. 1998. Systemic spread of sequence-specific transgene RNA degradation in plants is initiated by localized introduction of ectopic promoterless DNA. *Cell* **95:** 177–187.

Waterhouse P.M., Graham M.W., and Wang M.B. 1998. Virus resistance and gene silencing in plants can be induced by simultaneous expression of sense and antisense RNA. *Proc. Natl. Acad. Sci.* **95:** 13959–13964.

Waterhouse P.M., Wang M.-B., and Finnegan E.J. 2001. Role of short RNAs in gene silencing. *Trends Plant Sci.* **7:** 297–301.

Wassenegger M., Heimes S., Riedel L., and Sanger H.L. 1994. RNA-directed de novo methylation of genomic sequences. *Cell* **76:** 567–576.

# CHAPTER 2

# Transgene Cosuppression in Animals

James A. Birchler, Manika Pal-Bhadra, and Utpal Bhadra*

*Division of Biological Sciences, University of Missouri, Columbia, Missouri 65211*

WHEN TRANSGENES WERE FIRST INTRODUCED INTO PLANT SPECIES, it was discovered that their expression was often compromised. Matzke et al. (1989) described a situation in which transgenes, with sequence homology between the promoter regions, were silenced when combined together into tobacco. Napoli et al. (1990) and van der Krol et al. (1990) found that introduction of transgenes for anthocyanin pigment synthesis into petunia plants silenced not only multiple transgenes, but the endogenous gene as well. This process is referred to as cosuppression (Jorgensen 1995) or homology-dependent gene silencing (HDGS) (see also Chapter 1).

The numerous examples of cosuppression subsequently described were classified into two categories. Many cases in which homology existed between the RNA-encoding portions of the transgenes were found to have normal rates in transcriptional run-on assays (e.g., see Blokland et al. 1994). These results led to the concept that the silencing occurred posttranscriptionally, presumably by affecting RNA turnover (e.g., see Depicker et al. 1996; Metzlaff et al. 1997). The other class of silencing occurred between transgenes with homology in the promoter regions. This type was accompanied by increased methylation in the regions of homology and was correlated with reduced rates of transcription (e.g., see Matzke et al. 1989).

It was also known that transgenes expressing portions of plant viruses (most of which are of the double-stranded [ds] RNA type) would confer resistance to the homologous virus (e.g., see Goodwin et al. 1996). Subsequently, it was determined that the viral RNA was turned over in a manner similar to that found for posttranscriptional silencing (Baulcombe and English 1996). This led to the concept that one of the evolutionary pressures for cosuppression was as a defense against viruses (Baulcombe and English 1996). Another evolutionary rationale for silencing was described as a means of holding the expression of transposable elements at a low level to keep the mutation rate in check (Chabossier et al. 1998; Jensen et al. 1999; Ketting et al. 1999; Tabara et al. 1999).

Although good evidence exists that the process does function as a defense against foreign elements and transposons, the genes responsible for silencing in a variety of species also affect developmental processes. There is also an overlap of genes involved with the maturation of small temporal RNAs and posttranscriptional silencing (Grishok et al. 2001; Hutvagner et al. 2001; Ketting et al. 2001). A full appreciation of the cellular processes involved and all of their ramifications has yet to be realized.

For many years, transgene silencing was thought to be a phenomenon unique to plants. In the mid 1990s when the authors first attempted to publish an example of trans-

*Present address:* Centre for Cellular and Molecular Biology, Uppal Road, Hyderabad 500 007, India.

gene silencing in *Drosophila*, a reviewer, while admitting no criticism of the data, likened the phenomenon to "cold fusion." Subsequently, we have found that most transgenes examined will exhibit some level of silencing if the dosage is great enough. Transgene silencing in *Drosophila* differs from many cases in plants in that the degree of silencing is not as strong and seldom reaches complete null levels. This fact explains why transgene silencing was not recognized earlier in the animal kingdom.

## TRANSGENE SILENCING IN *DROSOPHILA*

In the course of studies on dosage-dependent gene regulation, our lab produced a promoter-reporter construct to test whether modifier effects operating on the *white* locus could be conferred onto another gene via the *white* promoter (Rabinow et al. 1991). The promoter region of *white* extending into the mRNA encoding sequences to the *Hph*I site (174 bp from the transcriptional initiation site) was fused with the structural part of the *Alcohol dehydrogenase* (*Adh*) gene beginning 36 bp 5′ to the initiation AUG (see Figure 2.1). This fusion (*w-Adh*) was transformed back into *Drosophila* carrying a null allele of the endogenous *Adh* gene (*Adh^{fn6}*). The homozygous constructs produced no ADH activity. However, one hemizygous copy on the single X chromosome in males was active. With this result, it was thought perhaps the explanation resided in the phenomenon of transvection, a situation in *Drosophila* in which the pairing (or lack thereof) of alleles on homologous chromosomes affects their expression (Wu and Goldberg 1989; Geyer et al. 1990). It should be noted that homologous chromosomes are paired in somatic cells in *Drosophila*, allowing genes at homologous sites to be in close association.

To examine this possibility, single *w-Adh* constructs present on different chromosomes were examined each in an unpaired hemizygous condition. These flies all produced significant levels of ADH activity. This result would be consistent with transvection as the

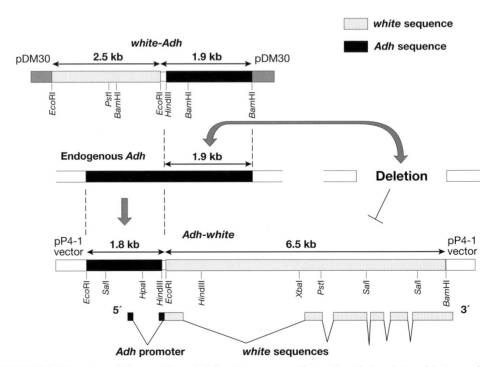

**FIGURE 2.1.** Structure of the *w-Adh* and *Adh-w* transgenes illustrating their relationship to endogenous *Adh*. Deletion of the endogenous gene removes any homologous connection between the reciprocally formed transgenes.

explanation, were it not for the fact that a control of two copies, represented by one unpaired insertion in two places in the genome, produced no more ADH than either single copy alone.

When higher copy numbers of the transgene were assayed by northern blots and RNase protection assays, the level of *Adh* message declined in quantity as the number of *w-Adh* constructs was increased from one to six copies (Pal-Bhadra et al. 1997). Flies with six copies had only about 15% of the *Adh* message level as flies with one copy. This result raised the issue of an analogy with the process of cosuppression, defined in plant species, in which transgenes introduced into the genome are silenced.

To test whether the analogy to cosuppression extended to an effect on the endogenous *Adh* gene, stocks carrying the *Adh* null allele and zero to four copies of the *w-Adh* transgene were crossed to the Canton S wild-type strain. The level of *Adh* endogenous mRNA was measured by RNase protection and northern blots. One copy of the transgene produced RNA levels that were additive with the endogenous *Adh*, but two to four copies lowered the total *Adh* message progressively to 16% of the level in flies with no transgenes. Thus, the *w-Adh* gene not only lowers its own expression with increasing doses, but also lowers that of the endogenous *Adh* gene held at one normal copy (Pal-Bhadra et al. 1997). In contrast, the copy number of *w-Adh* does not affect the endogenous *white* gene expression.

## Cosuppression with Other Constructs

To test whether silencing occurred with other constructs, three dosage series were produced: full-length *Adh*, full-length *white*, and the reciprocal construct to *w-Adh*—one carrying the *Adh* regulatory sequences fused to the *white* structural gene (see Figure 2.1). The *white-Adh* construct has 2.5 kb of *white* joined to 1.9 kb of *Adh* in vector DM30 (Rabinow et al. 1991). The *Adh-white* construct has 1.8 kb of *Adh* fused to 6.5 kb of *white* in vector P4-1 (Birchler et al. 1990). The *Adh* sequences begin 5′ to both of the two *Adh* enhancers and promoters (Benyajati et al. 1983) and end at the distally promoted RNA splice site, leaving only 34 bp of the proximally promoted RNA and 87 bp from the distally promoted RNA joined to the *Hph*I site 51 bp before the initiation AUG of *white*. Full-length *Adh* and *white* are present in the Carnegie 20 vector (Pal-Bhadra et al. 1997, 2002).

The expression of the full-length *white* construct is linear up to six copies—the highest tested. This result was consistent with the work of other investigators, which demonstrated that *mini-white*, a shortened version, exhibits a dosage effect. The *Adh* gene showed a dosage effect up to five copies, but beyond that level, the expression is reduced from the expected amount (Pal-Bhadra et al. 1997, 2002).

The *Adh-w* construct, however, shows silencing at lower dosage (M. Pal-Bhadra, U. Bhadra, and J.A. Birchler, unpubl.). This hybrid transgene can be scored phenotypically because it is present in a *white* deficiency background (*w^{67c23}*). At any one insertion site, *Adh-w* exhibits greater pigment as a homozygote compared to the respective hemizygote, as one might expect, but this result is in contrast to that observed with *w-Adh*. However, two unpaired copies show less expression than the additive effect predicted. Introduction of three copies into the genome causes a further decline in pigment. When an endogenous *white* gene is crossed into a multiple *Adh-w* stock, it is also reduced in expression. Thus, in this case, the endogenous gene is also drawn into the silencing pool.

In addition, a single *w-Adh* transgene will reduce the expression of a single *Adh-w* copy, even though they have no homologous sequences in common (see Figure 2.1) (Pal-Bhadra et al. 1999). The basis for this interaction appears to be that the two nonhomologous transgenes interact via the endogenous *Adh*. This hypothesis was tested by producing flies heterozygous for two overlapping deficiencies, which delete the *Adh* gene entirely, but produce a viable fly. Accordingly, a *w-Adh* insert on chromosome 3 and an *Adh-w* insert on the X chromosome were combined with both deficiencies on chromosome 2. When the two stocks with different *Adh* deficiencies were crossed together, the flies with both *Adh-w* and *w-Adh* but with the endogenous *Adh* deleted returned to the eye color characteristic of a single *Adh-w* insert alone. RNA analysis of *white* paralleled the phenotypic results. This led to the conclusion that the *Adh* structural region of *w-Adh* interacts with the part of the endogenous *Adh* in common (Pal-Bhadra et al. 1999). The signal might "spread" in the endogenous gene to the extreme 5′ leader region plus the promoter of *Adh*, followed by an interaction with the homologous sequences in the *Adh-w* insertion to cause the reduction in expression. Another possibility is that the three types of sequences might become associated in the nucleus to initiate silencing. Deletion of the endogenous *Adh* would fail to bring together *w-Adh* and *Adh-w*. Finally, the interaction might be mediated by ectopic transcription of the *Adh* promoter region, which triggers a chromatin change in *trans* via an interaction with the RNA or a doubled-stranded version of it. With the deletion of the endogenous copy, no homologous RNA could exist between the two types of transgenes.

Reintroduction of a full-length *Adh* transgene in the absence of endogenous *Adh* restores the interaction between *w-Adh* and *Adh-w*. Furthermore, a full-length *Adh* transgene alone was sufficient to silence the *Adh-w* construct in the absence of the endogenous *Adh*. This result allowed a test of a series of truncated *Adh* transgenes having deletions in the regulatory sequences of *Adh*. The smallest segment that is required for silencing surrounds the adult enhancer (Pal-Bhadra et al. 1999). This region of *Adh* is not known to be transcribed, leading to the suggestion that the silencing interaction is initiated by DNA-to-DNA association. However, some undetectable level of ectopic transcription of the regulatory sequences might foster an RNA-mediated mechanism, especially considering the impact of the *piwi* and *aubergine* mutations described below.

## Polycomb-Group Gene Effects

The setting of partially repressed states of gene expression suggested an involvement of the Polycomb-Group (Pc-G) genes (Kennison 1995). To test this, stocks carrying mutations of *Polycomb* (chromosome 3) and *Polycomblike* (chromosome 2) were crossed to the six-copy *w-Adh* stock and the $Adh^{fn6}$ *cn; ry*$^{506}$ transformation recipient. Neither *Pc* nor *Pcl* has an effect upon *Adh* RNA levels alone. However, when three *w-Adh* copies are included in the genotype, endogenous *Adh* is reduced, unless the *Pc* or *Pcl* mutations are also present, in which case the RNA levels return toward normal (Pal-Bhadra et al. 1997).

To examine this issue further, immunostaining with antibodies against *Pc-* and *polyhomeotic* (*ph*)-encoded proteins, both core components of the Pc-G complex, was conducted on salivary gland chromosome preparations from larvae with and without cosuppression. The Polycomb complex including these two proteins normally colocalizes to many sites on the chromosomes. Two of the three *w-Adh* insertions are not present at these sites and thus could be tested for accumulation of *Pc-* and *ph*-encoded proteins. The recipient strain and the single-copy genotypes do not show binding, but the six-copy *w-Adh* stock shows two additional bands of labeling that correspond to the testable sites (Figure 2.2). Thus, *Pc-* and *ph*-encoded proteins accumulate at the insertions under cosuppressing conditions but not otherwise. The mutational and binding studies provide evidence for an involvement of the Pc-G complex in cosuppression of this type. The association of the Pc-G on the transgene may act as a mechanism for maintenance of a set level of histone acetylation (Cavalli and Paro 1999; Cao et al. 2002; Czremin et al. 2002; Muller et al. 2002).

The cosuppressive effect of *w-Adh* transgenes on the *Adh-w* transposons also produces accumulation of Pc on the polytene chromosomes at the site of *Adh-w* insertion. Two dif-

## Chromosome 3

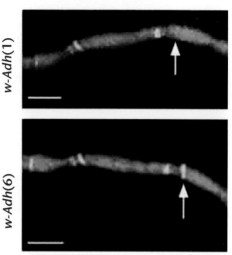

**FIGURE 2.2.** Accumulation of the Polycomb complex on a silenced transgene. A portion of chromosome 3 of *Drosophila* is shown that includes the site of a *w-Adh* transgene (arrow). The chromosomes were stained with propidium iodide and the antibody labeling against Polycomb is green. The merged image shows the sites of binding in yellow. Many locations accumulate the Polycomb complex normally, but none of these sites overlap the location of this transgene. When only a single transgene is active, there is no detectable Polycomb binding (*top*). When there is strong silencing in the presence of six *w-Adh* transgenes in the nucleus, there is accumulation of the Polycomb complex over the transgene (*bottom*).

ferent *Adh-w* insertions have no Pc labeling when present in a single copy in the genome. However, with the introduction of *w-Adh* transposons, the expression of *Adh-w* is reduced and Pc labeling occurs on the respective *Adh-w* sites. Deletion of the endogenous *Adh* gene, the presumed intermediary of this interaction as described above, eliminates the Pc binding.

The *Adh-w* transgene alone shows strong expression, as evidenced by in situ RNA analysis, as early as 2.5 hours of development and continues throughout the embryonic stages (Pal-Bhadra et al. 1999). However, when two *w-Adh* transgenes are also present, *Adh-w* again shows strong expression at 2.5 hours, but there is an obvious decline observed by 4 hours, which continues until about 13 hours, at which point the expression plateaus. By comparison, it should be noted that pairing of homologs after fertilization begins near the end of syncytial blastoderm (±3 hours) (Hiraoka et al. 1993). The initiation of Pc binding begins by cellular blastoderm and then spreads from the core association sites by the time of germ-band extension (Orlando et al. 1998). The time course of silencing is consistent with the possibility that the homologous copies associate during the homology search that establishes the homolog pairing in *Drosophila* and that Pc-G accumulation maintains the reduced level of gene expression. The time course cannot rule out the alternative possibility that expression of *w-Adh* ultimately triggers the *Adh-w* silencing, especially considering that posttranscriptional silencing readily occurs at this stage of development (see below).

## Interactions of PRE-containing Constructs

Hagstrom et al. (1997), Sigrist and Pirrotta (1997), and Muller et al. (1999) reported the *trans*-interaction of transposons carrying a Polycomb response element (PRE) and *mini-white* (*m-w*). PREs are DNA sequence motifs that attract the Polycomb complex (Mihaly et al. 1998) and were originally defined by their role in maintaining repressed states of gene expression of segment identity genes. When present in a transposon together with an *m-w* eye color reporter, the latter has altered expression. Typically, these *m-w* transgenes are active when hemizygous, but many insertions sites show strong silencing when homozygous. In addition, two unpaired copies in the genome are not additive in their expression and increasing copy number causes a decline in the total expression. We suspect that the interaction of PRE-containing transgenes is related to the cosuppression of the *w-Adh* and *Adh-w* examples described above. Cosuppression of *w-Adh* or *Adh-w* by *w-Adh* can involve *trans*-interactions of copies that do not label with Polycomb when active, but do label when the silencing becomes evident. However, when a bona fide PRE is present on the transposon, as was the case in the aforementioned studies, the silencing in *trans* appears to be much more severe and operates at a much lower dosage. Transgenes that carry a PRE together with *m-w* and *lacZ* (Muller 1995) or *m-w* and *yellow*+ (Mallin et al. 1998) exhibit silencing of both genes. Chromatin insulators placed around *yellow*+ can protect it from silencing by the PRE, even though the unprotected *m-w* is silenced (Mallin et al. 1998). By using a transformation cassette buffered from *y*+ silencing by chromatin insulators, Sigrist and Pirrotta (1997) found that many hemizygous transformants (20%) are basically null for *white* and would not have been recognized in previous experiments. This example illustrates that unless one is vigilant to the possibility of silencing, such transformants would not be recognized.

## Cosuppression and Pairing-sensitive Silencing

Transvection is the phenomenon whereby the expression of alleles is affected by the degree of pairing of the homologs. The change can be either a reduction in expression or complementation between alleles. This phenomenon was first discovered by Ed Lewis, using the *bithorax* locus (Lewis 1954; Wu and Goldberg 1989). Several subsequent examples have been found including the loci *white* (Jack and Judd 1979; Babu and Bhat 1980; Pirrotta 1991), *decapentaplegic* (Gelbart and Wu 1982), and *yellow* (Geyer et al. 1990). Although the mechanism is still unknown, a popular explanation is that transcriptional enhancers are shared between paired homologs (Geyer et al. 1990; Goldsborough and Kornberg 1996). In the case of complementation, one homolog might carry an allele with an enhancer that is inactivated and the other homolog, an allele in which the structural gene is impaired, but a nearly normal phenotype will still result. This occurs because the normal enhancer will work with the normal structural gene in *trans* across homologous pairs, as long as they can physically associate. If chromosomal aberrations or mutations are introduced that disrupt this association, a mutant phenotype will result. A pairing phenomenon referred to as "pairing-sensitive silencing" in which paired transgenes become inactive may or may not be related to transvection, as discussed below. The *w-Adh* transgene, which exhibits cosuppression, exhibits pairing silencing, so it would be of interest to determine whether there is a relationship between the two phenomena.

One mutation that will disrupt transvection at many loci is a "loss-of-function" allele at the X-linked *zeste* locus (Babu and Bhat 1980, 1981; Wu and Goldberg 1989). This allele of *zeste* is called *z-a* and disrupts transvection at *white* (Babu and Bhat 1980), *decapentaplegic* (Gelbart et al. 1985), and *bithorax* (Goldsborough and Kornberg 1996). Thus, a relationship between transvection and cosuppression was tested by examining whether cosuppression is eliminated in flies carrying the *z-a* allele. To do this, a stock homozygous for *z-a* and the *Adh* null allele was produced. Second, a stock was constructed that was homozygous for *z-a* and four copies of the *w-Adh* transgene. Once these were generated, the two lines were crossed together to produce *z-a* flies with one transgene on chromosome 2 and a second on chromosome 3. Then, comparisons for the level of *Adh* mRNA were made of the three types of flies, namely, (1) *z-a, Adh* null; (2) *z-a, Adh* null + hemizygous transgenes on two chromosomes; and (3) *z-a, Adh* null + two homozygous copies (= 4 copies). If *z-a* were to eliminate cosuppression, then the four copies of the transgene would have twice the level as two copies. If *z-a* does not eliminate cosuppression, the four copies would produce about one-third the level of mRNA as the two copies, based on our previous results. The latter result was found (Pal-Bhadra et al. 1997).

Despite the failure to find a relationship with *zeste*-mediated transvection, it should be noted that the *w-Adh* construct shows strong pairing-sensitive silencing. Its behavior is similar to that of the *engrailed-white* (*en-w*) transgenes that defined the phenomenon of pairing-sensitive silencing (Kassis et al. 1991, 1994). This transgene carries a portion of the nontranscribed regulatory region of the *engrailed* locus adjacent to *m-w*. The *en-w* transgenes express well as hemizygotes but are silenced as homozygotes. Interestingly, the original studies noted that two unpaired copies of *en-w* on different chromosomes very often were not expressed at greater levels than either single copy, just as with *w-Adh*. Furthermore, *z-a* had no effect upon the action of *en-w*. We have examined a dosage series of *en-w* for the presence of dispersed silencing (M. Pal-Bhadra, U. Bhadra, and J. Birchler, unpubl.). Indeed, double hemizygotes are not additive, and flies containing three and four copies of *en-w* progressively decline in eye color below the level seen in homozygous insertions. The flies with four copies of *en-w* show the most extreme reduction. Thus, the same type of silencing response occurs with *en-w* as was found with *w-Adh*. Further work is needed to understand the relationship of pairing-sensitive silencing between transgenes in identical positions on homologs versus transgene silencing among disperse locations in the genome (cosuppression). It is possible that the ability of the genes to associate enhances the silencing mechanism.

The question arises as to whether gene-to-gene encounters are involved in triggering silencing. There is very little known about homology searching in the nucleus. We reason that it occurs at meiosis, at the establishment of somatic pairing at some point following fertilization (Hiraoka et al. 1993), and in repair of P element excisions (Gloor et al. 1991). In addition, transposons containing PRE have a tendency to "home," in that they preferentially insert in chromosomal sites bound with the Polycomb complex (Kassis et al. 1992). Thus, it is possible that Pc-G-bound genes might foster gene-to-gene searches.

## POSTTRANSCRIPTIONAL TRANSGENE SILENCING IN *DROSOPHILA*

Work in *Caenorhabditis elegans* several years ago demonstrated that the injection of sense RNA into embryos would result in the extinction of homologous endogenous gene expression (Guo and Kemphues 1995). Subsequently, it was discovered that dsRNA was much more potent at silencing (Fire et al. 1998). This "RNA interference" or RNAi also occurs in *Paramecium* (Ruiz et al. 1998), trypanosomes (Ngo et al. 1998), *Drosophila* (Kennerdell and Carthew 1998; Misquitta and Paterson 1999), and mammals (Wianny and Zernicka-Goetz 2000; Elbashir et al. 2001b), as well as many other species. Extremely low levels of dsRNA can trigger interference, suggesting that there is an amplification of the signal. The silencing signal can act systemically, and, in some cases, the silenced state is transmitted to the subsequent generation (Tabara et al. 1998). Intron sequences are not effective, suggesting that RNA rather than DNA is the molecule targeted (Montgomery et al. 1998).

The 1–10-dosage series of full-length *Adh* described above shows a linear expression correlated with dosage up to five copies. After that level, the expression falls off and in some cases is quite low (Pal-Bhadra et al. 1997, 2002). In comparison, when a *heat-shock* (*hs*)-*Adh* construct is induced, an increase in the level of *Adh* RNA follows. However, when four copies of full-length *Adh* and the *hs-Adh* construct are combined, heat shock induction produces a collapse of the RNA levels as assayed by northern blots of adult RNA (Pal-Bhadra et al. 2002) or RNA in situ hybridizations of early embryos (Figure 2.3). The 45-minute heat shock has no effect on the levels in the five-*Adh*-copy flies and causes an increase with *hs-Adh* alone; but with the combination, the RNA levels are virtually extinguished. The rapid decline in *Adh* RNA under these circumstances suggests that an active degradation is occurring within a matter of minutes (Figure 2.4). The silenced con-

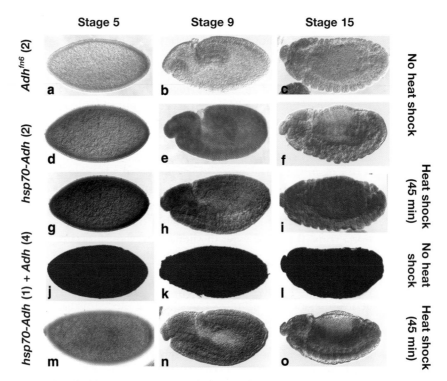

**FIGURE 2.3.** Threshold posttranscriptional gene silencing (PTGS) of *Adh* transgenes in early embryos. Silencing of five copies of *Adh* by the induction of *hsp70-Adh*. In situ analysis of RNA, using an *Adh* probe, was performed on mixed ages of embryos of different genotypes and treatments. (*a–c*) *Adh* null (*Adh^fn6^*); (*d–f*) two copies of *hs-Adh* uninduced; (*g–i*) two copies of *hs-Adh* induced; (*j–l*) four copies of *Adh* + one *hs-Adh* uninduced; (*m–o*) four copies of *Adh* + one *hs-Adh* induced. All embryos were fixed at the end of the 45-minute heat shock treatment.

dition persists at a reasonably low level for about 20 hours in adults, but will eventually recover to normal by 120 hours. This suggests that in *Drosophila*, a continuous overexpression of RNA is needed for silencing of this type. There is no influence of the *Polycomblike* mutation on the level of silencing as assayed immediately following heat shock or at 20 hours (Pal-Bhadra et al. 2002).

The dosage series of *Adh* and the heat-shock-induced silencing were examined for the presence of small 21–25-bp dsRNAs that are characteristic of RNAi and posttranscriptional gene silencing (Hamilton and Baulcombe 1999; Tuschl et al. 1999; Hammond et al. 2000; Yang et al. 2000; Zamore et al. 2000; Elbashir et al. 2001a) (see also Chapter 13). In the silencing doses, the quantity of such RNAs is quite abundant (Pal-Bhadra et al. 2002); in addition, with the threshold-induced silencing, they correlate with the collapse of *Adh* mRNA levels. Therefore, it appears that at a certain threshold of *Adh* mRNA, a double-stranded molecule is formed, presumably by an RNA-dependent RNA polymerase activity (Bass 2000; Lipardi et al. 2001) (see also Chapter 9). Such dsRNA structures would be cleaved to 21–25-bp units by an RNase type III nuclease (Bernstein et al. 2001) and subsequently incorporated into a larger RNase complex to target the homologous mRNA for digestion (Hammond et al. 2000).

## RUN-ON TRANSCRIPTION ANALYSIS

The ability to induce silencing of full-length *Adh* constructs suggested a posttranscriptional mechanism. To test this, run-on transcription experiments (So and Rosbash 1997)

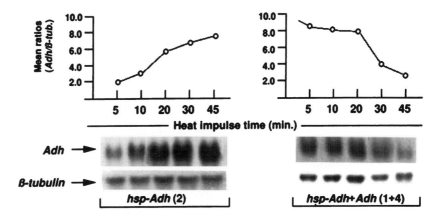

**FIGURE 2.4.** Time course for threshold-induced posttranscriptional silencing of *Adh*. Northern blot autoradiograms from two classes of adult RNAs carrying two copies of *hsp70-Adh* or one copy of *hsp70-Adh* and four copies of *Adh* were probed with radiolabeled *Adh* antisense RNA. The flies were heat-incubated for 45 minutes at 37°C. The total cellular RNA was isolated from each genotype at several intervals throughout the incubation period. The β-*tubulin* level acts as a gel-loading control. The mean *Adh*/β-*tubulin* ratios from triplicate measurements were plotted at time intervals during the heat incubation. At the time point that the *hsp-Adh* construct has significantly contributed *Adh* to the pool, the degradation of total RNA in the combined genotype begins to decline. Once the threshold for silencing is achieved, destruction of the RNA follows rapidly.

were performed with 0–10 copies of the *Adh* transgene, with 0–6 copies of *w-Adh*, and with the *Adh-w/w-Adh* combination (Pal-Bhadra et al. 2002). The amount of run-on transcription in the former case showed a linear relationship with transgene dosage, in contrast to the decline of the steady-state RNA level. The lack of linearity in the northern analysis indicates that the silencing in that case is indeed posttranscriptional. This result is consistent with the production of short interfering RNAs (siRNAs) in silenced doses. With *w-Adh* and *Adh-w/w-Adh*, the transcription followed a similar pattern with the northern data, indicating a transcriptional silencing (Pal-Bhadra et al. 2002). This result was foreshadowed by the finding of an association of the Pc-G complex with the silenced *w-Adh* transgenes.

## INFLUENCE OF *PIWI* AND *AUBERGINE* ON POSTTRANSCRIPTIONAL AND TRANSCRIPTIONAL SILENCING

Although transcriptional and posttranscriptional silencing might appear to be unrelated, some data demonstrate that RNA-mediated silencing can cause epigenetic changes in the homologous nuclear genes. The first indication that this might be the case was the finding that a transgene of a plant virus became methylated upon infection by the corresponding virus (Wassenegger et al. 1994; Jones et al. 1998, 1999; Pelissier and Wassenegger 2000). Second, the introduction into cells of DNA constructs with no means of expression can trigger the RNA degradation reaction (Voinnet et al. 1998). Third, mutations in plants that exhibit reduced DNA methylation will relieve gene silencing (Jeddeloh et al. 1998; Mittelsten-Scheid et al. 1998), and there is a relief of posttranscriptional silencing in some individuals of this genotype (Morel et al. 2000). Furthermore, Mette et al. (1999, 2000) showed that transcription of promoter regions to produce an aberrant RNA would cause methylation of the homologous sequences on a different transgene in the nucleus and silencing of the reporter gene being promoted. In addition, the introduction of a promoter into the plant cell on a virus will trigger transcriptional silencing of another reporter construct in the same cell with a homologous promoter (Jones et al. 2001). In animal species, silenced transgene arrays in the *C. elegans* germ line are activated (Tabara et al. 1999; Ketting and Plasterk 2000) and appear to be less condensed (Dernburg et al. 2000) in some mutant backgrounds defective for RNAi. Such mutations were initially recovered on the basis of their inability to support RNAi. These results raise the issue of whether a relationship exists between the two types of silencing (Jones et al. 1998, 1999; Vionnet et al. 1998; Mette et al. 1999, 2000; Birchler et al. 2000; Marathe et al. 2000; Sijen et al. 2001).

Genes have been identified in plants, *C. elegans*, and *Neurospora* that are defective in RNA silencing processes (Elmayan et al. 1998; Cogoni and Macino 1999a,b; Ketting et al. 1999, 2001; Tabara et al. 1999; Grishok et al. 2000; Smardon et al. 2000). The predicted functions of some of these genes include RNA-dependent RNA polymerase (Cogoni and Macino 1999a; Dalmay et al. 2000), RNA or DNA helicases (Cogoni and Macino 1999b; Ketting et al. 1999; Tabara et al. 1999; Wu-Scharf et al. 2000; Dalmay et al. 2001), and RNases (Ketting et al. 1999; Hammond et al. 2000; Bernstein et al. 2001). The thinking is that these gene products generate or amplify the dsRNA, which is cleaved to siRNA. They, in turn, act as a guide for targeting the homologous mRNA for destruction via an RNase complex (Bass 2000; Hammond et al. 2000; Bernstein et al. 2001). The mutations that are defective for RNAi (*rde*) in *C. elegans* are all viable, so there may be vital genes involved that have yet to be identified. Several *Drosophila* genes are homologs or gene family members of the *C. elegans rde* genes, including *sting* (= *aubergine*) (Schupbach and Wieschaus 1991; Wilson et al. 1996; Schmidt et al. 1999; Aravin et al. 2001), *piwi* (Cox et al. 1998, 2000; Tabara et al. 1999; Cerutti et al. 2000), *argonaute* (Katoaka et al. 2001), *argonaute2* (Hammond et al. 2001), and *homeless* (Aravin et al. 2001; Stapleton et al. 2001).

Given the parallels between posttranscriptional silencing of *Adh* and RNAi, *Drosophila* mutations in gene family members related to RNAi-defective mutations in *C. elegans* were tested for an effect. The *piwi*, but not the *aubergine*, mutation will block threshold-induced silencing and eliminate the production of 21–25-bp dsRNAs. It should be noted, however, that the *aubergine* mutation will block the germ-line silencing of the *Stellate* gene by *Suppressor of Stellate*, which operates via an RNAi-like mechanism (Avarin et al. 2001). Furthermore, given the potential relationship between RNAi-like and transcriptional silencing noted above, these two mutations were tested for an effect on *Adh-w/w-Adh* transcriptional silencing. Neither has a significant effect on *w-Adh* silencing of endogenous *Adh*, but will prevent rather strongly the extension of this silencing to *Adh-w* (Figure 2.5) (Pal-Bhadra et al. 2002). Therefore, it appears that in *Drosophila*, as in plants, RNA silencing in the cytoplasm can have an impact on transcriptional silencing. The basis of this connection is unknown, but it is presumed to involve siRNAs as the homology guide to the sequences in the nucleus. The recent finding that chromodomain proteins contain RNA interaction motifs (Akhtar et al. 2000; Muchardt et al. 2002) provides the interesting possibility that siRNAs direct the Polycomb complex (whose members possess chromodomains) to the transgenes for establishment of the transcriptionally silenced state.

## TRANSGENE EFFECTS ON TRANSPOSABLE ELEMENTS

One function of silencing mechanisms appears to be as a means to keep transposable elements from exhibiting extensive transpositions, which would increase the mutation rate. Several groups have reported silencing of the I transposable element in *Drosophila*. The I element is a type of retroposon that is responsible for one type of hybrid dysgenesis. When females devoid of functional elements are crossed by male inducer strains carry-

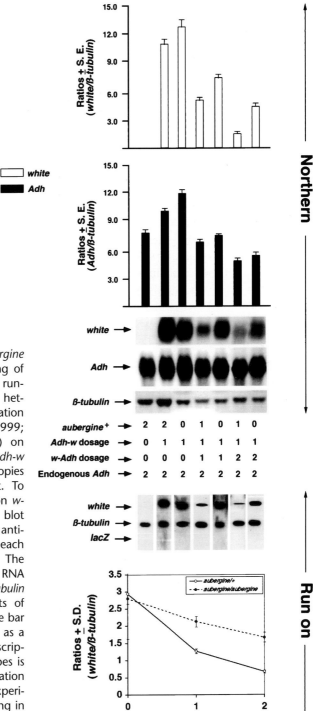

**FIGURE 2.5.** Effect of the *aubergine* mutation on transcriptional silencing of *Adh-w*. Northern and transcriptional run-on analyses showing the effect of a heteroallelic *aubergine* mutant combination (*ΔP-3a/QC42*) (Schmidt et al. 1999; Schupbach and Wieschaus 1999) on *white* transcripts of the reporter *Adh-w* transgene in the presence of 0–2 copies of the reciprocal *w-Adh* construct. To determine the effect of *aubergine* on *w-Adh/Adh* silencing, each northern blot was probed with a radiolabeled *Adh* antisense RNA. The copy number of each construct or allele is noted below. The relative amount of *white* and *Adh* RNA (*white*/β-*tubulin* ±S.E. and *Adh*/β-*tubulin* ±S.E.) from triplicate measurements of each genotype is represented by the bar diagrams. The β-*tubulin* probe acts as a gel-loading control. Run-on transcriptional analysis of the same genotypes is shown below. A graphical representation of the means (±S.D.) of triplicate experiments illustrates the relief of silencing in the *aubergine* mutant flies.

ing full-length elements in their genomes, sterility is observed as well as high rates of mutation due to new insertions of I. Transgenes carrying different forms of the I element are capable of reducing the reactivity of these elements over the course of several generations (Pritchard et al. 1988; Jensen et al. 1995).

To examine the relationship of this phenomenon to cosuppression, 186 bp of the 5′-untranslated region (5′UTR) of the I element were introduced into reactive flies

(Chaboissier et al. 1998). Strains with various numbers of this sequence ranging from 1 to 11 copies were produced, and the effect on an I element promoter-CAT (chloramphenicol acetyltransferase) reporter was determined. With increasing dosage of the UTR fragment, there was decreasing expression of the reporter. These transgenes were then tested for the fertility of dysgenic females. With increasing dosage of this segment, the fertility increased, suggesting a silencing of the elements in the genome. A regulatory mutation that decreases the expression of the I element transgene did not alter the silencing of the I element promoter-CAT construct significantly nor the effect on fertility.

In contrast, introduction of a different type of transgene transcribing internal portions of the I element had an effect on fertility, but when the transgene was not expressed, the response was absent (Jensen et al. 1999a). These constructs also showed a dosage effect with greater numbers of transgenes conditioning better fertility. Similar versions of the transgene that could or could not produce a protein were equally effective. The impact on fertility accumulates over several generations. The greater the number of transgenes, the fewer generations required to achieve the same level of effect. This accumulation only occurs through the female lineage and is reversed by one generation of passage through the male line. Different fragments of the I element were effective and the combined length was additive. Transgenes engineered not to express any RNA product were ineffective. Further analysis (Jensen et al. 1999b) demonstrated that the expression of sense or antisense fragments of the I element was equally effective in repressing incoming I elements. Both produced an accumulating silencing through the female germ line.

Gauthier et al. (2000) examined these issues further using antisense I element transgenes. With these constructs, the accumulation effect through the female lineage was also observed, although in this case, an effect could still be measured for two generations following the removal of the transgene. The strength of the effect of the transgene was dosage-dependent. Gauthier et al. also described a weaker paternal effect.

Malinsky et al. (2000) found that I element transgenes expressed by a heat shock promoter were quite effective at decreasing sterility. Sense or antisense constructs produced the same level of response. However, a transgene without the promoter was ineffective. Interestingly, when the same fragment of the I element was expressed by the I element promoter itself, there was no effect.

Jensen et al. (2002) reported that different nonhomologous portions of the I element were capable of producing the silencing interaction. The case is formally analogous to the *w-Adh/Adh-w* case described above. These authors suggested that pericentromeric copies of the I element that are normally present in the genome were capable of mediating the interaction of the two types of transgenes.

A fascinating aspect of the transgene silencing of the I element is the accumulating nature of the silencing. The transposition rate of the I factor progressively decreases over several generations as long as the transgene is inherited through the female germ line. It remains an open question whether the silencing operates at the posttranscriptional or transcriptional level. Determining the nature of this accumulating effect will reveal a process unique among cases of transgene silencing.

Data also exist which illustrate a *trans*-sensing mechanism for the P transposable element when present in telomeric-associated sequences (Ronsseray et al. 1996; 1998). At present, the relationship of this *trans*-interaction with the phenomenon of cosuppression is not known, although several parallels exist.

## STELLATE/SUPPRESSOR OF STELLATE INTERACTION IN DROSOPHILA

Although technically not an example of cosuppression, the *Stellate/Suppressor of Stellate* (*Su [Ste]*) interaction in *Drosophila* (Schmidt et al. 1999) is an example of homology-dependent silencing (Avarin et al. 2001; Stapleton et al. 2001) that appears to be related to cases of cosuppression. The *Stellate* gene is present on the X chromosome and the *Suppressor of Stellate* is present on the Y. In males without a Y chromosome, the *Stellate* gene (a casein kinase II) is overexpressed in the testis and conditions male sterility. The respective genes on the two chromosomes share a region of homology. The *Su(Ste)* gene cluster generates sense and antisense RNAs that have homology with *Stellate*. These enter the pool for the formation of siRNAs that can silence both *Stellate* and *Su(Ste)*. Thus, the system represents a case in which RNA silencing acts as an autogenous control mechanism.

## COSUPPRESSION IN *C. ELEGANS*

When transgenes are introduced into *C. elegans*, they typically form highly repeated arrays without integration into the chromosomes. In the soma, most such arrays are expressed as expected and are capable of complementing mutations in the corresponding genes. In contrast, these arrays in the germ line are not active and, in many cases, are capable of silencing the endogenous gene. Arrays of a transgene lacking a promoter are not effective. This fact has led some investigators to postulate that the silencing is post-transcriptional (Dernburg et al. 2000; Ketting and Plasterk 2000), and the term "cosuppression" has been applied. However, some transgene array silencing is affected by mutations in the *mes* genes, which are homologs to a class of Pc-G genes (see Kelly and Fire 1998). This fact would suggest that their silencing is chromatin-based, although an indirect effect of the *mes* genes is possible. Therefore, it is not presently clear whether the silencing of transgene arrays in worms is analogous to the dispersed gene-silencing type of cosuppression or to repeat-induced silencing of tandem arrays that resembles heterochromatin (see Dorer and Henikoff 1994). Of course, these two types of silencing may be related.

As noted above, mutations were recovered in *C. elegans* that would not support RNAi (RNAi-defective or *rde*). Some of these mutations overlap with mutator (*mut*) genes that derepress transposable elements. Tabara et al. (1999) noted that transgene arrays were desilenced in a background of some of these mutations. Ketting and Plasterk (2000) tested a collection of these mutations on the ability to desilence transgene arrays in the germ line. The mutations, *mut-2, mut-7, mut-8,* and *mut-9*, which are defective in transposon silencing, also released the silencing of the transgene arrays. In contrast, *mut-6* did not affect RNAi or eliminate cosuppression. Furthermore, the *rde1* mutation does not support RNAi but has no effect on cosuppression, thus documenting some differences in the two phenomena.

Dernburg et al. (2000) also observed that *rde1* had no effect on silencing of transgene arrays. They too found that *rde2* and *mut-7* would release the silencing of the array tested. Arrays generated in these backgrounds were not silenced, but they could become so when outcrossed to normal. These two mutations also made the transgene arrays appear less condensed than in a wild-type background. No evidence of association of the transgene array with the normal chromosomes was observed, although transient associations could not be ruled out.

The tissue-specific dichotomy of silencing of transgene arrays in worms raises an interesting question concerning its basis. RNAi is effective in somatic tissue and therefore must have distinctions with the phenomenon of cosuppression, which is restricted to the germ line. The mutational analysis is consistent with a relationship between the two phenomena, but they are not equivalent. It is possible that the initiation of cosuppression of transgene arrays shares mechanistic steps with RNAi, but thereafter a divergence in maintenance functions occurs.

## COSUPPRESSION IN MAMMALS

Transgenes introduced into mammalian cells are often integrated as tandem arrays and exhibit silencing in a mosaic fashion in a field of cells (Garrick et al. 1998). Whether this type of silencing is related to that of dispersed single copies is unknown. The ability to conduct gene replacement in mammalian cells eliminates the need for experiments using individual transgenes. Thus, little data are available about dispersed gene silencing in mammals. Given the presence of RNAi in mammals (Elbashir et al. 2001b) and a large number of antisense transcripts (Lehner et al. 2002), it is likely that similar processes are present. In addition, silencing by sense RNAs has been documented in mammalian cells (Cameron and Jennings 1991; zu Putlitz and Wands 1999).

Indeed, an example of transgene silencing in rodent cells has been reported (Bahramian and Zarbl 1999). Electroporation of rat fibroblast cells with a collagen transgene causes a rapid reduction of the endogenous RNA. The stable integration of the same plasmid would exhibit normal expression approximately equal to the endogenous copy. The thousands of transgenes in the transient introduction to the cells did not express an RNA, leading to the interpretation that they were transcriptionally silenced. The transgene RNA could have been distinguished from the endogenous RNA by RNase protection, but it was not detected. Because the endogenous RNA was present, although reduced, the transgene RNA should have been found if it had been expressed to any degree. Various methods of introducing the transgene copies all produced the silencing. These observations were interpreted as a reflection of transcriptional silencing of the transgenes.

A collagen promoter-CAT construct also caused the rapid depletion of the endogenous RNA, but in this case, there was expression of the reporter gene. Such constructs with different promoter strengths were equally effective at silencing the endogenous RNA. Another plasmid with the promoter region and a portion of the protein-encoding part of the gene extending through the first exon and intron showed greater effectiveness at reducing the level of endogenous RNA. Sequences in the protein-encoding portion of the transgene were postulated to be involved with posttranscriptional silencing.

Transgenes delivered to mice in lentiviral vectors apparently have a low rate of silencing (Lois et al. 2002). A green fluorescent protein (GFP) construct was expressed from most insertions and, when judged qualitatively, was correlated with gene copy number. The emphasis of this study, as is often the case, was on recovery of expressing transgenes. Therefore, the basis of the silenced cases was not investigated.

## SUMMARY

Gene silencing studies have led to the discovery of cellular mechanisms that were unknown until quite recently. The gene products identified to date indicate a group with implicated functions in RNA metabolism and another with chromatin functions. Many of these genes were previously identified for their role in developmental control. Further identification of the full spectrum of the participating gene products and their interactions is the challenge for understanding all of the processes involved and the evolution of their complex array of interactions.

Some transgenes show silencing at only a few copies, whereas others require much higher numbers, if they silence at all. The parameters that determine whether a particular transgene will silence and at what level are not known. In our experience, hybrid transgenes appear to be more susceptible, but the sample size is too small to make a generalization. It has long been recognized that transgenes vary considerably in expression, depending on the insertion site. Typically, these cases have been interpreted as reflecting the chromatin environment. Clearly, this would influence expression, but it is also possible that some cases may involve RNA metabolism steps that have yet to be recognized.

Moreover, it is not obvious what the determinants are that condition transcriptional silencing versus posttranscriptional silencing. The *Adh* gene described above exhibits either type in different transgenes that have only subtle differences in structure. The involvement of pairing-sensitive silencing is a related issue. This type of silencing seems to involve gene-to-gene associations that are recognized by the cell, which modify the chromatin configuration of the targeted gene. Whether dispersed transgene silencing also involves gene-to-gene associations and what mediates such interactions, if they occur, are more difficult questions to address experimentally.

Another important issue in the field is whether there is a connection between posttranscriptional and transcriptional silencing. Clearly, posttranscriptional effects can occur in the absence of transcriptional silencing. However, numerous studies in plant species have indicated methylation of the nuclear gene when posttranscriptional silencing is induced. Nevertheless, mutations defective for posttranscriptional silencing in plants appear to have little impact on transcriptional silencing, and viral inhibitors of RNA silencing do not affect transcriptional silencing (Marathe et al. 2000). In animal species, *piwi* and *aubergine* will diminish the Polycomb-dependent silencing of *Adh-w* in the *Adh-w/w-Adh* interaction, suggesting some shared function. In *C. elegans*, Tabara et al. (1999) noted that *rde* mutants would desilence tandem arrays of transgenes in the germ line, a phenomenon that could involve transcriptional silencing. The silencing of the tandem arrays is suppressed by mutations in the *mes* genes, which are homologs of some members of the Polycomb-Group (Kelly and Fire 1998). It is conceivable that two independent mechanisms are operative, but under some circumstances, they have a connection (Figure 2.6). Alternatively, it is possible that all dispersed gene silencing has a basis in dsRNA mechanisms. It is important to establish whether such a connection can ever occur because this impacts the proper conduct of reverse genetic procedures and gene therapy, as well as expanding our basic understanding of the cellular processes involved.

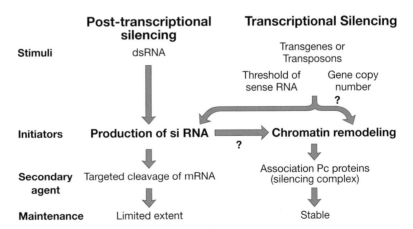

**FIGURE 2.6.** Silencing pathways in *Drosophila* for transgene cosuppression. Posttranscriptional silencing of transgenes in *Drosophila* exhibits characteristics typical of RNA-silencing processes including the production of siRNAs. The silencing can be strong if induced by rapid threshold induction, but it is not stable. Full recovery in the case of *Adh* silencing occurs within 120 hours. Transcriptional silencing is associated with chromatin modification involving the Polycomb-Group complex. Some evidence suggests connections with pairing-sensitive silencing and RNA-silencing processes but the nature of these potential interactions is unknown. Once transcriptional silencing is established early in embryogenesis, it is stable throughout development.

## ACKNOWLEDGMENTS

The authors' work on this topic was supported by a grant from the National Science Foundation (USA).

## REFERENCES

Akhtar A., Zink D., and Becker P.B. 2000. Chromodomains are protein-RNA interaction modules. *Nature* **407**: 405–409.

Aravin A.A., Naumova N.M., Tulin A.A., Rozovsky Y.M., and Gvozdev V.A. 2001. Double stranded RNA-mediated silencing of genomic tandem repeats and transposable elements in *Drosophila melanogaster* germline. *Curr. Biol.* **11**: 1017–1027.

Babu P. and Bhat S. 1980. Effect of *zeste* on *white* complementation. In *Developmental and neurobiology* of Drosophila (ed. O. Siddiqi et al.), pp. 35–40. Plenum Press, New York.

———. 1981. Role of *zeste* in transvection at the *bithorax* locus of *Drosophila*. *Mol. Gen. Genet.* **183**: 400–402.

Bass B.L. 2000. Doubled-stranded RNA as a template for gene silencing. *Cell* **101**: 235–238.

Bahramian M.B. and Zarbl H. 1999. Transcriptional and posttranscriptional silencing of rodent alpha1(i) collagen by a homologous transcriptionally self-silenced transgene. *Mol. Cell. Biol.* **19**: 274–283.

Baulcombe D.C. and English J.J. 1996. Ectopic pairing of homologous DNA and post transcriptional gene silencing in transgenic plants. *Curr. Opin. Biotech.* **7**: 173–180.

Benyajati C., Spoerel N., Haymerle H., and Ashburner M. 1983. The messenger RNA for *Alcohol dehydrogenase* in *Drosophila melanogaster* differs in its 5′ end in different developmental stages. *Cell* **33**: 125–133.

Bernstein E., Caudy A.A., Hammond S.M., and Hannon G.J. 2001. Role for a bidentate ribonuclease in the initiation step of RNA interference. *Nature* **409**: 295–296.

Birchler J.A., Hiebert J.C., and Paigen K. 1990. Analysis of autosomal dosage compensation involving the *Alcohol dehydrogenase* locus in *Drosophila melanogaster*. *Genetics* **124**: 677–686.

Birchler J.A., Pal-Bhadra M., and Bhadra U. 2000. Making noise about silence: Repression of repeated genes in animals. *Curr. Opin. Genet. Dev.* **10**: 211–216.

Blokland R.V., Geest N.V., Mol J.N.M., and Kooter J.M. 1994. Transgene mediated suppression of chalcone synthase expression in *Petunia hybrida* results in an increase in RNA turnover. *Plant J.* **6**: 861–877.

Cao R., Wang L., Wang H., Xia L., Erdjument-Bromage H., Tempst P., Jones R.S., and Zhang Y. 2002. Role of histone H3 lysine 27 methylation in Polycomb-group silencing. *Science* **298**: 1039–1043.

Cavalli G. and Paro R. 1999. Epigenetic inheritance of active chromatin after removal of the main transactivator. *Science* **286**: 955–958.

Cerutti L., Mian N., and Bateman A. 2000. Domains in gene silencing and cell differentiation proteins: The novel PAZ domain and redefinition of the *piwi* domain. *Trends Biochem. Sci.* **25**: 481–482.

Chabossier M.-C., Bucheton A., and Finnegan D.J. 1998. Copy number control of a transposable element, the I factor, a LINE-like element in *Drosophila*. *Proc. Natl. Acad. Sci.* **95**: 11781–11785.

Cogoni C. and Macino G. 1999a. Gene silencing in *Neurospora crassa* requires a protein homologous to RNA-dependent RNA polymerase. *Nature* **399**: 166–169.

———. 1999b. Posttranscriptional gene silencing in *Neurospora* by RecQ DNA helicase. *Science* **286**: 2342–2344.

Cox D.N., Chao A., and Lin H. 2000. *piwi* encodes a nucleoplasmic factor whose activity modulates the number and division rate of germline stem cells. *Development* **127**: 503–514.

Cox D.N., Chao A., Baker J., Chang L., Qiao D., and Lin H. 1998. A novel class of evolutionary conserved genes defined by *piwi* are essential for stem cell self-renewal. *Genes Dev.* **12**: 3715–3727.

Czermin B., Melfi R., McCabe D., Seitz V., Imhof A., and Pirrotta V. 2002. *Drosophila* enhancer of Zeste/ESC complexes have a histone H3 methyltransferase activity that marks chromosomal Polycomb sites. *Cell* **111**: 185–196.

Dalmay T., Horsefield R., Braunstein T.H., and Baulcombe D.C. 2001. SDE3 encodes an RNA helicase required for post-transcriptional gene silencing in *Arabidopsis*. *EMBO J.* **20:** 2069–2078.

Dalmay T., Hamilton A., Rudd S., Angell S., and Baulcombe D.A. 2000. An RNA-dependent RNA polymerase gene in *Arabidopsis* is required for posttranscriptional gene silencing mediated by a transgene but not by a virus. *Cell* **101:** 543–553.

Depicker A., Ingelbetcht I., van Houdt H., de Loose M., and van Montagu M. 1996. Post-transcriptional reporter transgene silencing in transgenic tobacco. In *Mechanisms and applications of gene silencing* (ed. D. Grierson et al.), pp. 71–84, Nottingham University Press, United Kingdom.

Dernburg A.F., Zalevsky J., Cloaiacovo M.P., and Villeneuve A.M. 2000. Transgene-mediated cosuppression in the *C. elegans* germ line. *Genes Dev.* **14:** 1578–1583.

Dorer D.R. and Henikoff S. 1994. Expansions of transgene repeats cause heterochromatin formation and gene silencing in *Drosophila*. *Cell* **77:** 993–1002.

Elbashir S.M., Leneckel W., and Tuschl T. 2001a. RNA interference is mediated by 21–22-nucleotide RNAs. *Genes Dev.* **15:** 188–200.

Elbashir S.M., Harborth J., Lendeckel W., Yalcin A., Weber K., and Tuschl T. 2001b. Duplexes of 21-nucleotide RNAs mediate RNA interference in cultured mammalian cells. *Nature* **411:** 494–498.

Elmayan T., Balzergue S., Beon F., Bourdon V., Daubrement J., Guenet Y., Mourrain P., Palauqui J.C., Vernhettes S., Vialle T., Wostrikoff K., and Vaucheret H. 1998. *Arabidopsis* mutants impaired in cosuppression. *Plant Cell* **10:** 1747–1758.

Fire A., Xu S., Montgomery M.K., Kostas S.A., Driver S.E., and Mello C.C. 1998. Potent and specific genetic interference by double-stranded RNA in *Caenorhabditis elegans*. *Nature* **391:** 806–811.

Garrick D., Fiering S., Martin D.I., and Whitelaw E. 1998. Repeat induced gene silencing in mammals. *Nat. Genet.* **18:** 56–59.

Gauthier E., Tatout C., and Pinon H. 2000. Artificial and epigenetic regulation of the I factor, a non-viral retrotransposon of *Drosophila melanogaster*. *Genetics* **156:** 1867–1878.

Gelbart W.M. and Wu C.T. 1982. Interactions of *zeste* mutations with loci exhibiting transvection effects in *Drosophila melanogaster*. *Genetics* **102:** 179–189.

Geyer P.K., Green M.M., and Corces V.G. 1990. Tissue specific transcriptional enhancers may act in trans on the gene located in the homologous chromosome: The molecular basis of transvection in *Drosophila*. *EMBO J.* **9:** 2247–2256.

Gloor G.B., Nassif N.A., John-Schlitz D.M., Peston C.R., and Engels W.R. 1991. Targeted gene replacement in *Drosophila* via P element-induced gap repair. *Science* **253:** 1110–1117.

Goldsborough A.S. and Kornberg T.B. 1996. Reduction of transcription by homologue asynapsis in *Drosophila* imaginal discs. *Nature* **381:** 807–810.

Goodwin J., Chapman K., Swaney S., Parks T.D., Wernsman E.A., and Doughtery W.G. 1996. Genetic and biochemical dissection of transgenic RNA-mediated virus resistance. *Plant Cell* **8:** 95–105.

Grishok A., Pasquinelli A.E., Conte D., Li N., Parrish S., Ha I., Baillie D.L., Fire A., Ruvkun G., and Mello C.C. 2001. Genes and mechanisms related to RNA interference regulate expression of the small temporal RNAs that control *C. elegans* developmental timing. *Cell* **106:** 23–34.

Guo S. and Kemphues K. 1995. *par-1*, a gene required for establishing polarity in *C. elegans* embryos, encodes a putative Ser/Thr kinase that is asymmetrically distributed. *Cell* **81:** 611–620.

Hagstrom K., Muller M., and Schedl P. 1997. A Polycomb and GAGA dependent silencer adjoins the *Fab-7* boundary in the *Drosophila bithorax* complex. *Genetics* **146:** 1365–1380.

Hamilton A.J. and Baulcombe D.C. 1999. A species of small antisense RNA in posttranscriptional gene silencing in plants. *Science* **286:** 950–952.

Hammond S.M., Bernstein E., Beach D., and Hannon G.J. 2000. An RNA-directed nuclease mediates post-transcriptional gene silencing in *Drosophila* cells. *Nature* **404:** 293–296.

Hammond S.C., Boettcher S., Caudy A.A., Kobayashi R., and Hannon G.J. 2001. *Argonaute2*, a link between genetic and biochemical analyses of RNAi. *Science* **293:** 1146–1150.

Hiraoka Y., Dernberg A.F., Parmelee S.J., Rykowski M.C., Agard D.A., and Sedat J.W. 1993. The onset of homologous chromosome pairing during *Drosophila melanogaster* embryogenesis. *J. Cell Biol.* **120:** 592–600.

Hutvagner G., McLachlan J., Pasquinelli A.E., Balint E., Tuschl T., and Zamore P.D. 2001. A cellular function for the RNA-interference enzyme Dicer in the maturation of the *let-7* small temporal RNA. *Science* **293:** 834–838.

Jack J.W. and Judd B.H. 1979. Allelic pairing and gene regulation: A model for the *zeste-white* interaction in *Drosophila melanogaster*. *Proc. Natl. Acad. Sci.* **76:** 1368–1372.

Jeddeloh J.A., Bender J., and Richards E.J. 1998. The DNA methylation locus DDM1 is required for maintenance of gene silencing in *Arabidopsis. Genes Dev.* **12:** 1714–1725.

Jensen S., Gassama M.-P., and Heidmann T. 1999. Taming of transposable elements by homology dependent gene silencing. *Nat. Genet.* **21:** 209–212.

———. 1999. Cosuppression of I transposon activity in *Drosophila* by I-containing sense and antisense transgenes. *Genetics* **153:** 1767–1774.

Jensen S., Cavarec L., Gassama M.P., and Heidman T. 1995. Defective I element introduced into *Drosophila* as transgenes can regulate reactivity and prevent I-R hybrid dysgenesis. *Mol. Gen. Genetics* **248:** 381–390.

Jensen S., Gassama M.-P., Dramand X., and Heidmann T. 2002. Regulation of I-transposon activity in *Drosophila*. Evidence for cosuppression of nonhomologous transgenes and possible role of ancestral I-related pericentromeric elements. *Genetics* **162:** 1197–1209.

Jones L., Ratcliff F., and Baulcombe D.C. 2001. RNA-directed transcriptional gene silencing in plants can be inherited independently of the RNA trigger and requires *Met1* for maintenance. *Curr. Biol.* **11:** 747–757.

Jones A.L., Thomas C.L., and Maule A.J. 1998. De novo methylation and co-suppression induced by a cytoplasmically replicating plant RNA virus. *EMBO J.* **17:** 6385–6393.

Jones L., Hamilton A.J., Vionnet O., Thomas C.L., Maule A.J., and Baulcombe D.C. 1999. RNA-DNA interactions and DNA methylation in post-transcriptional gene silencing. *Plant Cell* **11:** 2291–2301.

Jorgensen R.A. 1995. Cosuppression, flower color patterns, and metastable gene expression states. *Science* **268:** 686–691.

Kassis J.A. 1994. Unusual properties of regulatory DNA from the *Drosophila engrailed* gene: Three "pairing sensitive" sites within a 1.6 kb region. *Genetics* **136:** 1025–1038.

Kassis J.A., VanSickle E.P., and Sensabaugh S.M. 1991. A fragment of *engrailed* regulatory DNA can mediate transvection of the *white* gene in *Drosophila. Genetics* **128:** 751–761.

Kassis J.A., Noll E., VanSickle E.P., Odenwald W.F., and Perrimon N. 1992. Altering the insertional specificity of a *Drosophila* transposable element. *Proc. Natl. Acad. Sci.* **89:** 1919–1923.

Kataoka Y., Takeichi M., and Uemura T. 2001. Developmental roles and molecular characterization of a *Drosophila* homologue of *Arabidopsis Argonaute1*, the founder of a novel gene superfamily. *Genes Cells* **6:** 313–325.

Kelly W.G. and Fire A. 1998. Chromatin silencing and the maintenance of a functional germline in *Caenorhabditis elegans. Development* **125:** 2451–2456.

Kennerdell J.R. and Carthew R.W. 1998. Use of dsRNA-mediated genetic interference to demonstrate that *frizzled* and *frizzled2* act in the wingless pathway. *Cell* **95:** 1017–1026.

Kennison J.A. 1995. The Polycomb and trithorax Group proteins of *Drosophila:* Trans-regulators of homeotic gene function. *Ann. Rev. Genet.* **29:** 289–303.

Ketting R.F. and Plasterk R.H.A. 2000. A genetic link between co-suppression and RNA interference in *C. elegans. Nature* **404:** 296–298.

Ketting R.F., Haverkamp T.H., van Luenen H.G.A.M. and Plasterk H.A. 1999. *mut-7* of *C. elegans*, required for transposon silencing and RNA interference, is a homolog of Werner Syndrome helicase and RNaseD. *Cell* **99:** 133–141.

Ketting R.F., Fischer S.E.J., Bernstein E., Sijen T., Hannon G.J., and Plasterk R.H.A. 2001. Dicer functions in RNA interference and in synthesis of small RNA involved in developmental timing in *C. elegans. Genes Dev.* **15:** 2654–2659.

Lehner B., Williams G., Campbell R.D., and Sanderson C.M. 2002. Antisense transcripts in the human genome. *Trends Genet.* **18:** 63–65.

Lewis E.B. 1954. The theory and application of a new method of detecting chromosomal rearrangements in *Drosophila melanogaster. Am. Nat.* **88:** 225–239.

Lipardi C., Wei Q., and Paterson B.M. 2001. RNAi as random degradative PCR: siRNA primers convert mRNA into dsRNAs that are degraded to generate new siRNAs. *Cell* **107:** 297–307.

Malinsky S., Bucheton A., and Busseau I. 2000. New insights on homology-dependent silencing of I factor activity by transgenes containing ORF1 in *Drosophila melanogaster. Genetics* **156:** 1147–1155.

Mallin D.R., Myung J.S., Patton J.S., and Geyer P.K. 1998. Polycomb group repression is blocked by the *Drosophila suppressor of Hairy-wing* [*su(Hw)*] insulator. *Genetics* **148:** 331–339.

Marathe R., Smith T.H., Anandalakshmi R., Bowman L.H., Fagard M., Mourrain P., Vaucheret H., and Vance V.B. 2000. Plant viral suppressors of post-transcriptional silencing do not suppress transcriptional silencing. *Plant J.* **22:** 51–59.

Matzke M.A., Primig M., Trnovsky J., and Matzke A.J.M. 1989. Reversible methylation and inactivation of marker genes in sequentially transformed tobacco plants. *EMBO J.* **8:** 643–649.

Mette M.F., van der Winder J., Matzke M.A., and Matzke A.J.M. 1999. Production of aberrant promoter transcripts contributes to methylation and silencing of unlinked homologous promoters in trans. *EMBO J.* **18:** 241–248.

Mette M.F., Aufsatz W., van der Winder J., Matzke M.A., and Matzke A.J.M. 2000. Transcriptional silencing and promoter methylation triggered by double-stranded RNA. *EMBO J.* **19:** 5194–5201.

Metzlaff M., O'Dell M., Cluster P.D., and Flavell R.B. 1997. RNA-mediated RNA degradation and chalcone synthase A silencing in Petunia. *Cell* **88:** 845–854.

Mihaly J., Mishra R.K. and Karch F. 1998. A conserved sequence motif in Polycomb-response elements. *Mol. Cell* **1:** 1065–1066.

Misquitta L. and Paterson B.M. 1999. Targeted disruption of gene function in *Drosophila* by RNA interference (RNAi): A role for *nautilus* in embryonic somatic muscle formation. *Proc. Natl. Acad. Sci.* **96:** 1451–1456.

Mittelsten-Scheid O., Afsar K., and Paszkowski J. 1998. Release of epigenetic gene silencing by *trans*-acting mutations in *Arabidopsis. Proc. Natl. Acad. Sci.* **95:** 632–637.

Montgomery M.K., Xu S., and Fire A. 1998. RNA as a target of dsRNA-mediated genetic interference in *Caenorhabditis elegans. Proc. Natl. Acad. Sci.* **95:** 15502–15507.

Morel J., Mourrain P., Beclin C., and Vaucheret H. 2000. DNA methylation and chromatin structure affect transcriptional and post-transcriptional transgene silencing in *Arabidopsis. Curr. Biol.* **10:** 1591–1594.

Muchardt C., Guilleme M., Seeler J.S., Trouche D., Dejean A., and Yaniv M. 2002. Coordinated methyl and RNA binding is required for heterochromatin localization of mammalian HP1alpha. *EMBO Rep.* **3:** 975–981.

Muller J., Hart C.M., Francis N.J., Vargas M.L., Sengupta A., Wild B., Miller E.L., O'Connor M.B., Kingston R.E., and Simon J.A. 2002. Histone methyltransferase activity of a *Drosophila* Polycomb group repressor complex. *Cell* **111:** 197–208.

Muller M., Hagstrom K., Gyurkovics H., Pirrotta V., and Schedl P. 1999. The *Mcp* element from the *Drosophila melanogaster bithorax* complex mediates long-distance regulatory interactions. *Genetics* **153:** 1333–1356.

Napoli C., Lemieux C., and Jorgenson R. 1990. Introduction of a chimeric chalcone synthase gene in Petunia results in reversible co-suppression of homologous genes in trans. *Plant Cell* **2:** 279–289.

Ngo H., Tschudi C., Gull K., and Ullu E. 1998. Double-stranded RNA induces mRNA degradation in *Trypanosoma brucei. Proc. Natl. Acad. Sci.* **95:** 14687–14692.

Pal-Bhadra M., Bhadra U., and Birchler J.A. 1997. Cosuppression in *Drosophila:* Gene silencing of *Alcohol dehydrogenase* by *white-Adh* transgenes is Polycomb dependent. *Cell* **90:** 479–490.

———. 1999. Cosuppression of nonhomologous transgenes in *Drosophila* involves mutually related endogenous sequences. *Cell* **99:** 35–46.

———. 2002. RNAi related mechanisms affect both transcriptional and posttranscriptional transgene silencing in *Drosophila. Mol. Cell* **9:** 315–327.

Pelissier T. and Wassenegger M. 2000. A DNA target of 30 bp is sufficient for RNA-directed DNA methylation. *RNA* **6:** 55–65.

Pirrotta V. 1991. The genetics and molecular biology of *zeste* in *Drosophila melanogaster. Adv. Genet.* **29:** 301–348.

Pritchard M.A., Dura J.M., Pelisson A., Bucheton A., and Finnegan D.J. 1988. A cloned I-factor is fully functional in *Drosophila melanogaster. Mol. Gen. Genet.* **214:** 533–540.

Rabinow L., Nguyen-Huynh A.T., and Birchler J.A. 1991. A *trans*-acting regulatory gene that inversely affects the expression of the *white, brown* and *scarlet* loci in *Drosophila. Genetics* **129:** 463–480.

Ronsseray S., Lehmann M., Nouaud D., and Anxolabehere D. 1996. The regulatory properties of autonomous subtelomeric P elements are sensitive to a Suppressor of variegation in *Drosophila melanogaster. Genetics* **143:** 1663–1674.

Ronsseray S., Marin L., Lehmann M., and Anxolabehere D. 1998. Repression of hybrid dysgenesis in *Drosophila melanogaster* by combinations of telomeric P-element reporters and naturally occurring P elements. *Genetics* **149:** 1857–1866.

Ruiz F., Vassie L., Klotz K., Sperling L., and Madeddu L. 1998. Homology-dependent gene silencing in Paramecium. *Mol. Biol. Cell* **9:** 931–943.

Schmidt A., Palumbo G., Bozzetti M.P., Tritto P., Pimpinelli S., and Schafer U. 1999. Genetic and molecular characterization of *sting*, a gene involved in crystal formation and meiotic drive in the male germ line of *Drosophila melanogaster*. *Genetics* **151:** 749–760.

Schupbach T. and Wieschaus E. 1991. Female sterile mutations on the second chromosome of *Drosophila melanogaster*. *Genetics* **129:** 1119–1136.

Sigrist C.J.A. and Pirrotta V. 1997. Chromatin insulator elements block the silencing of a target gene by the *Drosophila* Polycomb response element (PRE) but allow trans interactions between PREs on different chromosomes. *Genetics* **147:** 209–221.

Sijen T., Vijn I., Rebocho A., van Blokland R., Roelofs D., Mol J.N., and Kooter J.M. 2001. Transcriptional and posttranscriptional gene silencing are mechanistically related. *Curr. Biol.* **11:** 436–440.

Smardon A., Spoerke J.M., Stacey S.C., Klein M.E., Mackin N., and Maine E.M. 2000. EGO-1 is related to RNA-directed RNA polymerase and functions in germ-line development and RNA interference in *C. elegans*. *Curr. Biol.* **10:** 169–178.

Stapleton W., Das S., and McKee B.D. 2001. A role of the *Drosophila* homeless gene in repression of Stellate in male meiosis. *Chromosoma* **110:** 228–240.

So W.V. and Rosbash M. 1997. Post-transcriptional regulation contributes to *Drosophila* clock gene mRNA cycling. *EMBO J.* **16:** 7146–7155.

Tabara H., Grishok A., and Mello C.C. 1998. RNAi in *C. elegans:* Soaking in the genome sequence. *Science* **282:** 430–431.

Tabara H., Sarkissian M., Kelly W.G., Fleenor J., Grishok A., Timmons L., Fire A., and Mello C.C. 1999. The *rde-1* gene, RNA interference, and transposon silencing in *C. elegans*. *Cell* **99:** 123–132.

Tuschl R., Zamore P.D., Lehman R., Bartel D.P., and Sharp P.A. 1999. Targeted mRNA degradation by double-stranded RNA in vitro. *Genes Dev.* **13:** 3191–3197.

van der Krol A.R., Mur L.A., Beld M., Mol J.N.M., and Stuitje A.R. 1990. Flavonoid genes in Petunia: Addition of a limited number of gene copies may lead to a suppression of gene expression. *Plant Cell* **2:** 291–299.

Voinnet O., Vain P., Angell S., and Baulcombe D.C. 1998. Systemic spread of sequence-specific transgene RNA degradation in plants is initiated by localized introduction of ectopic promoterless DNA. *Cell* **95:** 177–187.

Wassenegger M., Heimes S., Riedel L., and Sanger H.L. 1994. RNA-directed de novo methylation of genomic sequences in plants. *Cell* **76:** 567–576.

Wianny F. and Zernicka-Goetz M. 2000. Specific interference with gene function by double-stranded RNA in early mouse development. *Nat. Cell Biol.* **2:** 70–75.

Wilson J.E., Connell J.E., and Macdonald P.M. 1996. *aubergine* enhances *oskar* translation in the *Drosophila* ovary. *Development* **122:** 1631–1639.

Wu C.T. and Goldberg M.L. 1989. The *Drosophila zeste* gene and transvection. *Trends Genet.* **5:** 189–194.

Wu-Scharf D., Jeong B., Zhang C., and Cerutti H. 2000. Transgene and transposon silencing in *Chlamydomonas reinhardtii* by a DEAH-Box RNA helicase. *Science* **290:** 1159–1163.

Yang D., Lu H., and Erickson J.W. 2000. Evidence that processed small dsRNAs may mediate sequence-specific mRNA degradation during RNAi in *Drosophila* embryos. *Curr. Biol.* **10:** 1191–1200.

Zamore P.D., Tuschl T., Sharp P.A., and Bartel D.P. 2000. RNAi: Double-stranded RNA directs the ATP-dependent cleavage of mRNA at 21 to 23 nucleotide intervals. *Cell* **101:** 25–33.

zu Putlitz J. and Wands J.R. 1999. Specific inhibition of hepatitis B virus replication by sense RNA. *Antisense Nucleic Acid Drug Dev.* **9:** 241–252.

# CHAPTER 3

# Regulation of the Genome by Double-stranded RNA

Marjori Matzke, M. Florian Mette, Tatsuo Kanno, Werner Aufsatz, and Antonius J.M. Matzke

*Institute of Molecular Biology, Austrian Academy of Sciences, A-5020 Salzburg, Austria*

RNA SILENCING, OR EPIGENETIC GENE SILENCING EFFECTS that are triggered by double-stranded RNA (dsRNA), occurs predominantly in the cytoplasm and has been termed variously posttranscriptional gene silencing (PTGS) in plants, RNA interference (RNAi) in animals, and quelling in the filamentous fungus *Neurospora crassa*. Briefly, PTGS/RNAi involves the targeted elimination of mRNA by a homologous dsRNA that is processed into short interfering RNAs (siRNAs) approximately 21–24 nucleotides in length by an RNase-III-type enzyme termed Dicer. The antisense siRNAs are thought to guide an RNA-induced silencing complex (RISC) to the complementary mRNA, which is subsequently degraded (Figure 3.1) (Bernstein et al. 2001; Chicas and Macino 2001; Cogoni 2001; Tuschl 2001; Vance and Vaucheret 2001; Voinnet 2001; Hutvágner and Zamore 2002).

A second type of RNA silencing that is manifested at the genome level has been identified in plants and, to a growing extent, in other organisms (Matzke et al. 2001). The original process, recognized in plants, is termed RNA-directed DNA methylation (RdDM) (Wassenegger 2000, 2002). RdDM leads to de novo methylation of all cytosine residues within the region of sequence identity between the triggering RNA and the target DNA. Similar to the situation for PTGS/RNAi, RdDM involves a dsRNA that is processed to short RNAs slightly over 20 nucleotides in length. RdDM is presumed to be responsible for the DNA methylation observed in many cases of PTGS, and it has been implicated in a type of transcriptional gene silencing (TGS) that is elicited by dsRNAs containing promoter sequences. Although the phenomenon of RdDM is well established in plants, still unknown are the identity of the DNA methyltransferase(s) that catalyzes RdDM, the RNA species (short RNAs or dsRNA) that triggers RdDM, and the relationship between RdDM and changes in chromatin structure. Moreover, it is not yet clear whether RdDM occurs in organisms other than plants, although recent evidence suggests that chromatin modifications, even in organisms that lack DNA methylation, can be directed by guide RNAs. These issues are addressed in this chapter, with an emphasis on experimental work from plant systems.

## RNA-DIRECTED DNA METHYLATION

RdDM was first revealed in studies on viroid-infected transgenic plants. Viroids are plant pathogens that consist solely of a noncoding, closed circular RNA several hundred bases in length (Diener 2001; Flores 2001). In plants infected with replication-competent

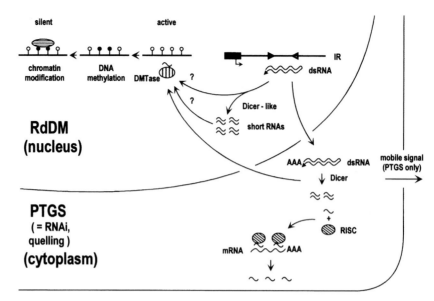

**FIGURE 3.1.** Model of RNA silencing in plants. dsRNA (*wavy lines*) can enter pathways that lead to RdDM in the nucleus and/or PTGS, the plant equivalent of RNAi in animals and quelling in *Neurospora*, in the cytoplasm. dsRNA transcribed from an inverted DNA repeat (IR) can be processed to short RNAs in the nucleus by a nuclear Dicer-like activity, or if polyadenylated (AAA) and transported in the cytoplasm, by a cytoplasmic form of Dicer. Nuclear and cytoplasmic short RNAs (or, possibly, dsRNA) can direct methylation of homologous DNA sequences (*closed circles*), perhaps by serving as a guide for a DNA methyltransferase (DMTase). Methylated DNA can attract chromatin-modifying proteins that lock in the silent state. In PTGS, antisense short RNAs are thought to associate with a RNA silencing complex "RISC" and target homologous mRNAs for degradation. PTGS, but not TGS, produces a mobile silencing signal that can trigger PTGS in adjacent cells and systemic tissues.

viroids, cDNA copies of the viroid that had been introduced as transgenes into plant chromosomes became methylated de novo (Wassenegger et al. 1994). These results indicated that the replicating viroid RNA was able to trigger methylation of the homologous cDNA copies in the genome.

Bisulfite-assisted genomic sequencing has demonstrated that RdDM induces dense de novo methylation of almost all cytosines (Cs) within the region of sequence identity between the triggering RNA and the target DNA (Pélissier et al. 1999; Wang et al. 2001; W. Aufsatz et al., in prep.). The minimum DNA target size of RdDM is approximately 30 bp (Pélissier and Wassenegger 2000). The methylation of cytosines in any sequence context, not just those in canonical CG dinucleotides, is a hallmark of RdDM. The restriction of methylation primarily to the region of RNA-DNA sequence identity strongly suggests that base-pairing interactions delimit the region of DNA modification. To place the discovery of RdDM into the context of plant gene-silencing research in the early 1990s, the range of homologous sequence interactions being invoked at that time to account for various gene silencing effects must be considered.

## RdDM AND HOMOLOGY-DEPENDENT GENE SILENCING

In contrast to the clear involvement of dsRNA in initiating RNAi in *Caenorhabditis elegans* (Fire et al. 1998), plant scientists working with transgenic systems a decade ago were confronted with a complex set of silencing phenomena that appeared to depend on various types of homologous nucleic acid sequence interactions. In addition to PTGS (which is called cosuppression when both a transgene and the homologous endogenous gene are coordinately silenced [Napoli et al. 1990; van der Krol et al. 1990]), homology-dependent gene silencing also included a type of TGS that affected unlinked transgenes sharing sequence identity predominantly in promoter regions (Matzke et al. 1989; Vaucheret 1992, 1993).

Both TGS and PTGS were frequently associated with sequence-specific DNA methylation, i.e., methylation was concentrated in protein-coding regions in cases of PTGS (Ingelbrecht et al. 1994) and in promoters in instances of TGS (Matzke et al. 1989; Vaucheret 1992, 1993). The discovery of RdDM provided a plausible explanation for the sequence-specific methylation associated with PTGS because the DNA sequence acquiring the modification was homologous to the RNA targeted for degradation by the PTGS mechanism. It was more difficult, however, to envision how promoter sequences, which are not normally thought to be transcribed, could become methylated via an RNA signal. Experiments designed to study the role of RNA signals in de novo methylation of promoter sequences demonstrated that promoters could indeed be targets of RdDM. Moreover, these studies provided some of the best evidence that RdDM is initiated by dsRNA.

## RNA-DIRECTED METHYLATION OF PROMOTER SEQUENCES AND TGS

*translational gene silencing*

Investigation of the role of RNA signals in promoter methylation and TGS followed from unusual cases of unidirectional transmission of methylation from one transgene promoter to an unlinked homologous copy in somatic cells (Matzke et al. 1989; Vaucheret 1992, 1993). The sequence specificity of methylation inspired models invoking DNA-DNA pairing to explain this process (Matzke et al. 1994; Bender 1998). Limitations in these models became apparent when methylation failed to be serially transmissible from one modified promoter to another, as would be expected if only DNA pairing were involved. As an alternative explanation for promoter-specific methylation and TGS induced by homologous sequence interactions, it was suggested that RNAs comprising promoter sequences might be synthesized unintentionally from scrambled, multicopy transgene inserts. The promoter RNAs were proposed to function as *trans*-acting signals that could trigger methylation and silencing of homologous promoters elsewhere in the genome (Park et al. 1996).

To test this hypothesis, experiments in which promoter sequences were deliberately transcribed in transgenic plants were carried out. These studies demonstrated that dsRNAs transcribed from a promoter inverted DNA repeat (IR), but not single-stranded promoter RNAs, were indeed able to induce TGS and methylation of homologous "target" promoters in *trans* (Figure 3.1) (Mette et al. 1999, 2000). The requirement for promoter dsRNA was confirmed in two ways: first, by epigenetically silencing the transcription of the promoter inverted repeat, which abolished synthesis of promoter dsRNAs without inducing a structural change in the inverted repeat region (Mette et al. 1999); and second, by using Cre/*lox*-mediated site-specific recombination to remove the promoter transcribing the inverted repeat (Aufsatz et al. 2002a). Both of these approaches led to loss of methylation and reactivation of the target promoter, thus implicating the promoter dsRNA in the *trans*-silencing and methylation phenomenon.

dsRNA transcribed from promoter inverted repeats has been used to silence a number of promoters in plants, including those of several endogenous plant genes (Sijen et al. 2001; W. Aufsatz and A. Matzke, unpubl.). Moreover, two previously characterized transgene silencing loci in tobacco, the $H_2$ locus, which induces methylation and silencing of the nopaline synthase promoter (Jakowitsch et al. 1999), and the *271* locus, which induces methylation and silencing of the cauliflower mosaic virus 35S and 19S promoters (Vaucheret 1992, 1993), have been shown to generate promoter short RNAs. The $H_2$ and *271* loci were not originally designed to produce promoter dsRNAs, and their structures do not reveal any obvious ways to synthesize dsRNAs containing promoter sequences. Therefore, the promoter short RNAs in both plant lines are presumably derived from unintended synthesis of promoter dsRNA from the respective complex transgene inserts (M.F. Mette, unpubl.; R. van Blokland, J. Kooter, and H. Vaucheret, pers. comm.).

## A SEPARATE PROCESSING PATHWAY FOR dsRNA IN PLANT NUCLEI?

Similar to the dsRNA involved in PTGS/RNAi, promoter dsRNAs transcribed from promoter inverted repeats in the nucleus are cleaved into short RNAs of both polarities approximately 21–24 nucleotides in length (Figure 3.1) (Mette et al. 2000; Sijen et al. 2001). Evidence that this processing step occurs via a distinct nuclear pathway in plants has been obtained using an RNA virus protein, called helper-component proteinase (HC-Pro), which is able to suppress PTGS and block the accumulation of siRNAs (Llave et al. 2000; Mallory et al. 2001). In contrast to its effects on PTGS, HC-Pro did not suppress dsRNA-mediated TGS (Marathe et al. 2000; Mette et al. 2001). Moreover, instead of preventing the accumulation of the promoter short RNAs, the abundance of these RNAs in one system actually increased approximately fivefold in HC-Pro-expressing plants (Mette et al. 2001). This observation was used to argue that the promoter dsRNA, which is not polyadenylated, is processed to short RNAs in the nucleus, because HC-Pro is an exclusively cytoplasmic protein. Further evidence for the nuclear localization of the promoter short RNAs was provided by their inability to efficiently induce degradation of homologous single-stranded promoter RNAs, a PTGS-like process that would take place primarily in the cytoplasm (Mette et al. 2001).

Following from these results, a current model of RNA silencing in plants postulates the existence of distinct nuclear and cytoplasmic pathways for dsRNA processing (Figure 3.1). According to this model, dsRNAs in the nucleus would be processed by a nuclear Dicer-like activity, whereas dsRNAs that are either synthesized in, or introduced into, the cytoplasm would be degraded by a cytoplasmic form of Dicer. Whether the nuclear or the cytoplasmic pathway is entered by a dsRNA that is synthesized in the nucleus probably depends in part on whether it is polyadenylated, which would presumably affect how efficiently it is transported to the cytoplasm (Carmo-Fonseca et al. 1999; Zhao et al. 1999).

## RNA SIGNALS FOR RdDM PRODUCED IN THE CYTOPLASM

Studies with promoter dsRNAs transcribed from promoter inverted repeats have demonstrated that RdDM in plants can be initiated by dsRNAs that are synthesized and probably processed in the nucleus. In a twist to this process, an RNA species that is produced in the cytoplasm as a consequence of PTGS has been shown to enter the nucleus and trigger RdDM of homologous DNA sequences (Figure 3.1). Silencing occurring in the cytoplasm can thus be imprinted on the genome level. This phenomenon has been shown most convincingly by infecting plants with RNA viruses that are engineered with sequences homologous to nuclear DNA. De novo methylation of the homologous DNA sequences is observed only upon the initiation of virus replication, which takes place exclusively in the cytoplasm (A.L. Jones et al. 1998; Jones et al. 1999, 2001).

Depending on the sequence introduced into an RNA viral genome, either PTGS or TGS of homologous host genes can be induced in infected plants. RNA viruses modified with protein-coding sequences are able to trigger PTGS in the cytoplasm and to direct methylation of DNA sequences homologous to the target mRNA. When RNA viruses are engineered to contain sequences homologous to promoters of host nuclear genes, they can induce TGS and methylation of host promoter regions (Jones et al. 1999, 2001). The nature of the RNAs that are produced in the cytoplasm and that subsequently enter the RdDM pathway is not yet clear. It is assumed that they are either dsRNAs or short RNAs similar to those implicated in the proposed nuclear pathway of dsRNA processing and RdDM (Figure 3.1).

## MOBILE SILENCING SIGNALS AND DNA METHYLATION

In plants, PTGS, but not TGS (Mlotshwa et al., 2002), is able to move from cell to cell through plasmodesmata and systemically through the vascular system to induce silencing at some distance from the site of local initiation (Figure 3.1). The mobile silencing signal has not been clearly defined, but it is hypothesized to contain an RNA molecule because of the sequence specificity of silencing and the resemblance of signal trafficking to viral movement (Mlotshwa et al. 2002). Mobile signals are able to induce not only PTGS, but also methylation of homologous DNA sequences at distant sites (Jones et al. 1999). Indeed, recent results suggest that the mobile silencing signal and the methylation signal might be the same RNA entity. The cucumber mosaic virus 2b protein (Cmv2b), which can suppress PTGS, has been shown to eliminate the mobile silencing signal and to reduce methylation of nuclear transgenes (Guo and Ding 2002; Mlotshwa et al. 2002).

## MECHANISM OF RdDM

Analyzing the mechanism of RdDM requires a well-defined, manipulable system that can be dissected by genetic and biochemical approaches. Additional tools available to plant biologists for studying RdDM consist of plant viral proteins that suppress the PTGS pathway at different steps.

One example of RdDM that has been well characterized at the molecular level is a TGS system involving the nopaline synthase promoter (NOSpro), a constitutive plant promoter with weak to moderate activity in most plant organs. The NOSpro TGS system is made up of two unlinked transgene loci: a target locus and a silencing locus (Figure 3.2). The target locus comprises a NOSpro-*NPTII* gene that is active and unmethylated when the silencing locus is not present. The silencing locus contains a NOSpro inverted repeat that is transcribed by the 35S promoter, a strong constitutive plant viral promoter. The silencing locus produces NOSpro dsRNA, which is processed into short RNAs approximately 21–24 nucleotides in length. Either the NOSpro dsRNA or the short RNA processing products are able to diffuse through the nucleoplasm and trigger de novo methylation and TGS of the target NOSpro in *trans* (Figure 3.2). Silencing and methylation of the target NOSpro are alleviated when the two loci segregate from each other in progeny plants. The reversibility of target locus methylation and inactivation demonstrates a strict requirement for the silencing locus in this promoter methylation-TGS phenomenon.

The NOSpro dsRNA-mediated TGS system was established initially in tobacco (Mette et al. 1999, 2000) and more recently in *Arabidopsis* (Aufsatz et al. 2002a,b). This silencing system is being used to investigate:

- the nature of the RNA molecule required for RdDM,
- the identity of the enzyme that cleaves promoter dsRNA to short RNAs,

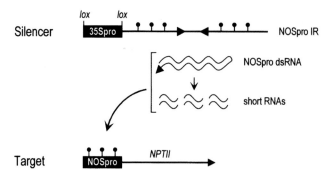

**FIGURE 3.2.** NOSpro dsRNA-mediated TGS system. The target locus encodes a NOSpro-*NPTII* gene. The silencing locus encodes NOSpro dsRNA that is transcribed from a NOSpro inverted repeat by the 35S promoter. Either the dsRNA or short RNA processing products can diffuse through the nucleoplasm and trigger de novo methylation (*closed circles*) and transcriptional silencing of the homologous NOSpro at the target locus. A requirement for NOSpro dsRNA in the *trans*-silencing and methylation phenomenon was demonstrated by removing the 35S promoter via Cre/*lox*-mediated recombination. This step abolished transcription of NOSpro dsRNA, resulting in loss of methylation and reactivation of the target NOSpro.

- the identity of the DNA methyltransferase(s) (MTase[s]) involved in RdDM, and
- the requirement for chromatin-modifying factors to either establish or maintain promoter methylation and TGS (Figure 3.3).

## RdDM: Short RNAs or dsRNA?

Although dsRNAs that are processed into short RNAs are associated with RdDM, it is not clear which of these two RNA species is actually required for directing modification of homologous DNA sequences (Figure 3.1). In addition, it is not clear whether the triggering RNA associates first with a protein catalyzing an epigenetic change, such as a DNA MTase or chromatin-modifying factor, and guides it to the homologous DNA sequence, or whether the RNA initially base pairs with DNA, creating a structural perturbation that attracts a modifying protein to that region of the genome (Matzke et al. 2001).

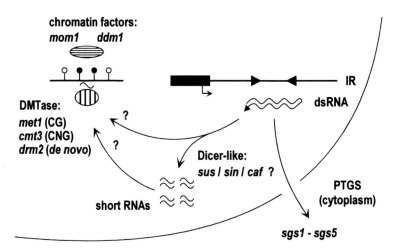

**FIGURE 3.3.** Genetic analysis of NOSpro dsRNA-mediated TGS and promoter RdDM. Several existing mutants that could affect this silencing system are possibly defective in dsRNA processing (*sus/sin/caf*), DNA methylation (*met1, cmt3, drm2*), chromatin remodeling (*mom1, ddm1*), and PTGS (*sgs1-sgs5*).

Intuitively, short RNAs are the more plausible candidates for serving a guide function and for having unimpeded access to DNA. Moreover, their size (~21–24 nucleotides) is close to the minimum target size (~30 bp) of RdDM (Pélissier and Wassenegger 2000). Two approaches are being followed to assess the involvement of short RNAs in RdDM: (1) the use of plant viral proteins that block the accumulation of siRNAs required for PTGS and (2) genetic screens to find mutations in proteins that produce or stabilize short RNAs.

HC-Pro, a viral suppressor of PTGS that prevents the accumulation of siRNAs (Llave et al. 2000; Mallory et al. 2001), provides a potential tool for studying the role of short RNAs in RdDM. In studies on PTGS systems, conflicting results on changes in methylation in HC-Pro-expressing plants have been reported. In two cases, HC-Pro had no effect on methylation in the protein-coding region of a silenced nuclear transgene (Jones et al. 1999; Mallory et al. 2001), whereas a third study suggested partial loss of methylation (Llave et al. 2000). In a TGS system, where HC-Pro failed to suppress silencing, a slight increase in target promoter methylation was observed in HC-Pro-expressing plants (Mette et al. 2001). Because the increased methylation was associated with enhanced accumulation of promoter short RNAs without a concomitant increase in the level of promoter dsRNA, it was suggested that short RNAs are involved in triggering promoter methylation (Mette et al. 2001). Similar support for short RNA involvement in RdDM emerged from two other studies (Llave et al. 2000; Wang et al. 2001).

The analysis of short RNAs is becoming increasingly complicated. Recent experiments using a variety of viral suppressors of PTGS have suggested that there are at least two distinct size classes of short RNAs, one shorter (~21–22 nucleotides) and one longer (~24–26 nucleotides), which can trigger either PTGS or RdDM, respectively (Hamilton et al. 2002). Different viral suppressors prevent accumulation of specific size classes of short RNAs. One of these, the p19 protein of tomato bushy stunt virus, blocks accumulation of all sizes of short RNAs (Hamilton et al. 2002), making it useful for future analyses of RdDM and the proposed nuclear pathway of dsRNA processing. The Cmv2b protein, which suppresses PTGS, is localized in the nucleus (Lucy et al. 2000). Because the Cmv2b protein affects both the mobile silencing signal and DNA methylation (Guo and Ding 2002), it also provides a potentially powerful tool for dissecting the RdDM pathway.

A second way to assess the RNA species involved in RdDM is to recover mutations in the Dicer-like enzyme(s) responsible for processing dsRNAs to short RNAs (Figure 3.3). With current models depicting separate nuclear and cytoplasmic pathways for dsRNA processing in plants (Figure 3.1), it might be anticipated that distinct nuclear and cytoplasmic versions of Dicer will be identified. Indeed, one of three possible Dicer homologs in *Arabidopsis*—a protein variously termed SUS, SIN, or CAF and recently renamed DICER-LIKE1 (DCL1) (Schauer et al. 2002)—has two putative bipartite nuclear localization signals, suggesting that a variant of this protein might localize to the nucleus and function there. Thus, DCL1 is an attractive candidate for producing promoter short RNAs by the proposed nuclear pathway for dsRNA processing. A nuclear form of a Dicer-like activity might be unique to plants; the human Dicer is encoded by a single gene, and the protein is localized exclusively in the cytoplasm (Billy et al. 2001).

Several mutations in the *Arabidopsis DCL1* gene are available: two missense mutations in the RNA helicase domain (*sin1* alleles; Golden et al. 2002) and a T-DNA insertion removing the second of two dsRNA-binding domains (*caf1* allele; Jacobsen et al. 1999). Mutations in the RNase III domain, which would probably have the biggest impact on dsRNA processing, have not yet been recovered. Tests of the effects of the *sin1* and *caf1* alleles on NOSpro dsRNA-mediated TGS are in progress. Mutants defective in the other two Dicer-like homologs in *Arabidopsis* have not yet been isolated.

Finally, despite the attractiveness of short RNAs and the diversity of their sizes and potential functions in gene silencing, it cannot yet be ruled out that unprocessed dsRNA interacts directly with DNA to induce RdDM. The nature of the sequence interactions between DNA and the triggering RNA (RNA-DNA heteroduplex? RNA-DNA triple helix?) also remains to be determined.

## DNA MTases and RdDM

RdDM must be considered with respect to both de novo and maintenance methylation steps. It is postulated that as long as the triggering RNA is present, methylation will be catalyzed continuously by a de novo DNA MTase activity at nearly all cytosines within the region of RNA-DNA sequence identity. If the RNA signal is removed, only methylation that can be maintained by maintenance DNA MTases will persist. The pattern of methylation resulting from RdDM at any given time is thus dependent on the availability of the RNA signal and on the presence of various DNA MTases.

In *Arabidopsis*, there are four known or putative classes of DNA MTases (Finnegan and Kovac 2000). The "domain-rearranged" (DRM) class, which is homologous to the Dnmt3 family in mammals, is probably the major de novo MTase (Cao et al. 2000; Cao and Jacobsen 2002). Two classes of DNA MTases appear to serve primarily maintenance functions. MET1, which is homologous to mammalian Dnmt1, maintains methylation predominantly in CG dinucleotides (Kishimoto et al. 2001), although similar to Dnmt1, it might have de novo activity in certain cases (Bestor 2000). CMT3, a member of the chromomethylase (CMT) family that is found only in plants (Henikoff and Comai 1998), maintains methylation primarily at CNG trinucleotides (Lindroth et al. 2001; Papa et al. 2001). The Dnmt2 class is an enigmatic type of putative DNA MTase that is found in vertebrates, plants, and two organisms not thought to methylate their DNA: *Schizosaccharomyces pombe* and *Drosophila melanogaster*. Although the type of DNA methylation, if any, catalyzed by Dnmt2 is unclear (Bestor 2000), non-CGs are possibly the preferred target sites (Lorincz and Groudine 2001). Mutants defective in three classes of DNA MTases in *Arabidopsis* MET1, CMT3, and DRM2, are available and are being tested for their effects on RdDM and their ability to release NOSpro dsRNA-mediated TGS (Figure 3.3).

MET1 is essential for maintaining methylation induced by RdDM, both in plants where the RNA signal is continuously synthesized and in cases where the RNA signal has been removed. In an experiment exemplifying the latter, MET1 was required to maintain CG methylation and TGS induced in a previous generation by an RNA virus engineered with sequences homologous to the promoter of a nuclear gene (Jones et al. 2001). MET1 was also needed to maintain full methylation of a target promoter in plants that continued to synthesize promoter dsRNA from a promoter inverted repeat (Aufsatz et al. 2002a). In a separate study, mutations in MET1 stochastically reduced silencing and methylation associated with *cis*-TGS (which is probably not triggered by RNA) and PTGS (Morel et al. 2000).

The CMTs are unique because they are plant-specific and they contain a chromodomain. Chromodomains are highly conserved sequence motifs present in proteins involved in remodeling chromatin structure in plants and animals (Jones et al. 2000). CMT was suggested as a promising candidate for RdDM (Habu et al. 2001; Matzke et al. 2001) following a report that the chromodomain of the *Drosophila* MOF histone acetylase is an RNA-protein interaction module (Akhtar et al. 2000). When crossed into the NOSpro dsRNA-mediated TGS system, however, the *cmt3* mutation had no effect on TGS in the $F_2$ generation, which is the first generation when a recessive mutation becomes homozygous and therefore effective (W. Aufsatz et al., unpubl.). Methylation of the target NOSpro must still be examined in the *cmt3* mutant background. Although silencing might be released in advanced generations, it appears unlikely that RdDM is initially catalyzed by CMT3, at least in the NOSpro TGS system. Methylation in CNG trinucleotides, however, should be maintained by CMT3. If CNG methylation is critical for silencing of a particular promoter, as appears to be the case for SUPERMAN epialleles (Lindroth et al. 2001) and members of the PAI gene family (Bartee et al. 2001) as well as certain transposable elements (Lindroth et al. 2001; Tompa et al. 2002), mutations in CMT3 can lead to reactivation of gene expression.

Experiments to test *drm2* mutants with the NOSpro dsRNA-mediated TGS system are in progress. In view of the questionable de novo activity of MET1 and the negligible effect of the *cmt3* mutation on this system, the DRM class remains a good candidate for the initiation of RdDM and silencing. However, the involvement of more than one de novo DNA MTase in RdDM cannot be ruled out (Aufsatz et al. 2002b).

## Chromatin Factors and RdDM

Given the close relationships between DNA methylation and chromatin remodeling factors (Jeddeloh et al. 1999; Gibbons et al. 2000; Dennis et al. 2001), and DNA methylation and histone modifications such as acetylation (Chen and Pikaard 1997; Nan et al. 1998; P. Jones et al. 1998; Fuks et al. 2000) and methylation (Tamaru and Selker 2001; Jackson et al. 2002), it might be anticipated that RdDM would be associated with chromatin modifications. Mutants defective in two chromatin remodeling factors are being assessed for their ability to release TGS and RdDM (Figure 3.3). In the NOSpro dsRNA-mediated TGS system, mutations in DDM1, a putative component of a SWI/SNF chromatin remodeling complex (Jeddeloh et al. 1999), induce a partial, stochastic release of TGS and reduction of promoter methylation, whereas a mutation in MOM1, another possible chromatin remodeling factor (Amedeo et al. 2000), has no effect on either silencing or methylation (Aufsatz et al. 2002a). A separate investigation showed that *ddm1* mutations sporadically affect silencing and methylation associated with transgene *cis*-TGS and PTGS (Morel et al. 2000).

Additional *Arabidopsis* silencing mutants available for testing on the NOSpro dsRNA-mediated TGS system include *hog1* and *sil1*, which encode proteins that have not yet been identified (Furner et al. 1998). Mutants defective in chromatin modifications, including histone deacetylation (Murfett et al. 2001) and histone methylation (Jackson et al. 2002), as well as various Polycomb-Group (Pc-G) proteins such as CURLY LEAF (Goodrich et al. 1997) and VERNALIZATION 2 (Gendall et al. 2001), have been identified in genetic screens (Li et al. 2002) and will be analyzed for their effects on the NOSpro dsRNA-mediated TGS system. In addition, mutants impaired in PTGS (*sgs1-sgs5*), some of which also reduce transgene methylation (Elmayan et al. 1998), are being evaluated to identify overlapping components in the two pathways (Figure 3.3) (W. Aufsatz and H. Vaucheret, unpubl.).

In summary, although genetic analyses of the RdDM mechanism using existing silencing mutants are not yet completed, results obtained so far indicate a requirement for MET1 to maintain methylation triggered by RdDM, and an indirect role for chromatin remodeling by DDM1 to establish or maintain methylation triggered by RdDM (Morel et al. 2000; Aufsatz et al. 2002a). Of particular interest in the future will be the identification of the DNA MTase involved in the de novo methylation step of RdDM and the complete set of chromatin factors associated with initiating and retaining methylation induced by RdDM.

## Genetic Screens to Recover Novel RdDM-defective Mutants

Genetic screens to recover novel mutations are under way using the NOSpro dsRNA-mediated TGS system. One mutation (*rts1* for RNA-directed transcriptional silencing) reduces target promoter methylation approximately 50%, primarily in symmetrical CG and CNG nucleotide groups, while allowing almost full recovery of target gene expression. The partial uncoupling of target promoter methylation from promoter activity hinted that RTS1 could be a chromatin-modifying factor that is critical for silencing. The *rts1* mutation has been mapped recently and found to encode a histone deacetylase, HDA6 (Aufsatz et al. 2002b). The observation that methylation was reduced primarily in symmetrical C(N)Gs in the *hda6^{rts1}* mutant, while methylation of asymmetric CNNs was largely unaffected, suggested a model in which HDA6 acted in between the CG de novo and CG maintenance (or reinforcement) methylation steps (Aufsatz et al. 2002b). This model is consistent with the general picture emerging from studies with plants, animals, and fungi, which indicates that chromatin structure and methylation are intertwined. Chromatin modifications help to maintain DNA methylation, which in turn probably contributes to the stability and heritability of silencing (Bird 2001; Dobosy and Selker 2001; Richards and Elgin 2002).

## DOES RdDM OCCUR IN PLANTS ONLY?

RdDM was discovered in plants in 1994 and all subsequent reports of RdDM since that time have involved plant systems. Given the length of time since the discovery of RdDM, it is noteworthy that no definitive data either demonstrating or ruling out a similar phenomenon in animals have been reported. Therefore, it is legitimate to ask whether RdDM is unique to plants.

One argument for RdDM being restricted to plants (Bird 2001) invokes the pattern of methylation, in which cytosines in any sequence context become methylated. This contrasts with the distribution of methylation in animals, which is present predominantly in CG dinucleotides. A second argument revolves around the possible involvement of CMTs, which are found exclusively in plants (Habu et al. 2001; Matzke et al. 2001). Both of these arguments are losing strength as additional data accumulate on RdDM and the dynamics of DNA methylation patterns in animals.

As discussed in the section on genetic analysis of RdDM, a promising candidate for the DNA MTase catalyzing RdDM in plants is not CMT, but DRM, the de novo MTase that is related to the mammalian Dnmt3 family. If the DRM/Dnmt3 class indeed catalyzes RdDM, then this process cannot be ruled out in mammals on the basis of the absence of the appropriate DNA MTase. With respect to patterns of methylation, a recent report indicates that substantial non-CG methylation might indeed be present in mammals early in development (Ramsahoye et al. 2000). The non-CG methylation is possibly catalyzed by Dnmt3a, which is the major de novo MTase that reestablishes methylation patterns in early stages of mammalian embryogenesis. It is conceivable that both CG and non-CG methylation are initially guided by RNA in postimplantation mammalian embryos; then, as the activity of Dnmt3a subsides later in development, a maintenance MTase activity takes over and only CG methylation is retained in adult somatic cells.

This scenario supports the argument that the pattern of methylation during a particular stage of development in mammals or plants is determined by the DNA MTases and the methylation signals that are available at a given time. Various types of DNA MTases differ in their activities during the course of plant and mammalian development. In plants, de novo MTase activity persists throughout development, potentially allowing continuous RdDM if suitable RNA signals are produced. Moreover, maintenance activities for both CG and CNG nucleotide groups are present in plants, giving rise to heavily methylated genomes in adult vegetative cells (Figure 3.4A). In contrast, since the activity of a major de novo MTase activity, Dnmt3a, diminishes as mammalian development

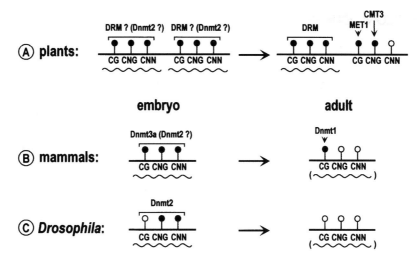

**FIGURE 3.4.** Potential role of RNA in establishing DNA methylation patterns in plants and animals. (*A*) In plants, de novo DNA MTases (DRM and possibly the Dnmt2 homolog) in conjunction with RNA (*wavy line*) are postulated to establish methylation (*closed circles*) at cytosines in any sequence context within the region of RNA-DNA sequence identity. During development, full methylation can be renewed continuously if the triggering RNA persists because DRM remains active throughout the plant life cycle. If the triggering RNA is withdrawn, maintenance MTases, MET1 and CMT3, retain methylation at CGs and CNGs, respectively. (*B*) In mammals, de novo DNA MTases (Dnmt3a and possibly Dnmt2) establish CG and non-CG methylation, postulated here to be targeted by RNA, early in embryonic development. As Dnmt3a activity decreases during development, non-CG methylation is lost (even if the triggering RNA persists), and only methylation in CGs is retained in adult somatic cells by the maintenance activity of Dnmt1. (*C*) In *Drosophila* embryos, Dnmt2 establishes non-CG methylation, postulated here to be targeted by RNA. When Dnmt2 activity declines in adult flies, methylation is lost (even if the triggering RNA persists) owing to the lack of a maintenance MTase activity.

unfolds (Ramsahoye et al. 2000), RdDM might not occur in adult somatic cells. Consequently, methylation would be present primarily in CG dinucleotides owing to a maintenance MTase activity such as Dnmt1 (Figure 3.4B).

In addition to the DRM/Dnmt3a class of de novo MTases, a second possible candidate for catalyzing RdDM is the enigmatic Dnmt2 class of putative DNA MTases that are present in vertebrates, plants, *D. melanogaster,* and, in a mutated form, in *S. pombe* (Bestor 2000). In an intriguing parallel to methylation patterns early in mammalian development, non-CG methylation has been observed in young *Drosophila* embryos (Lyko 2001). It has been proposed that this non-CG methylation, which would be consistent with the pattern of methylation induced by RdDM, is catalyzed by the *Drosophila* Dnmt2 (Lyko 2001). Because the activity of Dnmt2 declines during *Drosophila* development (Lyko et al. 2000) and no maintenance MTase is active, methylation disappears completely in adult flies and long-term silencing relies on chromatin modifications (Figure 3.4C). Dnmt2 in mammals is also active during early development (Okano et al. 1998), thus raising the interesting possibility that this DNA Mtase—either independently or in conjunction with Dnmt3a— is responsible for the non-CG methylation detected during that period (Figure 3.4B).

In summary, the occurrence of RdDM in organisms other than plants is suggested by the non-CG methylation present early in mammalian and *Drosophila* development. The identification of the de novo DNA MTase(s) required for RdDM will help to clarify the species distribution of RdDM. Current candidates for RdDM include the DRM/Dnmt3 class, which is present in plants and vertebrates, and the Dnmt2 class, a potentially active form present in plants, vertebrates, and *Drosophila*.

## DIFFERENT CONSEQUENCES OF METHYLATION IN PROTEIN-CODING REGIONS VS. PROMOTERS

RdDM leads to dense de novo methylation of almost all cytosines within a region of RNA-DNA sequence identity. This pattern presumably persists as long as the RNA signal and the appropriate de novo DNA MTase(s) are available. When the RNA signal is withdrawn, methylation is retained primarily at symmetrical cytosines (CG and CNG nucleotide groups in plants; CG dinucleotides in mammals) by DNA MTases that have maintenance activity (MET1 and CMT3 in plants; Dnmt1 in mammals). Owing to differences in the patterns of de novo and maintenance methylation, the consequences of RdDM might change over time.

Depending on the sequence composition of the triggering RNA, RdDM can lead to methylation of protein-coding regions, which is associated with PTGS, or of promoters, which leads to TGS. Methylation that is confined to symmetrical cytosines in protein-coding sequences probably does not interfere with transcription and hence is generally inconsequential for gene expression (Jones and Takai 2001). However, if dense methylation is present at all cytosines in a protein-coding region, transcription elongation can be inhibited, as has been observed with the methylation-induced premeiotically (MIP) phenomenon in the fungus *Ascobolus immersus* (Faugeron 2000). Prematurely terminated transcripts arising from highly methylated DNA templates might be sources of "aberrant" RNAs that are thought to be substrates for an RNA-directed RNA polymerase (RdRP), an enzyme required for PTGS in plants, quelling in *Neurospora*, and RNAi in *C. elegans* (Chicas and Macino 2001). Therefore, continuous RdDM of coding regions might fuel the PTGS pathway by leading to the production of templates for dsRNA synthesis via RdRP activity.

Depending on the sensitivity of a given promoter to cytosine methylation and the efficiency of establishing a repressive chromatin conformation at the promoter, silencing via promoter RNAs can potentially give rise to a highly stable, transcriptionally repressed state. Consequently, mitotic heritability and somatic stability of silencing can be achieved more reliably with TGS than with PTGS. Indeed, for one strong promoter in plants, RdDM and TGS were inherited through meiosis in the absence of the inducing dsRNA, which was transcribed either from a nuclear transgene (Park et al. 1996) or as part of an RNA virus genome (Jones et al. 2001). In both cases, the postmeiotic residual methylation was present primarily in CG and CNG nucleotide groups, as would be anticipated by the activity of maintenance DNA MTases. Thus, unlike maintenance methylation of symmetrical cytosines in protein-coding regions, which has negligible effects on gene expression, retention of CG and CNG methylation in promoters can still be associated with significant transcriptional silencing of the associated gene.

## RNA-GUIDED CHROMATIN MODIFICATIONS?

RdDM is well established in plants and possibly occurs early in mammalian and *Drosophila* development when an RNA-guided de novo DNA MTase is active. What about organisms that lack DNA methylation? Are alterations in chromatin structure guided by RNA? Suggestive evidence that components of the RNAi machinery have a role in chromosome-level regulation of transposons was obtained initially with several RNAi-defective mutants in *C. elegans*, which does not methylate its DNA (Ketting et al. 1999; Tabara et al. 1999; Dernburg et al. 2000). Recent work in *Drosophila*, *S. pombe*, and *C. elegans* has strengthened the idea that RNA is able to guide chromatin modifications in organisms that lack substantial DNA methylation.

Transgene TGS and PTGS in *Drosophila* were both found to be sensitive to mutations in the Piwi protein, which is a member of the Argonaute family of proteins required for

RNAi (Pal-Bhadra et al. 2002). This finding suggests a link between an RNAi-like mechanism and TGS. In the *Drosophila* transgene system, TGS was associated with Pc-G complexes at the silenced transgene promoter. One possibility is that the Pc-G proteins are targeted to the transgene promoters by guide RNAs with sequence identity to transcriptional regulatory regions (Pal-Bhadra et al. 2002).

Genetic analyses in both *S. pombe* and *C. elegans* have also suggested links between RNAi and genome modifications. In contrast to *Saccharomyces cerevisiae*, which appears to completely lack the RNAi pathway, the genome of *S. pombe* encodes homologs of three proteins essential for RNAi: Dicer, RdRP, and Argonaute (Grewal and Elgin 2002). Mutations in these proteins interfere with targeting of histone methyltransferase activity and Swi6, the *S. pombe* homolog of HP1 (heterochromatin protein 1), to centromeric repeats (Volpe et al. 2002) and to the silent mating-type locus (Hall et al. 2002). In an RNAi-based genetic screen to find genes required for RNAi in *C. elegans* (Dudley et al. 2002), five essential genes were identified that encode proteins predicted to associate with chromatin. It was speculated that chromatin-binding proteins have a direct role in RNAi (Dudley et al. 2002).

There are numerous other examples of noncoding RNA involvement in triggering genome modifications in mammals (Eddy 2001; Mattick 2001; Maison et al. 2002). Although there is no evidence that these noncoding RNAs participate in known RNA-silencing pathways involving processed dsRNAs (Figure 3.1), they are nevertheless candidates for further study in the context of RNA-silencing mechanisms.

## MEIOTIC SILENCING BY UNPAIRED DNA IN *NEUROSPORA*

A novel relationship between dsRNA and the genome has been revealed by the unusual phenomenon of "meiotic silencing by unpaired DNA" (MSUD) in *Neurospora* (Shiu et al. 2001). DNA that remains unpaired during meiosis is able to silence itself and all other homologous DNA in the genome, even if that DNA is paired. A semidominant mutant, *Sad-1*, which is defective in MSUD, encodes a putative RdRP (Shiu et al. 2001). These results suggest the involvement of a PTGS/RNAi-like mechanism in MSUD, which is believed to silence new transposon insertions that might remain unpaired when homologs come together at meiosis. Although MSUD is not yet known to be associated with epigenetic modifications of the genome, the phenomenon demonstrates a link between a genome surveillance mechanism that detects disruptions in DNA pairing and a silencing mechanism occurring at the RNA level. MSUD can thus be placed under the general heading of "homology effects" (Cogoni 2002).

## NATURAL SOURCES OF PROMOTER DsRNA

Most of the experimental work on RdDM and dsRNA-mediated TGS has been performed with artificial transgene systems in plants. Transgenes are easy to manipulate and useful for dissecting silencing mechanisms. The degree to which transgene silencing effects reflect natural silencing processes, however, is unclear. Recent evidence suggests that certain natural short RNAs (microRNAs or miRNAs) might indeed be able to target promoter sequences. Cloning and sequencing miRNAs from *Arabidopsis* have revealed that they are usually encoded in intergenic regions (Llave et al. 2002; Mette et al. 2002; Reinhart et al. 2002). Some of these miRNAs are homologous to DNA sequences that are within 0.1–1 kb of the ATGs of protein-coding regions, i.e., putative promoter regions (Park et al. 2002). It is conceivable that these miRNAs could target DNA methylation and/or chromatin modifications to the cognate promoter and thus induce TGS of the corresponding endogenous genes. In addition to studying natural miRNAs, another way to identify endogenous targets of RdDM is to examine genes that are reactivated in RdDM mutants as they become available.

The intergenic sequences giving rise to short RNAs might have evolved from a recently discovered group of transposable elements known as miniature inverted repeat transposable elements (MITEs). Found in a wide range of plants and animals, MITEs are short (several hundred base pairs), do not encode any protein, and are believed to be defective Class 2 (DNA) elements (Zhang et al. 2001). Unlike other DNA elements, they have achieved extraordinarily high copy numbers in host genomes by a transposition mechanism that is not yet completely understood. Due to their lack of protein-coding capacity, MITEs are not normally thought to be transcribed (S. Wessler, pers. comm.). When integrated next to an active host promoter, however, MITEs can be transcribed to produce an RNA with a stable secondary structure. Conceivably, MITE-derived dsRNAs can be processed by Dicer-like enzymes to produce short, *trans*-regulatory RNAs (Mette et al. 2002). Because MITEs preferentially integrate close to genes, they can furnish host genes with target sequences homologous to the cognate short RNAs. In this way, RNA-silencing mechanisms, which probably evolved to combat viruses and transposable elements (Voinnet 2001), could be imposed on host genes.

## VIROID PATHOGENICITY: RDDM OF HOST GENES?

The involvement of dsRNA in RdDM is consistent with the highly base-paired structure of viroids (Gross et al. 1978), the plant pathogens used originally to demonstrate RdDM (Wassenegger et al. 1994). Moreover, viroids produce dsRNA intermediates during replication (Branch and Robertson 1984), making them ideal targets for RNA silencing mechanisms. Despite more than two decades of research, the basis of viroid pathogenicity remains mysterious. The discovery of RdDM gave rise to the suggestion that viroids might inappropriately methylate and silence homologous plant genes. Initial screens for sequence homology between viroids and plant DNA concentrated on regions 100 bp or longer. However, the finding of an approximately 30-bp minimum DNA target size for RdDM (Pélissier and Wassenegger 2000) and the observation that viroids are processed to short RNAs (Papaefthimiou et al. 2001) reopened the search for much shorter regions of homology between viroids and plant genes. Although most viroids are replicated in the nucleus by the host RNA polymerase II, they can also exit the nucleus and move systemically through the plant vascular system in the manner of a virus. Viroids could therefore initiate RdDM in the nucleus and/or PTGS in the cytoplasm.

## USE OF RNA-MEDIATED TGS IN FUNCTIONAL GENOMICS

dsRNA-mediated TGS and RdDM of promoter sequences can be useful in functional genomics approaches in plants as an alternative to PTGS (Wang and Waterhouse 2001). The advantages of TGS are its stability and the ability to target individual members of gene families that differ primarily in the sequence of the promoter regions (W. Aufsatz and A. Matzke, unpubl.). It is not yet known, however, whether every plant promoter will be susceptible to this type of silencing. In addition, although the approach works well throughout plant development, presumably owing to the continuous activity of a de novo DNA MTase, dsRNA-mediated TGS and RdDM of promoter sequences in mammals might be restricted to early embryonic stages when the necessary de novo DNA MTase is active (Figure 3.4).

The basic requirement for dsRNA-mediated TGS is a dsRNA that contains promoter sequences. In a study designed to test different ways of producing promoter dsRNA, the only successful strategy involved transcribing a promoter inverted repeat introduced as a transgene into the host genome (Mette et al. 2000). TGS and RdDM of a target promoter were not triggered and promoter short RNAs were not detected when abundant sense and antisense single-stranded promoter RNAs were transcribed simultaneously from unlinked loci (Mette et al. 2000; M.F. Mette, unpubl.). TGS was also not achieved with constructs designed to produce overlapping transcripts of target promoter sequences by two opposing promoters (Mette et al. 2000). In assembling a transcribed promoter inverted repeat construct designed to synthesize promoter dsRNA, factors to consider include:

- Length and composition of promoter sequences in the inverted repeat.
- Size and sequence of the spacer used to separate the two halves of the inverted repeat.
- Nature of the promoter used to transcribe the inverted repeat.
- Need for polyadenylation signals.

## Length and Composition of the Promoter Inverted Repeat

Ideally, the promoter dsRNA should cover the transcription start site and include any cytosine residues that are known or suspected to be critical for silencing. Although 5′-untranslated regions (5′UTRs) in addition to promoter sequences have been present in some inverted repeat constructs, sequences upstream of the transcription start site are sufficient for TGS and target promoter methylation (Mette et al. 2000; T. Kanno, W. Aufsatz, and A. Matzke, unpubl.). Although no systematic analysis of length requirements has been carried out, a few hundred base pairs for each half of the promoter inverted repeat have efficiently induced TGS and target promoter methylation in all cases tested. dsRNA formed from an approximately 300-bp NOSpro inverted repeat with a 273-bp spacer in between was efficiently processed to short RNAs throughout the entire dsRNA region (M.F. Mette, unpubl.). Therefore, one can anticipate targeting methylation to the full DNA region that is homologous to the dsRNA. Notable down-regulation of transcription can be obtained with this method. A strong seed-specific promoter is almost completely silenced in the presence of a transcribed promoter inverted repeat, each half of which is about 250 bp in length (T. Kanno and A. Matzke, unpubl.).

A promoter inverted repeat can be either assembled directly in the plasmid used for plant transformation (in which case a spacer is needed for stability during bacterial cloning steps) or created *in planta* using Cre/*lox*-mediated recombination to convert a direct repeat of promoter sequences into an inverted repeat (Mette et al. 2000). An inverted repeat can also be generated in bacteria using a vector system containing two pairs of *attP* recombination sites arranged in inverse orientation and interrupted by an intron. Polymerase chain reaction (PCR) fragments flanked by *attB* sites can be recombined directionally into this vector in one step (Wang and Waterhouse 2002). Although not yet tested in our lab, an alternative to making promoter dsRNA is to transcribe directly short RNAs homologous to target promoter sequences using an RNA polymerase III promoter. Transcription by RNA polymerase III terminates at a stretch of Ts (Riedel et al. 1996) and does not produce polyadenylated transcripts, possibly favoring nuclear retention of the promoter short RNAs.

## Size and Sequence of Spacer

A spacer comprising sequences unrelated to the target promoter is required if cloning steps involving the inverted repeat are performed in bacteria. Although not yet analyzed systematically, spacer regions of about 250–300 bp have been used successfully in promoter inverted repeat constructs (Mette et al. 2000; T. Kanno and A. Matzke, unpubl.). For inverted repeats to be stable in the plant genome, spacer regions do not need to be too long. For example, the spacer in a promoter inverted repeat produced spontaneously at the transgene locus $H_{9NP}$ is 89 bp (Mette et al. 1999); the spacer in a Cre-generated inverted repeat is 149 bp (Mette et al. 2000).

## Transcribing Promoter

Strong, constitutive viral promoters have been used to transcribe promoter inverted repeats in plants (Mette et al. 2000; Sijen et al. 2001). This approach has induced dsRNA-mediated TGS and promoter RdDM at many stages of plant development, including embryos (T. Kanno and A. Matzke, unpubl.), seedlings and adult leaves (Mette et al. 2000; Aufsatz et al. 2002a,b), and flower petals (Sijen et al. 2001). In principle, it should be possible to transcribe promoter inverted repeats with tissue-specific promoters or promoters of varying strengths to achieve the desired level and cell-type specificity of TGS and target promoter methylation.

## Polyadenylation Signals

The issue of whether to put polyadenylation signals in promoter inverted repeat constructs is unsettled because the nature of the RNA triggering RdDM is unresolved. Depending on whether short RNA or dsRNA is involved in RdDM, the decision to include a polyadenylation site might differ depending on the experimental system used. If dsRNA is involved in RdDM, then a polyadenylation signal is not required because dsRNA forms rapidly by intramolecular folding when the entire inverted repeat is transcribed. Indeed, nonpolyadenylated dsRNAs might be retained in the nucleus and induce RdDM more efficiently than polyadenylated dsRNAs.

If short RNAs guide homologous DNA methylation, then the situation in plants and mammals differs. In plants, which probably possess a nuclear form of Dicer, nonpolyadenylated dsRNAs would still be optimal because they should feed preferentially into a nuclear pathway for dsRNA processing. In contrast, the sole Dicer protein in mammalian cells is located in the cytoplasm (Billy et al. 2001), and dsRNA processing presumably occurs only in this cellular compartment. Therefore, in mammalian systems, a polyadenylated dsRNA might be preferable because it would be transported more efficiently to the cytoplasm and be available for processing into short RNAs by Dicer. As has been shown in plants, an RNA species produced in the cytoplasm as a consequence of PTGS can enter the nucleus and induce RdDM (A.L. Jones et al. 1998; Jones et al. 1999). If short RNAs can be similarly transported in mammalian cells, then the lack of a nuclear form of Dicer will not preclude RdDM in mammals.

## RNA-GUIDED GENOME MODIFICATIONS

During the past year, the idea that short RNAs can target epigenetic modifications to specific regions of the genome has gained tremendous momentum. Although RdDM has been regarded as a plant-specific phenomenon, it is now clear that it was the first example of a general class of RNA-guided genome modifications. In retrospect, the exquisite sequence specificity of RdDM, which provided compelling evidence for direct RNA-DNA sequence interactions, and the occurrence of RdDM during the entire plant life cycle were probably critical features allowing its detection and characterization. As evidence accumulates for non-CG methylation in organisms other than plants, and for RNA-guided chromatin modifications in species that do not methylate their DNA, there is increasing awareness that RNA-directed genome alterations may be universal and diverse in their outcome. As genetic and biochemical analyses continue to reveal more about the mechanisms of RNA-directed genome modifications, the nuclear pathways of RNA silencing are approaching their cytoplasmic counterparts in importance.

## ACKNOWLEDGMENTS

We thank Stephen Schauer for a collaboration on *sin* and *caf* mutants and for providing us with unpublished information on Dicer-like proteins in *Arabidopsis*; Steve Jacobsen for a collaboration on *cmt3* and *drm2* mutants; Hervè Vaucheret for a collaboration on *sgs* mutants; Eric Richards for providing the *met1* mutant; Ortrun Mittelsten Scheid for providing a *som8/ddm1* mutant; and Jurek Paszkowski for supplying the *mom1* mutant. Research in our lab is supported by the Austrian Fonds zur Förderung der wissenschaftlchen Forschung (grant Z21-MED) and the European Union (Contract QLRT-2000-00078).

## REFERENCES

Akhtar A., Zink D., and Becker P.B. 2000. Chromodomains are protein-RNA interaction modules. *Nature* **407:** 405–409.

Amedeo P., Habu Y., Afsar K., Mittelsten Scheid O., and Paszkowski J. 2000. Disuption of the plant gene *MOM* releases transcriptional silencing of methylated genes. *Nature* **405:** 203–206.

Aufsatz W., Mette M.F., van der Winden J., Matzke A.J.M., and Matzke M.A. 2002a. RNA-directed DNA methylation in *Arabidopsis. Proc. Natl. Acad. Sci.* (suppl. 4) **99:** 16499–16506.

———. 2002b. HDA6, a putative histone deacetylase needed to enhance DNA methylation induced by double stranded RNA. *EMBO J.* **21:** 6832–6841.

Bartee L., Malagnac F., and Bender J. 2001. *Arabidopsis cmt3* chromomethylase mutations block non-CG methylation and silencing of an endogenous gene. *Genes Dev.* **15:** 1753–1758.

Bender J. 1998. Cytosine methylation of repeated sequences in eukaryotes: The role of DNA pairing. *Trends Biochem. Sci.* **23:** 252–256.

Bernstein E., Denli A.M., and Hannon G.J. 2001. The rest is silence. *RNA* **7:** 1509–1521.

Bestor T.H. 2000. The DNA methyltransferases of mammals. *Human Mol. Genet.* **9:** 2395–2402.

Billy E., Brondani V., Zhang H., Müller U., and Filipowicz W. 2001. Specific interference with gene expression induced by long, double-stranded RNA in mouse embryonal teratocarcinoma cell lines. *Proc. Natl. Acad. Sci.* **98:** 14428–14433.

Bird A. 2001. DNA methylation patterns and epigenetic memory. *Genes Dev.* **16:** 6–21.

Branch A.D. and Robertson H.D. 1984. A replication cycle for viroids and other small infectious RNAs. *Science* **223:** 450–455.

Cao X. and Jacobsen S. 2002. Role of the *Arabidopsis* DRM methyltransferases in de novo DNA methylation and gene silencing. *Curr. Biol.* **12:** 1138–1144.

Cao X., Springer N., Muszynski M.G., Phillips R.L., Kaeppler S., and Jacobsen S.E. 2000. Conserved plant genes with similarity to mammalian de novo DNA methyltransferases. *Proc. Natl. Acad. Sci.* **97:** 4979–4984.

Carmo-Fonseca M., Custodio N., and Calado A. 1999. Intranuclear trafficking of messenger RNA. *Crit. Rev. Eukaryot. Gene Expr.* **9:** 213–219.

Chen Z.J. and Pikaard C. 1997. Epigenetic silencing of RNA polymerase I transcription: A role for DNA methylation and histone modification in nucleolar dominance. *Genes Dev.* **11:** 2124–2136.

Chicas A. and Macino G. 2001. Characteristics of post-transcriptional gene silencing. *EMBO Rep.* **2:** 992–996.

Cogoni C. 2001. Homology-dependent gene silencing mechanisms in fungi. *Annu. Rev. Microbiol.* **55:** 381–406.

———. 2002. Unifying homology effects. *Nat. Genet.* **30:** 245–246.

Dennis K., Fan T., Geiman T., Yan Q., and Muegge K. 2001. Lsh, a member of the SNF2 family, is required for genome-wide methylation. *Genes Dev.* **15:** 2940–2944.

Dernburg A., Zalevsky J., Colaiácovo M., and Villeneuve A. 2000. Transgene-mediated cosuppression in the *C. elegans* germ line. *Genes Dev.* **14:** 1578–1583.

Diener T.O. 2001. The viroid: Biological oddity or evolutionary fossil? *Adv. Virus Res.* **57:** 137–184.

Dobosy J.R. and Selker E.U. 2001. Emerging connections between DNA methylation and histone acetylation. *Cell. Mol. Life Sci.* **58:** 721–727.

Dudley N., Labbé J.C., and Goldstein B. 2002. Using RNA interference to identify genes required for RNA interference. *Proc. Natl. Acad. Sci.* **99:** 4191–4196.

Eddy S. 2001. Non-coding RNA genes and the modern RNA world. *Nat. Rev. Genet.* **2:** 919–929.

Elmayan T., Balzergue S., Beon F., Bourdon V., Daubremet J., Guenet Y., Mourrain P., Palauqui J.C., Vernhettes S., Vialle T., Wostrikoff K., and Vaucheret H. 1998. *Arabidopsis* mutants impaired in cosuppression. *Plant Cell* **10:** 1747–1758.

Faugeron G. 2000. Diversity of homology-dependent gene silencing strategies in fungi. *Curr. Opin. Microbiol.* **3:** 144–148.

Finnegan E.J. and Kovac K.A. 2000. Plant DNA methyltransferases. *Plant Mol. Biol.* **43:** 189–201.

Fire A., Xu S., Montgomery M.K., Kostas S.A., Driver S.E., and Mello C.C. 1998. Potent and specific genetic interference by double-stranded RNA in *Caenorhabditis elegans*. *Nature* **391:** 806–811.

Flores R. 2001. A naked plant-specific RNA ten-fold smaller than the smallest known viral RNA: The viroid. *C.R. Acad. Sci. Paris* **324:** 943–952.

Fuks F., Burgers W., Brehm A., Hughes-Davies L., and Kouzarides T. 2000. DNA methyltransferase Dnmt1 associates with histone deacetylase activity. *Nat. Genet.* **24:** 88–91.

Furner I.J., Sheikh M.A., and Collett C. 1998. Gene silencing and homology-dependent gene silencing in *Arabidopsis:* Genetic modifiers and DNA methylation. *Genetics* **149:** 651–662.

Gendall A.R., Levy Y., Wilson A., and Dean C. 2001. The VERNALIZATION 2 gene mediates the epigenetic regulation of vernalization in *Arabidopsis. Cell* **107:** 525–535.

Gibbons R., McDowell T., Raman S., O'Rourke D., Garrick D., Ayyub H., and Higgs D. 2000. Mutations in *ATRX,* encoding a SWI/SNF-like protein, cause diverse changes in the pattern of DNA methylation. *Nat. Genet.* **24:** 368–371.

Golden T., Schauer S., Lang J., Pien S., Mushegian A., Grossniklaus U., Meinke D., and Ray A. 2002. SHORT INTEGUMENTS1/SUSPENSOR1/CARPEL FACTORY, a Dicer homolog, is a maternal effect gene required for embryo development in *Arabidopsis. Plant Physiol.* **130:** 808–822.

Goodrich J., Puangsomlee P., Martin M., Long D., Meyerowitz E., and Coupland G. 1997. A polycomb-group gene regulates homeotic gene expression in *Arabidopsis. Nature* **386:** 44–51.

Grewal S. and Elgin S. 2002. Heterochromatin: New possibilities for the inheritance of structure. *Curr. Opin. Genet. Dev.* **12:** 178–187.

Gross H.J., Domdey H., Lossow C., Jank P., Raba M., Alberty H., and Sänger H.L. 1978. Nucleotide sequence and secondary structure of potato spindle tuber viroid. *Nature* **18:** 203–208.

Guo H.S. and Ding S.W. 2002. A viral protein inhibits the long range signaling activity of the gene silencing signal. *EMBO J.* **21:** 398–407.

Habu Y., Kakutani T., and Paszkowski J. 2001. Epigenetic developmental mechanisms in plants: Molecules and targets of plant epigenetic regulation. *Curr. Opin. Genet. Dev.* **11:** 215–220.

Hall I., Shankaranarayana G., Noma K., Ayoud N., Cohen A., and Grewal S. 2002, Establishment and maintenance of a heterochromatin domain. *Science* **297:** 2232–2237.

Hamilton A., Voinnet O., Chappell L., and Baulcombe D. 2002. Two classes of short interfering RNA in RNA silencing. *EMBO J.* **21:** 4671–4679.

Henikoff S. and Comai L. 1998. A DNA methyltransferase homolog with a chromodomain exists in multiple polymorphic forms in *Arabidopsis. Genetics* **149:** 307–318.

Hutvágner G. and Zamore P.D. 2002. RNAi: Nature abhors a double-strand. *Curr. Opin. Genet. Dev.* **12:** 225–232.

Ingelbrecht I., Van Houdt H., Van Montagu M., and Depicker A. 1994. Posttranscriptional silencing of reporter transgenes in tobacco correlates with DNA methylation. *Proc. Natl. Acad. Sci.* **91:** 10502–10506.

Jackson J., Lindroth A., Cao X., and Jacobsen S.E. 2002. Control of CpNpG DNA methylation by the KRYPTONITE histone H3 methyltransferase. *Nature* **416:** 556–560.

Jacobsen S.E., Running M.P., and Meyerowitz E.M. 1999. Disruption of an RNA helicase/RNAse III gene in *Arabidopsis* causes unregulated cell division in floral meristems. *Development* **126:** 5231–5243.

Jakowitsch J., Papp I., Moscone E.A., van der Winden J., Matzke M.A., and Matzke A.J.M. 1999. Molecular and cytogenetic characterization of a transgene locus that can silence in *trans* via promoter homology. *Plant J.* **17:** 131–140.

Jeddeloh, J., Bender J., and Richards E.J. 1999. The DNA methylation locus *DDM1* is required for maintenance of gene silencing in *Arabidopsis. Genes Dev.* **12:** 1714–1725.

Jones A.L., Thomas C.L., and Maule A.J. 1998. De novo methylation and co-suppression induced by a cytoplasmically replicating plant RNA virus. *EMBO J.* **17:** 6385–6393.

Jones D., Cowell I., and Singh P. 2000. Mammalian chromodomain proteins: Their role in genome organization and expression. *BioEssays* **22:** 124–137.

Jones L., Ratcliff F., and Baulcombe D.C. 2001. RNA-directed transcriptional gene silencing in plants can be inherited independently of the RNA trigger and requires Met1 for maintenance. *Curr. Biol.* **11:** 747–757.

Jones L., Hamilton A.J., Voinnet O., Thomas C.L., Maule A.J., and Baulcombe D.C. 1999. RNA-DNA interactions and DNA methylation in post-transcriptional gene silencing. *Plant Cell* **11:** 2291–2301.

Jones P., Veenstra G., Wade P., Vermaak D., Kass S., Landsbeger N., Strouboulis J., and Wolffe A.P. 1998. Methylated DNA and MeCP2 recruit histone deacetylase to repress transcription. *Nat. Genet.* **19:** 187–191.

Jones P.A. and Takai D. 2001. The role of DNA methylation in mammalian epigenetics. *Science* **293:** 1068–1070.

Ketting R., Haverkamp T., van Luenen H., and Plasterk R. 1999. *mut-7* of *C. elegans*, required for transposon silencing and RNA interference, is a homolog of Werner syndrome helicase and RNaseD. *Cell* **99:** 133–141.

Kishimoto N., Sakai H., Jackson J., Jacobsen S.E., Meyerowitz E.M., Dennis E.S., and Finnegan E.J. 2001. Site specificity of the *Arabidopsis* MET1 DNA methyltransferase demonstrated through hypermethylation of the superman locus. *Plant Mol. Biol.* **46:** 171–183.

Li G., Hall T.C., and Holmes-Davis R. 2002. Plant chromatin: Development and gene control. *BioEssays* **24:** 234–243.

Lindroth A., Cao X., Jackson J., Zilberman D., McCallum C., Henikoff S., and Jacobsen S.E. 2001. Requirement of *CHROMOMETHYLASE3* for maintenance of CpXpG methylation. *Science* **292:** 2077–2080.

Llave C., Kasschau K.D., and Carrington J.C. 2000. Virus-encoded suppressor of posttranscriptional gene silencing targets a maintenance step in the silencing pathway. *Proc. Natl. Acad. Sci.* **97:** 13401–13406.

Llave C., Kasschau K., Rector M., and Carrington J. 2002. Endogenous and silencing-associated small RNAs in plants. *Plant Cell* **14:** 1605–1619.

Lorincz M.C. and Groudine M. 2001. C$^m$C(a/t)GG methylation: A new epigenetic mark in mammalian DNA? *Proc. Natl. Acad. Sci.* **98:** 10034–10036.

Lucy A.P., Guo H.S., Li W.X., and Ding S.W. 2000. Suppression of post-transcriptional gene silencing by a plant viral protein localized in the nucleus. *EMBO J.* **19:** 1672–1680.

Lyko F. 2001. DNA methylation learns to fly. *Trends Genet.* **17:** 169–172.

Lyko F., Whittaker A., Orr-Weaver T., and Jaenisch R. 2000. The putative *Drosophila* methyltransferase gene *dDnmt2* is contained in a transposon-like element and is expressed specifically in ovaries. *Mech. Dev.* **95:** 215–217.

Maison C., Bailly D., Peters A., Quivy J., Roche D., Taddei A., Lachner M., Jenuwein T., and Almouzni G. 2002. Higher-order structure in pericentric heterochromatin involves a distinct pattern of histone modification and an RNA component. *Nat. Genet.* **30:** 329–334.

Mallory A.C., Ely L., Smith T.H., Marathe R., Anandalakshmi R., Fagard M., Vaucheret H., Pruss G., Bowman L., and Vance V.B. 2001. HC-Pro suppression of transgene silencing eliminates the small RNAs but not transgene methylation or the mobile signal. *Plant Cell* **13:** 571–583.

Marathe R., Smith T.H., Anandalakshmi R., Bowman L.H., Fagard M., Mourrain P., Vaucheret H., and Vance V.B. 2000. Plant viral suppressors of post-transcriptional silencing do not suppress transcriptional silencing. *Plant J.* **22:** 51–59.

Mattick J.S. 2001. Noncoding RNAs: The architects of eukaryotic complexity. *EMBO Rep.* **2:** 986–991.

Matzke A., Neuhuber F., Park Y.D., Ambros P., and Matzke M.A. 1994. Homology-dependent gene silencing in transgenic plants: Epistatic silencing loci contain multiple copies of methylated transgenes. *Mol. Gen. Genet.* **244:** 219–229.

Matzke M., Matzke A.J.M., and Kooter J. 2001. RNA: Guiding gene silencing. *Science* **293:** 1080–1083.

Matzke M.A., Primig M., Trnovsky J., and Matzke A.J.M. 1989. Reversible methylation and inactivation of marker genes in sequentially transformed tobacco plants. *EMBO J.* **8:** 643–649.

Mette M.F., Matzke A.J.M., and Matzke M.A. 2001. Resistance of RNA-mediated transcriptional gene silencing to HC-Pro, a viral suppressor of PTGS, suggests alternative pathways of dsRNA processing. *Curr. Biol.* **11:** 1119–1123.

Mette M.F., van der Winden J., Matzke M.A., and Matzke A.J.M. 1999. Production of aberrant promoter transcripts contributes to methylation and silencing of unlinked homologous promoters *in trans*. *EMBO J.* **18:** 241–248.

———. 2002. Short RNAs can identify new candidate transposable element families in *Arabidopsis*. *Plant Physiol.* **130:** 6–9.

Mette M.F., Aufsatz W., van der Winden J., Matzke M.A., and Matzke A.J.M. 2000. Transcriptional gene silencing and promoter methylation triggered by double stranded RNA. *EMBO J.* **19:** 5194–5201.

Mlotshwa S., Voinnet O., Mette M.F., Matzke M., Vaucheret H., Ding S.W., Pruss G., and Vance V.B. 2002. RNA silencing and the mobile silencing signal. *Plant Cell* (suppl.) **14:** 289–301.

Morel J.B., Mourrain P., Béclin C., and Vaucheret H. 2000. DNA methylation and chromatin structure affect transcriptional and post-transcriptional transgene silencing in *Arabidopsis*. *Curr. Biol.* **10:** 1591–1594.

Murfett J., Wang X., Hagen G., and Guilfoyle T. 2001. Identification of *Arabidopsis* histone deacetylase HDA6 mutants that affect transgene expression. *Plant Cell* **13:** 1047–1061.

Nan X., Ng H., Johnson C., Laherty C., Turner B., Eisenman R., and Bird A. 1998. Transcriptional repression by the methyl-CpG-binding protein MeCP2 involves a histone deacetylase complex. *Nature* **393:** 386–389.

Napoli C., Lemieux C., and Jorgensen R.A. 1990. Introduction of a chimeric chalcone synthase gene into petunia results in reversible co-suppression of homologous genes in *trans*. *Plant Cell* **2:** 279–289.

Okano M., Shaoping X., and Li E. 1998. Dnmt2 is not required for de novo and maintenance methylation of viral DNA in embryonic stem cells. *Nucleic Acids Res.* **26:** 2536–2540.

Pal-Bhadra M., Bhada U., and Birchler J. 2002. RNAi related mechanisms affect both transcriptional and posttranscriptional transgene silencing in *Drosophila*. *Mol. Cell* **9:** 315–327.

Papa C.M., Springer N., Muszynski M., Meeley R., and Kappler S.M. 2001. Maize chromomethylase *Zea methyltransferase2* is required for CpNpG methylation. *Plant Cell* **13:** 1919–1928.

Papaefthimiou I., Hamilton A.J., Denti M.A., Baulcombe D.C., Tsagris M., and Tabler M. 2001. Replicating potato spindle tuber viroid RNA is accompanied by short RNA fragments that are characteristic of post-transcriptional gene silencing. *Nucleic Acids Res.* **29:** 2395–2400.

Park W., Li J., Song R., Messing J., and Chen X. 2002. CARPEL FACTORY, a Dicer homolog, and HEN1, a novel protein, act in microRNA metabolism. *Curr. Biol.* **12:** 1484–1495.

Park Y.D., Papp I., Moscone E.A., Iglesias V., Vaucheret H., Matzke A.J.M., and Matzke M. 1996. Gene silencing mediated by promoter homology occurs at the level of transcription and results in meiotically heritable alterations in methylation and gene activity. *Plant J.* **9:** 183–194.

Pélissier T. and Wassenegger M. 2000. A DNA target size of 30 bp is sufficient for RNA-directed DNA methylation. *RNA* **6:** 55–65.

Pélissier T., Thalmeir S., Kempe D., Sänger H.L., and Wassenegger M. 1999. Heavy de novo methylation at symmetrical and non-symmetrical sites is a hallmark of RNA-directed DNA methylation. *Nucleic Acids Res.* **27:** 1625–1634.

Ramsahoye B., Biniszkiewicz D., Lyko F., Clark V., Bird A., and Jaenisch R. 2000. Non-CpG methylation is prevalent in embryonic stem cells and may be mediated by DNA methyltransferase 3a. *Proc. Natl. Acad. Sci.* **97:** 5237–5242.

Reinhart B., Weinstein E., Rhoades M., Bartel B., and Bartel D. 2002. MicroRNAs in plants. *Genes Dev.* **16:** 1616–1626.

Richards E.J. and Elgin S.C.R. 2002. Epigenetic codes for heterochromatin formation and silencing. *Cell* **108:** 489–500.

Riedel L., Volger U., Luckinger R., Ptz A., Sänger H.L., and Wassenegger M. 1996. Molecular analysis of the gene family of the signal recognition particle (SRP) RNA of tomato. *Plant Mol. Biol.* **31:** 113–125.

Schauer S., Jacobsen S., Meinke D., and Ray A. 2002. DICER-LIKE1: Blind men and elephants in *Arabidopsis* development. *Trends Plant Sci.* **7:** 487–491.

Shiu P., Raju N., Zickler D., and Metzenberg R. 2001. Meiotic silencing by unpaired DNA. *Cell* **107:** 905–916.

Sijen T., Vijn I., Rebocho A., van Blokland R., Roelofs D., Mol J.N.M., and Kooter J.M. 2001. Transcriptional and posttranscriptional gene silencing are mechanistically related. *Curr. Biol.* **11:** 436–440.

Tabara H., Sarkissian M., Kelly W., Fleenor J., Grishok A., Timmons L., Fire A., and Mello C. 1999. The *rde-1* gene, RNA interference, and transposon silencing in *C. elegans*. *Cell* **99:** 123–132.

Tamaru H. and Selker E.U. 2001. A histone H3 methyltransferase controls DNA methylation in *Neurospora crassa*. *Nature* **414:** 277–283.

Tompa R., McCallum C., Delrow J., Henikoff J., van Steensel B., and Henikoff S. 2002. Genome-wide profiling of DNA methylation reveals transposon targets of CHROMOMETHYLASE3. *Curr. Biol.* **12:** 65–68.

Tuschl T. 2001. RNA interference and small interfering RNAs. *Chembiochem.* **2:** 239–245.

Van der Krol A.R., Mur L., Beld M., Mol J.N.M., and Stuitje A.R. 1990. Flavonoid genes in petunia: Addition of a limited number of gene copies may lead to a suppression of gene expression. *Plant Cell* **2:** 291–299.

Vance V. and Vaucheret H. 2001. RNA silencing in plants—Defense and counterdefense. *Science* **292:** 2277–2280.

Vaucheret H. 1992. Promoter-dependent *trans*-inactivation in transgenic tobacco plants: Kinetic aspects of gene silencing and gene reactivation. *C.R. Acad. Sci.* **317:** 310–323.

———. 1993. Identification of a general silencer of 19S and 35S promoters in transgenic tobacco plants: 90 bp of homology in the promoter sequence are sufficient for *trans*-inactivation. *C.R. Acad. Sci.* **316:** 1471–1483.

Voinnet O. 2001. RNA silencing as a plant immune system against viruses. *Trends Genet.* **17:** 449–459.

Volpe T., Kidner C., Hall I., Teng G., Grewal S., and Martienssen M. 2002. Regulation of heterochromatic silencing and histone H3 lysine-9 methylation by RNAi. *Science* **297:** 1833–1837.

Wang M.B. and Waterhouse P.M. 2001. Application of gene silencing in plants. *Curr. Opin. Plant Biol.* **5:** 146–150.

Wang M.B., Wesley S., Finnegan E.J., Smith N.A., and Waterhouse P.M. 2001. Replicating satellite RNA induces sequence-specific DNA methylation and truncated transcripts in plants. *RNA* **7:** 16–28.

Wassenegger M. 2000. RNA-directed DNA methylation. *Plant Mol. Biol.* **43:** 203–220.

———. 2002. Gene silencing. *Int. Rev. Cytol.* **219:** 61–113.

Wassenegger M., Heimes S., Riedel L., and Sänger H.L. 1994. RNA-directed de novo methylation of genomic sequences in plants. *Cell* **76:** 567–576.

Zhang X., Feschotte C., Zhang Q., Jiang N., Eggleston W., and Wessler S. 2001. P instability factor: An active maize transposon system associated with the amplification of *Tourist*-like MITEs and a new superfamily of transposases. *Proc. Natl. Acad. Sci.* **98:** 12572–12577.

Zhao J., Hyman L., and Moore C. 1999. Formation of mRNA 3′ ends in eukaryotes: Mechanism, regulation, and interrelationships with other steps in mRNA synthesis. *Microbiol. Mol. Biol. Rev.* **63:** 405–445.

# CHAPTER 4

# RNAi in *Caenorhabditis elegans*

René F. Ketting, Marcel Tijsterman, and Ronald H.A. Plasterk

*Hubrecht Laboratory, Department of Functional Genomics, 3584 CT Utrecht, The Netherlands*

I N 1995, KEMPHUES' GROUP MADE A CONFUSING OBSERVATION. They had tried to down-regulate genes in *Caenorhabditis elegans* by injecting antisense RNA molecules. Indeed, they observed that injection of antisense RNA led to a reduction of gene function. However, one of the control experiments resulted in a big surprise—the sense RNA injection also resulted in almost the same levels of inhibition of gene function (Guo and Kemphues 1995). This was in contrast to the expectation that only the antisense RNA would be able to base pair with the endogenous target mRNA, thus inhibiting its translation. The Fire and Mello labs provided an explanation for this paradox when they showed in 1998 that the protocol routinely used for the preparation of single-stranded RNA (ssRNA) resulted in the formation of low levels of double-stranded RNA (dsRNA) (Fire et al. 1998). When this dsRNA species was purified away from the ssRNA, the antisense ssRNA lost most of its interfering activity, and the sense ssRNA was totally inactive. In contrast, the dsRNA was a very potent inducer of the so-called interference effect. Thus, a new phenomenon had been discovered, and it was named RNA interference, or RNAi for short. Since then, the number of research groups working on RNAi has exploded, and the RNAi technique has inspired many laboratories to embark on large genome-wide screens for gene functions.

Studies performed shortly after the discovery of the potency of dsRNA led to a number of observations that can now be regarded as basic characteristics of the RNAi response (Montgomery and Fire 1998).

- It was established that the endogenous sequence of the targeted gene was not changed, a condition that correlates with the fact that RNAi does not result in stable genetic changes. However, despite the fact that no sequence changes are induced at the chromosomal loci, some RNAi effects can be inherited for one or two generations (discussed below).

- The levels of targeted mRNA were found to be reduced in the cytoplasm, but the amount of pre-mRNA in the nucleus was not affected. Accordingly, it was found that genes located in operons usually do not cross-react in RNAi experiments. These results suggest that RNAi acts posttranscriptionally by increasing the rate of mRNA turnover sequence specifically. This observation made the link between RNAi and the process of posttranscriptional gene silencing (PTGS) in plants.

- Only a few molecules per cell are required to affect a large pool of mRNA, implying that some aspect of amplification and/or catalytic activity had to be at work.

- The RNAi effect is capable of spreading within the nematode: dsRNA can be injected in the gonad and exert an effect elsewhere in the body. A striking case of this spreading effect is that the dsRNA can be delivered by soaking the animals in a solution of dsRNA

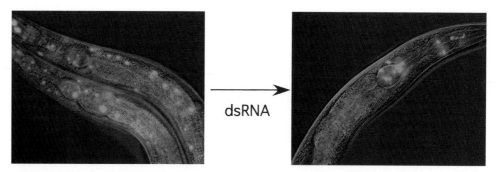

**FIGURE 4.1.** RNAi in *C. elegans*. Detection of the RNAi effect in *C. elegans*. (*Left panel*) These two animals show the green fluorescence protein (GFP) pattern of a strain harboring a ubiquitously expressed GFP reporter construct, with nuclear localization signal. (*Right panel*) After feeding bacteria that express dsRNA directed against GFP, the GFP signal is lost throughout the animal. Note that some neurons in the head region are resistant to this effect (see text, Spreading of RNAi).

(Tabara et al. 1998) or even by feeding the worms on *Escherichia coli* that produce the dsRNA (Timmons and Fire 1998; Timmons et al. 2001) (see also Protocol 1, below). Apparently, the dsRNA produced inside the *E. coli* cell is able to reach virtually all tissues within the fed animal and induce an RNAi response (Figure 4.1).

This chapter focuses on the genetic studies performed to elucidate the mechanism by which dsRNA induces breakdown of homologous mRNA molecules and by which the RNAi effect can spread throughout the organism, i.e., *C. elegans*. Also discussed are the presently known natural functions of RNAi. At the end of this chapter, a number of protocols regarding the synthesis, administration, and detection of RNA used in RNAi experiments in *C. elegans* are described.

## CONCEPTS AND STRATEGIES

### RNAi-defective Mutants

In *C. elegans*, mutant screens have been performed from which mutants have been isolated with a defective RNAi response. One screen was developed specifically for the isolation of RNAi-defective (Rde) mutants, and the loci coming from this screen have been named Rde (Tabara et al. 1999). This screen takes advantage of the fact that when *C. elegans* is fed on bacteria that express dsRNA directed against an essential gene, the progeny fail to develop and they die, making the selection for RNAi-resistant mutants straightforward. Using this screen, several loci have been identified that are required for RNAi: *rde-1* through *rde-4* (see Table 4.1). Some of these loci do not display other obvious phenotypes besides the RNAi defect (*rde-1* and *rde-4*), but other loci do show additional effects, the most notable being the activation of transposable elements (*rde-2* and *rde-3*).

A second type of screen from which RNAi-resistant mutants have been isolated was aimed at the isolation of mutants in which transposable elements had been activated (Collins et al. 1987; Ketting et al. 1999). In *C. elegans*, transposons are not active in the germ line, whereas these same elements are active in the somatic tissues, demonstrating that the elements are specifically silenced in the germ line (Emmons and Yesner 1984). To understand this germ-line-specific silencing better, a screen was done to isolate mutants in which germ-line transposition had been activated. The isolated loci were named *mut*, for the mutant alleles resulting in a *mut*ator phenotype. Interestingly, approximately one half of the *mut* loci resulted in a defective RNAi response (Ketting et al. 1999): *mut-7, mut-8* (=*rde-2*), *mut-9, mut-14, mut-15* (=*rde-5*), and *mut-16* (=*rde-6*) (see

**TABLE 4.1.** Genes involved in RNAi in *C. elegans*

| Locus | RNAi function | | Protein domains | References |
|---|---|---|---|---|
| | soma | germ line | | |
| *rde-1* | √ | √ | PAZ; PIWI | Tabara et al. (1999) |
| *rde-3* | √ | √ | not cloned | Tabara et al. (1999) |
| *rde-4* | √ | √ | dsRNA binding | Tabara et al. (2002 and unpubl.) |
| *mut-7* | – | √ | RNase D | Ketting et al. (1999) |
| *mut-8/rde-2* | – | √ | none detected | B. Tops et al. (unpubl.) |
| *mut-14* | – | √ | DEAD-box RNA helicase | Tijsterman et al. (2002) |
| *mut-15* | √ | √ | none detected | R.F. Ketting et al. (unpubl.) |
| *ego-1* | – | √ | RdRP | Smardon et al. (2000) |
| *rrf-1* | √ | – | RdRP | Sijen et al. (2001) |
| *rrf-3* | √[a] | √[a] | RdRP | Sijen et al. (2001) |
| *dcr-1* | √ | √ | dsRNA binding; helicase; nuclease; PAZ | Grishok et al. (2001); Knight et al. (2001); Ketting et al. (2001) |
| *sid-1* | √ | √ | transmembrane | Winston et al. (2002) |

The locations for where RNAi defects are observed in mutant strains is indicated for each gene. The one exception in the list is *rrf-2*. This protein is one of the four RdRP family members in *C. elegans*, but no role in RNAi has yet been observed.

[a]Loss of *rrf-3* leads to hypersensitivity to RNAi.

Table 4.1). Although these loci are required for RNAi in the germ-line tissue, somatic RNAi is usually not affected in these mutants, which is in contrast to the *rde-1* and *rde-4* mutant alleles, that affect both germ-line and somatic RNAi to the same extent. Apparently, differences exist between RNAi in the soma and RNAi in the germ line that are so far not well understood.

In addition to screens performed in *C. elegans*, studies in other organisms have led to the discovery of genes involved in RNAi (Table 4.1). A putative RNA-directed RNA polymerase (RdRP) protein was first shown to be involved in PTGS in *Neurospora*, in an analogous process called quelling (Cogoni and Macino 1999). The *C. elegans* genome harbors four RdRP loci: *ego-1, rrf-1, rrf-2,* and *rrf-3*. Indeed, *ego-1* (Smardon et al. 2000) and *rrf-1* (Sijen et al. 2001) are required for RNAi in *C. elegans*. Interestingly, the EGO-1 protein is required for RNAi in the germ line only, and the RRF-1 protein is specific for the soma. No RNAi involvement has been found for *rrf-2*, and *rrf-3* appears to be an inhibitor of RNAi, because RNAi acts more efficiently in *rrf-3* mutants than in wild-type animals (Sijen et al. 2001; Simmer et al. 2002).

Dicer, an evolutionarily conserved RNase-III-like enzyme, was first characterized in *Drosophila* (Bernstein et al. 2001), and was shown to be involved in RNAi in *Drosophila* (Bernstein et al. 2001), and *C. elegans*, where the protein is named DCR-1 (Grishok et al. 2001; Ketting et al. 2001; Knight and Bass 2001). Like EGO-1, DCR-1 is also required for normal germ-line development, and disruption of the corresponding genes leads to complete sterility. This may indicate either that these proteins are involved in other processes besides RNAi or that the process of RNAi itself is essential for proper germ-line development.

## Mechanism of RNAi

Several proteins and protein complexes that have a role in the RNAi reaction have been identified through biochemical experiments in *Drosophila* (Hammond et al. 2000, 2001; Nykanen et al. 2001) (see Table 4.1), and some reaction intermediates have been detected (Elbashir et al. 2001; Nykanen et al. 2001). The dsRNA is first cleaved into short RNA

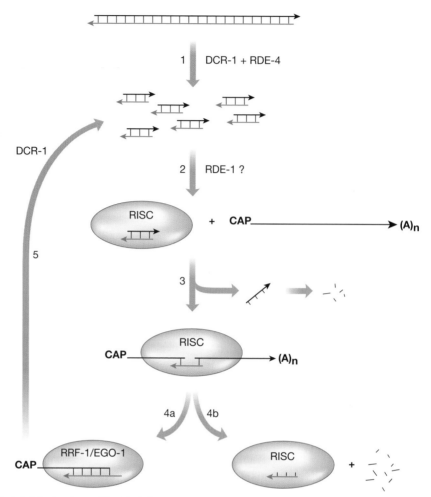

**FIGURE 4.2.** Mechanism of RNAi. Schematic drawing of the mechanism of RNAi. This model results from the combination of genetic data from *C. elegans* and biochemical data from *Drosophila*. The double-stranded RNA (the sense strand is black and the antisense strand is red; the direction of the arrow indicates 5´→3´ polarity) is processed by Dicer into siRNA fragments. (*Step 1*) The RDE-4 protein is required for full activity. (*Step 2*) The siRNA molecules are incorporated into the protein complex RISC. RDE-1 is possibly involved in this step of the reaction. (*Step 3*) The double-stranded siRNA in RISC then becomes single-stranded, and this single-stranded siRNA recognizes a homologous target mRNA. The siRNA strand that is removed is most likely degraded, for in vivo, mainly siRNA of antisense polarity is detected. RISC then cleaves the mRNA in the region covered by the siRNA. After this, two things may happen: The siRNA is elongated by RdRP activity (*Step 4a*), and the resulting dsRNA again enters the RNAi pathway (*Step 5*). Alternatively, the nicked mRNA is further degraded (*Step 4b*). After the nicking of the mRNA, the RISC complex may be recycled to perform additional rounds of mRNA degradation. The genetically identified proteins MUT-7, MUT-8, MUT-14, and MUT-15 are not depicted in this figure, as their roles in this process are not yet known.

fragments of approximately 23 bp by Dicer (Figure 4.2, Step 1) (Bernstein et al. 2001). Dicer starts at the end of a dsRNA molecule and successively cleaves a fragment of 21–23 bp from the dsRNA molecule (Ketting et al. 2001). In vitro and in vivo studies in *Drosophila* and *C. elegans* have shown that this activity is essential for RNAi (Bernstein et al. 2001; Ketting et al. 2001; Knight and Bass 2001). In addition, in vivo studies in *C. elegans* indicate that an additional protein, RDE-4, is required for efficient processing of injected dsRNA (Parrish and Fire 2001). Indeed, the RDE-4 protein is found in complex with DCR-1 in vivo in *C. elegans*, together with RDE-1 and an RNA helicase, DRH-1

(Tabara et al. 2002), indicating that these proteins together are required for short interfering RNA (siRNA) production during RNAi in *C. elegans*.

The siRNA fragments produced by Dicer were first observed in plants undergoing PTGS (Hamilton and Baulcombe 1999). Since then, siRNAs have been detected in all species undergoing RNAi-like phenomena (Hammond et al. 2000; Parrish et al. 2000; Zamore et al. 2000). The two strands of the siRNA are not functionally equivalent, and a number of studies have reported differences in reactivity and/or abundance between the sense and antisense strands (Parrish et al. 2000; Elbashir et al. 2001; Tijsterman et al. 2002b).

The siRNA is incorporated into a protein complex named RISC (RNA-induced silencing complex) (Hammond et al. 2000), which induces a break in the target mRNA in the region covered by the siRNA molecule (Figure 4.2, Steps 2–3) (Elbashir et al. 2001). One of the components of RISC has been identified as Ago2 (Hammond et al. 2001). This protein is one of the fly homologs of the *C. elegans* RDE-1 gene product (Tabara et al. 1999). These proteins are members of a large family, with other members being the *Arabidopsis* Argonaute (Bohmert et al. 1998; Fagard et al. 2000) and the *Neurospora crassa* QDE-2 proteins (Catalanotto et al. 2000), both of which are also required for RNA-induced gene silencing. *C. elegans* has approximately 25 *rde-1* homologs in its genome, but not all are required for RNAi (Grishok et al. 2001). A biochemical function for the RDE-1 protein family has not yet been determined, but it has been suggested that they might function as an adaptor between Dicer and another protein complex, for example, RISC (Hammond et al. 2001).

Other protein factors that have been found through genetics in *C. elegans* so far are an exoribonuclease, MUT-7 (Ketting et al. 1999); an RNA helicase, MUT-14 (Tijsterman et al. 2002b); an RNA-directed RNA polymerase, EGO-1 (Smardon et al. 2000); and two proteins with no recognizable protein motifs, MUT-8 (B. Tops et al., unpubl.) and MUT-15 (Table 4.1) (R.K. Ketting et al., unpubl.). Most of these factors are required for RNAi, but only in the germ-line tissue of *C. elegans*. MUT-15 is an exception, because this protein is required both for somatic RNAi and for germ-line RNAi. EGO-1 has a somatically acting counterpart, RRF-1 (Sijen et al. 2001), but such somatically acting orthologs are not known for MUT-7, MUT-8, and MUT-14. At present, nothing is known about the precise roles of these proteins during the RNAi reaction.

## Heritability of RNAi

As stated above, RNAi does not usually result in a meiotically stable down-regulation of gene expression. However, when RNAi is performed against some genes, effects can be seen two or three generations later, meaning that a heritable agent is formed upon the dsRNA treatment that can still trigger RNAi in later generations. Elegant genetic experiments have shown that this heritable effect can be divided into two steps, with distinct genetic requirements (Grishok et al. 2000). First, an inheritable agent must be formed, which requires the *rde-1* and *rde-4* gene products. This agent is inherited extrachromosomally, because it can be transmitted through sperm that lacks the targeted locus. In the second generation, these transmitted agents must instigate an RNAi response, which requires the *mut-7, rde-2,* and *rde-3* genes. At this step, *rde-1* and *rde-4* are dispensable. From these studies, it has been concluded that *rde-1* and *rde-4* are required for the formation of an inherited agent and that the mutator/RNAi genes are active in the execution step.

With regard to the nature of the inherited agent, it is of interest that short stretches of antisense RNA are on their own capable of triggering a potent RNAi response that can be inherited (Tijsterman et al. 2002b). These ssRNA species are most likely elongated by an RdRP activity, using the endogenous target RNA as a template (also see below, Transitive RNAi and the Immune System). The dsRNA that is formed is then probably processed by Dicer, and the resulting siRNA enters the RNAi pathway. This process requires the MUT-7 and MUT-14 proteins, but not RDE-1 and RDE-4. These requirements reflect those for inheritable RNAi described above, suggesting that the inherited species may be antisense ssRNA.

## Transitive RNAi and the Immune System

Soon after the discovery of RNAi, it was found that enzymes that have RdRP activity were essential for RNAi (and related processes) to occur in several organisms (Cogoni and Macino 1999; Dalmay et al. 2000; Mourrain et al. 2000; Smardon et al. 2000). What are the functions of these enzymes in RNAi?

In *C. elegans*, the following appears to be one function of the RdRP enzymes (Sijen et al. 2001): The siRNA molecules formed by Dicer could be elongated by RdRP enzymes, using the homologous template RNA as a target. This, in turn, produces new dsRNA that can again enter the RNAi pathway (Figure 4.2, Steps 4a–5), leading to three outcomes:

1. Some of the target mRNA will be degraded by Dicer, contributing to the RNAi effect.

2. The siRNA population is amplified, but only if there is a suitable target RNA that can be used as a template.

3. When an siRNA molecule that is close to the 5′ end of the original dsRNA is used as a primer for RdRP systhesis, dsRNA will be produced that is not present in the originally introduced dsRNA. This dsRNA will also be processed by Dicer, leading to so-called secondary siRNAs, which again mediate gene silencing. This has been demonstrated using gene fusions: Introduction of dsRNA directed against the 3′ part of a gene fusion leads to an RNAi response of the fusion gene, but also to the silencing of the endogenous copy of the upstream gene (see Figure 4.3). This effect has been named "transitive RNAi" (Sijen et al. 2001). dsRNA directed against the 5′ part of the gene fusion does not result in transitive RNAi, a result that can be explained by the one-directionality of the RdRP reaction. It is noteworthy that transitivity is also observed in plants. Interestingly, here the effect is not unidirectional, but is also observed in 5′ to 3′ direction (Vaistij et al. 2002). It is not clear what the basis of this 5′ to 3′ directionality is, but it could involve template switching by the RdRP enzyme or primer-independent polymerase activity.

The following is another interesting aspect of siRNAs in *C. elegans:* When RNAi occurs, only siRNAs of antisense polarity are observed (Tijsterman et al. 2002b). Apparently, the system selects only the one strand of the siRNA that has a homologous mRNA target and discards the other. Combined with the finding of template-dependent amplification, this system strikingly resembles the mammalian immune system, where antibody titers are only then amplified when a recognizable antigen is present. In this light, it is interesting to note that one of the natural functions of RNAi is to protect the genome from molecular parasites such as transposable elements and RNA viruses (Lindbo and Dougherty 1992; van der Vlugt et al. 1992; Ketting et al. 1999; Tabara et al 1999), a typical task for an immune system.

Taking all of this evidence together, it seems that the RNAi process is capable of building up a powerful response when both dsRNA and homologous target RNA are present, all of which bears a striking resemblance to the vertebrate immune response. Of course, the immune system, as we know it in vertebrates, depends on extensive recombination that is catalyzed by a transposon-derived protein pair: RAG-1 and RAG-2. Most likely, early in vertebrate evolution, a transposon inserted into a particular gene, setting the

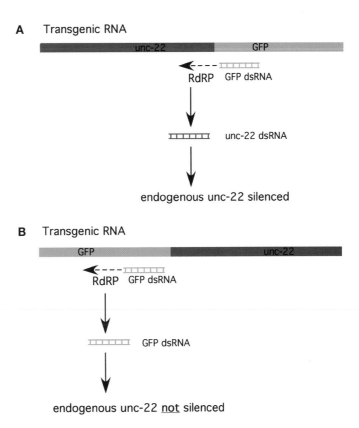

**A** Transgenic RNA

**B** Transgenic RNA

endogenous unc-22 <u>not</u> silenced

**FIGURE 4.3.** Transitive RNAi. When dsRNA directed against GFP is introduced into an animal that expresses an *unc-22*::GFP fusion construct, the action of RdRP enzymes will result in the formation of dsRNA specific for *unc-22*. This *unc-22* dsRNA will be processed into secondary siRNAs, which will lead to the silencing of the endogenously encoded *unc-22* mRNA. This will only happen when the *unc-22* fragment is located 5′ of the GFP part of the transgenic RNA (*A*). When the relative orientations of the *unc-22* and GFP parts are inverted, the RdRP activity will not result in the formation of *unc-22* dsRNA, but in the formation of more GFP dsRNA (*B*).

scene for the *V(D)J* recombination process that is essential for the development of a proper immune system. It is a funny coincidence that our immune system appears to be based on the insertion of a transposable element (Plasterk 1998), a process that was supposed to be inhibited by the primitive "immune system" we now know as RNAi.

## Spreading of RNAi

As described elsewhere, PTGS in plants can be a systemic effect, i.e., the silencing information is transmitted throughout the plant. When the PTGS effect is induced by a virus, this process of spreading leads to resistance to the virus in regions of the plant where the virus is not yet active, protecting the plant from further virus infection. Interestingly, plants are not the only phylum in which systemic silencing effects have been observed.

The effects of dsRNA can also spread to tissues in which it has not been directly introduced in *C. elegans* (Fire et al. 1998). For example, one can feed *E. coli* that produces dsRNA encoded by a plasmid to *C. elegans*. This will lead to a very robust RNAi response not only in the fed animal itself, but also in its progeny (Timmons and Fire 1998). Note that this systemic RNAi response usually does not affect neuronal cells (Figure 4.1); however, when dsRNA is expressed directly in neuronal cells, this does trigger an RNAi response (Tavernarakis et al. 2000), which shows that although the RNAi machinery is present in neurons, systemic RNAi does not reach these cells. Interestingly, similar effects have been seen in plants, where the stomata do not respond to systemic silencing (Voinnet et al. 1998). Apparently, like the neuronal cells in *C. elegans*, these cells are isolated from the rest of the organism.

Obviously, the question regarding systemic RNAi is: How does the dsRNA spread throughout the animal? Are there specialized transport routes for dsRNA, or is it taken up through more general pathways that are also used for the uptake of nutrients? Mutant screens have been aimed at the isolation of mutations that display an RNAi defect when the dsRNA is introduced through feeding, but respond normally to injected or transgene-driven dsRNA (Winston et al. 2002; M. Tijsterman et al., unpubl.). One of the loci isolated from such screens, *sid-1* (systemic RNAi-deficient), has been cloned and encodes a transmembrane protein involved in the uptake of an undefined spreading signal into the cell (Winston et al. 2002). Further molecular characterization of *sid-1* and other loci will undoubtedly lead to a better understanding of this systemic silencing process.

Although differential sensitivity of *C. elegans* to injected versus consumed dsRNA may indeed reflect a defect in an RNA transport process, one should also consider the possibility that such differences may be caused by mutations that render the RNAi reaction in general less efficient. An example of such a scenario is a naturally occurring mutation in the natural isolate CB4856. This Hawaiian strain carries a defective allele of an *rde-1* homolog, *ppw-1*, making the strain resistant to dsRNA introduced through feeding, but not to dsRNA introduced through injection (Tijsterman et al. 2002a). Upon injecting a concentration range of the dsRNA, it appeared that there was a general reduction of RNAi efficiency, rather than a defect in RNA transport.

## Cosuppression and RNAi

Another process in *C. elegans* that closely resembles RNAi is cosuppression. Cosuppression was first described in plants, where the introduction of transgenic copies of a gene led to the down-regulation of those transgenic copies as well as the endogenous gene (Napoli et al. 1990; van der Krol et al. 1990) (see also Chapter 1). Since then, this phenomenon has also been observed in animals (Pal-Bhadra et al. 1997; Ketting and Plasterk 2000; Dernburg et al. 2000). In contrast to RNAi, the cosuppression effect in *C. elegans* does not spread throughout the animal, because genes that are cosuppressed in the germ line may not be silenced in the soma, and, in addition, one of both gonads of *C. elegans* may be cosuppressed and the other may be wild-type (R.F. Ketting and R.H. Plasterk, unpubl.). Both the requirements of the transgene itself and some of the genetic requirements for cosuppression have been studied (Ketting and Plasterk 2000; Dernburg et al. 2000).

1. Cosuppression in *C. elegans* is frequently observed for germ-line-expressed genes. It has been shown that the transgene must express RNA to induce the silencing and that the sequence of the RNA determines the specificity of the cosuppression effect. The promoter must be active in the germ-line tissue, but it does not contribute to the gene specificity.

2. Genetically, cosuppression depends on the presence of genes that are involved in RNAi and transposon silencing: *mut-7, mut-8, mut-14, rde-2,* and *rde-3. rde-1* and *rde-4* are not required. This shows that cosuppression most likely shares aspects with RNAi, but that the two processes are not identical.

Besides posttranscriptional processes, cosuppression may involve silencing at the transcriptional level. Studies from plants have clearly demonstrated that cosuppression is accompanied by changes at the chromosomal loci; when promoter sequences are targeted, this may lead to transcriptional down-regulation of that gene (for review, see Sijen and Kooter 2000). More recently, the labs of Martienssen, Grewal, and Gorovsky showed that siRNA molecules and proteins involved in RNAi are required for centromere function in *Schizosaccharomyces pombe* and DNA rearrangements in *Tetrahymena* (Hall et al. 2002; Mochizuki et al. 2002; Volpe et al. 2002).

Does this link between posttranscriptional and DNA-related processes also exist in metazoans? Studies on cosuppression in *Drosophila* have shown that posttranscriptional and transcriptional silencing processes may indeed be mechanistically connected (Pal-Bhadra et al. 1997, 2002). Also observed in *C. elegans* is evidence for changes in chromatin conformation induced by transgenes and RNAi. The first evidence has been obtained by studying germ-line-specific transgene silencing. This process, in which repetitive transgenes are silenced in the germ line, depends on the presence of Polycomb-Group proteins (Kelly et al. 1997; Kelly and Fire 1998) and other chromatin-related proteins such as the histone H1.1 isoform (Jedrusik and Schulze 2001) and HP1 (Couteau et al. 2002). Interestingly, also in RNAi-resistant mutator mutants, such as *mut-7*, *mut-14*, *rde-2*, and *rde-3*, this repetitive transgene silencing effect is compromised (Tabara et al. 1999).

In addition, the chromosome segregation defects in the mutator mutants could hint at the involvement of RNAi in chromatin-related processes (Ketting et al. 1999; Tabara et al. 1999). A requirement for chromatin components in RNAi was suggested by Dudley et al. (2002), who show that knocking down Polycomb-like factors in *C. elegans* can lead to an RNAi-deficient phenotype. Taken together, these data suggest that the processes of RNAi and chromatin-mediated silencing may also be mechanistically related in metazoans.

## Transposon Silencing, Cosuppression, and RNAi: Three Phenomena, One Mechanism?

How are transposon silencing, cosuppression, and RNAi related? As described above, mutants that have been selected for a defective RNAi response often show release of transposition in the germ line, and, when mutants are specifically selected for this latter phenotype, RNAi resistance is often observed (Ketting et al. 1999; Tabara et al. 1999), revealing a clear relationship between RNAi and transposon silencing. Such a relationship is also observed in plants, not between RNAi and transposon silencing but between cosuppression and virus resistance. It is thus clear that the silencing of molecular parasites, like transposons and viruses, and RNAi has a mechanistic link, but how exactly RNAi is involved in the taming of the mobile elements is subject to speculation. Here, we discuss three alternative hypotheses.

One possibility that has been suggested is that the transposons and repetitive transgenes make dsRNA. In principle, there could be several ways in which transposons can make dsRNA. First, many copies of most transposons are present in the genome. These elements have inserted randomly with regard to the direction of transcription in the area where the transposon integrated, and thus different transposon insertion alleles could produce RNA species of either sense or antisense polarity. When both are produced, dsRNA can be formed, and the RNAi reaction would be initiated.

A second hypothesis for generating transposon-derived dsRNA relies on the presence of inverted repeat sequences within the silenced elements. These inverted repeats are indeed transcribed (T. Sijen et al., unpubl.) and may fold into double-stranded structures. The transposon inverted repeats vary greatly in length, from 54 to more than 300 bp, depending on the specific element. Whether such molecules are indeed good substrates for entering the RNAi pathway is not known.

A third possibility involves transposon-derived aberrant RNA molecules, which could trigger the RNAi response. The molecular nature of aberrant RNAs is unknown, but they

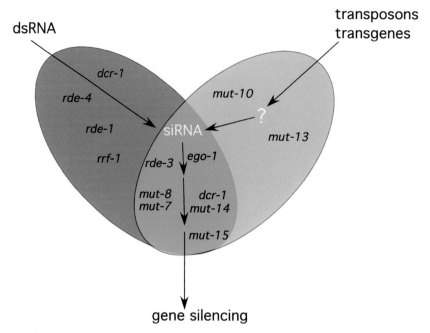

**FIGURE 4.4.** Classes of genes involved in RNAi. Schematic overview of the three classes of genes described in this chapter. (*Blue area*) Genes are indicated that are involved in the RNAi process; (*green area*) genes involved in transposon silencing are displayed. The overlap between the two areas shows the genes that have a role in both phenomena. Most of the genes in the overlap region have also been shown to be involved in gene silencing triggered by antisense RNA and cosuppression, whereas genes in the blue region are not. *dcr-1* is depicted both in the blue area and in the overlap to indicate that *dcr-1* acts both on the introduced dsRNA and on the dsRNA produced by RdRP enzymes such as *ego-1* and *rrf-1*. Involvement of *dcr-1* and *ego-1* in transposon silencing has not been demonstrated, due to the sterility associated with mutations in these genes. However, both genes are required for the dsRNA-mediated silencing. The RNA silencing trigger produced by transposons is unknown, but the involvement of RNAi enzymes suggests that transposon-derived siRNA will be produced.

could very well resemble the before-mentioned antisense RNA oligonucleotides (Tijsterman et al. 2002b) because the same protein requirements have been found: *rde-1* and *rde-4* are dispensable, whereas the mutator/RNAi genes *mut-7* and *mut-14* are required (also see Figure 4.4).

In all cases, an RNAi-like response is triggered, with the mRNA of the transposase as a target, resulting in the breakdown of transposon-encoded mRNA. This, in turn, prevents the production of the transposase protein, thus effectively silencing the transposition process.

One may come up with more models, explaining the link between RNAi and transposon silencing. However, it should not be forgotten that not all RNAi-resistant mutants have activated transposon activity and that not all mutants in which transposons are desilenced display RNAi resistance. These facts hint at the existence of additional regulatory mechanisms involved in the silencing of transposons. As already discussed with respect to cosuppression, this may involve silencing of transposons through changes in their chromatin structure that would effectively prevent their transcription, or perhaps cause the production of transposon-derived aberrant RNA molecules, which would then trigger the RNAi response.

## Natural Functions of RNAi

From many studies performed in different organisms from several phyla, it has become clear that a major role for the RNAi response is the protection of the genome against invading sequences, both exogenous (viruses) and endogenous (transposable elements) (Plasterk and Ketting 2000). In *C. elegans*, however, many of the RNAi-resistant mutator mutants show additional defects, mainly in the germ line. At elevated temperatures, these animals become sterile. One of the reasons for this sterility is a defect in sperm development (Ketting et al. 1999). Another germ-line defect frequently observed in mutator mutants involves the segregation of chromosomes: All mutator mutants display an elevated loss of the X chromosome, leading to an increased percentage of spontaneous males in hermaphrodite cultures (Ketting et al. 1999; Tabara et al. 1999). Also in other mutants, germ-line development is impaired: Mutations in *dcr-1* and *ego-1* lead to complete sterility (Smardon et al. 2000; Grishok et al. 2001; Ketting et al. 2001; Knight and Bass 2001), indicating that the process of RNAi, or at least parts of it, are essential for proper germ-line development.

*C. elegans* is not unique regarding links between RNAi and development. In plants, evidence also exists that the PTGS response requires genes that also function in developmental processes (Bohmert et al. 1998; Jacobsen et al. 1999). Thus, it seems that RNAi-like processes have a crucial role in development. At present, there is one clear example in which the role of an RNAi protein, Dicer, in development is clearly established. As discussed above, Dicer processes dsRNA molecules into the 21-bp siRNA molecules, which subsequently target the homologous mRNA. In addition, it has been shown that Dicer is involved in the maturation of endogenously encoded small RNA molecules (microRNAs or miRNAs) (Grishok et al. 2001; Hutvagner et al. 2001; Ketting et al. 2001), some of which are involved in development, but most of which have no known function at present (Lagos-Quintana et al. 2001; Lau et al. 2001; Lee and Ambros 2001). In addition, it has been shown that proteins of the RDE-1/Argonaute family are also involved in the mechanisms by which these miRNAs function (Grishok et al. 2001). miRNAs may not lead to an increase of mRNA turnover as do the siRNAs. In fact, for the two most-studied miRNAs, *let-7* (Reinhart et al. 2000) and *lin-4* (Lee et al. 1993), the mechanism of action most likely involves the translational inhibition of target genes (Olsen and Ambros 1999; Slack et al. 2000). It is interesting to note that *let-7* and *lin-4* RNAs require a mismatch with their target mRNA (Ha et al. 1996) at the position where the RISC enzyme would induce the break (Elbashir et al. 2001). This plus the fact RDE-1-like proteins are involved suggest that a RISC-like complex is involved in miRNA-mediated silencing. Indeed, when HeLa cell extracts containing protein complexes with *let-7* miRNA are offered an RNA substrate with perfect complementarity to *let-7* sequence, an endonucleolytic activity can be observed, resembling that of RISC activity in RNAi (Hutvagner and Zamore 2002). In addition, in plants, miRNA molecules with perfect target complementarity have been found, and these miRNAs direct cleavage of that target RNA molecule (Llave et al. 2002). Thus, the Dicer protein has a dual role: one in the maturation of miRNAs, independent of the DCR-1-binding proteins RDE-1, RDE-4, and DRH-1, and one in the process that we call RNAi. Most likely, Dicer is not the only example of an RNAi protein with such a dual role, and other proteins required for both RNAi and other processes will probably be found. Of course, worms were never designed to eat high concentrations of dsRNA or to be injected with it. Nevertheless, the strong evolutionary conservation of the reaction triggered by the dsRNA, and the strong conservation of at least some of the protein factors carrying out the reaction, suggests that this pathway does have important cellular functions. As described above, such functions are indeed being found, and without doubt, future research will further clarify the role of RNAi during normal life.

# TECHNIQUES

**▼ CAUTION**

*See Appendix for appropriate handling of materials marked with <!>.*

## MATERIALS

### BUFFERS AND MEDIA

The following buffers and media are used in the protocols detailed below.

**10× M9**

30 g of $KH_2PO_4$ <!>
60 g of $K_2HPO_4$ <!>
50 g of NaCl

    After autoclaving, add 10 ml of 1 M $MgSO_4$. <!>

**NGM (Worm Culturing)**

3 g of NaCl
17 g of agar
2.5 g of peptone
1 ml of 5 mg/ml cholesterol

After autoclaving add:

    1 ml of 1 M $CaCl_2$
    1 ml of 1 M $MgSO_4$ <!>
    25 ml of 1 M $K_2HPO_4/KH_2PO_4$ (pH 6) <!>
    Seed plates with the *E. coli* strain OP50, as a food source for the nematodes.

**Hybridization Buffer (RNase Protection)**

50 mM PIPES (pH 6.4)
1.3 mM EDTA
0.5 M NaCl
67% formamide <!>

**High-salt Buffer (RNase Protection)**

0.3 M NaCl
10 mM Tris (pH 7.5) <!>
5 mM EDTA

**Low-stringency Hybridization Buffer (Northern Blotting)**

0.36 M $Na_2HPO_4$ <!>
0.14 M $NaH_2PO_4$ <!>
1 mM EDTA
7% SDS <!>

    The pH of entire mix should be 7.2. Do not autoclave.

## PROTOCOL 1: INTRODUCTION OF DsRNA

dsRNA can be introduced in *C. elegans* in various ways. The most direct way is microinjection into the gonad, the gut, or the body fluid. Another method that is easily scalable is feeding with bacteria that express dsRNA. When using an RNase-III-negative *E. coli* strain (HT115), the effectiveness of this method is comparable to injection. A third way is to soak the animals in a solution of dsRNA. Routinely, we use either injection or feeding as means to induce RNAi; a soaking protocol will be given for completeness.

## Procedures

**▼ CAUTION**
*See Appendix for appropriate handling of materials marked with <!>.*

### MATERIALS

**REAGENTS**

Antibiotics

*C. elegans* culture (well-growing) of the desired genotype

Culture media

DNase

*E. coli:* HT115 (RNase-III-negative)

$H_2O$ (RNase-free)

In vitro transcription system (T7, T3, or SP6)

IPTG (1 μM)

M9 buffer (see Buffers and Media, above), 10 μl supplemented with 1 mg/ml dsRNA <!>

NGM medium (see Buffers and Media, above)

PCR products with T7/T3/SP6 transcription start sites

Phenol:chloroform <!>

RNA purification kit

RNase-free PCR products

> Transcription start sites can be attatched to the primers. These can be present on both ends of the PCR fragment.

**EQUIPMENT**

Culture plates (standard)

Eppendorf tube

Microinjection equipment

### Method 1: Injection of dsRNA

1. Synthesize the RNA strands by placing the desired PCR products with T7/T3/SP6 transcription start sites on both ends into standard transcription reactions. Follow instructions provided by the manufacturer.

2. Treat the reaction with DNase, and then heat the transcription reaction to 95°C. Slowly cool the reaction down to room temperature.

3. Purify the RNA from the mixture by phenol:chloroform extraction and precipitation, or by using RNA purification kits. Dissolve the RNA in $H_2O$.

> Keep in mind that when the dsRNA is dissolved in $H_2O$, it will melt at low temperatures and will not reanneal.

4. Inject the dsRNA into the animals at concentrations ranging from 100 to 1000 ng/μl.

> The RNA can be injected into the animal at any anatomical location, but most often it is injected into the gonad or the gut. No systematic differences in effectiveness have been reported for different injection sites.

## Method 2: Soaking

1. Place animals at the L4 stage of development in an Eppendorf tube containing 10 μl of M9 buffer supplemented with 1 mg/ml of dsRNA.

2. Incubate for 24 hours at 20°C.

3. Single out the animals onto standard culture plates and score the progeny for phenotypes.

## Method 3: Feeding

1. Clone a fragment of the gene of interest into a vector in between two T7 promoters. Transform this plasmid into the *E. coli* strain HT115.

2. Grow overnight cultures in standard media containing the appropriate selection markers.

3. Introduce the dsRNA by seeding the bacteria onto standard NGM plates containing the appropriate antibiotics and 1 mM IPTG.

4. Store the plates overnight at room temperature to induce the dsRNA.

5. Place L4-stage animals onto the food. Score the $F_1$ progeny for phenotypes.

> The P0-stage animals can sometimes show effects of the RNAi, but it is usually at a level lower than that of the $F_1$ animals.
>
> The RNAi plates can be stored for 2–3 weeks at 4°C without significantly loosing potency.

## PROTOCOL 2: COSUPPRESSION

Cosuppression in *C. elegans* can be triggered by highly repetitive transgenes that contain gene constructs that are expressed in the germ line. Examples of such genes are *fem-1, gld-1, mrt-2, him-14,* and *dpy-30.*

## Procedure

---

**MATERIALS**

**REAGENTS**

*C. elegans* culture

This culture should be well-growing and of the desired genotype (Bristol N2, or *mut-7 [pk204]*).

Plasmid preparations of high quality

We routinely purify the plasmid DNA with commercially available plasmid isolation kits. It is best to dissolve the DNA in $H_2O$, rather than TE.

**EQUIPMENT**

Micro-injection setup (standard *C. elegans*)

---

1. Make constructs containing the genes described above at concentrations of 10–50 ng/μl in a mixture with selection and carrier plasmids at a total concentration of ~100 ng/μl. For example, mix 50 ng/μl pAJ37 (Jones and Schedl 1995) and 50 ng/μl pRF4 for cosuppression of the *gld-1* gene.

2. Inject the DNA mix from Step 1 into the gonad arms of the animal, using standard transgenesis protocols (Mello et al. 1991).

3. Select $F_1$ transgenic animals, and score the $F_2$ for transmission of the transgene.

   When a cosuppression effect will lead to sterility, no or few transmitting lines will be obtained. In this case, inject animals that have a defective cosuppression response, such as NL917 (*mut-7[pk204]*III). When a transmitting line is established, the *mut-7(pk204)* can be crossed out to yield the cosuppression effect. The establishment of cosuppression may occur either in the heterozygous $F_1$ generation or not until the $F_2$ generation.

## PROTOCOL 3: RNA ISOLATION

RNA can be isolated from *C. elegans* in several ways. An efficient approach is to grind the animals in liquid nitrogen, and then extract the RNA from the powder using Trizol (Invitrogen). Alternatively, the animals can be lysed using proteinase K, followed by phenol:chloroform or Trizol extraction.

## Procedure

▼ **CAUTION**
*See Appendix for appropriate handling of materials marked with <!>.*

---

**MATERIALS**

**REAGENTS**

    *C. elegans* culture (well-growing)

    Liquid nitrogen <!>

    M9 buffer (see Buffers and Media, above) <!>

    Trizol (Invitrogen)

**EQUIPMENT**

    Mortar and pestle

    Pipette

---

1. Collect worms from the culture. The more animals, the easier the procedure, but 500 μl of animals is sufficient for 500–1000 μg of total RNA. Keep the nematode suspension as concentrated as possible, but the slurry should be able to be pipetted.

2. Drop little droplets of the nematode slurry into liquid nitrogen. This is most easily done in a mortar. The pestle should be chilled.

3. Grind the frozen balls to a fine powder.

4. Dissolve the powder in Trizol. Use 10 volumes of Trizol, compared to the original worm slurry.

5. Follow the instructions for RNA isolation provided with the Trizol. Dissolve the RNA in RNase-free $H_2O$.

## PROTOCOL 4: SI/MIRNA DETECTION

In *C. elegans*, siRNA can be most easily detected using RNase protection. siRNAs can also be detected on northern blot, but in our hands, the RNase protection is much more sensitive.

## Procedures

### Method 1: RNase Protection

---

**MATERIALS**

**REAGENTS**

$[\alpha\text{-}^{32}P]UTP$ <!>

Denaturing polyacrylamide gel (12%)

DNA-free RNA

  After a Trizol isolation, we usually perform a DNase digestion.

Ethanol <!>

Formamide loading dye <!>

High-salt buffer (see Buffers and Media, above) <!>

Hybridization buffer (see Buffers and Media, above) <!>

In vitro transcription kit

Phenol:chloroform <!>

Proteinase K (10 mg/ml)

SDS (10%) <!>

---

1. Prepare a $[\alpha\text{-}^{32}P]UTP$-labeled RNA probe using a standard in vitro transcription reaction. To a 20-µl in vitro transcription reaction, add 20 µCi $[\alpha\text{-}^{32}P]UTP$ and lower the concentration of the cold UTP tenfold. Gel-purify the full-length RNA.

2. Mix 4 ml (10–100 µg) of RNA with 15 µl of hybridization buffer and 1 µl of one $[\alpha\text{-}^{32}P]UTP$-labeled probe RNA (50–100 cps).

3. Store the mixture for 5 minutes at 95°C and then overnight at 42°C.

4. Add 300 µl of high-salt buffer containing the appropriate RNases. For each reaction, add:

   0.6 µl of RNase A (10 mg/ml)

   0.6 µl of RNase I (10 units/µl)

   0.03 µl of RNase T1 (80 units/µl)

5. Incubate the mixture for 45 minutes at 30°C and for 45 minutes at 37°C.

6. Add 15 µl of 10% SDS and 5 µl of 10 mg/ml proteinase K. Incubate the mixture for 15 minutes at 37°C.

7. Extract the proteinase K reaction with phenol:chloroform and precipitate (using tRNA or glycogen as a carrier) with 1 ml of ethanol.

8. Dissolve the pellet in formamide loading dye and run the RNA on a 12% denaturing polyacrylamide gel. Expose it to detect the RNA signals.

## Method 2: Northern Blot

We usually use this method to detect microRNAs in *C. elegans*.

**MATERIALS**

**REAGENTS**

Denaturing acrylamide gel (15%) (1x TBE)

Low-stringency hybridization buffer (see Buffers and Media, above) <!>

Radiolabeled probe <!>

RNA (50 µg)

1x TBE

**EQUIPMENT**

Nylon membrane (positively charged)

Semidry blotting apparatus

1. Separate the RNA (50 µg) on a 15% denaturing acrylamide gel (1x TBE). The 21-nucleotide RNA species runs approximately halfway between the XC and BFB dyes. As a marker, use a radiolabeled oligonucleotide (preferably RNA) or an oligonucleotide that will hybridize with the probe (do not use too much; ~10 fmoles should give a decent signal).

2. Blot the gel onto the membrane. In 1x TBE, blot at 6 V for 20 minutes, depending on the specific kind of membrane used. Cross-link the RNA to the membrane before further use.

3. Prehybridize the blot for 1–2 hours in low-stringency hybridization buffer, and then add the radiolabeled probe.

   The probe should be a radiolabeled single-stranded RNA or DNA. Also effective are 5′-labeled oligonucleotides.

4. Perform the hybridization overnight at 37°C.

5. Wash the blot twice with 2x SSC/0.2% SDS for 20 minutes.

| Troubleshooting | **Complementary ssRNAs Will Not Anneal** |
|---|---|

- Check complementarity.
- Make sure that there is salt in the RNA solutions.
- Check for hairpin structures in the RNA. Remember that G-U is also a stable base pair in RNA.

**High Background on Northern Blots**

- Try adding less probe in the hybridization.
- Make sure that the SDS is of good quality and is well dissolved in the buffers (heat if necessary).
- Remove free nucleotides from the probe.
- Wash for longer periods of time.

**No Signal on Northern or in RNase Protection (RPA)**

- Check the probe (well labeled? sequence OK?).
- Use more input RNA.
- For the RPA, make sure that the RNases have been inactivated properly.
- For the northern, try optimizing the blotting time. Too long a time may take the RNA through the membrane; too short a time will give a low transfer yield.

**No RNAi Phenotype**

- Knocking down the gene does not result in a detectable phenotype.
- When an effect is expected: try different sequences.
- Try doing RNAi in the *rrf-3* mutant background. This background is more sensitive to RNAi than the wild-type Bristol N2 strain.

## ACKNOWLEDGMENTS

The authors thank the members of our laboratory for helpful discussions: Titia Sijen for critical reading of the manuscript and Titia Sijen and Bastiaan Tops for permission to cite unpublished work.

## REFERENCES

Bernstein E., Caudy A.A., Hammond S.M., and Hannon G.J. 2001. Role for a bidentate ribonuclease in the initiation step of RNA interference. *Nature* **409:** 363–366.

Bohmert K., Camus I., Bellini C., Bouchez D., Caboche M., and Benning C. 1998. AGO1 defines a novel locus of *Arabidopsis* controlling leaf development. *EMBO J.* **17:** 170–180.

Catalanotto C., Azzalin G., Macino G., and Cogoni C. 2000. Transcription: Gene silencing in worms and fungi. *Nature* **404:** 245.

Cogoni C. and Macino G. 1999. Gene silencing in *Neurospora crassa* requires a protein homologous to RNA-dependent RNA polymerase. *Nature* **399:** 166–169.

Collins J., Saari B., and Anderson P. 1987. Activation of a transposable element in the germ line but not the soma of *Caenorhabditis elegans. Nature* **328:** 726–728.

Couteau F., Guerry F., Muller F., and Palladino F. 2002. A heterochromatin protein 1 homologue in *Caenorhabditis elegans* acts in germline and vulval development. *EMBO Rep.* **3.**

Dalmay T., Hamilton A., Rudd S., Angell S., and Baulcombe D.C. 2000. An RNA-dependent RNA polymerase gene in *Arabidopsis* is required for posttranscriptional gene silencing mediated by a transgene but not by a virus. *Cell* **101:** 543–553.

Dernburg A.F., Zalevsky J., Colaiacovo M.P., and Villeneuve A.M. 2000. Transgene-mediated cosuppression in the *C. elegans* germ line. *Genes Dev.* **14:** 1578–1583.

Dudley N.R., Labbe J.-C., and Goldstein B. 2002. Using RNA interference to identify genes required for RNA interference. *Proc. Natl. Acad. Sci.* **99:** 4191–4196.

Elbashir S.M., Lendeckel W., and Tuschl T. 2001. RNA interference is mediated by 21- and 22-nucleotide RNAs. *Genes Dev.* **15:** 188–200.

Emmons S.W. and Yesner L. 1984. High-frequency excision of transposable element Tc 1 in the nematode *Caenorhabditis elegans* is limited to somatic cells. *Cell* **36:** 599–605.

Fagard M., Boutet S., Morel J.B., Bellini C., and Vaucheret H. 2000. AGO1, QDE-2, and RDE-1 are related proteins required for post-transcriptional gene silencing in plants, quelling in fungi, and RNA interference in animals. *Proc. Natl. Acad. Sci.* **97:** 11650–11654.

Fire A., Xu S., Montgomery M.K., Kostas S.A., Driver S.E., and Mello C.C. 1998. Potent and specific genetic interference by double-stranded RNA in *Caenorhabditis elegans*. *Nature* **391:** 806–811.

Grishok A., Tabara H., and Mello C.C. 2000. Genetic requirements for inheritance of RNAi in *C. elegans*. *Science* **287:** 2494–2497.

Grishok A., Pasquinelli A.E., Conte D., Li N., Parrish S., Ha I., Baillie D.L., Fire A., Ruvkun G., and Mello C.C. 2001. Genes and mechanisms related to RNA interference regulate expression of the small temporal RNAs that control *C. elegans* developmental timing. *Cell* **106:** 23–34.

Guo S. and Kemphues K.J. 1995. *par-1*, a gene required for establishing polarity in *C. elegans* embryos, encodes a putative Ser/Thr kinase that is asymmetrically distributed. *Cell* **81:** 611–620.

Ha I., Wightman B., and Ruvkun G. 1996. A bulged lin-4/lin-14 RNA duplex is sufficient for *Caenorhabditis elegans* lin-14 temporal gradient formation. *Genes Dev.* **10:** 3041–3050.

Hall I.M., Shankaranarayana G.D., Noma K., Ayoub N., Cohen A., and Grewal S.I. 2002. Establishment and maintenance of a heterochromatin domain. *Science* **297:** 2215–2218.

Hamilton A.J. and Baulcombe D.C. 1999. A species of small antisense RNA in posttranscriptional gene silencing in plants. *Science* **286:** 950–952.

Hammond S.M., Bernstein E., Beach D., and Hannon G.J. 2000. An RNA-directed nuclease mediates post-transcriptional gene silencing in *Drosophila* cells. *Nature* **404:** 293–296.

Hammond S.M., Boettcher S., Caudy A.A., Kobayashi R., and Hannon G.J. 2001. Argonaute2, a link between genetic and biochemical analyses of RNAi. *Science* **293:** 1146–1150.

Hutvagner G. and Zamore P.D. 2002. A microRNA in a multiple-turnover RNAi enzyme complex. *Science* **297:** 2056–2060.

Hutvagner G., McLachlan J., Pasquinelli A.E., Balint E., Tuschl T., and Zamore P.D. 2001. A cellular function for the RNA-interference enzyme Dicer in the maturation of the let-7 small temporal RNA. *Science* **293:** 834–838.

Jacobsen S.E., Running M.P., and Meyerowitz E.M. 1999. Disruption of an RNA helicase/RNAse III gene in *Arabidopsis* causes unregulated cell division in floral meristems. *Development* **126:** 5231–5243.

Jedrusik M.A. and Schulze E. 2001. A single histone H1 isoform (H1.1) is essential for chromatin silencing and germline development in *Caenorhabditis elegans*. *Development* **128:** 1069–1080.

Jones A.R. and Schedl T. 1995. Mutations in *gld-1*, a female germ cell-specific tumor suppressor gene in *Caenorhabditis elegans*, affect a conserved domain also found in Src-associated protein Sam68. *Genes Dev.* **9:** 1491–1504.

Kelly W.G. and Fire A. 1998. Chromatin silencing and the maintenance of a functional germline in *Caenorhabditis elegans*. *Development* **125:** 2451–2456.

Kelly W.G., Xu S., Montgomery M.K., and Fire A. 1997. Distinct requirements for somatic and germline expression of a generally expressed *Caernorhabditis elegans* gene. *Genetics* **146:** 227–238.

Ketting R.F. and Plasterk R.H. 2000. A genetic link between co-suppression and RNA interference in *C. elegans*. *Nature* **404:** 296–298.

Ketting R.F., Haverkamp T.H., van Luenen H.G., and Plasterk R.H. 1999. Mut-7 of *C. elegans*, required for transposon silencing and RNA interference, is a homolog of Werner syndrome helicase and RNaseD. *Cell* **99:** 133–141.

Ketting R.F., Fischer S.E., Bernstein E., Sijen T., Hannon G.J., and Plasterk R.H. 2001. Dicer functions in RNA interference and in synthesis of small RNA involved in developmental timing in *C. elegans*. *Genes Dev.* **15:** 2654–2659.

Knight S.W. and Bass B.L. 2001. A role for the RNase III enzyme DCR-1 in RNA interference and germ line development in *Caenorhabditis elegans*. *Science* **293:** 2269–2271.

Lagos-Quintana M., Rauhut R., Lendeckel W., and Tuschl T. 2001. Identification of novel genes coding for small expressed RNAs. *Science* **294:** 853–858.

Lau N.C., Lim L.P., Weinstein E.G., and Bartel D.P. 2001. An abundant class of tiny RNAs with probable regulatory roles in *Caenorhabditis elegans*. *Science* **294:** 858–862.

Lee R.C. and Ambros V. 2001. An extensive class of small RNAs in *Caenorhabditis elegans*. *Science* **294:** 862–864.

Lee R.C., Feinbaum R.L., and Ambros V. 1993. The *C. elegans* heterochronic gene *lin-4* encodes small RNAs with antisense complementarity to *lin-14*. *Cell* **75:** 843–854.

Lindbo J.A. and Dougherty W.G. 1992. Pathogen-derived resistance to a polyvirus: Immune and resistant phenotypes in transgenic tobacco expressing altered forms of a polyvirus coat protein nucleotide sequence. *Mol. Plant-Microbe Interact* **5:** 144–153.

Llave C., Xie Z., Kasschau K.D., and Carrington J.C. 2002. Cleavage of Scarecrow-like mRNA targets directed by a class of *Arabidopsis* miRNA. *Science* **297:** 2053–2056.

Mello C.C., Kramer J.M., Stinchcomb D., and Ambros V. 1991. Efficient gene transfer in *C. elegans:* Extrachromosomal maintenance and integration of transforming sequences. *EMBO J.* **10:** 3959–3970.

Mochizuki K., Fine N.A., Fujisawa T., and Gorovsky M.A. 2002. Analysis of a piwi-related gene implicates small RNAs in genome rearrangement in *Tetrahymena*. *Cell* **110:** 689–699.

Montgomery M.K. and Fire A. 1998. Double-stranded RNA as a mediator in sequence-specific genetic silencing and co-suppression. *Trends Genet.* **14:** 255–258.

Mourrain P., Béclin C., Elmayan T., Feuerbach F., Godon C., Morel J.-B., Jouette D., Lacombe A.-M., Nikic S., Picault N., Rémoué K., Sanial M., Vo T.-A., and Vaucheret H. 2000. *Arabidopsis* SGS2 and SGS3 genes are required for posttranscriptional gene silencing and natural virus resistance. *Cell* **101:** 533–542.

Napoli C., Lemieux C., and Jorgensen R.A. 1990. Introduction of a chimeric chalcone synthase gene into *Petunia* results in reversible co-suppression of homologous genes *in trans*. *Plant Cell* **2:** 279–289.

Nykanen A., Haley B., and Zamore P.D. 2001. ATP requirements and small interfering RNA structure in the RNA interference pathway. *Cell* **107:** 309–321.

Olsen P.H. and Ambros V. 1999. The lin-4 regulatory RNA controls developmental timing in *Caenorhabditis elegans* by blocking LIN-14 protein synthesis after the initiation of translation. *Dev. Biol.* **216:** 671–680.

Pal-Bhadra M., Bhadra U., and Birchler J.A. 1997. Cosuppression in *Drosophila:* Gene silencing of Alcohol dehydrogenase by white-Adh transgenes is Polycomb dependent. *Cell* **90:** 479–490.

———. 2002. RNAi related mechanisms affect both transcriptional and posttranscriptional transgene silencing in *Drosophila*. *Mol. Cell* **9:** 315–327.

Parrish S.N. and Fire A. 2001. Distinct roles for RDE-1 and RDE-4 during RNA interference in *C. elegans*. *RNA* **7:** 1397–1402.

Parrish S., Fleenor J., Xu S., Mello C., and Fire A. 2000. Functional anatomy of a dsRNA trigger. Differential requirement for the two trigger strands in RNA interference. *Mol. Cell* **6:** 1077–1087.

Plasterk R. 1998. V(D)J recombination. Ragtime jumping. *Nature* **394:** 718–719.

Plasterk R.H. and Ketting R.F. 2000. The silence of the genes. *Curr. Opin. Genet. Dev.* **10:** 562–567.

Reinhart B.J., Slack F.J., Basson M., Pasquinelli A.E., Bettinger J.C., Rougvie A.E., Horvitz H.R., and Ruvkun G. 2000. The 21-nucleotide *let-7* RNA regulates developmental timing in *Caenorhabditis elegans*. *Nature* **403:** 901–906.

Sijen T. and Kooter J.M. 2000. Post-transcriptional gene-silencing: RNAs on the attack or on the defense? *BioEssays* **22:** 520–531.

Sijen T., Fleenor J., Simmer F., Thijssen K.L., Parrish S., Timmons L., Plasterk R.H., and Fire A. 2001. On the role of RNA amplification in dsRNA-triggered gene silencing. *Cell* **107:** 465–476.

Simmer F., Tijsterman M., Parrish S., Koushika S., Nonet M., Fire A., Ahringer J., and Plasterk R. 2002. Loss of the putative RNA-directed RNA polymerase RRF-3 makes *C. elegans* hypersensitive to RNAi. *Curr. Biol.* **12:** 1317–1319.

Slack F.J., Basson M., Liu Z., Ambros V., Horvitz H.R., and Ruvkun G. 2000. The *lin-41* RBCC gene acts in the *C. elegans* heterochromic pathway between the *let-7* regulatory RNA and the LIN-29 transcription factor. *Mol. Cell* **5:** 659–669.

Smardon A., Spoerke J.M., Stacey S.C., Klein M.E., Mackin N., and Maine E.M. 2000. EGO-1 is related to RNA-directed RNA polymerase and functions in germ-line development and RNA interference in *C. elegans*. *Curr. Biol.* **10:** 169–178.

Tabara H., Grishok A., and Mello C.C. 1998. RNAi in *C. elegans:* Soaking in the genome sequence. *Science* **282:** 430–431.

Tabara H., Yigit E., Siomi H., and Mello C.C. 2002. The dsRNA binding protein RDE-4 interacts with RDE-1, DCR-1, and a DExH-box helicase to direct RNAi in *C. elegans*. *Cell* **109:** 861–871.

Tabara H., Sarkissian M., Kelly W.G., Fleenor J., Grishok A., Timmons L., Fire A., and Mello C.C. 1999. The *rde-1* gene, RNA interference, and transposon silencing in *C. elegans. Cell* **99:** 123–132.

Tavernarakis N., Wang S.L., Dorovkov M., Ryazanov A., and Driscoll M. 2000. Heritable and inducible genetic interference by double-stranded RNA encoded by transgenes. *Nat. Genet.* **24:** 180–183.

Tijsterman M., Okihara K., Thijssen K., and Plasterk R. 2002a. PPW-1, a PAZ/PIWI protein required for efficient germline RNAi, is defective in a natural isolate of *C. elegans. Curr. Biol.* **12:** 1535.

Tijsterman M., Ketting R.F., Okihara K.L., Sijen T., and Plasterk R.H. 2002b. RNA helicase MUT-14-dependent gene silencing triggered in *C. elegans* by short antisense RNAs. *Science* **295:** 694–697.

Timmons L. and Fire A. 1998. Specific interference by ingested dsRNA. *Nature* **395:** 854.

Timmons L., Court D.L., and Fire A. 2001. Ingestion of bacterially expressed dsRNAs can produce specific and potent genetic interference in *Caenorhabditis elegans. Gene* **263:** 103–112.

Vaistij F.E., Jones L., and Baulcombe D.C. 2002. Spreading of RNA targeting and DNA methylation in RNA silencing requires transcription of the target gene and a putative RNA-dependent RNA polymerase. *Plant Cell* **14:** 857–867.

van der Krol A.R., Mur L.A., Beld M., Mol J.N., and Stuitje A.R. 1990. Flavonoid genes in *Petunia:* Addition of a limited number of gene copies may lead to a suppression of gene expression. *Plant Cell* **2:** 291–299.

van der Vlugt R.A., Ruiter R.K., and Goldbach R. 1992. Evidence for sense RNA-mediated protection to PVYN in tobacco plants transformed with the viral coat protein cistron. *Plant Mol. Biol.* **20:** 631–639.

Voinnet O., Vain P., Angell S., and Baulcombe D.C. 1998. Systemic spread of sequence-specific transgene RNA degradation in plants is initiated by localized introduction of ectopic promoterless DNA. *Cell* **95:** 177–187.

Volpe T.A., Kidner C., Hall I.M., Teng G., Grewal S.I., and Martienssen R.A. 2002. Regulation of heterochromatic silencing and histone H3 lysine-9 methylation by RNAi. *Science* **297:** 1833–1837.

Winston W.M., Molodowitch C., and Hunter C.P. 2002. Systemic RNAi in *C. elegans* requires the putative transmembrane protein SID-1. *Science* **295:** 2456–2459.

Zamore P.D., Tuschl T., Sharp P.A., and Bartel D.P. 2000. RNAi: Double-stranded RNA directs the ATP-dependent cleavage of mRNA at 21 to 23 nucleotide intervals. *Cell* **101:** 25–33.

# CHAPTER 5

# Small RNAs, Big Biology: Biochemical Studies of RNA Interference

Gregory J. Hannon* and Phillip D. Zamore†

*Watson School of Biological Sciences, Cold Spring Harbor Laboratory, Cold Spring Harbor, New York 11724; †Department of Biochemistry and Molecular Pharmacology, University of Massachusetts Medical School, Worcester, Massachusetts 01605

As DESCRIBED IN THE PREVIOUS CHAPTERS, genetic experiments built the foundation of our model for RNA interference (RNAi). Not long after the discovery of RNAi, genetic screens in *Caenorhabditis elegans*, *Arabidopsis thaliana*, *Neurospora crassa*, *Chlamydomonas reinhardtii*, and *Drosophila melanogaster* identified genes required for post-transcriptional gene silencing (PTGS). These studies helped to define the RNAi pathway, but a detailed understanding of the precise roles individual proteins have in the interference process can come only from biochemical analysis. The ultimate goal of such studies is the reconstitution from purified components of each step of the interference pathway, coupled with a detailed understanding of the precise functions of the structural domains of constituent proteins. A complete biochemical description of RNAi is a prerequisite for a full appreciation of the remarkable abilities of this machinery—its exquisite specificity in seeking targets and its efficiency in regulating the expression of those targets by a surprisingly diverse set of mechanisms.

Biochemical studies of RNAi also have biological and technological implications. In fact, study of the mechanism of this process has already helped to uncover novel biological roles for the RNAi pathway. Mechanistic studies have also been critical to the development of RNAi as a tool, particularly in mammals, where an intermediate in the RNAi pathway is used to trigger interference without provoking nonsequence-specific effects. Without the mechanistic studies described in this chapter, our ability to harness the RNAi machinery as an experimental tool in mammalian cells and in animals would likely be restricted to cell lines with considerable embryonic or stem cell character and to the earliest developmental stages of animal embryos. No doubt, mechanistic studies of RNAi will continue to have an important role in further honing this adept gene-silencing tool for mammalian cells.

## THE BIOCHEMISTRY OF RNAi

The core RNAi machinery carries out numerous cellular functions, ranging from gene regulation to virus resistance to chromatin remodeling. As described elsewhere in this volume, these functions operate through numerous, apparently distinct mechanisms. For example, introduction of double-stranded RNA (dsRNA) into *C. elegans*, *Drosophila*, plants, or mammals can elicit a silencing response that operates at a posttranscriptional level. However, in plants, fungi, protozoa, and probably even animals, dsRNA can also induce changes in chromatin structure that result in the repression of transcription of an individual gene or regions of a chromosome. Thus far, biochemical approaches have been restricted to but one aspect of silencing—posttranscriptional silencing. Although genetic studies are leading the way toward a model in which closely related complexes are also involved in silencing at the genomic level, this chapter deals mainly with the biochemical analysis of posttranscriptional silencing.

## FIRST INSIGHTS INTO MECHANISM FROM IN VIVO STUDIES

Although the basic properties of RNA silencing were first described in plants (see Chapter 1), it was the experiments of Fire and Mello in *C. elegans* which first demonstrated that dsRNA was the silencing trigger (Fire et al. 1998). Despite the early and continued role of *C. elegans* in the genetic dissection of RNAi, nematodes have not proved useful for the biochemical dissection of RNA silencing. Plant extracts that recapitulate some aspects of posttranscriptional silencing have been introduced only recently (Tang et al. 2003). Although they promise to accelerate our understanding of the mechanism by which endogenous and transgenic RNAs trigger silencing in plants, genetic analysis has been the major strategy for the study of silencing in plants. Nonetheless, detailed descriptions of the phenomenology of silencing in these model systems framed the key questions to be addressed by biochemical studies of the RNAi mechanism.

In both plants and *C. elegans*, exposure to dsRNA results in a sequence-specific silencing response that occurs mainly at the posttranscriptional level (Fire et al. 1998; Waterhouse et al. 1998). Loss of gene expression in both cases correlates with the disappearance of the targeted mRNA (Fire et al. 1998; Vaucheret et al. 1998). In plants, nuclear run-on experiments have demonstrated that this occurs without effect on transcription of the target loci (Jones et al. 2001). Indeed, in *C. elegans*, dsRNA sequences competent to trigger silencing are restricted to sequences in mature mRNAs. In plants, intronic dsRNAs are similarly less potent, although genomic methylation and transcriptional gene silencing can be triggered by dsRNAs homologous to promoter regions (see Chapter 3). These observations suggested that dsRNAs destabilize homologous mRNAs, probably through a process that acted in the cytoplasm. How the mRNA was destabilized and how the target mRNA was identified by the silencing machinery remained unclear until the development of cell-free systems that recapitulated RNAi.

## BIOCHEMICAL MODELS FOR THE STUDY OF RNAi

Successful biochemical study of any process has two important requirements: One requires an in vitro assay that recapitulates elements of the process, and the other requires that large amounts of relatively homogeneous material be made for cell-free extract preparation. Although extracts of monocot (wheat) embryos were among the very first in vitro extracts for studying protein synthesis, neither *C. elegans* nor dicot plants have historically been considered convenient biochemical models. Therefore, the first biochemical approaches to RNAi awaited the demonstration of RNAi in an organism that was biochemically tractable. The key breakthrough was Carthew's demonstration that dsRNA could induce sequence-specific silencing upon introduction into *Drosophila* embryos (Kennerdell and Carthew 1998).

Since that discovery, the broad outlines of a biochemical pathway for RNAi have emerged from experiments using cell-free lysates of *Drosophila* syncytial blastoderm embryos or extracts from cultured S2 cells and, more recently, human HeLa cells. The pathway has been teased apart by examining the ATP dependence of each step, by using intermediates in the pathway to initiate RNAi, and by classical biochemical fractionation of the protein enzymes and protein-RNA complexes required for RNAi. These studies have revealed a number of key steps. First, experiments using in vitro systems demonstrated that dsRNA did, in fact, target corresponding mRNAs for destruction (Tuschl et al. 1999). Second, in vitro studies established that fragmentation of long dsRNA produces approximately 22 nucleotides of short interfering RNAs (siRNAs) (Zamore et al. 2000). Third, in vitro experiments demonstrated that siRNAs are bona fide intermediates in the RNAi pathway (Nykänen et al. 2001; Elbashir et al. 2001b) and that they direct an RNA-protein complex, the RNA-induced silencing complex (RISC), to destroy the target mRNA (Hammond et al. 2000). Fourth, in vitro studies showed that the siRNA-programmed RISC directs the endonucleolytic cleavage of the target RNA at a single phosphodiester bond, whose position on the target is determined by its distance from the 5´ end of the siRNA (Zamore et al. 2000; Elbashir et al. 2001a). Finally, studies in cell-free extracts demonstrated that RISC-associated activities can completely destroy the target mRNA (Hammond et al. 2000; A.A. Caudy et al., in prep.).

Perhaps the most important breakthrough to emerge from in vitro studies was the identification of the structure and function of siRNAs. In a series of remarkable papers, Tuschl and co-workers sequenced the small RNAs produced upon incubation of long dsRNA in a *Drosophila* embryo lysate, deduced their architecture, and then demonstrated that synthetic siRNA duplexes direct target cleavage in vitro (Elbashir et al. 2001a). This work led directly to the discovery that synthetic siRNAs can trigger RNAi ex vivo in cultured mammalian cells (Caplen et al. 2001; Elbashir et al. 2001b) and in vivo in mice (Lewis et al. 2002; McCaffrey et al. 2002; Song et al. 2003) and plants (Klahre et al. 2002). Beyond the revolutionary implications of these studies for functional genomics and reverse genetics, these authors demonstrated that siRNA-mediated RNAi is conserved in mammals, an idea suggested by earlier studies in mouse embryos (Svoboda et al. 2000; Wianny and Zernicka-Goetz 2000). In fact, RISC-based destruction of target RNAs, directed by small RNAs, is likely to be as ancient as the eukaryotic lineage itself, occurring in animals, plants, fungi, and basal eukaryotes.

Although few of the steps observed in vitro have been confirmed in vivo, the cardinal prediction of the in vitro work—that target mRNAs are first cleaved by an siRNA-directed endonuclease and then destroyed wholesale by other ribonucleases—has received in vivo support. Prydz and colleagues showed that in HeLa cells, highly effective siRNAs cause the accumulation of truncated target mRNAs whose length suggests they are 3´ cleavage products generated by siRNA-directed endonucleolytic cleavage (Holen et al. 2002). In contrast, such cleavage products do not accumulate with inefficient siRNAs. The authors interpreted these results as reflecting the relative rates of endonucleolytic cleavage and subsequent degradation of cleavage products. Thus, for efficient siRNAs, $k_{endo} \gg k_{degradation}$, whereas for inefficient siRNAs, $k_{endo} \ll k_{degradation}$. Putative target cleavage products consistent with siRNA-directed endonucleolytic cleavage have also been observed in plants in which overexpression of a transgenic reporter construct triggered silencing (Klahre et al. 2002). However, in most cases of RNA silencing in vivo in *C. elegans*, *Drosophila*, plants, and mammals, the targeted mRNA appears to disappear wholesale, implying that destruction of the cleavage products is tightly coupled to the initial endonucleolytic event.

It is important to note that formally, initial target cleavage has not been proved to be endonucleolytic, because only 5´ cleavage products have been reported. However, 3´ cleavage products can be detected when the enzyme(s) responsible for degrading them is suppressed (B. Haley et al., in prep.), lending support to the idea that siRNAs direct the cleavage of a single phosphodiester bond in their mRNA targets. Furthermore, siRNA-directed endonucleolytic cleavage has been detected using *Drosophila* embryo lysates and human HeLa cell cytoplasmic extracts, but not extracts derived from *Drosophila* cultured S2 cells. Although this may reflect technical differences in the experiments, a more likely explanation is that RISC-associated ribonucleases responsible for cleavage product degradation are either less abundant or less active in extracts from *Drosophila* embryos or HeLa cells than from S2 cells or that their interaction with RISC in these systems is less stable. The RNAi machinery may be intimately associated with general ribonucleases that destroy the products of endonucleolytic cleavage because (1) the RNAi effect is very short lived in worms mutant for some components of the nonsense-mediated decay pathway (Domeier et al. 2000), (2) a putative exonuclease, Mut-7, is required for RNAi in worms (Ketting et al. 1999), and (3) at least one general ribonuclease that is unlikely to be the target-cleaving endonuclease copurifies with the RISC (A.A. Caudy et al., in prep.).

## RISC: THE CATALYTIC ENGINE OF RNAi

siRNAs were the first subunit of the RISC enzyme to be revealed by biochemical purification (Hamilton and Baulcombe 1999; Hammond et al. 2000). The development of in vitro assays for RISC from *Drosophila* embryo lysates and S2 cell extracts opened the way for the isolation of this enzyme complex and for the identification of its components. Work in these two systems has proceeded in parallel, and each continues to generate distinct and complementary insights into RNAi.

Studies in S2 cell extracts rely on the assembly of the RISC enzyme in vivo following the exposure of cells to silencing triggers, typically, approximately 500-bp dsRNAs. In S2 cell extracts, RISC is characterized by its ability to completely degrade synthetic mRNA substrates homologous to the silencing trigger. In contrast, studies in embryo extracts rely on the assembly of the RISC enzyme in vitro by programming naïve lysates with in-vitro-synthesized dsRNAs or directly with siRNAs. With reconstituted enzyme, homologous mRNAs are also cleaved. However, in these assays, RISC mediates endonucleolytic cleavage of the target RNA, not its complete destruction.

The differences between these model systems and those more recently developed from mammals probably reflect the degree to which each enzyme assembles with accessory factors in vitro and in vivo. Consistent with this idea, RISC enzymes formed in *Drosophila* embryo lysates, S2 cell extracts, and, more recently, human cell extracts (see below) have distinct and characteristic sizes. RISC has been isolated from S2 cells as a complex of approximately 500 kD (Hammond et al. 2000). In embryo extracts, siRNAs form an approximately 250-kD complex, proposed to be an inactive precursor to the RISC, that is converted in the presence of ATP to an active enzyme of about 100 kD (Nykänen et al. 2001). To date, four protein subunits of the effector machinery have emerged from purification of the approximately 500-kD enzyme from S2 cell extracts.

A significant fraction of the RISC emerges from *Drosophila* S2 cells associated with ribosomes. This was first evidenced by the insoluble nature of the enzyme during centrifugation steps designed to clarify extracts for column chromatography (Hammond et al. 2000). In these experiments, RISC was restricted to the P100 pellet, but it could be extracted from the insoluble material by high salt. Furthermore, both RISC activity and RISC components comigrate with polysomes during gradient fractionation of cell-free extracts. Of course, the association between RISC and ribosomes could be an artifactual interaction that occurs during cell lysis; however, several lines of evidence suggest a functional association of at least a population of RISC with polysomes. For example, immunolocalization of several RISC components (see below) indicates elevated concentrations of these proteins near the endoplasmic reticulum, a site of active protein synthesis in the cell (Cikaluk et al. 1999). Recent studies from trypanosomes suggest that interference occurs much less effectively if initiation of translation is perturbed by treatment of cells with specific inhibitors (E. Ullu, pers. comm.). It is important to note that dsRNA-directed mRNA degradation does not require protein synthesis in vitro (Tuschl et al. 1999). Considered together, these studies suggest that RISC exploits ribosomal localization as a mechanism to improve the efficiency with which it scans for potential substrates. The association between RISC and the translation machinery is unlikely simply to reflect the coincident action of both of these multicomponent ribonucleoproteins (RNPs) on mRNA, because components of the RNAi machinery appear to associate directly with ribosomal proteins (Ishizuka et al. 2002). Recent evidence suggests that animal microRNAs (miRNAs; see below), a class of endogenous small RNAs, mediate translational control as components of RISC complexes (see below). Thus, the association of RISC with ribosomes may reflect an inherent capacity of RISC to stall translation. This speculative model has not yet received direct experimental support.

The RISC enzyme has been purified from *Drosophila* S2 cells to near homogeneity, revealing a number of protein components that consistently cofractionate with specific mRNA-degrading activity. Four of these proteins have been identified by mass spectrometry as Argonaute2 (Hammond et al. 2001b), VIG, dFXR (Caudy et al. 2002), and a novel protein, named TSN, composed of multiple staphylococcal nuclease motifs and a tudor domain (A.A. Caudy et al., in prep.). Argonaute proteins are evolutionarily conserved proteins represented in nearly all eukaryotic lineages. Relatives of these proteins can be found in several archaeal genomes. Argonaute protein sequences are defined by the presence of two conserved regions of homology. The biochemical function of these two domains is not yet known. Near the amino termini of Argonaute proteins lies the PAZ motif (so named for founding members of this family, Piwi, Argonaute, and Zwille), while carboxy-terminal regions contain the Piwi domain (Cerutti et al. 2000). Argonaute proteins have been implicated in RNAi and related processes in animals, plants, fungi, and basal eukaryotes.

The first link between the Argonaute family and RNAi came from genetic studies in *C. elegans*, which defined mutations in the *rde-1* gene as causing resistance to silencing by dsRNA (Tabara et al. 1999). RDE-1 is one of 27 such proteins in the *C. elegans* genome, and Argonaute proteins in many organisms are represented by multiprotein families. To date, reverse genetic experiments have also linked two other *C. elegans* Argonaute family members, ALG-1 and ALG-2, to RNAi-related processes (Grishok et al. 2001). In *Arabidopsis*, Argonaute1 and Argonaute4 also participate in dsRNA-mediated silencing (Morel et al. 2002; Zilberman et al. 2003). In *N. crassa*, Qde-2 (one of two Argonaute proteins) is essential for transgene cosuppression (Catalanotto et al. 2000). In *Drosophila*, four of five family members have been linked to gene silencing either by biochemical studies in cell-free systems or by genetic analyses (Hammond et al. 2001b; Kennerdell et al. 2002; Pal-Bhadra et al. 2002; Williams and Rubin 2002). In mammalian cells, at least two of seven or eight (human and mouse, respectively) of these proteins have been implicated in RNAi by in vitro studies (Hutvagner and Zamore 2002a; Martinez et al. 2002; Mourelatos et al. 2002; A.A. Caudy et al., in prep.). Mutation of a single *Argonaute* gene in *Schizosaccharomyces pombe* relieves silencing of heterochromatic regions, the formation of which is thought to be directed by RNAi-related mechanisms (Volpe et al. 2002). Argonaute has also been implicated in heterochromatin formation in *Tetrahymena* (Mochizuki et al. 2002).

Given the mounting evidence for the participation of Argonaute proteins in a wide variety of homology-dependent silencing phenomena, it is likely that these proteins are key components of the RNAi machinery and that many more Argonaute proteins will also be found to participate in similar silencing phenomena. Perhaps many of these family members will find tissue-specific roles, but it is also possible that Argonaute proteins may act at multiple distinct steps in RNAi. Indeed, evidence to this effect has already emerged. Although biochemical experiments show that Argonaute2 is an essential component of the RNAi effector machinery, genetic studies from *C. elegans* have indicated that RDE-1 is essential only for the initiation of RNAi (Grishok et al. 2000). Given the complexity of Argonaute family members in worms (27 in total), tissue-specific or other types of developmental specializations may account for the discrepancy between conclusions from the genetics in worms and that from the biochemistry in flies. A resolution may come from biochemical studies of the RDE-1 complex in *C. elegans*. Affinity purification of RDE-1-containing complexes in worms shows that RDE-1 is associated with a set of proteins distinct from those found in RISC. Remarkably, these proteins are Dicer (DCR-1), which initiates RNAi by converting long dsRNA into siRNAs, and a putative RNA-dependent RNA polymerase (RdRP), proposed to amplify and propagate the silencing signal (Tabara et al. 2002; see below). Thus, it now seems possible that Argonaute proteins can have multiple distinct roles in the silencing process, raising significant questions about what precise mechanistic roles these family members and their constituent domains might assume. However, an alternative view is that the Dicer-RDE1 complex represents a subset of *C. elegans* RISC. This would be in agreement with previously demonstrated interactions between Dicer and Argonaute proteins (Hammond et al. 2001b; Caudy et al. 2002), which led to the proposal that siRNA production and RISC assembly on siRNAs might be physically coupled in *Drosophila* (Hammond et al. 2001b).

Virtually nothing is known about the precise biochemical functions of the PAZ and Piwi domains that define Argonaute proteins. The PAZ domain is approximately 130 amino acids long and is found not only in Argonaute family members, but also in the RNAi-initiating enzyme, Dicer (see below). The Piwi domain is unique to the Argonaute family. Although only a combination of biochemical studies, mutation analysis, and structural studies will reveal the exact functions of these domains, we will indulge in a few predictions. It has been suggested that the PAZ and Piwi domains might represent protein-protein interaction domains (Cerutti et al. 2000). Indeed, Dicer interacts with Argonaute family members in several organisms (Hammond et al. 2000; Caudy et al. 2002; Ishizuka et al. 2002; Tabara et al. 2002; J. Silva and G.J. Hannon, unpubl.). However, the highly basic nature of both the PAZ and Piwi domains defies the conventional notion that interaction domains most often present surfaces containing patches of complementary charges and of hydrophobic residues. Considering their basic nature, it seems more likely that the Piwi and PAZ domains might function as nucleic-acid-binding domains. Nonspecific RNA binding to homopolymers has been demonstrated for several members of the Argonaute family (Kataoka et al. 2001; Deng and Lin 2002).

One common feature of Piwi- and PAZ-containing proteins is that they interact in some fashion with siRNAs, either in complexes that contain siRNAs as specificity determinants or in complexes that generate siRNAs. It is possible that one or both of these domains recognize specific features of these RNAs, distinguishing them from other small RNAs, for example, random degradation products, of mRNAs in cells, and allowing only those RNAs with specific characteristics to enter the silencing pathway. In this regard, the 5′ phosphate of both strands of an siRNA is required to "license" siRNAs to enter the RNAi pathway (Nykänen et al. 2001). Of the two domains characteristic of Argonaute proteins, the PAZ domain seems to be a likely candidate for siRNA binding, based on its presence in multiple RNAi-related proteins. Although the basic nature of the Piwi domain could imply that it also interacts with siRNAs, the presence of this domain mainly in complexes that involve interactions between siRNAs and their targets might suggest a function in either target identification/binding or perhaps even a catalytic role in silencing.

In summary, Argonaute proteins lie at the core of the silencing mechanism and are likely to be the primary proteins that interact with siRNAs and use them to guide RISC complexes to their targets. Yet cells likely contain many small RNAs that occur as random degradation products of cellular mRNAs. How can RISC and, according to our proposal, Argonaute proteins in RISC distinguish siRNAs from such degradation products? Mounting evidence suggests that the specific structure of siRNAs is critical for this discrimination.

## THE PRODUCTION OF siRNAs BY DICER

siRNA duplexes are 21–25 nucleotides long, double-stranded, and bear 2-nucleotide 3′ overhangs. Each siRNA strand has 5′ phosphate and 3′ hydroxyl termini (Zamore et al. 2000). Indeed, studies with synthetic siRNAs indicate that these characteristics are critical to the function of siRNAs in the silencing pathway (Elbashir 2001a,b; Nykänen 2001). The structure of siRNAs provided strong clues to their origins, in that one family of dsRNA-specific ribonucleases, the RNase III family, produces cleavage products with characteristics very similar to those of siRNAs. This family of enzymes is reviewed in detail in Chapter 8.

The discovery of Dicer was presaged by the demonstration that small RNAs—now known as siRNAs—correlated with PTGS in plants (Hamilton and Baulcombe 1999) and that these siRNAs were incorporated into the effector complex of RNAi (Hammond et al. 2000). Such small RNAs could be produced by fragmentation of dsRNA into siRNAs in *Drosophila* embryo lysates, and this reaction required ATP (Zamore et al. 2000).

The identification of the enzyme that produces siRNAs was a triumph of using the enzyme's properties to predict its domain structure. Because Dicer cut dsRNA, Bass (2000) hypothesized that it would contain one or more dsRNA-specific nuclease domains. Because siRNA production was enhanced by ATP, an ATP-binding motif was also proposed to be present in the enzyme. Fungi, plants, and animals contain several classes of RNase III enzymes, enzymes that cut dsRNA. Chief among candidates for Dicer were class II and class III enzymes (Filippov et al. 2000) that contain tandem nuclease domains. Class II enzymes contain two RNase III domains near their carboxyl termini. A similar arrangement is seen in class III enzymes, but these also have additional domains near their amino termini. Notably, class III enzymes contain a putative ATP-dependent helicase domain and the PAZ domain that is also found in Argonaute family proteins (see above).

A survey of RNase III family members demonstrated that immunoaffinity-purified class III ribonucleases from *Drosophila* and humans could produce siRNAs from dsRNA substrates (Bernstein et al. 2001). Furthermore, the *Drosophila* enzyme was required in cultured cells for RNAi triggered by long dsRNA. This class was aptly named "Dicer"; subsequent studies have established that Dicer is required for RNAi in a wide variety of organisms (for review, see Hannon 2002).

All members of the RNase III family cleave dsRNA, leaving 3′ overhanging ends with 5′ phosphates and 3′ hydroxyls—the precise structure of the termini of siRNAs. Bacterial RNase III enzymes, which contain a single nuclease domain and a dsRNA-binding motif, generate fragments of 9–11 nucleotides in each cycle of dsRNA cleavage (see Chapter 8). This is accomplished by coordinated digestion of dsRNAs by an RNase III dimer containing four catalytic half-sites (one for each strand scission reaction) (Blaszczyk et al. 2001). Catalysis requires a divalent cation ($Mg^{++}$), consistent with the release of products with 5′ phosphates (see Chapter 8). In contrast, Dicer proteins produce discrete products of about 22 nucleotides. (The precise length depends on the species from which the enzyme is derived.) Insight into how precisely measured cleavages are achieved has been provided by a recent crystallographic study of an RNase III family member from *Aquifex aeolicus* (Blaszczyk et al. 2001). These structural studies suggest that two of the four catalytic half-sites in Dicer are defective. If Dicer indeed acts as an antiparallel dimer, like its prokaryotic counterparts, then the central pair of active sites would be poisoned by noncanonical residues, leaving only the 5′- and 3′-terminal active sites competent for cutting. This assertion has been supported by the conversion of residues in the *Aquifex* RNase III to their Dicer counterparts, demonstrating that the resulting mutant enzymes are indeed inactive in vivo (Blaszczyk et al. 2001). The fact that Dicer carries two potentially catalytic domains also raises an alternative possibility—that the two Dicer RNase III domains associate to form an intramolecular antiparallel pseudodimer. Indeed, although bacterial RNase III can form a dimer in the absence of substrate, we have not yet detected a similar interaction with Dicer (E. Bernstein, unpubl.). This raises the possibility that Dicer may act as a monomer and that the approximately 22-nucleotide periodicity of Dicer cleavage may be determined in another fashion. For example, Dicer may cleave in a semiprocessive fashion, with the periodicity being determined by the translocation of the enzyme down its substrate (Bernstein et al. 2001). This might explain the function of the helicase domain and the dependence of Dicer on ATP hydrolysis for mediating multiple cycles of dsRNA cleavage (Bernstein et al. 2001; Hutvagner et al. 2001; Ketting et al. 2001; Nykänen et al. 2001; Provost et al. 2002; Zhang et al. 2002; Myers et al. 2003). Alternatively, Dicer may use a different method for determining the size of siRNAs. For example, occupancy of the RNA by another resident RNA-binding domain (e.g., the dsRBM [RNA-binding motif]) may position catalytic motifs at the appropriate positions for cleavage. In this model, ATP hydrolysis and the helicase domain would be proposed to serve a different function. For example, the helicase domain might be involved in enhancing product release or in facilitating removal of dsRNA-binding proteins from Dicer substrates. Of course, Dicer could also utilize different cleavage modes depending on the substrate, with the dependency of cleavage on different domains varying with the structure of the dsRNA. For example, in processing miRNAs from their approximately 70-nucleotide hairpin precursors (see below), Dicer may employ a different cleavage mode than it does for long dsRNA silencing triggers.

## DICER AND THE DISCOVERY OF miRNAs

Biochemical studies provide substantial support for Dicer as the initiating enzyme of RNAi, because it converts long dsRNA into siRNAs. However, definitive evidence could only come from genetic analyses. Following its discovery, four groups decided to probe the role of Dicer in RNAi and related phenomena by generating hypomorphic or null mutations in the gene. Three groups worked in *C. elegans* while the other group used the siRNAs themselves to attack the Dicer mRNA in HeLa cells (Grishok et al. 2001; Hutvagner et al. 2001; Ketting et al. 2001; Knight and Bass 2001).

Work in *C. elegans* generated a number of alleles in its single Dicer gene that were predicted to be nulls (see Chapter 4). However, Dicer mutant worms were still capable of mounting an RNAi response in the soma, at least against some genes (Grishok et al. 2001; Ketting et al. 2001; Knight and Bass 2001). RNAi in the germ line of these animals was defective, leading to the hypothesis that maternal contributions of Dicer might be sufficient to rescue somatic RNA in many cases but that maternal stores (P0) became depleted in the $F_1$ germ line, revealing the silencing defect. Although these studies suggested that Dicer was required for RNAi, the intriguing aspect of the story was that Dicer null worms displayed remarkable phenotypic abnormalities. Previous studies had identified several genes required only for RNAi; null mutations in these genes give rise to normal worms. Yet Dicer mutant worms are sterile, with pleiotropic defects in the generation of mature oocytes. Furthermore, adult $F_1$ worms show a number of somatic defects characteristic of problems in the timing of development. For example, at the L4-adult transition, specific structures (alae) are formed by the fusion of specialized hypodermal cells called seam cells. These cells fail to fuse in the Dicer mutant worms, and the mutants reiterate a larval fate by undergoing an additional round of division.

Phenotypes similar to those noted in the Dicer nulls had previously been reported by Ambros (Chalfie et al. 1981; Lee et al. 1993) and Ruvkun (Wightman et al. 1993) in their studies of developmental timing and had been mapped to two unusual loci, *lin-4* and *let-7*. Both of these encode small RNA molecules that are transcribed from the genome as longer precursors that have a double-stranded character and are processed posttranscriptionally into mature, approximately 21-nucleotide RNAs (Lee et al. 1993; Reinhardt et al. 2000). These small RNAs are known collectively as small temporal RNAs (stRNAs), in view of their role in regulating the timing of developmental events. Remarkably, both are conserved (*let-7*, strictly so) in other organisms, including *Drosophila* and mammals.

Because these small RNAs are transcribed as precursor RNAs thought to contain a small double-stranded region, several groups investigated whether Dicer might process pre-*let-7* and pre-*lin-4* into their mature forms. Indeed, Dicer can generate mature stRNAs from synthetic precursors in vitro (Hutvagner et al. 2001; Ketting et al. 2001). Dicer null worms accumulate both *lin-4* and *let-7* precursors with a corresponding loss of mature stRNAs (Ketting et al. 2001; Knight and Bass 2001). Similarly, in HeLa cells, depletion of Dicer by RNAi causes a buildup of pre-*let-7* and a loss of mature *let-7* (Hutvagner et al. 2001). *lin-4* and *let-7* RNAs regulate the expression of endogenous protein-coding genes, raising the possibility that RNAi might be used as a general mechanism to control programs of gene expression. Two members of the Argonaute gene family in worms, ALG-1 and ALG-2, are essential for stRNA-mediated regulation, strengthening the hypothesis that stRNAs might act through the RNAi pathway on their regulatory targets (Grishok et al. 2001). However, neither *lin-4* nor *let-7* triggers degradation of its target mRNA, *lin-14* and *lin-28*, respectively. Instead, these small RNAs act at the level of protein synthesis, interfering in some way with the translational machinery, possibly translational elongation (Olsen et al. 1999).

The discovery of two endogenous small RNAs linked to the RNAi machinery raised numerous questions concerning the extent to which cells use this paradigm to control gene expression generally. This fueled an already intensive search for endogenously encoded small RNAs that might function via the RNAi pathway. Efforts were centered on cloning of RNAs with the structure of siRNAs: a size of about 22 nucleotides and 5′ phosphate and 3′ hydroxyl termini. Three publications appearing in *Science* in 2001 presented approximately 100 small RNAs from *Drosophila, C. elegans,* and mammals that shared structural similarity with *lin-4* and *let-7* (Lagos-Quintana et al. 2001; Lau et al. 2001; Lee and Ambros 2001). The genomic loci corresponding to these small RNAs predict a precursor transcript that can form a hairpin structure. Many of these precursors have now been shown to be transcribed and to be cleaved by Dicer into siRNA-like approximately 22-nucleotide RNAs. Such small, noncoding RNAs, including *lin-4* and *let-7*, are now collectively called miRNAs.

The identification of numerous miRNAs from several organisms led to wild speculation regarding the prevalence of miRNA-mediated gene regulation. Estimates of the number of miRNAs in various genomes ranged into the tens of thousands. More recently, a combination of bioinformatics and exhaustive cloning efforts suggests that about 1% of a metazoan organism's genes encode miRNAs. Approximately 88 miRNA genes have been identified in *C. elegans* (Lim et al. 2003b) and fewer than 200 for humans (Lim et al. 2003a).

In contrast to the success in identifying potential miRNAs, scant progress has been made in deciphering their individual functions in animals. Many miRNAs are evolutionarily conserved. For example, mature *let-7* RNA is identical in worms, flies, and mammals and is developmentally regulated in both worms and flies (Reinhart et al. 2000; Hutvágner et al. 2001; Sempere et al. 2002). Considering that miRNAs are thought to direct RISC-related complexes to their targets on the basis of sequence complementarity, one might imagine that locating target genes should be a simple matter of homology searches. However, most, if not all, animal miRNAs appear to recognize their targets by imperfect pairing. For example, the interaction between *let-7* and its target *lin-41* involves multiple binding sites, each of which pairs imperfectly with the small RNA. This problem does not exist in plants, in which many, perhaps even all, miRNAs are extensively complementary to their targets (Llave et al. 2002; Rhoades et al. 2002) and act as siRNAs to direct endonucleolytic cleavage of target mRNAs (Llave et al. 2002; Kasschau et al. 2003; Tang et al. 2003). In animals, the biologically relevant targets for nearly all miRNAs remain to be identified.

A variety of data from both plants and animals suggest that miRNAs function as key developmental regulators. Since miRNAs have been detected only in metazoans, they may represent a key adaptation on the route to multicellular life. *lin-4* and *let-7* control events at the L1–L2 and L4–adult transitions, respectively, and many mammalian, fly, and worm miRNAs are expressed in a developmentally regulated manner. In plants, the presumptive targets of many miRNAs are transcription factors that are known to be involved in patterning during development or the control of cell differentiation (Rhoades et al. 2002). This observation correlates well with the phenotypes in *Arabidopsis* caused by mutations in Argonaute family proteins, which likely use miRNAs as a guide to the selection of targets for silencing (for review, see Carmell et al. 2002). Argonaute1 is generally expressed throughout the development and life of the plant. Mutation of *AGO1* causes numerous defects in overall plant architecture, including radicalized leaves and abnormal flowers (Bohmert et al. 1998). In some mutants, shoot apical meristem defects are also noted with variable penetrance. Pinhead/Zwille, a second *Arabidopsis* Argonaute family member, is critical for maintaining undifferentiated stem cells in the shoot apical meristem (Lynn et al. 1999). Although it is not clear whether these defects can be traced to a failure to regulate certain miRNA-targeted genes, the demonstrated role of some of these family members, notably, *AGO1*, in RNAi makes this a logical assumption. Furthermore, strong hypomorphic alleles of *Arabidopsis* Dicer homologs, particularly DCL-1 (formerly CAF-1/SIN1/SUS-1), cause embryonic-lethal phenotypes, whereas weak alleles can produce phenotypes similar to those that arise from alterations in certain Argonaute family members (for review, see Hammond et al. 2001a) and are defective in PTGS but not miRNA biogenesis (Finnegan et al. 2003).

## DICER IN VIVO

The dual role of Dicer in generating both miRNA and siRNAs has confounded analysis of its function in vivo. What role does Dicer have in initiating RNAi in vivo? Clearly, Dicer is required for RNAi triggered by long dsRNA, because it generates siRNAs (Grishok et al. 2001; Ketting et al. 2001; Knight and Bass 2001). However, the role of Dicer in RNAi initiated directly by siRNAs is less clear. In vitro, Dicer is not required for siRNA-directed target cleavage (Martinez et al. 2002), although small RNA-directed target cleavage by a Dicer-dependent mechanism has been proposed to be more efficient than the Dicer-independent route (Hutvágner and Zamore 2002a). Recent evidence suggests that Dicer is required for siRNA-triggered RNAi in cultured mammalian cells, because Dicer depletion (by RNAi!) reduces the efficacy of siRNAs directed against a luciferase reporter gene (Doi et al. 2003). Several studies show that Dicer and components of the RISC coimmunoprecipitate, suggesting that they are associated in a common complex (Hammond et al.

2001b; Tabara et al. 2002; Doi et al. 2003). It is possible that Dicer depletion blocks RNAi in vivo simply because it destabilizes this complex, leading to the depletion of core components of the RISC. Alternatively, siRNA production by Dicer may be directly coupled to RISC assembly. In this view, Dicer may pass siRNAs directly to the RISC, without siRNAs diffusing freely in the cytoplasm after their production, but before joining the RISC. Support for this view comes from recent studies demonstrating that *Drosophila* Dicer and at least one of the Dicer-like activities in wheat bind siRNAs tightly, probably in the Dicer active site (Tang et al. 2003). Thus, synthetic siRNAs might enter the RISC in vivo by first binding Dicer, simulating a natural pathway in which they are produced from dsRNA, and then passing from the Dicer active site to the RISC. This would aid in the solution to the problem of discriminating bona fide siRNAs from various RNA degradation products in the cell. High siRNA concentrations in vitro and perhaps even in vivo may bypass the coupling of Dicer to RISC formation.

## BIOCHEMISTRY OF RNAi IN MAMMALIAN CELLS

Biochemical dissection of the mammalian RNAi pathway has been carried out using HeLa cell S100 extracts, which recapitulate siRNA-directed target cleavage in vitro. Using these extracts, several components of the mammalian RISC have been identified, including the Argonaute homologs eIF2C1 and eIF2C2, the putative RNA helicase Gemin3, and a protein of unknown function, Gemin4 (Hutvágner and Zamore 2002a; Martinez et al. 2002). Like the RISC detected in *Drosophila* embryo lysates, the mammalian RISC mediates cleavage of the target RNA at a single phosphodiester bond, near the center of the antisense siRNA strand. Gem3 and Gem4 are also components of the SMN (survival of motor neurons) complex, which is required for small nuclear ribonucleoprotein (snRNP) biogenesis. The presence of the SMN protein, rather than one of the two eIF2C paralogs, distinguishes this latter complex from the RISC. The eiF2C/Gem3/Gem4 complex was first identified by Dreyfus and colleagues as an RNP containing numerous miRNAs (Mourelatos et al. 2002). Because the same complex can cleave a complementary target RNA at a single phosphodiester bond near the center of the small RNA guide, we refer to it as a RISC.

The demonstration that an endogenous miRNA was a component of a functional RISC led to the proposal that only a single complex exists to mediate the diverse functions of small RNAs (Hutvágner and Zamore 2002a). This model proposes that small RNA functional diversity is achieved not by multiple effector complexes, but by a single complex that can carry out at least two types of posttranscriptional regulation: target cleavage and translational repression. The one-complex, two-functions hypothesis was formulated to avoid postulating two evolutionarily distinct complexes that shared a single remarkable property—the ability to hold on tightly to a small RNA guide, irrespective of its sequence, in single-stranded (before binding its mRNA target) and double-stranded (after binding its target) states. Which type of regulation occurs was proposed to be determined solely by the degree of complementarity between the small RNA guide and its target. Additional support from this model comes from studies which show that siRNA can mediate translational control when cleavage is blocked by a lack of complementarity with the target RNA at the center of the siRNA guide (Doench et al. 2003).

Recently, a complex similar in composition to that of *Drosophila* RISC has been identified in mammalian cells in which RNAi has been initiated by transfection with siRNAs (A.A. Caudy et al., in prep.). This complex includes siRNAs or miRNAs, an Argonaute family member (eIF2C-1/Ago-1), the mammalian homolog of VIG (PAI-RBP1; Heaton et al. 2001), the Fragile X mental retardation protein, and the mammalian homolog of the

RISC-associated micrococcal nuclease family member (P100) (Callebaut and Mornon 1997; Ponting 1997). This complex is present in low amounts in naïve cells, but it can be induced to assemble by transfection with siRNAs. Furthermore, it is related in composition to a similar complex that cofractionates with siRNAs and that participates in miRNA-mediated repression in *C. elegans* (A.A. Caudy et al., in prep.). The precise relationship of this complex to the Gem3/Gem4-containing enzyme previously characterized in HeLa cells remains unclear. However, it is possible that these are the same complex or closely related complexes that differ only in their association with accessory factors.

## VARIATIONS ON THE THEME?

Although Dicer may have a role in RISC assembly, in vitro studies show that it is not involved in target destruction by the RISC (Hammond et al. 2001b; Martinez et al. 2002). Remarkably, RISC activity has not yet been demonstrated in *C. elegans*, the organism in which RNAi was first discovered (Guo and Kemphues 1995; Fire et al. 1998). In fact, the model of siRNA-programmed RISC as the primary executioner of target mRNAs, which has proved so powerful in explaining RNAi in *Drosophila* and mammals, fails to account for much of what we know about RNAi in *C. elegans*. Two observations in worms are difficult to reconcile with the RISC model. First, siRNAs are poor triggers for silencing in *C. elegans* (Caplen et al. 2001). Second, nematodes, as well as plants, *Dictyostelium*, and *Neurospora*, contain members of a conserved family of RdRPs that are required for RNAi. Such RdRP homologs are absent from both *Drosophila* and mammals, insofar as can be ascertained by searching the most current releases of the *Drosophila*, human, and mouse genome sequences.

To understand the significance of these two observations, we need first to review the properties of RNAi in *C. elegans*. In contrast to flies, where RNAi is strictly cell-autonomous (Roignant et al. 2003), RNAi seems to spread from cell to cell in worms. In fact, RNAi can be initiated simply by soaking worms in solutions containing dsRNA (Tabara et al. 1998) or by feeding worms *Escherichia coli* expressing dsRNA (Timmons and Fire 1998). RNAi in worms, but not in cultured *Drosophila* or human cells nor in flies or mice, displays "transitivity" (Sijen et al. 2001). Transitivity occurs when dsRNA (Y) corresponding to the 3′ portion of a hybrid gene (XY) triggers silencing of a separate gene encompassing only X sequences. Such silencing is surprising because the trigger dsRNA shares no sequence with the X target. Instead, the silencing of X requires the intercession of the hybrid gene, XY, to generate X-specific siRNAs. Because silencing of X is not triggered by Y dsRNA when the chimeric gene is configured Y-X, transitive RNAi has been proposed to be mediated by an RdRP that uses an antisense siRNA strand as a primer to generate new dsRNA by the direct copying of sequence 5′ to the Y sequences. Transitivity in worms is short-range, however, and is not detected more than about 100 nucleotides 5′ to the position of the input dsRNA along the chimera. This observation is surprising, because RdRP activities in other organisms (e.g., wheat) are highly processive, capable of synthesizing many hundreds of nucleotides of complementary RNA without pause.

RNAi triggered by ingestion of dsRNA requires a putative membrane protein, Sid-1 (Winston et al. 2002), but little else is known of how ingested dsRNA causes systemic RNAi. Even more remarkable is the finding that RNAi can be passed from generation to generation (Grishok et al. 2000). In principle, intergenerational silencing might reflect the inheritance of siRNA-programmed RISCs. In cultured mammalian cells, RISC inheritance is thought to explain the persistence of silencing through multiple rounds of cell division. However, siRNA-triggered RNAi is relatively short-lived in dividing mammalian cells (Amarzguioui et al. 2003). RNAi in mammalian cells likely lasts for days, despite the dilution of siRNA-programmed RISC at each cell division, because each RISC is a multiple-turnover enzyme capable of repeated cycles of mRNA target cleavage (Hutvágner and Zamore 2002a). In worms, where RNAi persists through multiple generations, it is assumed that a cellular mechanism amplifies the original trigger dsRNA or the siRNAs generated from it, counterbalancing the effects of dilution at each cell cycle. The role of amplifying either the trigger dsRNA or the siRNAs themselves has been ascribed to two enzymes, a somatic RdRP, Rrf-1 (Sijen et al. 2001), and a germ-line-specific RdRP, Ego-1 (Smardon et al. 2000). Support for this idea comes from the difficulty in detecting in worms "primary" siRNAs (i.e., the siRNAs derived directly from the trigger dsRNA), but the relative ease with which "secondary" siRNAs (those whose production requires the presence of a corresponding RNA target) can be found. Several models for amplification have been proposed, but none reconcile amplification of the trigger dsRNA with the lack of RdRP processivity observed in transitive RNAi experiments.

In plants, "spreading" of the silencing signal, rather than transitive RNAi, is observed. Spreading, like transitivity, reflects the production of siRNAs 5′ to the site of the original trigger dsRNA (Fabian et al. 2002). Unlike transitive RNAi, spreading also occurs 3′ to the target region corresponding to the dsRNA. Both 5′ and 3′ spreading occur even when a single siRNA is used to trigger silencing. For the production of new siRNAs 3′ to the site corresponding to the original siRNA trigger to be catalyzed by an RdRP using the original antisense siRNA strand as a primer, the target RNA would have to be circular. An alternative model is that the RdRP does not use a primer, but rather initiates RNA synthesis at the extreme 3′ end of the RNA template. In this view, the two target cleavage products generated by an siRNA-programmed RISC would each serve as templates for the RdRP. Although this model solves the problem of bidirectional spreading, it fails to explain how the RdRP recognizes the products of RISC cleavage as substrates while ignoring other cellular RNAs. Furthermore, endogenous plant RISC complexes, programmed with miRNAs (see above), cleave RNA targets, but do not trigger spreading. To resolve this paradox, some investigators have proposed that the RdRP senses the intracellular concentration of an RNA, only copying it into new dsRNA when it accumulates past a threshold amount (Hutvágner and Zamore 2002b; Tang et al. 2003).

Of course, such a model merely defers the question because it does not explain how an enzyme can distinguish a high concentration of a single RNA species from the aggregate concentration of many species. Furthermore, the model does not explain why transgenic mRNA itself, rather than just mRNA cleavage products, does not trigger spreading. A possible resolution to this paradox is that plant cells perceive an RNA as "aberrant" not merely because it is too abundant, but also because it is both too abundant and fails to be productively translated. Thus, the cleavage products produced by an siRNA-programmed RISC would trigger the "aberrancy" sensor (an RdRP?), but highly expressed mRNA would not. Normal cellular mRNAs can be cleaved when targeted for destruction by miRNA-programmed RISC complexes, but the cleavage products are never sufficiently abundant to qualify as aberrant. This proposal makes several predictions. First, the probability that a silenced transgene triggers spreading should be a function of the strength of the transgenic promoter. Second, chimeric RNAs containing long internal poly(A) stretches in their 3′-untranslated regions (3′UTRs) should not induce silencing when the downstream portion of the 3′UTR is cleaved off in response to an exogenous siRNA. Finally, in the absence of a corresponding transgene, exogenous siRNAs should trigger cleavage of endogenous mRNAs, but not spreading, i.e., such siRNAs should behave like plant miRNAs.

The mechanisms of both transitive RNAi in worms and spreading in plants are obscure, in part because suitable in vitro systems remain to be developed. Recent attempts to produce in vitro extracts that recapitulate plant RNA silencing have seen success for the function of miRNAs (Tang et al. 2003), but do not recapitulate the entire PTGS pathway. Wheat-germ extracts do recapitulate several key features of plant silencing—Dicer and RdRP activity—and also contain endogenous miRNA-programmed RISC complexes that cleave target RNAs, but they cannot yet be programmed with exogenous single-stranded RNA (ssRNA), dsRNA, or siRNAs to cleave target RNAs. A deeper mechanistic understanding of plant silencing is likely to be closely allied with the solution to this technical problem.

## LINKS TO HUMAN DISEASE?

The association of the miRNAs *lin-4* and *let-7* with the RNAi machinery suggested that RNAi regulates developmental processes. Parallel discoveries may also link RNAi to the molecular defects that underlie Fragile X mental retardation (Caudy et al. 2002; Ishizuka et al. 2002). Starting from the purification of apparently different complexes, two labs independently arrived at the same conclusion: RNAi components associate with dFXR (Fragile-X-related), the *Drosophila* homolog of Fragile X mental retardation protein (FMRP). The loss of FMRP expression in humans causes the most common form of mental retardation.

One group discovered dFXR in association with purified RISC complexes (Caudy et al. 2002), whereas the second group identified Ago-2 in attempts to purify proteins that associate with endogenous dFXR (Ishizuka et al. 2002). Consistent with this reciprocal interaction, antibodies to dFXR immunoprecipitate RISC activity, and essentially all of the dFXR protein in *Drosophila* S2 cells cofractionate with RISC activity and with the RISC components Ago-2, siRNAs, VIG, and TSN (Caudy et al. 2002, and unpubl.). In addition to Ago-2, Ishizuka et al. (2002) found that dFXR purified with 5S ribosomal RNA, ribosomal proteins L5 and L11, and Dmp68, a DEAD-box helicase. Both RISC complexes and the dFXR protein bind to RNA, but the complexes were RNase-resistant, suggesting that the proteins were interacting directly, rather being associated via mutual interaction with cellular RNAs in extracts. The two studies disagreed on the requirement of dFXR for RNAi: Caudy et al. found that depletion of dFXR by RNAi suppressed RNAi slightly, whereas Ishizuka et al. did not observe an effect. These conflicting RNAi-of-RNAi results indicate that dFXR is probably an accessory factor, rather than an essential component of RISC for targeting of mRNAs as directed by exogenous dsRNAs. However, the lack of substantial effects could also be reflective of incomplete depletion of dFXR by RNAi.

Not only siRNAs but also miRNAs are present in complexes with dFXR. This is again consistent with the hypothesis that siRNAs and miRNAs are not sorted into distinct complexes, but that both types of small RNAs enter multifunctional complexes capable of mRNA degradation and translational regulation. Both siRNAs and miRNAs copurify with RISC activity, although the majority of miRNAs (and a minority of the siRNAs) in S2 cells associate in a smaller, distinct complex with Ago-1 (Caudy et al. 2002). miRNAs are present in immunoprecipitates of dFXR, Ago-2, VIG, and TSN. dFXR and its family members are RNA-binding proteins (see below) and may represent the first example of an general mechanism by which RISC complexes effectively target endogenous messages by partnering with other RNA-binding proteins.

Recent data in *C. elegans* support the hypothesis that siRNAs and miRNAs are both present in similar complexes (A.A. Caudy et al., in prep.). In RNP complexes purified from adult worms and eggs, miRNAs were present in the complexes that also contained the *C. elegans* homologs of VIG and the nuclease TSN. In addition, depletion of VIG and TSN by RNAi prevents the proper down-regulation of a *lin-41* 3′UTR reporter gene that normally occurs at the L4 to adult transition. This *lin-41* down-regulation is dependent on proper function of *let-7*, suggesting that the VIG and TSN proteins not only have a role in mRNA degradation as found in *Drosophila*, but also are important for the function of the *let-7* miRNA pathway. As an additional example of miRNAs associated with proteins originally identified in degradation complexes, *let-7* is present in immunoprecipitates from human cells using PAI-RBP1, the human VIG homolog, and p-100, the TSN homolog. These immunoprecipitates also contain FMRP, suggesting that complexes present in human cells are similar to those originally identified in fly cells.

The Fragile X family of RNA-binding proteins contain two KH domains and an RGG box, all of which bind RNA. Many RNAs have been found to interact with FMRP (Brown et al. 1998, 2001), including the binding of FMRP to its own message. Both SELEX (systematic evolution of ligands by exponential enrichment) experiments (Darnell et al. 2001) and a directed analysis of the binding of FMRP to its message (Schaeffer et al. 2001) indicate that the RGG box selectively binds G-quartet structures. However, the RNAs resulting from the SELEX experiments are not only potential G-quartets, but also short hairpins with a 14-nucleotide stem. Some mutations that disrupt binding also disrupt this stem loop (although not the potential G-quartet). Thus, some type of secondary structure, perhaps a duplex, might also be important for recognition of target RNAs by FMRP.

Although there is a single Fragile X family member in *Drosophila*, mammals have three Fragile-X-related proteins, including FMRP, FXR1P, and FXR2P. These proteins are closely related in sequence and are expressed in most tissue types. Knockout mice for FXR2P show a number of neurological and behavioral phenotypes, some similar to those in FMRP knockout animals (Bontekoe et al. 2002). FXR1P knockout mice develop normally but die shortly after birth (cited as a personal communication in Bontekoe et al. 2002). In addition to mental retardation, individuals with Fragile X syndrome have a number of other phenotypes, including craniofacial abnormalities, macro-orchidism, and hyperextensible joints. This suggests that the Fragile X protein family may be important for gene regulation in a number of cell types and not just the brain.

Fragile X family members associate with polyribosomes in an RNA-dependent manner (Eberhart et al. 1996; Khandjian et al. 1996; Tamanini et al. 1996; Feng et al. 1997). The I304N point mutation in the FMRP KH domain is associated with disease (De Boulle et al. 1993) and disrupts the binding of FMRP to polysomes (Tamanini et al. 1996; Feng et al. 1997). The analogous mutation in *Drosophila* also disrupts dFXR association with ribosomes (Caudy et al. 2002). FMRP has been observed trafficking along dendrites in hippocampal and PC-12 cells (De Diego Otero et al. 2002). FMRP and the family member FXR1P colocalize with RNA and 60S ribosomes in these dendrites, suggesting that FMRP could have a role in regulating the expression of target mRNAs, perhaps in an activity-dependent manner.

Recent work has implicated another RNA-binding partner for FMRP in neurons. The nontranslated BC1 RNA is an approximately 150-nucleotide RNA expressed in murine dendrites. It was found to associate specifically with FMRP (Zalfa et al. 2003). The BC1 RNA contains regions of base-pairing complementarity ranging between 15 and 23 nucleotides to a number of the known targets of FMRP. Addition of competing complementary oligoribonucleotides could disrupt the interaction of FMRP with the mRNA targets (and, to some extent, the BC1 RNA, perhaps due to structural changes). BC1 represents an interesting new example of homology-directed gene regulation in an RNP complex. The length of complementary sequence between BC1 and some of its target RNAs may be short enough that regulation requires additional forms of molecular recognition to achieve specificity.

Alignment of the *lin-28* 3′UTRs from *C. elegans* and *Caenorhabditis briggsae* reveals strong conservation at the *lin-4*-binding site, but there is also a large amount of sequence conservation beyond the miRNA-binding site (Seggerson et al. 2002). It may be the case that a variety of RNA-binding proteins or mRNP complexes such as FMRP/BC1 partner with RISC complexes to target miRNAs specifically and efficiently to their messages for regulation. These additional RNA-binding partners may also further tune the translational regulation or localization of the target proteins.

## RNAi AND THE GENOME

In addition to their posttranscriptional functions, small RNAs also direct chromatin remodeling, leading to changes in DNA transcription and stability. The year 2002 saw the discovery in both *S. pombe* (Volpe et al. 2002) and *Tetrahymena* (Mochizuki 2002; Taverna et al. 2002) that the presence of small RNAs correlates with methylation of histone H3 lysine 9 (H3K9) on genes corresponding to the small RNA sequence. H3K9 methylation signals the formation of heterochromatin (see Chapter 10). In fission yeast, heterochromatin formation represses gene transcription. In *Tetrahymena*, methyl-H3K9 serves as a signal for the physical deletion of specific DNA sequences in the macronucleus (an intact copy of these genes is retained in the transcriptionally quiescent micronucleus). The dramatic consequences of H3K9 methylation in *Tetrahymena* probably reflect a macronuclear response to heterochromatin, rather than a distinct mechanism of methyl-H3K9 activity.

Although it remains to be formally demonstrated that these small RNAs are bona fide siRNAs, a strong circumstantial case has been assembled: Both H3K9 methylation and small RNA production require homologs of Dicer, an Argonaute family member, and a putative RdRP (Volpe et al. 2002). The involvement of these "RNAi genes" in pathways required for heterochromatin formation is by no means unprecedented. Transcriptional gene silencing can be triggered by dsRNA corresponding to promoter sequences in plants and requires Argonaute homologs in flies. Furthermore, a distinct class of small RNAs, approximately 25-nucleotide-long siRNAs, correlates with PTGS in plants (Hamilton et al. 2002). The involvement of both small RNAs and Argonaute homologs in transcriptional silencing suggests that the "single complex" hypothesis may be extended further. Might not the same RISC complex proposed to mediate target cleavage and translational control also catalyze the targeted production of heterochromatin? If so, how?

One simple idea is that the small RNA guide, as a component of the core RISC, binds to DNA, delivering an H3K9-specific methyltransferase to the chromatin. Alternatively, the complex might bind to rare RNA transcripts crossing promoter elements of the gene targeted for transcriptional repression. The existence of such RNA can be inferred from endogenous transcripts, likely dsRNA, detected in fission yeast from the genomic regions subject to silencing by an RNAi-related mechanism. Gene-specific targeting of H3K9 methylation can be achieved artificially by targeting the methyltransferase to the DNA by fusing it to an engineered zinc finger transcription factor (ZFT) that binds specifically to the promoter of the target gene (Snowden et al. 2002). Models in which small RNAs direct an H3K9-specific methyltransferase to specific chromatin regions predict that such H3K9 methyltransferase-ZFT chimeras will not require genes in the RNAi pathway to initiate formation of repressive heterochromatin on a target gene.

A direct interaction between the siRNA and putative RNA-binding domains of the H3K9 methyltransferase itself (Hall et al. 2002) seems unlikely, given the intimate association of Argonaute proteins with the siRNA, and the requirement for Ago-1 to initiate silencing of the *cenH* locus in fission yeast. However, the RISC is likely to be decorated with a variety of proteins that extend its core function of cleaving mRNA targets. Proteins such as the Fragile X protein, which is required for translational control of several neuron-specific transcripts, are intimately associated with the RISC, so it is not difficult to imagine that H3K9 methyltransferases such as Clr4 in fission yeast might also be peripherally bound to a subset of RISCs that initiate heterochromatin formation in the nucleus.

A more speculative idea is that the small RNA guides associated with heterochromatin production do not directly interact with a histone methyltransferase, but rather act as "old fashioned" RISCs to cleave RNA transcripts in the nucleus. Britten and Davidson (1969) originally postulated that low levels of "background" transcription might occur on both strands throughout the entire genome. This model holds a certain appeal when one considers the copy-number-dependent cosuppression that is triggered by unlinked transgenes (see Chapter 2). A threshold dependence might be achieved by accumulation of transcripts from both strands of the genome to a certain level at which point sufficient dsRNA would be formed to trigger RNAi.

That such RNAs might have a role in specifying the "state" of chromatin is not terribly far-fetched, considering the roles of Xist and Tsix in X inactivation (for review, see Lee 2002). In this case, we would postulate a positive role not only for transcription throughout what are traditionally thought of as nontranscribed regions, but also for the RNAs themselves. Elimination of such RNAs by RISC cleavage would remove a positive signal for open chromatin and favor formation of a heterochromatic state. In this model, H3K9 methylation would be a consequence, but not the direct cause of such RISC action. The appeal of this outlandish model is that it retains the simplicity of one complex with many functions and eliminates the question of how a single complex might recognize both ssRNA and dsDNA. A key prediction of this model is that an siRNA-directed, target-cleaving RISC activity remains to be discovered in fission yeast.

## SUMMARY

The past several years have seen remarkable progress in our understanding of the underlying mechanisms of RNAi and related silencing processes. Five years ago, the mechanisms through which dsRNA could induce sequence-specific gene silencing were a mystery. Through a combination of biochemical and genetic approaches in various model systems, we now have the outlines of a coherent model for the interference process and know many of the players involved. Still, many central questions remain unanswered, particularly as they relate to the ability of dsRNA to cause heritable changes in gene expression. Additionally, we know very little about the details of the effector enzyme and how it finds its targets and catalyzes the cleavage event. Perhaps one of the most important outcomes of biochemical studies of RNAi has been the remarkable complexity of both the pathway itself and the proteins comprising the RNAi machinery. This unexpected complexity has given us a new appreciation of the biological roles of the dsRNA-induced silencing machinery.

## REFERENCES

Amarzguioui M., Holen T., Babaie E., and Prydz H. 2003. Tolerance for mutations and chemical modifications in a siRNA. *Nucleic Acids Res.* **31:** 589–595.

Bass B.L. 2000. Double-stranded RNA as a template for gene silencing. *Cell* **101:** 235–238.

Bernstein E., Caudy A.A., Hammond S.M., and Hannon G.J. 2001. Role for a bidentate ribonuclease in the initiation step of RNA interference. *Nature* **409:** 363–366.

Blaszczyk J., Tropea J.E., Bubunenko M., Routzahn K.M., Waugh D.S., Court D.L., and Ji X. 2001. Crystallographic and modeling studies of RNase III suggest a mechanism for double-stranded RNA cleavage. *Structure* **9:** 1225–1236.

Bohmert K., Camus I., Bellini C., Bouchez D., Caboche M., and Benning C. 1998. AGO1 defines a novel locus of *Arabidopsis* controlling leaf development. *EMBO J.* **17:** 170–180.

Bontekoe C.J., McIlwain K.L., Nieuwenhuizen I.M., Yuva-Paylor L.A., Nellis A., Willemsen R., Fang Z., Kirkpatrick L., Bakker C.E., McAninch R., et al. 2002. Knockout mouse model for Fxr2: A model for mental retardation. *Hum. Mol. Genet.* **11:** 487–498.

Britten R.J. and Davidson E.H. 1969. Gene regulation for higher cells: A theory. *Science* **165:** 349–357.

Brown V., Small K., Lakkis L., Feng Y., Gunter C., Wilkinson K.D., and Warren S.T. 1998. Purified recombinant Fmrp exhibits selective RNA binding as an intrinsic property of the fragile X mental retardation protein. *J. Biol. Chem.* **273:** 15521–15527.

Brown V., Jin P., Ceman S., Darnell J.C., O'Donnell W.T., Tenenbaum S.A., Jin X., Feng Y., Wilkinson K.D., Keene J.D., Darnell R.B., and Warren S.T. 2001. Microarray identification of FMRP-associated brain mRNAs and altered mRNA translational profiles in fragile X syndrome. *Cell* **107:** 477–487.

Callebaut I. and Mornon J.P. 1997. The human EBNA-2 coactivator p100: Multidomain organization and relationship to the staphylococcal nuclease fold and to the tudor protein involved in *Drosophila melanogaster* development. *Biochem. J.* **321:** 125–132.

Caplen N.J., Parrish S., Imani F., Fire A., and Morgan R.A. 2001. Specific inhibition of gene expression by small double-stranded RNAs in invertebrate and vertebrate systems. *Proc. Natl. Acad. Sci.* **98:** 9742–9747.

Carmell M.A., Xuan Z., Zhang M.Q., and Hannon G.J. 2002. The Argonaute family: Tentacles that reach into RNAi, developmental control, stem cell maintenance, and tumorigenesis. *Genes Dev.* **16:** 2733–2742.

Catalanotto C., Azzalin G., Macino G., and Cogoni C. 2000. Gene silencing in worms and fungi. *Nature* **404:** 245.

Caudy A.A., Myers M., Hannon G.J., and Hammond S.M. 2002. Fragile X-related protein and VIG associate with the RNA interference machinery. *Genes Dev.* **16:** 2491–2496.

Cerutti L., Mian N., and Bateman A. 2000. Domains in gene silencing and cell differentiation proteins: The novel PAZ domain and redefinition of the Piwi domain. *Trends Biochem. Sci.* **25:** 481–482.

Chalfie M., Horvitz H.R., and Sulston J.E. 1981. Mutations that lead to reiterations in the cell lineages of *C. elegans. Cell* **24:** 59–69.

Cikaluk D.E., Tahbaz N., Hendricks L.C., DiMattia G.E., Hansen D., Pilgrim D., and Hobman T.C. 1999. GERp95, a membrane-associated protein that belongs to a family of proteins involved in stem cell differentiation. *Mol. Biol. Cell.* **10:** 3357–3372.

Darnell J.C., Jensen K.B., Jin P., Brown V., Warren S.T., and Darnell R.B. 2001. Fragile X mental retardation protein targets G quartet mRNAs important for neuronal function. *Cell* **107:** 489–499.

De Boulle K., Verkerk A.J., Reyniers E., Vits L., Hendrickx J., Van Roy B., Van den Bos F., de Graaff E., Oostra B.A., and Willems P.J. 1993. A point mutation in the FMR-1 gene associated with fragile X mental retardation. *Nat. Genet.* **3:** 31–35.

De Diego Otero Y., Severijnen L.A., van Cappellen G., Schrier M., Oostra B., and Willemsen R. 2002. Transport of fragile X mental retardation protein via granules in neurites of PC12 cells. *Mol. Cell. Biol.* **22:** 8332–8341.

Deng W. and Lin H. 2002. *miwi*, a murine homolog of *piwi*, encodes a cytoplasmic protein essential for spermatogenesis. *Dev. Cell.* **2:** 819–830.

Doench J.G., Petersen C.P., and Sharp P.A. 2003. siRNAs can function as miRNAs. *Genes Dev.* **17:** 438–442.

Doi N., Zenno S., Ueda R., Ohki-Hamazaki H., Ui-Tei K., and Saigo K. 2003. Short-interfering-RNA-mediated gene silencing in mammalian cells requires Dicer and eIF2C translation initiation factors. *Curr. Biol.* **13:** 41–46.

Domeier M.E., Morse D.P., Knight S.W., Portereiko M., Bass B.L., and Mango S.E. 2000. A link between RNA interference and nonsense-mediated decay in *Caenorhabditis elegans. Science* **289:** 1928–1931.

Eberhart D.E., Malter H.E., Feng Y., and Warren S.T. 1996. The fragile X mental retardation protein is a ribonucleoprotein containing both nuclear localization and nuclear export signals. *Hum. Mol. Genet.* **5:** 1083–1091.

Elbashir S.M., Martinez J., Patkaniowska A., Lendeckel W., and Tuschl T. 2001a. Functional anatomy of siRNAs for mediating efficient RNAi in *Drosophila melanogaster* embryo lysate. *EMBO J.*

**20:** 6877–6888.

Elbashir S.M., Harborth J., Lendeckel W., Yalcin A., Weber K., and Tuschl T. 2001b. Duplexes of 21-nucleotide RNAs mediate RNA interference in cultured mammalian cells. *Nature* **411:** 494–498.

Fabian E., Jones L., and Baulcombe D.C. 2002. Spreading of RNA targeting and DNA methylaion in RNA silencing requires transcription of the target gene and a putative RNA dependent RNA polymerase. *Plant Cell* **14:** 857–867.

Feng Y., Gutekunst C.A., Eberhart D.E., Yi H., Warren S.T., and Hersch S.M. 1997. Fragile X mental retardation protein: Nucleocytoplasmic shuttling and association with somatodendritic ribosomes. *J. Neurosci.* **17:** 1539–1547.

Filippov V., Solovyev V., Filippova M., and Gill S.S. 2000. A novel type of RNase III family proteins in eukaryotes. *Gene* **245:** 213–221.

Finnegan E.J., Margis R., and Waterhouse P.M. 2003. Posttranscriptional gene silencing is not compromised in the *Arabidopsis* CARPEL FACTORY (DICER-LIKE1) mutant, a homolog of Dicer-1 from *Drosophila. Curr. Biol.* **13:** 236–240.

Fire A., Xu S., Montgomery M.K., Kostas S.A., Driver S.E., and Mello C.C. 1998. Potent and specific genetic interference by double-stranded RNA in *Caenorhabditis elegans. Nature* **391:** 806–811.

Grishok A., Tabara H., and Mello C.C. 2000. Genetic requirements for inheritance of RNAi in *C. elegans. Science* **287:** 2494–2497.

Grishok A., Pasquinelli A.E., Conte D., Li N., Parrish S., Ha I., Baillie D.L., Fire A., Ruvkun G., and Mello C.C. 2001. Genes and mechanisms related to RNA interference regulate expression of the small temporal RNAs that control *C. elegans* developmental timing. *Cell* **106:** 23–34.

Guo S. and Kemphues K.J. 1995. *par-1*, a gene required for establishing polarity in *C. elegans* embryos, encodes a putative Ser/Thr kinase that is asymmetrically distributed. *Cell* **81:** 611–620.

Hall I.M., Shankaranarayana G.D., Noma K., Ayoub N., Cohen A., and Grewal S.I. 2002. Establishment and maintenance of a heterochromatin domain. *Science* **297:** 2232–2237.

Hamilton A.J. and Baulcombe D.C. 1999. A species of small antisense RNA in posttranscriptional gene silencing in plants [see comments]. *Science* **286:** 950–952.

Hamilton A., Voinnet O., Chappell L., and Baulcombe D. 2002. Two classes of short interfering RNA in RNA silencing. *EMBO J.* **21:** 4671–4579.

Hammond S.M., Bernstein E., Beach D., and Hannon G.J. 2000. An RNA-directed nuclease mediates post-transcriptional gene silencing in *Drosophila* cells. *Nature* **404:** 293–296.

Hammond S.M., Caudy A.A., and Hannon G.J. 2001a. Post-transcriptional gene silencing by double-stranded RNA. *Nat. Rev. Genet.* **2:** 110–119.

Hammond S.M., Boettcher S., Caudy A.A., Kobayashi R., and Hannon G.J. 2001b. Argonaute2, a link between genetic and biochemical analyses of RNAi. *Science* **293:** 1146–1150.

Hannon G.J. 2002. RNAi. *Nature* **418:** 244–251.

Heaton J.H., Dlakic W.M., Dlakic M., and Gelehrter T.D. 2001. Identification and cDNA cloning of a novel RNA-binding protein that interacts with the cyclic nucleotide-responsive sequence in the Type-1 plasminogen activator inhibitor mRNA. *J. Biol. Chem.* **276:** 3341–3347.

Holen T., Amarzguioui M., Wiiger M.T., Babaie E., and Prydz H. 2002. Positional effects of short interfering RNAs targeting the human coagulation trigger tissue factor. *Nucleic Acids Res.* **30:** 1757–1766.

Hutvágner G. and Zamore P.D. 2002a. A microRNA in a multiple-turnover RNAi enzyme complex. *Science* **1:** 1.

———. 2002b. RNAi: Nature abhors a double-strand. *Curr. Opin. Genet. Dev.* **12:** 225–232.

Hutvágner G., McLachlan J., Pasquinelli A.E., Balint E., Tuschl T., and Zamore P.D. 2001. A cellular function for the RNA-interference enzyme Dicer in the maturation of the *let-7* small temporal RNA. *Science* **293:** 834–838.

Ishizuka A., Siomi M.C., and Siomi H. 2002. A *Drosophila* fragile X protein interacts with components of RNAi and ribosomal proteins. *Genes Dev.* **16:** 2497–2508.

Jones L., Ratcliff F., and Baulcombe D.C. 2001. RNA-directed transcriptional gene silencing in plants can be inherited independently of the RNA trigger and requires Met1 for maintenance. *Curr. Biol.* **11:** 747–757.

Kasschau K.D., Xie Z., Allen E., Llave C., Chapman E.J., Krizan K.A., and Carrington J.C. 2003. P1/HC-Pro, a viral suppressor of RNA silencing, interferes with *Arabidopsis* development and miRNA function. *Dev. Cell* **4:** 205–217.

Kataoka Y., Takeichi M., and Uemura T. 2001. Developmental roles and molecular characterization of a *Drosophila* homologue of *Arabidopsis* Argonaute1, the founder of a novel gene superfamily. *Genes Cells* **6:** 313–325.

Kennerdell J.R. and Carthew R.W. 1998. Use of dsRNA-mediated genetic interference to demonstrate that frizzled and frizzled 2 act in the wingless pathway. *Cell* **95:** 1017–1026.

Kennerdell J.R., Yamaguchi S., and Carthew R.W. 2002. RNAi is activated during *Drosophila* oocyte maturation in a manner dependent on aubergine and spindle-E. *Genes Dev.* **16:** 1884–1889.

Ketting R.F., Haverkamp T.H., van Luenen H.G., and Plasterk R.H. 1999. Mut-7 of *C. elegans*, required for transposon silencing and RNA interference, is a homolog of Werner syndrome helicase and RNaseD. *Cell* **99:** 133–141.

Ketting R.F., Fischer S.E., Bernstein E., Sijen T., Hannon G.J., and Plasterk R.H. 2001. Dicer functions in RNA interference and in synthesis of small RNA involved in developmental timing in *C. elegans. Genes Dev.* **15:** 2654–2659.

Khandjian E.W., Corbin F., Woerly S., and Rousseau F. 1996. The fragile X mental retardation protein is associated with ribosomes. *Nat. Genet.* **12:** 91–93.

Klahre U., Crete P., Leuenberger S.A., Iglesias V.A., and Meins F., Jr. 2002. High molecular weight RNAs and small interfering RNAs induce systemic posttranscriptional gene silencing in plants. *Proc. Natl. Acad. Sci.* **99:** 11981–11986.

Knight S.W. and Bass B.L. 2001. A role for the RNase III enzyme DCR-1 in RNA interference and germ line development in *Caenorhabditis elegans. Science* **293:** 2269–2271.

Lagos-Quintana M., Rauhut R., Lendeckel W., and Tuschl T. 2001. Identification of novel genes coding for small expressed RNAs. *Science* **294:** 853–858.

Lau N.C., Lim L.P., Weinstein E.G., and Bartel D.P. 2001. An abundant class of tiny RNAs with probable regulatory roles in *Caenorhabditis elegans. Science* **294:** 858–862.

Lee J.T. 2002. Is X-chromosome inactivation a homology effect? *Adv. Genet.* **46:** 25–48.

Lee R.C. and Ambros V. 2001. An extensive class of small RNAs in *Caenorhabditis elegans. Science* **294:** 862–864.

Lee R.C., Feinbaum R.L., and Ambros V. 1993. The *C. elegans* heterochronic gene *lin-4* encodes small RNAs with antisense complementarity to *lin-14. Cell* **75:** 843–854.

Lewis D.L., Hagstrom J.E., Loomis A.G., Wolff J.A., and Herweijer H. 2002. Efficient delivery of siRNA for inhibition of gene expression in postnatal mice. *Nat. Genet.* **32:** 107–108.

Lim L.P., Glasner M.E., Yekta S., Burge C.B., and Bartel D.P. 2003a. Vertebrate microRNA genes. *Science* **299:** 1540.

Lim L.P., Lau N.C., Weinstein E.G., Abdelhakim A., Yekta S., Rhoades M.W., Burge C.B., and Bartel D.P. 2003b. The microRNAs of *Caenorhabditis elegans. Genes Dev.* **17:** 99–108.

Llave C., Xie Z., Kasschau K.D., and Carrington J.C. 2002. Cleavage of Scarecrow-like mRNA targets directed by a class of *Arabidopsis* miRNA. *Science* **297:** 2053–2056.

Lynn K., Fernandez A., Aida M., Sedbrook J., Tasaka M., Masson P., and Barton M.K. 1999. The *PINHEAD/ZWILLE* gene acts pleiotropically in *Arabidopsis* development and has overlapping functions with the *ARGONAUTE1* gene. *Development* **126:** 469–481.

Martinez J., Patkaniowska A., Urlaub H., Luhrmann R., and Tuschl T. 2002. Single-stranded antisense siRNAs guide target RNA cleavage in RNAi. *Cell* **110:** 563–574.

McCaffrey A.P., Meuse L., Pham T.T., Conklin D.S., Hannon G.J., and Kay M.A. 2002. RNA interference in adult mice. *Nature* **418:** 38–39.

Mochizuki K., Fine N.A, Fujisawa T., and Gorovsky M.A. 2002. Analysis of a *piwi*-related gene implicates small RNAs in genome rearrangement in *Tetrahymena. Cell* **110:** 689–699.

Morel J.B., Godon C., Mourrain P., Beclin C., Boutet S., Feuerbach F., Proux F., and Vaucheret H. 2002. Fertile hypomorphic ARGONAUTE (ago1) mutants impaired in post-transcriptional gene silencing and virus resistance. *Plant Cell* **14:** 629–639.

Mourelatos Z., Dostie J., Paushkin S., Sharma A., Charroux B., Abel L., Rappsilber J., Mann M., and Dreyfuss G. 2002. miRNPs: A novel class of ribonucleoproteins containing numerous microRNAs. *Genes Dev.* **16:** 720–728.

Myers J.W., Jones J.T., Meyer T., and Ferrell J.E. 2003. Recombinant Dicer efficiently converts large dsRNAs into siRNAs suitable for gene silencing. *Nat. Biotechnol.* **21:** 324–328.

Nykänen A., Haley B., and Zamore P.D. 2001. ATP requirements and small interfering RNA structure in the RNA interference pathway. *Cell* **107:** 309–321.

Olsen P.H. and Ambros V. 1999. The *lin-4* regulatory RNA controls developmental timing in *Caenorhabditis elegans* by blocking LIN-14 protein synthesis after the initiation of translation.

*Dev. Biol.* **216:** 671–680.

Pal-Bhadra M., Bhadra U., and Birchler J.A. 2002. RNAi related mechanisms affect both transcriptional and posttranscriptional transgene silencing in *Drosophila. Mol. Cell* **9:** 315–327.

Ponting C.P. 1997. P100, a transcriptional coactivator, is a human homologue of staphylococcal nuclease. *Protein Sci.* **6:** 459–463.

Provost P., Dishart D., Doucet J., Frendewey D., Samuelsson B., and Radmark O. 2002. Ribonuclease activity and RNA binding of recombinant human Dicer. *EMBO J.* **21:** 5864–5874.

Reinhart B.J., Slack F.J., Basson M., Pasquinelli A.E., Bettinger J.C., Rougvie A.E., Horvitz H.R., and Ruvkun G. 2000. The 21-nucleotide *let-7* RNA regulates developmental timing in *Caenorhabditis elegans. Nature* **403:** 901–906.

Rhoades M.W., Reinhart B.J., Lim L.P., Burge C.B., Bartel B., and Bartel D.P. 2002. Prediction of plant microRNA targets. *Cell* **110:** 513–520.

Roignant J.Y., Carre C., Mugat B., Szymczak D., Lepesant J.A., and Antoniewski C. 2003. Absence of transitive and systemic pathways allows cell-specific and isoform-specific RNAi in *Drosophila. RNA* **9:** 299–308.

Schaeffer C., Bardoni B., Mandel J.L., Ehresmann B., Ehresmann C., and Moine H. 2001. The fragile X mental retardation protein binds specifically to its mRNA via a purine quartet motif. *EMBO J.* **20:** 4803–4813.

Seggerson K., Tang L., and Moss E.G. 2002. Two genetic circuits repress the *Caenorhabditis elegans* heterochronic gene *lin-28* after translation initiation. *Dev. Biol.* **243:** 215–225.

Sempere L.F., Dubrovsky E.B, Dubrovskaya V.A., Berger E.M., and Ambros V. 2002. The expression of the *let-7* small regulatory RNA is controlled by ecdysone during metamorphosis in *Drosophila melanogaster. Dev. Biol.* **244:** 170–179.

Sijen T., Fleenor J., Simmer F., Thijssen K.L., Parrish S., Timmons L., Plasterk R.H., and Fire A. 2001. On the role of RNA amplification in dsRNA-triggered gene silencing. *Cell* **107:** 465–476.

Smardon A., Spoerke J.M., Stacey S.C., Klein M.E., Mackin N., and Maine E.M. 2000. EGO-1 is related to RNA-directed RNA polymerase and functions in germ-line development and RNA interference in *C. elegans. Curr. Biol.* **10:** 169–178. (Published erratum appears in *Curr. Biol.* 2000 May 18; 10(10), R393–394.)

Snowden A.W., Gregory P.D., Case C.C., and Pabo C.O. 2002. Gene-specific targeting of H3K9 methylation is sufficient for initiating repression in vivo. *Curr. Biol.* **12:** 2159–2166.

Song E., Lee S.K., Wang J., Ince N., Ouyang N., Min J., Chen J., Shankar P., and Lieberman J. 2003. RNA interference targeting Fas protects mice from fulminant hepatitis. *Nat. Med.* **9:** 347–351.

Svoboda P., Stein P., Hayashi H., and Schultz R.M. 2000. Selective reduction of dormant maternal mRNAs in mouse oocytes by RNA interference. *Development* **127:** 4147–4156.

Tabara H., Grishok A., and Mello C.C. 1998. RNAi in *C. elegans:* Soaking in the genome sequence. *Science* **282:** 430–431.

Tabara H., Yigit E., Siomi H., and Mello C.C. 2002. The dsRNA binding protein RDE-4 interacts with RDE-1, DCR-1, and a DExH-box helicase to direct RNAi in *C. elegans. Cell* **109:** 861–871.

Tabara H., Sarkissian M., Kelly W.G., Fleenor J., Grishok A., Timmons L., Fire A., and Mello C.C. 1999. The *rde-1* gene, RNA interference, and transposon silencing in *C. elegans. Cell* **99:** 123–132.

Tamanini F., Meijer N., Verheij C., Willems P.J., Galjaard H., Oostra B.A., and Hoogeveen A.T. 1996. FMRP is associated to the ribosomes via RNA. *Hum. Mol. Genet.* **5:** 809–813.

Tang G., Reinhart B.J., Bartel D.P., and Zamore P.D. 2003. A biochemical framework for RNA silencing in plants. *Genes Dev.* **17:** 49–63.

Taverna S.D., Coyne R.S., and Allis C.D. 2002. Methylation of histone H3 at lysine 9 targets programmed DNA elimination in *Tetrahymena. Cell* **110:** 701–711.

Timmons L. and Fire A. 1998. Specific interference by ingested dsRNA. *Nature* **395:** 854.

Tuschl T., Zamore P.D., Lehmann R., Bartel D.P., and Sharp P.A. 1999. Targeted mRNA degradation by double-stranded RNA in vitro. *Genes Dev.* **13:** 3191–3197.

Vaucheret H., Beclin C., Elmayan T., Feuerbach F., Godon C., Morel J.B., Mourrain P., Palauqui J.C., and Vernhettes S. 1998. Transgene-induced gene silencing in plants. *Plant J.* **16:** 651–659.

Volpe T., Kidner C., Hall I.M., Teng G., Grewal S.I., and Martienssen R. 2002. Regulation of heterochromatic silencing and histone H3 lysine-9 methylation by RNAi. *Science* **297:** 1833–1837.

Waterhouse P.M., Graham M.W., and Wang M.B. 1998. Virus resistance and gene silencing in plants can be induced by simultaneous expression of sense and antisense RNA. *Proc. Natl. Acad. Sci.* **95:** 13959–13964.

Wianny F. and Zernicka-Goetz M. 2000. Specific interference with gene function by double-stranded RNA in early mouse development. *Nat. Cell. Biol.* **2:** 70–75.

Wightman B., Ha I., and Ruvkun G. 1993. Posttranscriptional regulation of the heterochronic gene *lin-14* by *lin-4* mediates temporal pattern formation in *C. elegans. Cell* **75:** 855–862.

Williams R.W. and Rubin G.M. 2002. ARGONAUTE1 is required for efficient RNA interference in *Drosophila* embryos. *Proc. Natl. Acad. Sci.* **99:** 6889–6894.

Winston W.M., Molodowitch C., and Hunter C.P. 2002. Systemic RNAi in *C. elegans* requires the putative transmembrane protein SID-1. *Science* **295:** 2456–2459.

Zalfa F., Giorgi M., Primerano B., Moro A., Di Penta A., Reis S., Oostra B., and Bagni C. 2003. The fragile X syndrome protein FMRP associates with BC1 RNA and regulates the translation of specific mRNAs at synapses. *Cell* **112:** 317–327.

Zamore P.D., Tuschl T., Sharp P.A., and Bartel D.P. 2000. RNAi: Double-stranded RNA directs the ATP-dependent cleavage of mRNA at 21 to 23 nucleotide intervals. *Cell* **101:** 25–33.

Zhang H., Kolb F.A., Brondani V., Billy E., and Filipowicz W. 2002. Human Dicer preferentially cleaves dsRNAs at their termini without a requirement for ATP. *EMBO J.* **21:** 5875–5885.

Zilberman D., Cao X., and Jacobsen S.E. 2003. ARGONAUTE4 control of locus-specific siRNA accumulation and DNA and histone methylation. *Science* **299:** 716–719.

# shRNA-mediated Silencing of Mammalian Gene Expression

Michael T. McManus* and Douglas S. Conklin†

*Center for Cancer Research, Massachusetts Institute of Technology, Cambridge, Massachusetts 02139;
†Cancer Genome Research Center, Cold Spring Harbor Laboratory, Woodbury, New York 11797

THE RNA HAIRPIN STRUCTURE IS PERHAPS THE SIMPLEST of all RNA secondary structures. It is formed when a single-stranded RNA folds back onto itself, making a double-stranded RNA (dsRNA) stem joined at one end by a single-stranded loop. The hairpin, also referred to as a "stem-loop," is found ubiquitously throughout RNAs of all classes. Its fundamental architectural roles in RNAs such as tRNAs or rRNAs are well recognized, but hairpin RNAs also function in a diversity of other cellular roles. They define nucleation sites for folding, determine tertiary interactions in RNA enzymes, protect mRNAs from degradation, and are recognized by RNA-binding proteins. Recently, a special class of hairpin RNAs has been shown to be responsible for the posttranscriptional regulation of gene expression. These RNAs are central to hairpin-mediated stable gene silencing, a powerful technology for the genetic manipulation of mammalian cells in vitro and in vivo.

## RNA HAIRPINS THAT REGULATE GENE EXPRESSION

A novel strategy to regulate gene expression was discovered when *lin-4* and *let-7* hairpin RNAs were identified in *Caenorhabditis elegans* (Lee et al. 1993; Pasquinelli et al. 2000; Reinhart et al. 2000). The *C. elegans lin-4* regulatory gene, a 22-nucleotide RNA processed from a precursor hairpin RNA (Figure 6.1), was identified in a screen for mutations that

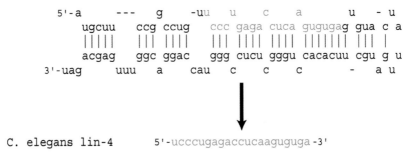

**FIGURE 6.1.** *lin-4* is processed from a hairpin RNA precursor. Although the exact structure of the precursor has not been determined, the *lin-4* precursor RNA is predicted to form a hairpin precursor RNA containing multiple bulges. The RNase III enzyme Dicer is believed to process the stable *lin-4* small temporal RNA (shown in *red*) from the hairpin.

**109**

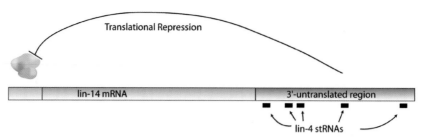

**FIGURE 6.2.** *lin-4* binds to multiple regions of the untranslated region of the *lin-14* gene. This schematic illustrates that multiple *lin-4* small temporal RNAs may bind to the targeted message. The exact number and placement of all *lin-4*-binding sites have not yet been determined.

affected the timing and sequence of postembryonic development in *C. elegans* (Wightman et al. 1993; Moss et al. 1997). *lin-4* was positionally cloned by isolating a 693-bp DNA fragment that could rescue the phenotype of mutant animals (Lee et al. 1993). Worms that are mutant for *lin-4* reiterate the L1 larval stage, rather than progressing to later stages of development. These worms exhibit a heterochronic phenotype, which is a defect in developmental timing. It has been observed that *lin-4* acts as a negative regulator of heterochronic protein-coding *lin-14* and *lin-28* genes (Lee et al. 1993; Wightman et al. 1993; Moss et al. 1997; Olsen and Ambros 1999). *lin-4* RNA contains homology with specific regions of the 3´-untranslated regions (3´UTRs) of *lin-14* and *lin-28* and perhaps other developmentally important genes (Lee et al. 1993; Wightman et al. 1993; Moss et al. 1997; Olsen and Ambros 1999). Correspondingly, deletion of the *lin-4* target sequences causes an unregulated gain-of-function phenotype. The exact mechanism by which *lin-4* inhibits expression of the *lin-14* and *lin-28* genes is unknown. Northern blot analysis of the target mRNAs indicates that the message remains stably associated with polysomes, and a model was suggested in which *lin-4* acted via translational repression (Figure 6.2) (Olsen and Ambros 1999).

The *lin-4* gene remained a peculiar and novel RNA until the discovery of a second heterochronic gene, *lethal-7* (*let-7*), which also encoded a 21-nucleotide product. The *let-7* gene is also processed from a larger approximately 70-nucleotide hairpin RNA. Data supported a role for *let-7* as a posttranscriptional negative regulator targeting the 3´UTR of *lin-41* and *lin-14*, two other known heterochronic genes. At that time, genomic databases were appearing, and Pasquinelli et al. (Reinhart et al. 2000) showed that the sequence of *let-7* was remarkably conserved among bilaterally symmetrical animals. Northern blots indicated that *let-7* was indeed expressed in mature tissues from humans, mice, chickens, worms, and flies, but not in cnidarians (jellyfish) or poriferans (sponges). Although the precise function of *let-7* is also unknown, it is clear that this gene shows a developmentally regulated timing of expression similar to the previously discovered *lin-4* gene. On this basis, it was proposed that the name small temporal RNAs (stRNAs) was appropriate for genes such as *lin-4* and *let-7* and suggested that others might be found (Pasquinelli et al. 2000; Reinhart et al. 2000).

## SMALL INTERFERING RNAs

Meanwhile, the RNA interference (RNAi) field was rapidly developing. RNAi has received considerable attention not only because it has proven to be an effective means to "knockdown" specific genes, but also because it is particularly useful in organisms in which traditional gene knockout analysis is not feasible. In fact, several groups have initiated genome-scale RNAi-mediated gene-knockout analysis with the hope of quickly ascribing functions to the thousands of genes that poured out from genome sequencing projects (Fraser et al. 2000; Gonczy et al. 2000; O'Neil et al. 2001; Walhout et al. 2002). RNAi gained popularity in *C. elegans* first, and *C. elegans* has proven to be an organism highly amenable to RNAi studies. Because *C. elegans* has an efficient means of taking up the dsRNA, automation for large-scale analysis is easily achieved simply by soaking the worms in a solution of dsRNA, and gross phenotypes can be scored under a dissecting microscope.

In *C. elegans*, the dsRNA is also distributed throughout the tissues, with the notable exception of the neuronal tissues. The entire organism, which comprises approximately <1000 different cells, can be observed at once through its semitransparent cuticle. This is particularly advantageous for observing genes that are functional in only certain cell types and that sometimes display different functions in different cells.

Although most studies of RNAi have been performed in diverse lower eukaryotes, such as *C. elegans*, *Drosophila*, *Arabidopsis*, *Neurospora*, and trypanosomes, the initial studies of mammalian RNAi were carried out by injecting long dsRNA into mouse oocytes and zygotes. Using this method, the inhibition of three different genes was demonstrated. In these studies, the RNAi response lasted for 6.5 days during the development of the mouse zygote, a time length that corresponded to an approximately 100-fold increase in the number of cells. On the one hand, the observation of RNAi in vertebrates was not altogether surprising, given its nearly universal description across phyla. On the other hand, it is well known that mammalian cells exhibit profound physiological responses to dsRNA, often leading to cell death. As little as one molecule of dsRNA longer than 30 nucleotides is sufficient to trigger the interferon response, part of which involves the activation of the dsRNA-responsive protein kinase (Marcus and Sekellick 1985; Clemens and Elia 1997; Williams 1999; Gil and Esteban 2000). The consequence is the nearly global inhibition of translation and nonspecific degradation of mRNA, which in turn triggers apoptosis. So then how can mammalian cells elicit an RNAi response? One possibility is that not all mammalian cells respond in the same way to dsRNA. For example, embryonal carcinoma (EC) cells, such as the murine F9 and P19 cell lines, are unresponsive in the dsRNA-interferon pathway and only become responsive upon induced differentiation (Francis and Lehman 1989; Harada et al. 1990; Stark et al. 1998).

Since RNAi was discovered, investigators have been homing in on its mechanisms. *C. elegans* genetics indicated that RNA occurred in at least two major steps, and the first likely involved the generation of a sequence-specific silencing agent (Tabara et al. 1999; Grishok et al. 2000). A strong candidate for this sequence-specific silencing agent is a special class of tiny RNAs. Hamilton and Baulcombe (1999) originally described such RNAs when they found that *Arabidopsis* plants undergoing posttranscriptional silencing manifested 21–25-nucleotide size RNAs complementary to both strands of the silenced gene (Jones et al. 1999). Since that article, reports that small 21–23-nucleotide RNAs were associated with RNAi silencing were beginning to appear in *C. elegans* and *Drosophila*. It was shown that when these RNAs are isolated from cells undergoing silencing, they could be used to silence expression in naïve *Drosophila* S2 cells and embryo extracts (Yang et al. 2000; Zamore et al. 2000; Elbashir et al. 2001). These small RNAs were then sequenced, and found to be 21–23-nucleotide dsRNAs that contain two nucleotide 3´ overhangs (Elbashir et al. 2001). This work represented a major advance in biology. These RNAs were named short interfering RNAs (siRNAs).

Part of the importance of the discovery of siRNAs was that they were proven to be an effective trigger for gene silencing in mammalian cell culture. siRNA-mediated gene silencing is now considered a standard laboratory practice. One major difference between

RNAi in *C. elegans* and mammalian cells is that, in general, mammalian cells do not appear to take up exogenously applied dsRNA efficiently. Thus, to perform siRNA-mediated silencing in mammalian cell culture, cationic lipid or electroporation must be used for transfecting the siRNAs, and these are not generally effective in vivo. A second major distinction between *C. elegans* and mammalian cells is that of the persistence of the RNAi response. It seems that mammalian cells lack the ability to amplify the RNAi response, and thus RNAi is limited to approximately four to eight cell doublings (McManus et al. 2002b; Stein et al. 2003). For experiments that require prolonged RNAi, transfection must be repeated multiple times. Clearly, mammalian siRNA-mediated silencing is limited as compared to *C. elegans*; however, as discussed in this chapter, new methods have been developed to overcome some of these disadvantages.

## LINKS BETWEEN TWO RNA SILENCING PATHWAYS

Many investigators noted the commonality in size between the siRNAs and stRNAs; however, the relationship between siRNA- and stRNA-mediated silencing remained unclear. As genetic experiments in the RNAi field began to uncover protein factors important for RNAi, links between siRNA- and stRNA-mediated silencing processes began to appear. The nuclease responsible for siRNA processing was identified as Dicer, an RNase-III-related enzyme that processes precursor dsRNAs into 21–23-nucleotide siRNAs (Bernstein et al. 2001; Elbashir et al. 2001). The human Dicer enzyme was knocked down in human cultured cells with siRNAs, and the accumulation of the 72-nucleotide unprocessed human *let-7* precursor was observed (Hutvagner et al. 2001). This result provided the biochemical link. When Dicer was mutated in worms, heterochronic phenotypes were observed (Grishok et al. 2001; Ketting et al. 2001; Knight and Bass 2001), providing a genetic link between these two processes. Another class of proteins related to the rabbit eIF2C/Argonaute family has provided another link between RNAi and stRNA-mediated silencing (Hammond et al. 2001; Morel et al. 2002; Williams and Rubin 2002). Worms that are deficient for certain proteins in this class are deficient in RNAi or exhibit a heterochronic phenotype (Grishok et al. 2001). These worms also exhibit an accumulation of unprocessed *lin-4* and *let-7* precursors.

The link between siRNA- and stRNA-mediated silencing became stronger when endogenous 21–23-nucleotide RNAs were identified. Originally, in the course of cloning and analyzing 21–23-nucleotide RNAs produced in a *Drosophila* in vitro RNAi assay, endogenous 21- and 22-mers were identified, and the suggestion was made that perhaps there were naturally occurring siRNAs (Elbashir et al. 2001; Harborth et al. 2001). Since that report, three groups have undertaken an effort to identify endogenous small RNA regulators (Lagos-Quintana et al. 2001; Lau et al. 2001; Lee and Ambros 2001). These early efforts have resulted in the identification of more than 90 new hairpin-structured RNAs encoding potential regulator 21–23-nucleotide RNAs. Because not all of these RNAs were temporally expressed like *lin-4* and *let-7*, the term microRNAs (miRNAs) was given to this large class of hairpin RNAs. Even though many miRNAs are actually produced in a stage-specific and/or tissue-specific manner, the definition of stRNAs has been expanded so that *lin-4* and *let-7* are considered miRNAs (Ambros et al. 2003). Northern blot analyses verified the 21–23-nucleotide size distribution of the mature miRNA, and a less abundant approximately 70-nucleotide precursor form. Most of the miRNAs were identified just once, suggesting that the sequencing screens have not been saturating. At that time, it seemed likely that the number of miRNA riboregulators might be far larger than the approximately 90 sequences thus identified.

After these three reports, a fourth group described about 40 more miRNAs, a few of which were identical to those reported in the earlier articles (Mourelatos et al. 2002).

This new batch was identified from immunoprecipitations of Gemin3 and Gemin4, two core components of the Survival of Motor Neurons (SMN) complex. The SMN complex has important roles in the assembly/restructuring and function of diverse ribonucleoprotein (RNP) complexes, including spliceosomal small nuclear RNPs (snRNPs) (Fischer et al. 1997; Pellizzoni et al. 1998; Meister et al. 2001), small nucleolar RNPs (snoRNPs) (Jones et al. 2001; Pellizzoni et al. 2001a), heterogeneous nuclear RNPs (hnRNPs) (Mourelatos et al. 2001), and transcriptosomes (Pellizzoni et al. 2001b). The link between the SMN complex and siRNA/stRNA-mediated silencing became apparent when it was discovered that the eIF2C/Argonaute family member was part of the SMN complex. Together, these studies add additional links between RNAi and the endogenous miRNA hairpins.

The identification of miRNA sequences is continuing, and successes are being made in both the sequencing and bioinformatics prediction of miRNA genes (Brennecke et al. 2003; Dostie et al. 2003; Lagos-Quintana et al. 2003; Lim et al. 2003). It is likely that large sequencing projects will define and annotate miRNA genes. It has been suggested that in humans and worms, miRNAs constitute approximately 1% of the gene content, a remarkable number considering that they are encoded in the intergenic and intronic regions that were once considered by many to be the "junk" of the genome.

## THE STRUCTURE OF A SILENCING RNA

Although miRNAs are derived from dsRNA hairpin precursors, typically only the miRNA single strand of the precursor stem is cloned. The precusor miRNAs appear to be transcribed as an approximately 70-nucleotide precursor hairpin RNA containing a 4–15-nucleotide loop. Sometimes the 21–23-nucleotide miRNA forms a perfect duplex within the hairpin, but more often, multiple bulges disrupt the perfect 21–23-nucleotide duplex. Similar to other RNase III enzymes, Dicer is active at processing complex hairpin structures that can contain multiple mismatches in the helical stem (Hutvagner et al. 2001). Little is known about the structural determinants necessary for processing of miRNAs into approximately 21-nucleotide RNAs. Studies performed on RNase III enzymes in other organisms have shown that dsRNA cleavage relies on antideterminants in the double-stranded stem, as well as sequence determinants in the terminal loop of the stem-loop RNA substrate (Zhang and Nicholson 1997; Chanfreau et al. 2000; Wu et al. 2001). However, comparison of the large class of miRNAs that are likely to be processed by Dicer has yet to reveal any obvious features that might guide Dicer recognition or processing.

## TARGETS OF MICRORNAS

Significantly, when the miRNA is paired to the mRNA, it does not exhibit perfect complementarity; this situation is in contrast to siRNA-mediated degradation. In the existing models for the pairing, typically 50–85% of the miRNA residues are base-paired to the mRNA 3´UTR (Figure 6.3). However, after the identification of *lin-4* and *let-7* targets, none of the additional cloned miRNAs have indicated potential mRNA targets. This is probably because the homology between miRNAs and the target mRNA comprises only 5–14 nucleotides, making it difficult to find candidate targets using bioinformatic searches. Several groups are continuing to search for targets, and in *Drosophila*, two 3´UTR sequence motifs, the K box (cUGUGAUa) and the Brd box (AGCUUUA), have been implicated as miRNA targets. They are partially complementary to several reported miRNAs and also because these sequences are known to mediate negative posttranscriptional regulation (Lai and Posakony 1997; Lai et al. 1998). It is interesting to note that among the miRNAs and their cognate targets, only the 5´-most stretch of miRNA nucleotides is perfectly

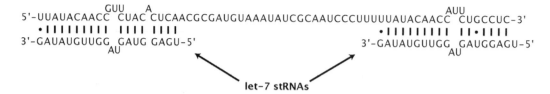

**section of the lin-41 UTR**

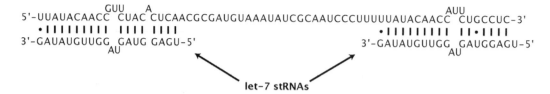

let-7 stRNAs

**FIGURE 6.3.** Anatomy of a miRNA:mRNA interaction. The *C. elegans let-7* stRNA is shown here duplexed to a small section of the 3′UTR of *lin-41*. The interaction between these two RNAs is imperfect and multiple bulges occur. Shown are Watson-Crick pairing (*vertical lines*) and G:U wobble pairing (*closed circles*). Two predicted 3′UTR-binding sites are shown, but there may be several more along the length of the 3′UTR.

paired to the target sequence. Elbashir and colleagues have demonstrated that nearly perfect duplexing between the target mRNA and the siRNA antisense strand is a requirement for mRNA cleavage in RNAi (Elbashir et al. 2001; Harborth et al. 2001). One possibility is that the miRNAs cannot activate the degradation pathway because they cannot form a perfect duplex with the target mRNA.

## ENGINEERING HAIRPIN RNAs FOR DIRECTED GENE SILENCING

RNAi analysis of gene function in lower organisms has been revolutionized by the development of in vivo systems that express long dsRNA hairpins that are typically 500–1000 nucleotides in length (Chuang and Meyerowitz 2000; Kennerdell and Carthew 2000; Smith et al. 2000; Tavernarakis et al. 2000). The primary advantages of in vivo transcription for the generation of siRNAs are uniform delivery and duration of the silencing as compared to the transient delivery schemes. First developed in *C. elegans*, in vivo transcription of dsRNA enabled the silencing of genes that were otherwise difficult to silence with standard dsRNA delivery (Tavernarakis et al. 2000). In vivo transcription was an enabling technology in cases in which genes were required for viability or reproduction, large populations of phenocopy mutants were required for biochemical analysis, or the phenotype required silencing of genes in relatively inaccessible tissues such as the nervous system. Similar benefits have been realized in other organisms. In *Drosophila*, Kennerdell and Carthew (2000) used long dsRNAs to silence gene expression specifically in adult tissues. In plants, Waterhouse and colleagues have developed long dsRNA expression vectors that offer improvements in both the proportion of transgenic plants in which silencing occurs and the extent of silencing when compared to standard methods (Wesley et al. 2001).

In mammals, silencing using long dsRNA has been accomplished in mouse ES cells and EC cells (Billy et al. 2001; Svoboda et al. 2001; Yang et al. 2001; Paddison et al. 2002a). However, attempts to develop a stable long dsRNA hairpin-based expression system in differentiated somatic mammalian cells have been unsuccessful. The nonspecific effects of long (>30 nucleotides) dsRNA expression in eliciting the interferon response (Marcus and Sekellick 1985; Gil and Esteban 2000) necessitated a different approach to stable dsRNA-based silencing in these cells.

As was the case in developing transient RNAi in mammalian cells, the key development in methodology for stable silencing in mammalian cells was to produce dsRNAs with short double-stranded regions, termed short hairpin RNAs (shRNA), that would fail to trigger dsRNA responses and yet still induce RNAi-type silencing (Bitko and Barik 2001; Elbashir et al. 2001). Several groups accomplished this at about the same time by expressing dsRNAs less than 30 nucleotides in length (Brummelkamp et al. 2002; McManus et al. 2002a; Paddison et al. 2002b; Paul et al. 2002; Sui et al. 2002; Yu et al. 2002). In most cases, these constructs contain sufficient sequence information to reduce specifically the expression of a cognate target gene to less than 10% of control levels. Nonspecific toxic effects such as induction of dsRNA-mediated apoptosis are not seen.

The structure of miRNAs served as a blueprint for many groups in the design of short double-stranded silencing RNAs. By fashioning constructs that mimicked the overall structure of miRNAs yet contained sequence homology with the targeted gene sequence, the sequence specificity of the silencing effect could be reprogrammed. This approach was successful in a number of instances (McManus et al. 2002a; Paddison et al. 2002b; Zeng et al. 2002). It became apparent, however, that simple stem structures in which dsRNA had perfect identity to the targeted gene were most effective at silencing. This implied that these molecules acted through a degradative rather than a translational repression pathway. Confirmation of this came with the finding that short hairpins were substrates for the Dicer nuclease and effected the degradation of target sequences (Paddison et al. 2002b). These data can be put into a general model whereby both miRNAs and shRNAs utilize components of a general RNA silencing pathway (Figure 6.4).

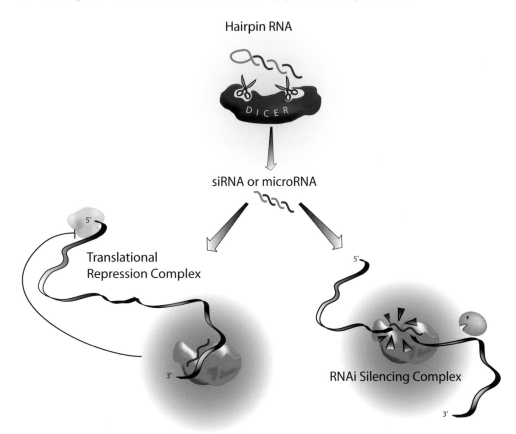

**FIGURE 6.4.** A shared pathway for miRNAs and silencing hairpin RNA transgenes. miRNA precursors and transgene shRNAs both may be processed by Dicer, giving a short 21–23-nucleotide silencing RNA. In this model, shRNAs give rise to siRNA, which silence via mRNA degradation. In the case of the miRNA precursor, a dsRNA may be processed from the hairpin RNA, but only one strand appears to be stable. The components of the silencing complexes may be shared between the RNA silencing complex and the translational repression complex. The entrance into either pathway is independent of the origin of the 21–23-nucleotide RNA (Hutvagner and Zamore 2002; Doench et al. 2003).

At present, a variety of workable strategies have been developed that effect stable silencing in mammalian cells. In each case, the expression of the interfering RNA is driven by mammalian promoters from DNA vectors that are introduced into the cell by transfection or infection methods and either propagated episomally or integrated into the genome. Most express dsRNA as a very short, simple hairpin with 19–29-nucleotide double-stranded stems. Because RNA polymerase III initiates and terminates small highly structured RNA transcripts precisely (Goomer and Kunkel 1992), many groups have employed polymerase-III-dependent promoters to drive the expression of these molecules. Mouse (Yu et al. 2002) and human U6-snRNA promoters (Paddison et al. 2002b; Paul et al. 2002; Sui et al. 2002), the human RNase P (H1) RNA promoter (Brummelkamp et al. 2002; McManus et al. 2002a), and the human Val-tRNA promoter (Kawasaki and Taira 2003) have all been used successfully. These promoters are active in most if not all embryonal and somatic cell types and offer similar levels of constitutive expression.

Although most approaches have been based on the expression of polymerase-III-transcribed dsRNA in the form of short hairpin, other successful strategies have yielded similarly effective results. As mentioned above, dsRNA can be produced in a form that mimics the structure of miRNAs and is therefore likely to serve as a substrate for the Dicer enzyme. In one strategy, expression of an miRNA-like construct is driven from the RNA polymerase II cytomegalovirus (CMV) immediate-early promoter (Zeng et al. 2002). Dicer recognizes the double-stranded region of this transcript and produces functional siRNAs. In another version, separate U6 promoters have been used to drive the transcription of single-stranded 21-nucleotide RNAs (Miyagishi and Taira 2002). These transcripts presumably anneal within the cell to form structures identical to siRNAs.

That short-hairpin-based silencing was developed independently in a number of laboratories speaks to the robust nature of the effect. Despite its widespread use, however, it is probably too early to expect that all of the parameters in these methods have been optimized. In the somewhat limited number of published studies, it appears that a variety of design elements have at best modest effects on the efficacy of the encoded shRNAs. Figure 6.5 outlines the various components of effective shRNA constructs for stable silencing in mammals. Provided the sequence is large enough to be incorporated into the RNAi silencing complex (RISC) and yet small enough to evade the protein kinase PKR and interferon pathways, the size of the targeting sequence is not crucial. Double-stranded stems between 18 and 29 nucleotides in length are roughly equivalent in efficacy (Yu et al. 2003). Although increasing the length of a short stem will improve the efficacy of an inefficient hairpin, it will not increase the efficacy of a hairpin that already silences a gene at greater than 90%. This may be explained by the greater number of potential siRNAs that can be processed from a 29-nucleotide stem. Whether the 5′ stem strand or 3′ stem strand is complementary to the sense strand of the mRNA is not important, provided one stem is completely complementary to the targeted sense mRNA sequence (Brummelkamp et al. 2002; McManus et al. 2002a; Paddison et al. 2002b; Paul et al. 2002; Sui et al. 2002; Yu et al. 2002). The sequence of the loop is fairly unimportant. Successful silencing has been observed with a variety of loop sequences between 3 and 9 nucleotides in length. Some longer loops have had negative effects on the strength of the silencing effect (Yu et al. 2003).

With the relative flexibility of shRNA design given above, the main determinant of silencing efficiency appears to reside in the target sequence. The crucial event in RNAi-mediated gene silencing is the interaction of the 21-nucleotide siRNA contained in RISC with its complementary sequence within an mRNA. There is, however, considerable variability in the degree of suppression from gene to gene and between target sequences within a single gene (McManus et al. 2002a; Kapadia et al. 2003). Unknown intrinsic factors related to mRNA abundance, structure, translation rate, or other features of the

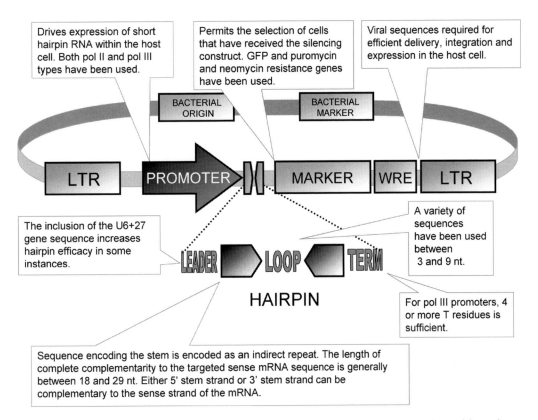

Drives expression of short hairpin RNA within the host cell. Both pol II and pol III types have been used.

Permits the selection of cells that have received the silencing construct. GFP and puromycin and neomycin resistance genes have been used.

Viral sequences required for efficient delivery, integration and expression in the host cell.

BACTERIAL ORIGIN

BACTERIAL MARKER

LTR    PROMOTER    )(    MARKER    WRE    LTR

The inclusion of the U6+27 gene sequence increases hairpin efficacy in some instances.

A variety of sequences have been used between 3 and 9 nt.

LEADER    LOOP    TERM

HAIRPIN

For pol III promoters, 4 or more T residues is sufficient.

Sequence encoding the stem is encoded as an indirect repeat. The length of complete complementarity to the targeted sense mRNA sequence is generally between 18 and 29 nt. Either 5' stem strand or 3' stem strand can be complementary to the sense strand of the mRNA.

**FIGURE 6.5.** Anatomy of a short hairpin dsRNA expression vector. Elements indicated have been included in a variety of vectors to improve the gene transfer efficiency and stability of the silencing effect. Although several vectors share the general design, vectors without markers, viral sequence elements, and leader sequences have been successful in some instances. (LTR) Long terminal repeat required for retrovirus or lentivirus replication; (WRE) woodchuck hepatitis regulatory element, a *cis*-acting sequence that enhances expression of transgenes delivered by viral vectors.

RNAi mechanism are expected to be responsible. Until the mechanism of RISC action is elucidated, our abilities to predict accurately the suitability of a specific target sequence will be limited. Not surprisingly, the published data on rules with which to select hairpin target sites are somewhat imprecise. The available general guidelines suggest a target sequence near the 5′ end of the gene, with a length of 18–29 nucleotides, and GC content of approximately 50%. However, many target sites that do not share these criteria are highly effective. In fact, there are several cases where targeting of the 3′ end of a gene was more effective (McManus et al. 2002b; Ge et al. 2003).

Another aspect of the RNAi mechanism that may lead to improved silencing technology when it is understood concerns the intracellular localization of RNAi. Although this question has been examined by a few groups, a complete picture has yet to emerge. Paul et al. (2003) have used the first 27 bases of the U6 promoter to localize transcribed hairpins targeting lamin A/C to the nucleus. These U6+27 hairpins are effective in silencing. When, however, the same hairpin sequences are expressed from a 5S ribosomal cassette, which causes them to be retained in the nucleolus, or the 7SL ribosomal particle cassette, which causes them to be exported to the cytoplasm, silencing is abolished. This suggests that at least some important step in the RNAi process occurs in the nucleus. Two other studies suggest that cytoplasmic localization may be important. Zeng and Cullen (2003) have used the retroviral Rev/Rev-responsive element (RRE) system to control the nuclear retention or cytoplasmic export of target mRNAs. RRE-containing target mRNAs, which were retained in the nucleus, were not affected by activation of RNAi. These same mRNAs were specifically degraded, however, when nuclear export was induced by expression of Rev. Similar results were obtained in a study in which transcripts expressed from a valine tRNA promoter were localized in the cytoplasm and were efficiently processed by Dicer (Kawasaki and Taira 2003). Transcripts from a U6 promoter were primarily found in the nucleus and were not efficiently processed. Taken together, these results suggest that subcellular localization can be important to the activity of transcribed hairpins, although a complete understanding of both the mechanism of RNAi and transcript transport may be required to drastically impact the efficacy of hairpin-based silencing.

In practice, the production of encoded hairpin constructs is relatively straightforward. In the simplest form, a hairpin target site sequence is chosen. Although this can be done "by hand," a number of design tools are available on the Internet that simplify this mind-bending process (http://katahdin.cshl.org:9331/RNAi/, http://jura.wi.mit.edu/bioc/siRNA/, http://www.dharmacon.com/, http://www.ambion.com/techlib/misc/siRNA_finder.html). Chemically synthesized DNA oligonucleotides that encode the chosen sequence are annealed to form a double-stranded fragment, which is then cloned into a hairpin expression vector downstream from a eukaryotic promoter. To facilitate cloning, G-U base pairs, which are permitted in duplexed RNA but not DNA, have been substituted in some cases for G-C or A-U encoded in the oligonucleotides. These substitutions reduce the stability of DNA hairpins that might form during cloning and propagation, but theoretically have no effect on the structure and no empirical effect on the activity of the resulting RNA hairpins. Another method of hairpin construction makes use of PCR (polymerase chain reaction) to affix the hairpin to the promoter (Paddison and Hannon 2002; Hemann et al. 2003). In any event, since oligonculeotides that encode hairpins are relatively long (60–90 nucleotides, depending on the method) and possess a high degree of secondary structure, DNA sequence confirmation of the constructs is always advisable. There is still much to be learned about the design of a hairpin target that is efficient at down-regulating specific gene expression. As mentioned above, this is largely in part because not much is known about target site selection of siRNAs. Because of this, we suggest that the researcher make at least two to three different constructs for each gene that is being targeted.

The introduction of shRNA constructs to cells can be accomplished using any of the various gene transfer techniques. Provided the delivery is efficient, the cognate gene will be suppressed to levels dictated by the efficacy of the hairpin. The ultimate utility of shRNAs as a genetic tool, however, lies in the generation of stably transfected shRNA-expressing cells. For this reason, gene transfer methodologies that are inherently stable are better platforms for the expression of shRNAs than transiently transfected vectors or PCR fragments (Castanotto et al. 2002). A number of well-characterized stable expression technologies have been used to effect permanent expression of shRNAs in target cells. These include systems based on retroviral and lentiviral integration and adenoviral expression. A growing list of genes has been silenced using these vectors.

## shRNA-MEDIATED SILENCING IN ANIMALS

The ultimate use of a genetic technology in mammals is to affect the phenotype of the organism. In this way, the effects of altered gene function on developmental, behavioral, and other complex phenotypes can be assessed. Although conventional knockout technologies have proven to be invaluable in the elucidation of countless biological processes, this approach is not without limitations. The significant investment of time and resources required can be prohibitive. Moreover, classical gene knockouts are of limited use because they eliminate gene function universally in the embryo. Thus, the phenotypes of mutations in genes essential for general cell viability cannot be studied. This effect also leads to problems with the interpretation of phenotypes caused by the mutation of genes with multiple spatially and temporally distinct functions during development. Complete disruption of such a gene by knockout only implicates it in the first crucial event or locale without providing any information regarding other events.

The first demonstration of the utility of dsRNA in silencing genes in mammals at the organismal level was provided by Wianny and Zernicka-Goetz (2000). In their work, 500–700-bp dsRNAs were microinjected into mouse oocytes and zygotes. Exogenously expressed green fluorescent protein (GFP) or endogenous target gene expression was found to be specifically reduced in embryos when targeted by these RNAs. The phenotypes caused by the silenced endogenous genes were indistinguishable from those exhibited by embryos carrying null alleles. Injection of double-stranded E-cadherin RNA resulted in the same developmental blastocyst abnormalities that are observed in E-cadherin knockout mice. The effect of injected c-*mos* dsRNA was a phenocopy loss of maternal expression of the c-*mos* gene in the oocyte, as evidenced by progression through the normal arrest of meiosis at metaphase II in most oocytes.

Although these experiments required the production of long dsRNA in vitro, an improved method of delivery to these cell types was developed by Svoboda and Schultz, who constructed vectors for the expression of long dsRNAs in mouse oocytes and preimplantation embryos (Svoboda et al. 2001). These vectors contained enhanced (E) GFP (EGFP) as a visual marker and an inverted repeat that, when expressed, would form a long dsRNA. In this vector, a long dsRNA hairpin that targeted *mos* was just as effective at silencing as dsRNA constructed from annealed sense and antisense RNA in vitro.

Despite these advances, studies with long dsRNA, as mentioned previously, are restricted to embryonal cells that do not possess appreciable levels of the interferon and PKR responses (Castanotto et al. 2002). As is the case with cells in vitro, the usage of short dsRNA is more broadly applicable and was key to developing gene silencing as a tool at the organismal level. McCaffery et al. (2002) were the first to report that gene expression could be silenced in animals with synthetic siRNAs. A modified hydrodynamic transfection method was used that delivered naked siRNAs and a firefly luciferase reporter gene to the livers of adult mice. Expression of the luciferase transgene was monitored using quantitative whole-body imaging. In these studies, expression of the luciferase reporter was specifically silenced to levels less than 5% of controls by a co-injected siRNA. Of greater importance for the future of this technology was the finding that shRNAs expressed from co-injected plasmids were found to have the same effect. Because constructs that contained a transcriptionally dead shRNA cassette failed to silence, it was clear that the silencing effect was mediated by transcription of the shRNA. This suggested that stable RNAi could be used effectively at the organismal level in mammals to create loss-of-function alleles.

## STABLE RNAi TRANSGENIC ANIMALS

Perhaps the best test of the utility of stable RNAi-based silencing in mammals is the generation of transgenic animals. Conventional knockout technologies require a significant investment of time and resources that prohibit large-scale functional genomic studies. A major impediment is that mammalian cells are diploid. This problem is minimized because shRNAs create what are effectively dominant loss-of-function mutations. The first demonstration of functional silencing constructs being passed through the germ line was by Hasuwa et al. (2002). Using shRNA vectors to engineer rodent ES cells, ubiquitously expressed GFP was silenced by a cognate hairpin in all tissues in mice and rats. Suppression in mice was to levels 4% of untreated controls. Importantly, shRNA-mediated suppression of GFP levels was stable, occurring in $F_1$ transgenic animals several weeks after introduction of the transgene. In contrast, silencing mediated by siRNAs was short-lived and was only observed in ES-cell-derived embryos that were less than 10.5 days old. Although these workers targeted an exogenously expressed reporter gene, this demonstrated two important concepts: (1) RNAi was active in a variety of differentiated cells in live animals and (2) transgenic RNAi could serve as an alternative method to traditional homologous recombination-based knockouts by using hairpin expression constructs in ES cells.

This approach was shown to work with endogenous targets by Carmell et al. (2003), who targeted the murine DNA glycosylase encoding the *Neil1* gene. An shRNA construct that effectively silenced this gene in ES cells was used to generate transgenic mouse lines. $F_1$ animals that carried the *Neil1*-targeted shRNA expression vector all displayed silencing and contained siRNAs corresponding to a Dicer-processed form of the shRNA. Similar levels of suppression, as reflected by decreased levels of mRNA and protein, were found in both the original ES cells and the transgenic $F_1$ animals. Consistent with a role for *Neil1* in DNA repair, cells from either source exhibited increased sensitivity to ionizing radiation. Aside from demonstrating the utility of this approach in generating mutant animals, these experiments also suggested that the level of silencing was similar in all cell types affected.

The fact that genes can be silenced efficiently and relatively inexpensively has enabled larger-scale analysis of gene function in transgenic mammals. Improved vector systems have been developed with this in mind (Rubinson et al. 2003; Stewart et al. 2003; Tiscornia et al. 2003). On the basis of self-inactivating lentiviruses, these vectors contain an shRNA expression cassette and express EGFP as a reporter that allows infected cells to be selected by flow cytometry. This system takes advantage of the efficient methods that have been optimized for the generation of transgenic animals through infection of ES cells or single-cell embryos (Lois et al. 2002). The relative resistance of lentiviruses to transgene silencing during mammalian development makes them an excellent choice for this application. Rubinson et al. (2003) have used this vector to silence a variety of genes in hematopoietic stem cells (HSCs) and differentiated progeny as well as in transgenic animals. Splenocytes from reconstituted mice that received HSCs infected with the virus that targeted CD8 had 90% fewer CD8$^+$ T cells when compared with those carrying the vector control. Transgenic chimeras generated from lentivirus-infected ES cells or single-cell embryos infected with lentiviruses targeted at the CD8, CD25, p53, and Mena$^+$ genes produced offspring that showed expression of GFP and siRNAs in all tissues tested. In most cases, the level of targeted silencing corresponded to the appropriate cellular phenotypes. Although somewhat variable penetrance of the phenotypes was noted in these experiments, the authors reasoned that it was likely due to the number of integrated lentiviruses in each of the transgenic lines.

## CONCLUSIONS

The initial motivation for developing RNAi in mammalian cells was that it enabled rapid reverse genetic production of knockdowns. An unexpected development in the use of shRNAs, however, was that hairpins which were less than completely effective could be equally valuable. Using retroviral shRNA constructs with intrinsic differences in silencing activity, Hemann et al. (2003) demonstrated that targeting a single gene with different hairpins can produce phenotypes of varying severity at both the cellular and organismal levels. In these experiments, the mouse *p53* gene was targeted with three different hairpin constructs each with different effects on p53 protein levels. These constructs were introduced by infection into hematopoietic stem cells derived from E1-Myc transgenic mice. The forced expression of Myc that occurs in these cells leads to greatly accelerated lymphomagenesis in recipient irradiated mice when coupled with deletion of the *p53* gene (Schmitt et al. 1999). Surprisingly, each *p53* shRNA produced a distinct phenotype in vivo in reconstituted mice. The phenotypes ranged from benign lymphoid hyperplasias to highly disseminated lymphomas similar to those found in the E1-Myc mouse with homozygous *p53* deletions. In each case, the severity and type of disease were correlated with the levels of suppression of the *p53* gene by the specific shRNAs. That this type of "epiallelic series" of silencing constructs can be created adds value to an already useful technique. It indicates that subtle phenotypes caused by incompletely silenced genes can be created using shRNAs. This will undoubtedly aid in elucidating gene function in vivo.

A number of papers have pointed to the potential utility of stable silencing as a form of gene therapy (Abbas-Terki et al. 2002; Agami 2002; Borkhardt 2002; Caplen et al. 2002; Lawrence 2002; Martinez et al. 2002; Wilda et al. 2002; Xia et al. 2002; Aoki et al. 2003; Cioca et al. 2003; Zhang et al. 2003). The concept is that in contrast to standard gene therapy, which normally relies on the ectopic expression of proteins, shRNAs could potentially be used to silence disease-related genes. The potential for RNAi as a therapeutic has been demonstrated by Lieberman and colleagues in protecting mice from fulminant hepatitis (Song et al. 2003). siRNAs introduced by hydrodynamic transfection that targeted the Fas receptor specifically reduced its mRNA and protein levels in hepatocytes. In a severe disease model caused by the injection of agonistic Fas-specific antibody, 82% of mice treated with these siRNAs survived for 10 days, whereas all control mice died within 3 days. These results show that siRNAs could be used therapeutically in a tissue type that is targeted effectively. With what is now known from in vivo studies, shRNAs would undoubtedly have the same effect and provide even longer-term silencing. However, delivery of shRNAs remains a major obstacle as it does in current gene therapy approaches.

Although stable silencing of genes using expressed hairpin-shaped siRNAs has already been put to work in a variety of applications (Figure 6.6), improvements and more ambitious applications of this technology are sure to occur. The commonly used promoters and the cellular silencing machinery are both active in most tissues. As a consequence, hypomorphic alleles of most genes can be generated in most tissues. In a number of instances, however, greater control over the timing and localization of the silencing effect would be useful. The development of promoters that express shRNAs only in response to small-molecule inducers or in specific tissue types would expand the range of potential experiments. This approach would be especially useful for genes that are essential or have multiple roles in development or that are related to behavior.

As detailed above, expressed shRNAs are already useful in generating knockdowns. In most cases, the effort required to introduce engineered cells into blastocysts and to propagate the recipient animals is the same in either approach, and because the RNAi constructs act in a dominant fashion, there is no need to carry out crosses to produce homozygous mutants. The dominant nature of shRNAs provides a further acceleration in mutant generation by making the simultaneous silencing of two or more genes in a transgenic animal a possibility. To date, only two genes have been silenced simultaneously in cells in vitro (Yu et al. 2003). It is likely, however, that the number of genes that can be silenced at one time will be significantly larger. Such combinatorial silencing will significantly increase the speed of analysis of gene product interactions in both cellular and animal models as multiply mutant animals can be generated in the time that it takes to gen-

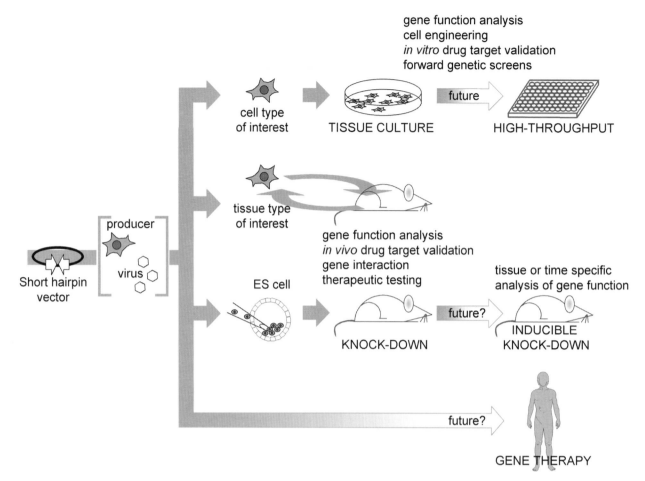

**FIGURE 6.6.** Applications of stable silencing in mammalian cells. Short hairpin vectors have been used to generate stable hypomorphic alleles of genes in tissue culture, in specific cell types within mice, and in all cells in an animal. High-throughput inducible knockouts and gene therapy are under development at the time of this writing.

erate a single heterozygous founder. This, coupled with the ability to generate alleles of varying degrees of severity, makes stable RNAi an even more significant improvement in the genetic analysis of mammals.

With the genome sequences of a number of mammals now known, genome-wide analysis of gene function has become a priority. The relative ease with which stable silencing vectors can be constructed makes functional genomics another promising application of stable RNAi (Agami 2002; Hannon 2002). Arrayed libraries of hairpins targeting each of the open reading frames (ORFs) in a genome can be used in phenotype-based, mid- to high-throughput screens in vitro and in vivo. The advantage of this type of approach is that information can be gained regarding the function of any gene in a given biological process with few if any presuppositions. Although this approach can also be taken with siRNAs, the benefits of stable silencing are similar to those found in other organisms in that populations of cells or animals exhibiting the same phenotype can be produced.

Stable silencing based on hairpin RNAs is possible in mammalian cells because it takes advantage of the cellular RNAi machinery—a collection of enzymes that has been honed by millions of years of evolution to silence genes. It is clear that there are an unlimited number of potential applications. The limitations appear to be only in the nature of the phenotypes and the relative ease of their assay, since the robust nature of this process is likely to make stable silencing a useful technique for the foreseeable future.

## ACKNOWLEDGMENTS

We thank Phillip A. Sharp, Chris Petersen, Carl Novina, Shola Aruleba, Greg Hannon, Patrick Paddison, Ravi Sachidanadam, Kim Scobie, and the rest of the Cold Spring Harbor RNAi Group for comments and helpful discussions. M.T.M. is partially supported by the Cancer Research Institute and by the U.S. Health Service MERIT Award (R37-GM34277) from the National Institutes of Health and P01-CA42063 from the National Cancer Institute. D.S.C. is supported by the U.S. Army Breast Cancer Research Program.

## REFERENCES

Abbas-Terki T., Blanco-Bose W., Deglon N., Pralong W., and Aebischer P. 2002. Lentiviral-mediated RNA interference. *Hum. Gene Ther.* **13:** 2197–201.

Agami R. 2002. RNAi and related mechanisms and their potential use for therapy. *Curr. Opin. Chem. Biol.* **6:** 829–834.

Ambros V., Bartel B., Bartel D.P., Burge C.B., Carrington J.C., Chen X., Dreyfuss G., Eddy S.R., Griffiths-Jones S., Marshall M., Matzke M., Ruvkun G., and Tuschl T. 2003. A uniform system for microRNA annotation. *RNA* **9:** 277–279.

Aoki Y., Cioca D.P., Oidaira H., Kamiya J., and Kiyosawa K. 2003. RNA interference may be more potent than antisense RNA in human cancer cell lines. *Clin. Exp. Pharmacol. Physiol.* **30:** 96–102.

Bernstein E., Caudy A.A., Hammond S.M., and Hannon G.J. 2001. Role for a bidentate ribonuclease in the initiation step of RNA interference. *Nature* **409:** 363–366.

Billy E., Brondani V., Zhang H., Muller U., and Filipowicz W. 2001. Specific interference with gene expression induced by long, double-stranded RNA in mouse embryonal teratocarcinoma cell lines. *Proc. Natl. Acad. Sci.* **98:** 14428–14433.

Bitko V. and Barik S. 2001. Phenotypic silencing of cytoplasmic genes using sequence-specific double-stranded short interfering RNA and its application in the reverse genetics of wild type negative-strand RNA viruses. *BMC Microbiol.* **1:** 34.

Borkhardt A. 2002. Blocking oncogenes in malignant cells by RNA interference—New hope for a highly specific cancer treatment? *Cancer Cell* **2:** 167–168.

Brennecke J., Hipfner D.R., Stark A., Russell R.B., and Cohen S.M. 2003. bantam encodes a developmentally regulated microRNA that controls cell proliferation and regulates the proapoptotic gene hid in *Drosophila*. *Cell* **113:** 25–36.

Brummelkamp T.R., Bernards R., and Agami R. 2002. A system for stable expression of short interfering RNAs in mammalian cells. *Science* **296:** 550–553.

Caplen N.J., Taylor J.P., Statham V.S., Tanaka F., Fire A., and Morgan R.A. 2002. Rescue of polyglutamine-mediated cytotoxicity by double-stranded RNA-mediated RNA interference. *Hum. Mol. Genet.* **11:** 175–184.

Carmell M.A., Zhang L., Conklin D.S., Hannon G.J., and Rosenquist T.A. 2003. Germline transmission of RNAi in mice. *Nat. Struct. Biol.* **10:** 91–92.

Castanotto D., Li H., and Rossi J.J. 2002. Functional siRNA expression from transfected PCR products. *RNA* **8:** 1454–1460.

Chanfreau G., Buckle M., and Jacquier A. 2000. Recognition of a conserved class of RNA tetraloops by *Saccharomyces cerevisiae* RNase III. *Proc. Natl. Acad. Sci.* **97:** 3142–3147.

Chuang C.F. and Meyerowitz E.M. 2000. Specific and heritable genetic interference by double-stranded RNA in *Arabidopsis thaliana*. *Proc. Natl. Acad. Sci.* **97:** 4985–4890.

Cioca D.P., Aoki Y., and Kiyosawa K. 2003. RNA interference is a functional pathway with therapeutic potential in human myeloid leukemia cell lines. *Cancer Gene Ther.* **10:** 125–133.

Clemens M.J. and Elia A. 1997. The double-stranded RNA-dependent protein kinase PKR: Structure and function. *J. Interferon Cytokine Res.* **17:** 503–524.

Doench J.G., Petersen C.P., and Sharp P.A. 2003. siRNAs can function as miRNAs. *Genes Dev.* 17: 438–442.

Dostie J., Mourelatos Z., Yang M., Sharma A., and Dreyfuss G. 2003. Numerous microRNPs in neuronal cells containing novel microRNAs. *RNA* **9:** 180–186.

Elbashir S.M., Lendeckel W., and Tuschl T. 2001. RNA interference is mediated by 21- and 22-nucleotide RNAs. *Genes Dev.* **15:** 188–200.

Elbashir S.M., Harborth J., Lendeckel W., Yalcin A., Weber K., and Tuschl T. 2001. Duplexes of 21-nucleotide RNAs mediate RNA interference in cultured mammalian cells. *Nature* **411:** 494–498.

Fischer U., Liu Q., and Dreyfuss G. 1997. The SMN-SIP1 complex has an essential role in spliceosomal snRNP biogenesis. *Cell* **90:** 1023–1029.

Francis M.K. and Lehman J.M. 1989. Control of beta-interferon expression in murine embryonal carcinoma F9 cells. *Mol. Cell. Biol.* **9:** 3553–3556.

Fraser A.G., Kamath R.S., Zipperlen P., Martinez-Campos M., Sohrmann M., and Ahringer J. 2000. Functional genomic analysis of *C. elegans* chromosome I by systematic RNA interference. *Nature* **408:** 325–330.

Ge Q., McManus M.T., Nguyen T., Shen C.H., Sharp P.A., Eisen H.N., and Chen J. 2003. RNA interference of influenza virus production by directly targeting mRNA for degradation and indirectly inhibiting all viral RNA transcription. *Proc. Natl. Acad. Sci.* **100:** 2718–2723.

Gil J. and Esteban M. 2000. Induction of apoptosis by the dsRNA-dependent protein kinase (PKR): Mechanism of action. *Apoptosis* **5:** 107–114.

Gonczy P., Echeverri C., Oegema K., Coulson A., Jones S.J., Copley R.R., Duperon J., Oegema J., Brehm M., Cassin E., Hannak E., Kirkham M., Pichler S., Flohrs K., Goessen A., Leidel S., Alleaume A.M., Martin C., Ozlu N., Bork P., and Hyman A.A. 2000. Functional genomic analysis of cell division in *C. elegans* using RNAi of genes on chromosome III. *Nature* **408:** 331–336.

Goomer R.S. and Kunkel G.R. 1992. The transcriptional start site for a human U6 small nuclear RNA gene is dictated by a compound promoter element consisting of the PSE and the TATA box. *Nucleic Acids Res.* **20:** 4903–4912.

Grishok A., Tabara H., and Mello C.C. 2000. Genetic requirements for inheritance of RNAi in *C. elegans*. *Science* **287:** 2494–2497.

Grishok A., Pasquinelli A.E., Conte D., Li N., Parrish S., Ha I., Baillie D.L., Fire A., Ruvkun G., and Mello C.C. 2001. Genes and mechanisms related to RNA interference regulate expression of the small temporal RNAs that control *C. elegans* developmental timing. *Cell* **106:** 23–34.

Hamilton A.J. and Baulcombe D.C. 1999. A species of small antisense RNA in posttranscriptional gene silencing in plants [see comments]. *Science* **286:** 950–952.

Hammond S.M., Boettcher S., Caudy A.A., Kobayashi R., and Hannon G.J. 2001. Argonaute2, a link between genetic and biochemical analyses of RNAi. *Science* **293:** 1146–1150.

Hannon G.J. 2002. RNA interference. *Nature* **418:** 244–251.

Harada H., Willison K., Sakakibara J., Miyamoto M., Fujita T., and Taniguchi T. 1990. Absence of the type I IFN system in EC cells: Transcriptional activator (IRF-1) and repressor (IRF-2) genes are developmentally regulated. *Cell* **63:** 303–312.

Harborth J., Elbashir S.M., Bechert K., Tuschl T., and Weber K. 2001. Identification of essential genes in cultured mammalian cells using small interfering RNAs. *J. Cell Sci.* **114:** 4557–4565.

Hasuwa H., Kaseda K., Einarsdottir T., and Okabe M. 2002. Small interfering RNA and gene silencing in transgenic mice and rats. *FEBS Lett.* **532:** 227–230.

Hemann M.T., Fridman J.S., Zilfou J.T., Hernando E., Paddison P.J., Cordon-Cardo C., Hannon G.J., and Lowe S.W. 2003. An epi-allelic series of p53 hypomorphs created by stable RNAi produces distinct tumor phenotypes in vivo. *Nat. Genet.* **3:** 3.

Hutvagner G. and Zamore P.D. 2002. A microRNA in a multiple-turnover RNAi enzyme complex. *Science* **297:** 2056–2060.

Hutvagner G., McLachlan J., Pasquinelli A.E., Balint E., Tuschl T., and Zamore P.D. 2001. A cellular function for the RNA-interference enzyme Dicer in the maturation of the *let-7* small temporal RNA. *Science* **293:** 834–838.

Jones K.W., Gorzynski K., Hales C.M., Fischer U., Badbanchi F., Terns R.M., and Terns M.P. 2001. Direct interaction of the spinal muscular atrophy disease protein SMN with the small nucleolar RNA-associated protein fibrillarin. *J. Biol. Chem.* **276:** 38645–38651.

Jones L., Hamilton A.J., Voinnet O., Thomas C.L., Maule A.J., and Baulcombe D.C. 1999. RNA-DNA interactions and DNA methylation in post-transcriptional gene silencing. *Plant Cell* **11:** 2291–2302.

Kapadia S.B., Brideau-Andersen A., and Chisari F.V. 2003. Interference of hepatitis C virus RNA replication by short interfering RNAs. *Proc. Natl. Acad. Sci.* **100:** 2014–2018.

Kawasaki H. and Taira K. 2003. Short hairpin type of dsRNAs that are controlled by tRNA(Val) promoter significantly induce RNAi-mediated gene silencing in the cytoplasm of human cells. *Nucleic Acids Res.* **31:** 700–707.

Kennerdell J.R. and Carthew R.W. 2000. Heritable gene silencing in *Drosophila* using double-stranded RNA. *Nat. Biotechnol.* **18:** 896–898.

Ketting R.F., Fischer S.E., Bernstein E., Sijen T., Hannon G.J., and Plasterk R.H. 2001. Dicer functions in RNA interference and in synthesis of small RNA involved in developmental timing in *C. elegans*. *Genes Dev.* **15:** 2654–2659.

Knight S.W. and Bass B.L. 2001. A role for the RNase III enzyme DCR-1 in RNA interference and germ line development in *Caenorhabditis elegans*. *Science* **293:** 2269–2271.

Lagos-Quintana M., Rauhut R., Lendeckel W., and Tuschl T. 2001. Identification of novel genes coding for small expressed RNAs. *Science* **294:** 853–858.

Lagos-Quintana M., Rauhut R., Meyer J., Borkhardt A., and Tuschl T. 2003. New microRNAs from mouse and human. *RNA* **9:** 175–179.

Lai E.C. and Posakony J.W. 1997. The Bearded box, a novel 3′UTR sequence motif, mediates negative post-transcriptional regulation of Bearded and Enhancer of split Complex gene expression. *Development* **124:** 4847–4856.

Lai E.C., Burks C., and Posakony J.W. 1998. The K box, a conserved 3′ UTR sequence motif, negatively regulates accumulation of enhancer of split complex transcripts. *Development* **125:** 4077–4088.

Lau N.C., Lim L.P., Weinstein E.G., and Bartel D.P. 2001. An abundant class of tiny RNAs with probable regulatory roles in *Caenorhabditis elegans*. *Science* **294:** 858–862.

Lawrence D. 2002. RNAi could hold promise in the treatment of HIV. *Lancet* **359:** 2007.

Lee R.C. and Ambros V. 2001. An extensive class of small RNAs in *Caenorhabditis elegans*. *Science* **294:** 862–864.

Lee R.C., Feinbaum R.L., and Ambros V. 1993. The *C. elegans* heterochronic gene *lin-4* encodes small RNAs with antisense complementarity to *lin-14*. *Cell* **75:** 843–854.

Lim L.P., Lau N.C., Weinstein E.G., Abdelhakim A., Yekta S., Rhoades M.W., Burge C.B., and Bartel D.P. 2003. The microRNAs of *Caenorhabditis elegans*. *Genes Dev.* **17:** 991–1008.

Lois C., Hong E.J., Pease S., Brown E.J., and Baltimore D. 2002. Germline transmission and tissue-specific expression of transgenes delivered by lentiviral vectors. *Science* **295:** 868–872.

Marcus P.I. and Sekellick M.J. 1985. Interferon induction by viruses. XIII. Detection and assay of interferon induction-suppressing particles. *Virology* **142:** 411–415.

Martinez M.A., Clotet B., and Este J.A. 2002. RNA interference of HIV replication. *Trends Immunol.* **23:** 559–561.

McCaffrey A.P., Meuse L., Pham T.T., Conklin D.S., Hannon G.J., and Kay M.A. 2002. RNA interference in adult mice. *Nature* **418:** 38–39.

McManus M.T., Petersen C.P., Haines B.B., Chen J., and Sharp P.A. 2002a. Gene silencing using micro-RNA designed hairpins. *RNA* **8:** 842–850.

McManus M.T., Haines B.B., Dillon C.P., Whitehurst C.E., van Parijs L., Chen J., and Sharp P.A. 2002b. Small interfering RNA-mediated gene silencing in T lymphocytes. *J. Immunol.* **169:** 5754–5760.

Meister G., Buhler D., Pillai R., Lottspeich F., and Fischer U. 2001. A multiprotein complex mediates the ATP-dependent assembly of spliceosomal U snRNPs. *Nat. Cell Biol.* **3:** 945–949.

Miyagishi M. and Taira K. 2002. U6 promoter-driven siRNAs with four uridine 3′ overhangs efficiently suppress targeted gene expression in mammalian cells. *Nat. Biotechnol.* **20:** 497–500.

Morel J.B., Godon C., Mourrain P., Beclin C., Boutet S., Feuerbach F., Proux F., and Vaucheret H. 2002. Fertile hypomorphic ARGONAUTE (ago1) mutants impaired in post-transcriptional gene silencing and virus resistance. *Plant Cell* **14:** 629–639.

Moss E.G., Lee R.C., and Ambros V. 1997. The cold shock domain protein LIN-28 controls developmental timing in *C. elegans* and is regulated by the *lin-4* RNA. *Cell* **88:** 637–646.

Mourelatos Z., Abel L., Yong J., Kataoka N., and Dreyfuss G. 2001. SMN interacts with a novel family of hnRNP and spliceosomal proteins. *EMBO J.* **20:** 5443–5452.

Mourelatos Z., Dostie J., Paushkin S., Sharma A., Charroux B., Abel L., Rappsilber J., Mann M., and Dreyfuss G. 2002. miRNPs: A novel class of ribonucleoproteins containing numerous microRNAs. *Genes Dev.* **16:** 720–728.

Olsen P.H. and Ambros V. 1999. The *lin-4* regulatory RNA controls developmental timing in *Caenorhabditis elegans* by blocking LIN-14 protein synthesis after the initiation of translation. *Dev. Biol.* **216:** 671–680.

O'Neil N.J., Martin R.L., Tomlinson M.L., Jones M.R., Coulson A., and Kuwabara P.E. 2001. RNA-mediated interference as a tool for identifying drug targets. *Am. J. Pharmacogenomics* **1:** 45–53.

Paddison P.J. and Hannon G.J. 2002. RNA interference: The new somatic cell genetics? *Cancer Cell* **2:** 17–23.

Paddison P.J., Caudy A.A., and Hannon G.J. 2002a. Stable suppression of gene expression by RNAi in mammalian cells. *Proc. Natl. Acad. Sci.* **99:** 1443–1448.

Paddison P.J., Caudy A.A., Bernstein E., Hannon G.J., and Conklin D.S. 2002b. Short hairpin RNAs (shRNAs) induce sequence-specific silencing in mammalian cells. *Genes Dev.* **16:** 948–958.

Pasquinelli A.E., Reinhart B.J., Slack F., Martindale M.Q., Kuroda M.I., Maller B., Hayward D.C., Ball E.E., Degnan B., Muller P., Spring J., Srinivasan A., Fishman M., Finnerty J., Corbo J., Levine M., Leahy P., Davidson E., and Ruvkun G. 2000. Conservation of the sequence and temporal expression of *let-7* heterochronic regulatory RNA. *Nature* **408:** 86–89.

Paul C.P., Good P.D., Winer I., and Engelke D.R. 2002. Effective expression of small interfering RNA in human cells. *Nat. Biotechnol.* **20:** 505–508.

Paul C.P., Good P.D., Li S.X., Kleihauer A., Rossi J.J., and Engelke D.R. 2003. Localized expression of small RNA inhibitors in human cells. *Mol. Ther.* **7:** 237–247.

Pellizzoni L., Baccon J., Charroux B., and Dreyfuss G. 2001a. The survival of motor neurons (SMN) protein interacts with the snoRNP proteins fibrillarin and GAR1. *Curr. Biol.* **11:** 1079–1088.

Pellizzoni L., Kataoka N., Charroux B., and Dreyfuss G. 1998. A novel function for SMN, the spinal muscular atrophy disease gene product, in pre-mRNA splicing. *Cell* **95:** 615–624.

Pellizzoni L., Charroux B., Rappsilber J., Mann M., and Dreyfuss G. 2001b. A functional interaction between the survival motor neuron complex and RNA polymerase II. *J. Cell. Biol.* **152:** 75–85.

Reinhart B.J., Slack F.J., Basson M., Pasquinelli A.E., Bettinger J.C., Rougvie A.E., Horvitz H.R., and Ruvkun G. 2000. The 21-nucleotide *let-7* RNA regulates developmental timing in *Caenorhabditis elegans*. *Nature* **403:** 901–906.

Rubinson D.A., Dillon C.P., Kwiatkowski A.V., Sievers C., Yang L., Kopinja J., Zhang M., McManus M.T., Gertler F.B., Scott M.L., and Van Parijs L. 2003. A lentivirus-based system to functionally silence genes in primary mammalian cells, stem cells and transgenic mice by RNA interference. *Nat. Genet.* **33:** 401–406.

Schmitt C.A., McCurrach M.E., de Stanchina E., Wallace-Brodeur R.R., and Lowe S.W. 1999. INK4a/ARF mutations accelerate lymphomagenesis and promote chemoresistance by disabling p53. *Genes Dev.* **13:** 2670–2677.

Smith N.A., Singh S.P., Wang M.B., Stoutjesdijk P.A., Green A.G., and Waterhouse P.M. 2000. Total silencing by intron-spliced hairpin RNAs. *Nature* **407:** 319–320.

Song E., Lee S.K., Wang J., Ince N., Ouyang N., Min J., Chen J., Shankar P., and Lieberman J. 2003. RNA interference targeting Fas protects mice from fulminant hepatitis. *Nat. Med.* **9:** 347–351.

Stark G.R., Kerr I.M., Williams B.R., Silverman R.H., and Schreiber R.D. 1998. How cells respond to interferons. *Annu. Rev. Biochem.* **67:** 227–264.

Stein P., Svoboda P., Anger M., and Schultz R.M. 2003. RNAi: Mammalian oocytes do it without RNA-dependent RNA polymerase. *RNA* **9:** 187–192.

Stewart S.A., Dykxhoorn D.M., Palliser D., Mizuno H., Yu E.Y., An D.S., Sabatini D.M., Chen I.S., Hahn W.C., Sharp P.A., Weinberg R.A., and Novina C.D. 2003. Lentivirus-delivered stable gene silencing by RNAi in primary cells. *RNA* **9:** 493–501.

Sui G., Soohoo C., Affar el B., Gay F., Shi Y., and Forrester W.C. 2002. A DNA vector-based RNAi technology to suppress gene expression in mammalian cells. *Proc. Natl. Acad. Sci.* **99:** 5515–5520.

Svoboda P., Stein P., and Schultz R.M. 2001. RNAi in mouse oocytes and preimplantation embryos: Effectiveness of hairpin dsRNA. *Biochem. Biophys. Res. Commun.* **287:** 1099–1104.

Tabara H., Sarkissian M., Kelly W.G., Fleenor J., Grishok A., Timmons L., Fire A., and Mello C.C. 1999. The *rde-1* gene, RNA interference, and transposon silencing in *C. elegans. Cell* **99:** 123–132.

Tavernarakis N., Wang S.L., Dorovkov M., Ryazanov A., and Driscoll M. 2000. Heritable and inducible genetic interference by double-stranded RNA encoded by transgenes. *Nat. Genet.* **24:** 180–183.

Tiscornia G., Singer O., Ikawa M., and Verma I.M. 2003. A general method for gene knockdown in mice by using lentiviral vectors expressing small interfering RNA. *Proc. Natl. Acad. Sci.* **100:**

1844–1848.

Walhout A.J., Reboul J., Shtanko O., Bertin N., Vaglio P., Ge H., Lee H., Doucette-Stamm L., Gunsalus K.C., Schetter A.J., Morton D.G., Kemphues K.J., Reinke V., Kim S.K., Piano F., and Vidal M. 2002. Integrating interactome, phenome, and transcriptome mapping data for the *C. elegans* germline. *Curr. Biol.* **12:** 1952–1958.

Wesley S.V., Helliwell C.A., Smith N.A., Wang M.B., Rouse D.T., Liu Q., Gooding P.S., Singh S.P., Abbott D., Stoutjesdijk P.A., Robinson S.P., Gleave A.P., Green A.G., and Waterhouse P.M. 2001. Construct design for efficient, effective and high-throughput gene silencing in plants. *Plant J.* **27:** 581–590.

Wianny F. and Zernicka-Goetz M. 2000. Specific interference with gene function by double-stranded RNA in early mouse development. *Nat. Cell. Biol.* **2:** 70–75.

Wightman B., Ha I., and Ruvkun G. 1993. Posttranscriptional regulation of the heterochronic gene *lin-14* by *lin-4* mediates temporal pattern formation in *C. elegans*. *Cell* **75:** 855–862.

Wilda M., Fuchs U., Wossmann W., and Borkhardt A. 2002. Killing of leukemic cells with a BCR/ABL fusion gene by RNA interference (RNAi). *Oncogene* **21:** 5716–5724.

Williams B.R. 1999. PKR; a sentinel kinase for cellular stress. *Oncogene* **18:** 6112–6120.

Williams R.W. and Rubin G.M. 2002. ARGONAUTE1 is required for efficient RNA interference in *Drosophila* embryos. *Proc. Natl. Acad. Sci.* **99:** 6889–6894.

Wu H., Yang P.K., Butcher S.E., Kang S., Chanfreau G., and Feigon J. 2001. A novel family of RNA tetraloop structure forms the recognition site for *Saccharomyces cerevisiae* RNase III. *EMBO J.* **20:** 7240–7249.

Xia H., Mao Q., Paulson H.L., and Davidson B.L. 2002. siRNA-mediated gene silencing in vitro and in vivo. *Nat. Biotechnol.* **20:** 1006–1010.

Yang D., Lu H., and Erickson J.W. 2000. Evidence that processed small dsRNAs may mediate sequence-specific mRNA degradation during RNAi in *Drosophila* embryos. *Curr. Biol.* **10:** 1191–1200.

Yang S., Tutton S., Pierce E., and Yoon K. 2001. Specific double-stranded RNA interference in undifferentiated mouse embryonic stem cells. *Mol. Cell. Biol.* **21:** 7807–7816.

Yu J.Y., DeRuiter S.L., and Turner D.L. 2002. RNA interference by expression of short-interfering RNAs and hairpin RNAs in mammalian cells. *Proc. Natl. Acad. Sci.* **99:** 6047–6052.

Yu J.Y., Taylor J., DeRuiter S.L., Vojtek A.B., and Turner D.L. 2003. Simultaneous inhibition of GSK3alpha and GSK3beta using hairpin siRNA expression vectors. *Mol. Ther.* **7:** 228–236.

Zamore P.D., Tuschl T., Sharp P.A., and Bartel D.P. 2000. RNAi: Double-stranded RNA directs the ATP-dependent cleavage of mRNA at 21 to 23 nucleotide intervals. *Cell* **101:** 25–33.

Zeng Y. and Cullen B.R. 2003. Sequence requirements for micro RNA processing and function in human cells. *RNA* **9:** 112–123.

Zeng Y., Wagner E.J., and Cullen B.R. 2002. Both natural and designed micro RNAs can inhibit the expression of cognate mRNAs when expressed in human cells. *Mol. Cell.* **9:** 1327–1333.

Zhang K. and Nicholson A.W. 1997. Regulation of ribonuclease III processing by double-helical sequence antideterminants. *Proc. Natl. Acad. Sci.* **94:** 13437–13441.

Zhang L., Yang N., Mohamed-Hadley A., Rubin S.C., and Coukos G. 2003. Vector-based RNAi, a novel tool for isoform-specific knock-down of VEGF and anti-angiogenesis gene therapy of cancer. *Biochem. Biophys. Res. Commun.* **303:** 1169–1178.

# CHAPTER 7

# Transposable Elements, RNA Interference, and the Origin of Heterochromatin

Rob Martienssen, Tom Volpe, Zach Lippman, Ann-Valerie Gendrel,*
Catherine Kidner, Pablo Rabinowicz, and Vincent Colot*

*Cold Spring Harbor Laboratory, Cold Spring Harbor, New York 11724; *IGRV, Evry, France*

TRANSPOSABLE ELEMENTS ARE MAJOR COMPONENTS of eukaryotic genomes, in some cases comprising by far the largest portion of the nuclear complement of DNA. Their prevalence, especially in heterochromatin, has led to the resurrection of McClintock's notion that gene regulation by transposon insertion might have a genome-wide role in gene regulation, participating in a variety of epigenetic phenomena from position effect variegation to gene silencing and imprinting (McClintock 1951; Martienssen et al. 1990).

Transposable elements fall into two major classes: retroelements that replicate and transpose via an RNA intermediate and those elements that transpose via a DNA-based mechanism. DNA transposons were discovered first, but retrotransposons comprise a much larger portion of complex eukaryotic genomes. Transposons are typically thought of as parasitic elements that have only survived genomic selection pressure because of their "selfish" nature. However, their role in chromosome structure was apparent from the time of their discovery in maize. For example, transposons can rearrange chromosomal segments by breakage, translocation, inversion, and, of course, transposition.

On the basis of the map position of certain autonomous elements (Figure 7.1), and on their properties, McClintock (1951) proposed that transposons were heterochromatic ele-

**FIGURE 7.1.** Heterochromatic knob on maize chromosome 9S. This deeply staining cytogenetic feature was used to physically map crossing over in maize. The mutability factor *Dotted* (*Dt*) mapped to this region. A pachytene spread is shown. (From McClintock 1951)

ments that upon redistribution throughout the genome could regulate gene expression during development. As a model for differentiation, this was perhaps before its time, but the role of transposons and heterochromatin in chromosomal organization and function has become widely recognized since then. Here, we review recent advances in understanding the epigenetic properties of transposons and heterochromatic repeats, their role in epigenetic mechanisms of gene regulation, and their role in chromosome organization and function.

## TRANSPOSONS AS CONTROLLING ELEMENTS

McClintock described transposons as "controlling elements" because of their ability to bring neighboring genes under their control. Such control was mediated in *trans* by autonomous elements located elsewhere in the genome (McClintock 1954). These studies provided some of the first evidence for gene regulation. Importantly, gene regulation mediated by transposons was frequently epigenetic, i.e., changes in gene expression induced by an autonomous transposon could be inherited in its absence. Such "preset" changes were typically erased after one or two further generations. However, the autonomous transposons themselves could also undergo epigenetic switches ("cycling") between active and inactive states. These states could be inherited for multiple generations.

In yeast, maize, mice, flies, *Neurospora*, and probably every organism in which they are found, retrotransposons and DNA transposons can regulate neighboring genes (Colot and Martienssen 2001). This is because transposons recruit transcriptional machinery to the chromosome as part of the mechanism of transposition. For example, Robertson's *Mutator* (*Mu*) and McClintock's *Suppressor-Mutator* (*Spm*) transposons control the anthocyanin biosynthesis gene *A1* when integrated in the promoter. In each case, the presence of the corresponding transposase, provided in *trans* from another location, regulates the gene. Interestingly, this regulation occurs in opposite directions even though the two transposons are integrated at the same nucleotide in each allele: *Spm* up-regulates *A1*, whereas *Mu* down-regulates it (for review, see Martienssen 1996).

In transgenic tobacco, the *TnpA* gene from *Spm* behaves like a transcription factor (Cui and Fedoroff 2002). Furthermore, outward-reading promoters were found in *Mu*, *Activator (Ac)*, and several animal transposons, such as the P element in *Drosophila* (for review, see Martienssen and Colot 2001). This led to a simple model for controlling element activity in which transposase genes regulate neighboring gene expression from promoters within the element. This model is already well-established for retrotransposons, which have outward-reading promoters in their 3′ long terminal repeat (LTR) and direct neighboring gene expression in yeast and *Drosophila* (for review, see Martienssen and Colot 2001). The situation with DNA transposons is more complex because transposase promoters typically read inward rather than outward, but transcriptional interference is one possible mechanism by which transposases could act as repressors and activators, depending on additional factors (Barkan and Martienssen 1991).

Sequencing of the *Arabidopsis* genome, as well as genome-wide surveys in other plants, has revealed that transposons are widespread components of the plant genome (*Arabidopsis* Genome Initiative 2000). In addition, in both maize and *Arabidopsis*, transposable elements are subject to extensive methylation (Rabinowitz et al. 1999). This raises the possibility that epigenetic regulation of gene expression mediated by transposons may be an inherent property of plant genomes. Such regulation would be predicted to be highly polymorphic, given the ability of transposons to move from one location in the genome to another, and could account for the properties of some hybrids, such as hybrid vigor (Colot and Martienssen 2001).

## HETEROCHROMATIN AND THE TRANSPOSON LANDSCAPE

In the "pregenomic" era, transposons were discovered only when they interrupted genes (May and Martienssen 2003). However, with the advent of genomic sequencing, the extent to which transposons contribute to the genome was realized. Up to 80% of the maize genome and 20% of the *Arabidopsis* genome are composed of transposons, in each case making up the majority of the repetitive fraction estimated by hybridization kinetics (San Miguel et al. 1996; *Arabidopsis* Genome Initiative 2000). In *Arabidopsis*, the majority of transposons do not interrupt genes. In contrast, mammalian genomes are structured very differently; mammalian introns can be 100–1000 times larger than plant introns, and they typically contain multiple transposon insertions. The origin of these differences is not clear, but they may reflect differences in splicing, transposon targeting, or epigenetic regulation.

*Arabidopsis* has a full complement of transposons, including *copia* and *gypsy* LTR retrotransposons, non-LTR long interspersed nuclear element (LINE) and short interspersed nuclear element (SINE) retrotransposons, as well as DNA transposons (Le et al. 2000). In addition, several new classes of transposable elements have been found. Terminal repeat retrotransposons in miniature (TRIM) elements are LTR retrotransposons in which genes encoding polyproteins have been replaced by noncoding sequences only 1–300 bp long (Witte et al. 2001). TRIMs resemble nonautonomous DNA elements in that they can be mobilized by full-length retroelements. Of the transposons found in *Arabidopsis*, 25% are "Basho" elements, which do not belong to any known structurally defined family of elements (*Arabidopsis* Genome Initiative 2000). Another 10% are thought to be helitrons, which replicate via a rolling circle mechanism (Kapitonov and Jurka 2001). In prokaryotes, related insertion sequences utilize host-encoded helicases and ligases, as well as a transposon-encoded replication protein. In *Arabidopsis*, helitrons encode helicases and replication proteins (Rpa1-like), whereas only helicases are encoded in *Caenorhabditis elegans*. Most copies in eukaryotes are deletion derivatives (Kapitonov and Jurka 2001).

The distribution of transposons in *Arabidopsis* is far from random (Figure 7.2), and the picture that emerges is one of a highly structured genome. Euchromatic sequences have a high frequency of matches to cDNA sequences, indicating that many of the genes are expressed. These regions have relatively few transposons. In contrast, heterochromatic

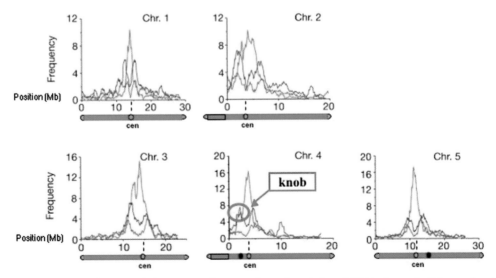

**FIGURE 7.2.** Distribution of class I retroelements (*green*) and class II DNA elements (*blue* and *red*) in the *Arabidopsis* genome. NOR, centromeres, and telomeres are shown in blue. Knobs are in *black*. (Modified, with permission, from the *Arabidopsis* Genome Initiative 2002.)

regions have few cDNA matches and many more transposons, with densities up to 20 times those found in euchromatin (*Arabidopsis* Genome Initiative 2000). The distribution bias varies greatly from one transposon family to another. *MULE* and *CACTA* class II elements, for example, are enriched in heterochromatin, whereas *Ac* and miniature inverted repeat transposable element (MITE) classes are more widely distributed, although they almost never come within 1 kb of a gene. Retroelements also differ, with LTR-class retroelements (especially of the Ty3/*gypsy* class) clustered in pericentromeric regions, and SINE-like elements excluded from them (Lenoir et al. 2001). Interestingly, some elements from all four classes in maize have been found to disrupt genes and to target genes specifically rather than repeat sequences. Maize *Ac/Ds* elements introduced into *Arabidopsis* also have a pronounced asymmetric distribution in the genome. In this case, they accumulate adjacent to rDNA and are excluded from pericentromeric domains (Parinov et al. 1999; CSHWU Consortium 2000).

Occasionally, entire chromosomal territories, with little to distinguish them in terms of sequence composition, have very few transposon insertions relative to the rest of the genome. For example, the distal portion of the long arm of chromosome 4 is almost devoid of class I and class II elements (*Arabidopsis* Genome Initiative 2000). There are several possible explanations. One is that these regions are single copy and cannot tolerate insertions into genes. This is not the case for the distal portion of chromosome 4L of *Arabidopsis*, however, which is extensively duplicated on chromosome 2. Another possibility is that these regions were recently introgressed into *Arabidopsis* from a related species. Transposons are highly polymorphic with large differences in copy number and chromosomal location between related genomes. This species-specific distribution is the basis for chromosome painting, where total genomic DNA from one species can be used to hybridize exclusively to its own chromosomes in hybrids with another species (Heslop-Harrison 2000). Thus, the "chromosomal landscape" of transposon insertions becomes diagnostic of a given species.

*Athila* and *del*-like retrotransposons have a unique distribution: They are almost all found in centromeric regions, internal to the regions populated by *MULE* and *CACTA* elements. This may indicate a functional role for these elements in chromosome structure and segregation. In yeast, Ty1 (*copia*) and Ty3/Ty5 (*gypsy*) elements are specifically targeted to distinct genomic features. It is possible that *Athila* and *del*-like elements are also targeted in *Arabidopsis*, although the nature of the target sequence remains a mystery (Wright and Voytas 2002). Certainly, 180-bp centromeric repeats can be interrupted by *Athila* elements in this way. Targeting to specific genomic regions may reflect higher-order nuclear structure. Alternatively, this highly organized distribution may impact chromosome behavior, as transposons may themselves recruit chromatin modifications. This is discussed in more detail below.

Heterochromatic sequences on *Arabidopsis* chromosome 4 have been mapped by fluorescent in situ hybridization (FISH) and sequence analysis (Fransz et al. 2000; CSHWU Consortium 2000). There are four heterochromatic regions: the nucleolar organizing region (NOR), telomeric domain, pericentromeric domain, and an interstitial heterochromatic knob or chromomere. Unlike pericentromeric sequences, the 0.5-Mb knob is small enough to have been completely cloned into overlapping fingerprinted bacterial artificial chromosomes (BACs) and was the first completely sequenced region of heterochromatin in a higher eukaryote (Figure 7.3).

The knob has about 35 genes, very few of which correspond to expressed sequence tags (ESTs), indicating that they are largely silent. More than 50 transposons include *Athila* and many DNA transposons, including members of the *MULE* and *CACTA* families. Most strikingly, a tandem array of 22 copies of a 2-kb repeat resembles those known to induce heterochromatic gene silencing in *Drosophila*. The heterochromatic repeats and *Athila* transposons are heavily methylated (CSHWU Consortium 2000).

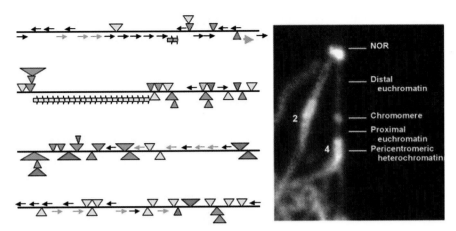

FIGURE 7.3. *Arabidopsis* heterochromatin. DAPI staining reveals the pericentromeric heterochromatin of chromosomes 2 and 4 at pachytene, the NOR, and the interstitial knob or chromomere (Fransz et al. 2000). The sequence of the knob is represented by *black* (hypothetical genes), *red* (known genes), and *yellow* arrows (tandem repeats) as well as triangles (transposons).

## MECHANISMS OF TRANSPOSON SILENCING

Although transposon silencing mechanisms were the first to be described genetically, transgene silencing mechanisms have been studied in more detail. Transgene silencing mechanisms are typically divided into those with transcriptional and posttranscriptional consequences. In the last few years, transposon silencing has been shown to involve both classes of silencing mechanisms via chromatin remodeling and DNA methylation, on the one hand, and RNA interference, on the other.

### DNA Methylation

The relationship between transposons and DNA methylation in eukaryotes was first demonstrated in maize, although it had been previously shown that retroviruses were methylated upon integration into the mouse genome. In maize, transposable elements were modified by cytosine methylation, and the pattern of methylation was correlated with activity (Bennetzen 1996). For *Ac, Spm,* and Robertson's *Mu* class I transposons, activity was associated with the loss of 5-methyl cytosine (5mC) from the promoter of the transposase gene(s) encoded by the element, which was located at one or both ends of the element. Therefore, a popular hypothesis was that DNA methylation might be a host response to restrict transposon mobilization.

Parallels have been drawn between DNA methylation of transposons and bacterial restriction modification systems because of their role in host defense, although, of course, these mechanisms function in the opposite fashion. DNA modification in bacteria protects the methylated sequences from restriction, whereas in transposons, it is proposed to act to restrict element transposition. Although eukaryotic and prokaryotic DNA methyltransferases share critical motifs, the derivation of one from the other is far from clear (Holliday 1996). Nonetheless, DNA methylation alters the binding properties of several DNA-binding proteins that impact gene regulation.

In maize and other eukaryotes, DNA methylation is mostly restricted to 5mC. We investigated the distribution of DNA methylation between transposons and the remainder of the genome using bacterial strains that could distinguish methylated DNA from unmethylated DNA. We found that repetitive, transposable element DNA was heavily methylated in plant genomes, whereas genic DNA was largely, if not entirely, unmethylated (Rabinowicz et al. 1999).

To investigate the role of DNA methylation in transposon silencing and other phenomena, pools of mutagenized *Arabidopsis* plants were screened for demethylated high-copy centromeric repeats, allowing detection of rare mutants by Southern blotting (Vongs et al. 1993). Several recessive alleles were isolated at two loci. The *MET1* gene encodes the most highly expressed DNMT1-class maintenance methyltransferase in *Arabidopsis* (Finnegan and Dennis 1993), and two *met1* alleles were recovered (Kankel et al. 2003). The second major *Arabidopsis* locus required for centromeric methylation, *DDM1*, encodes a SWI2/SNF2 superfamily protein (Jeddeloh et al. 1999).

*met1-2* and *ddm1-2* homozygotes both exhibit a 70% reduction in genome methylation. In *ddm1*, this reduction is found initially in 5S rDNA, NOR, centromeric repeats, interstitial knobs, and subtelomeric regions (Vongs et al. 1993; CSHWU Consortium 2000), but it becomes more widespread in subsequent inbred generations. Residual DNA methylation in *met1-2* may result from the activity of the *MET1* homologs, *MET2, MET2a, MET2b,* and *MET3*, as well as the chromomethylases *CMT1-3* (Henikoff and Comai 1998) and the de novo methyltransferase genes *Domains Rearranged* (*DRM1* and *DRM2*), which closely resemble the mouse *DNMT3* gene and have a similar function (Cao and Jacobsen 2002).

Several alleles of *cmt3* were recovered in screens for reversion of endogenous gene silencing (Bartee et al. 2001; Lindroth et al. 2001). In each case, spontaneous reversion was suppressed by an unlinked inverted repeat comprising the gene and its promoter, raising the interesting possibility that double-stranded RNA (dsRNA) is the trigger for methylation. Non-CG and especially C-X-G methylation in the target gene required *CMT3*. No other phenotypic defects were observed. Targeted screens have recovered several other mutants in both transcriptional and posttranscriptional silencing in plants (for review, see Martienssen and Colot 2001; Matzke et al. 2001). Transcriptional gene silencing (TGS) screens have recovered mutants at more than five loci. Two of these (*ddm1* and *mom1*) encode potential Swi/Snf superfamily members, but only *ddm1* results in reduced DNA methylation (Amedeo et al. 2000).

## Histone Modification

Chromatin can be modified through histone acetylation, phosphorylation, ubiquitination, and methylation (Allfrey et al. 1964). Reversible modifications of core histones are catalyzed by histone acetyltransferase (HAT) and histone deacetylase (HD or HDAC). In general, deacetylation is related to TGS, whereas acetylation correlates with activation, but many exceptions have been documented (Braunstein et al. 1993). In organisms that have both DNA and histone modifications, HDACs either exist within a complex containing methyl-cytosine-binding proteins MeCP2 or MBD3 (Nan et al. 1998; Ng et al. 1999), or they interact directly with Dnmt1 (Fuks et al. 2000; Robertson et al. 2000; Rountree et al. 2000). This implies that DNA methylation may influence gene activity, at least partly through changes in histone acetylation (Cameron et al. 1999).

Many organisms, such as *Drosophila*, yeast, and *C. elegans* have little or no DNA methylation, but heterochromatin is instead characterized by association with modified histones. Methylation of residues Lys-4 (K4) and Lys-9 (K9) of the histone H3 tail have been correlated with euchromatic and heterochromatic regions, respectively, in many different organisms (Jenuwein and Allis 2001). Macronuclear repeats in *Tetrahymena*, silent mating-type loci in *Schizosaccharomyces pombe*, and the inactive X chromosome in mouse are examples of silent chromosomal domains whose silencing depends on the methylation of histone H3 (Strahl et al. 1999; Rea et al. 2000; Litt et al. 2001; Nakayama et al. 2001; Noma et al. 2001; Peters et al. 2001).

Dimethyl histone H3 Lys-9 (H3mK9) and dimethyl histone H3 Lys-4 (H3mK4) appear to be complementary states, leading in part to the proposition that silent chromatin is marked by a "histone code" (Jenuwein and Allis 2001). This code is "read" by adaptor domains such as bromodomains (which bind acetylated histones) and some chromo-domains (which bind methylated histones). Multifunctional proteins that have these domains can interact in turn with DNA, RNA, and protein components to either promote or repress gene expression. H3mK9 recruits the highly conserved protein, heterochro-matin protein-1 (HP-1). Silencing at the mating-type region in *S. pombe*, which lacks DNA methylation, also depends genetically on histone modification, in that mutants in many of the *clr* (cryptic loci repression) genes encode histone modification enzymes (*clr3, clr4, clr6*). Furthermore, the HP-1 homolog Swi6 is recruited to the silent mating-type loci and participates in silencing.

The dependence of DNA methylation on histone H3 Lys-9 methylation was demon-strated in *Neurospora* and in *Arabidopsis* (Jackson et al. 2001; Tamaru and Selker 2001). In *Neurospora*, direct screens were performed for demethylation mutants. *dim-2 (decrease in methylation)* encodes a *dnmt1* homolog responsible for DNA methyltransferase activity, but otherwise has no reproducible phenotype (Kouzminova and Selker 2001). Another locus, *dim-5*, encodes a SET domain histone methyltransferase that resembles yeast *clr4 +* and animal *Su(var)3-9* in having a cysteine-rich domain, but it lacks the chromodomain found in these proteins (Tamaru and Selker 2001). DNA methylation is largely absent from this strain, as well as from strains transfected with histone genes in which Lys-9 has been substituted with leucine. These results suggest that DNA methylation may be guid-ed by histone modification.

This conclusion has been supported by isolation of mutations in a SET domain gene (*KRYPTONITE/SUVH4*) in *Arabidopsis* in the same silencing screens that recovered mutants in the chromomethylase *cmt3* (Jackson et al. 2002; Malagnac et al. 2002). However, demethylation was modest in this case and did not extend to the genome as a whole. More than 30 SET domain genes are encoded in the *Arabidopsis* genome, and each might effect a different subset of genes or chromosomal domains. *Arabidopsis* TGS mutants have also been recovered in the histone deacetylase HDAC6, which encodes a homolog of RPD3 in yeast (Vidal and Gaber 1991; Runlett et al. 1996; Murfett et al. 2001).

## Chromatin Remodeling

*DDM1* maintains genomic DNA methylation levels (Vongs et al. 1993) and is responsible for transcriptional silencing of transgenes and certain endogenous genes in *Arabidopsis* (Jeddeloh et al. 1998; Mittlesten-Scheid et al. 1999). In *ddm1* mutants, transposons from the *MULE* and *CACTA* families are demethylated and activated, resulting in transcription and transposition around the genome (CSHWU Consortium 2000; Miura et al. 2001; Singer et al. 2001). Both elements are normally found clustered in the pericentromeric heterochromatin. However, when transposed elements were mapped in *ddm1* mutants, they were found to be scattered. It is therefore possible that *DDM1* not only silences these transposons, but also contributes to their targeting (Martienssen and Colot 2001).

Rather than encoding a DNA methyltransferase, *DDM1* has similarity to the SWI/SNF family of ATP-dependent chromatin remodeling genes (Jeddeloh et al. 1999; Verbsky and Richards 2001), suggesting it affects DNA methylation indirectly, presumably by remod-eling nucleosomes (Brzeski and Jerzmanowski 2002). Methylation of Lys-4 and Lys-9 of histone H3 is typically associated with transcriptionally active and inactive nucleosomes, respectively (Jenuwein and Allis 2001). Given that *DDM1* primarily impacts heterochromatin, we set out to test whether heterochromatic methylation of histone H3 is affected in *ddm1*. We used antibodies raised against H3mK9 and H3mK4 peptides to show that *Arabidopsis* histone H3 has both modifications and that overall levels do not change appreciably in *ddm1-2* by western blotting (Gendrel et al. 2002).

We analyzed the pattern of histone H3 methylation in heterochromatic DNA from the knob of chromosome IV. Chromatin immunoprecipitation (ChIP) was carried out using 52 primer pairs from genes, upstream regions, and transposons from the heterochromatic knob. In wild-type plants, all 15 transposons were associated with H3mK9, whereas 4 out of 6 known genes were associated with H3mK4. In *ddm1*, a dramatic shift in the pattern of histone methylation was observed. Of the 41 sequences whose association with methylated histones could be detected, 32 either lost H3mK9 or gained H3mK4, or both. These included both known and hypothetical genes, transposons, and upstream regions, indicating that heterochromatin underwent a major restructuring (Gendrel et al. 2002).

We also determined whether changes in histone H3 methylation correlated with changes in expression. Up-regulated sequences included *Athila*, *Cinful*, *del-like*, and novel retrotransposons, as well as *MULE* and *CACTA* class II DNA transposable elements and several silent genes. Thus, heterochromatin reverted to H3mK4 in *ddm1*, perhaps because H3mK9 and H3mK4 were redistributed throughout the genome. We speculated that either *DDM1* actively eliminates H3mK4 from heterochromatin or it maintains a high concentration of heterochromatic H3mK9 (Gendrel et al. 2002).

Most DNA methylation in plants is confined to transposons and repeats (Rabinowicz et al. 1999), and given that transposons are also associated with high levels of H3mK9, *DDM1* may act primarily on DNA methylation, impacting histone methylation indirectly. Alternatively, DNA methylation during replication may depend on a high concentration of H3mK9. According to this scenario, redistribution of methylated histones in *ddm1* would lead to dilution of heterochromatic H3mK9 relative to H3mK4 and subsequent loss of DNA methylation. However, relative to *ddm1*, overall DNA methylation losses are modest in the histone H3 Lys-9 methyltransferase *kyp-1*. This may be because of gene redundancy: *Arabidopsis* has up to 15 genes that potentially encode this class of proteins (Jackson et al. 2002; Johnson et al. 2002).

The loss of DNA methylation in *ddm1* may thus be a consequence of the reduced association of heterochromatin with H3mK9. However, the mechanism by which this is achieved is unclear. The SWI2/SNF2 complex in yeast forms loops in naked or nucleosomal DNA by bringing distant sites together (Bazett-Jones et al. 1999). These loops have been observed in *Arabidopsis* and depend on *DDM1* (P. Fransz, pers. comm.). SWI/SNF complexes may interact specifically with modified histones (Sudarsanam and Winston 2000). Complexes related to Mi-2 are associated with the NURD histone deacetylase and the methylated DNA-binding protein MBD3. NURD recruits the MeCP1 complex to methylated DNA (Feng and Zhang 2001), whereas Mi-2/CHD SWI/SNF family proteins have chromodomains that might promote binding to methylated histones. These domains are absent from *DDM1*, which represents a novel, but highly conserved, subfamily of SWI/SNF ATPases (see Verbsky and Richards 2001). It remains to be seen whether *DDM1* and other SWI/SNF subfamilies interact specifically with methylated and other modified nucleosomes. However, it has been demonstrated that *DDM1* has no specificity for methylated DNA (see Brzeski and Jerzmanowski 2002).

Conversely, DNA methylation can reinforce histone methylation patterns via chromatin remodeling complexes that bind methylated DNA. For example, DNA methyltransferases and methyl-binding proteins both associate with histone deacetylase (Robertson et al. 2000; Rountree et al. 2000). This could account for the loss of H3mK9 from a subset of transposons in *met1* mutant strains of *Arabidopsis* (Johnson et al. 2002). Thus, both *MET1* and *DDM1* are required to maintain both DNA methylation and histone H3 Lys-9 methylation associated with transposable elements. However, the mechanism by which transposons are targeted for these epigenetic marks is still unclear.

## RNA Interference

It has long been suspected that transposon silencing mechanisms could somehow sense the presence of multiple copies of identical or near-identical DNA sequences in the genome. These sequences would be recognized as distinct from genes and therefore likely to represent parasitic or invasive elements. Until recently, the most popular mechanism proposed to account for this recognition was DNA-DNA pairing (Bennetzen 1996; Bender 1998). It was proposed that heteroduplexes between unlinked repeats would be recognized by hemimethylation-dependent DNA methyltransferases, which would methylate and thus silence the transposons. Such a mechanism for homology-dependent gene silencing (HDGS) was assumed to be responsible for transgene silencing as well.

Evidence for the transfer of DNA methylation in a heteroduplex was obtained from the filamentous ascomycete *Ascobolus immersus*. At meiosis, methylation could be transferred from a methylated to an unmethylated allele, by a process reminiscent of gene conversion (Colot et al. 1996).

As mutants in transposon silencing were recovered, other mechanisms for homology-dependent silencing emerged. In *C. elegans*, both forward and reverse screens uncovered a wide variety of genes, known as *Mutator* (*mut*) genes, required to silence *Tc* elements. *Tc* elements are transposons of the Mariner DNA class of transposable elements (for review, see Plasterk 2002) and are active in somatic cells, but silenced in the germ line. Characterization of *mut* genes and tests of RNA interference (RNAi) revealed that RNAi is a major pathway for transposon silencing in *C. elegans*.

It should be noted, however, that transposons are only silenced in the germ line, not in somatic cells. Yet RNAi is very active in somatic cells. Furthermore, several genes critically required for RNAi, such as the *argonaute* homolog *rde1*, have no impact on transposon silencing. Therefore, it is possible that the role of RNAi in transposon silencing in *C. elegans* is indirect. For example, perhaps transposons are targets of RNAi in all cells, but only in the germ line does the RNAi lead to silencing at the transcriptional level (see below).

In *Chlamydomonas*, trypanosomes, and *Drosophila*, retrotransposons are silenced by the RNAi machinery (Aravin et al. 2001; Djikeng et al. 2001; Wu-Scharf et al. 2002). Furthermore, in *Drosophila*, the *argonaute* homolog *aubergine/sting* is also involved in processing transcripts derived from heterochromatic repeats (Aravin et al. 2001). These repeat transcripts, known as *Suppressor of Stellate*, target the homologous *Stellate* gene for silencing. *Stellate* encodes an unusual protein that forms crystalline structures in spermatids and impacts male fertility.

In plants, a role for RNA silencing in transgene regulation has been well-established (see Chapter 1), but a similar role in transposon regulation has been harder to demonstrate, potentially because of redundant mechanisms of transposon regulation, such as DNA methylation. The SINE element AtSN1 is methylated in wild-type cells, but not in some RNAi mutants (Hamilton et al. 2002). Importantly, Llave et al. (2002) identified numerous small RNA species that corresponded to transposable elements, including *Athila*, *CACTA*, and *MULE* elements. This may reflect the existence of antiparallel transcripts arising from inward reading promoters or to some form of read-through transcription. *CACTA*, *MULE*, and *Athila* elements are all concentrated in pericentromeric regions.

## RNAi AND TRANSCRIPTIONAL SILENCING ARE MECHANISTICALLY RELATED

A connection between RNA silencing mechanisms and DNA silencing mechanisms was first recognized in plant viruses when viral transgenes within the tobacco genome were found to become methylated on infection with a cognate RNA virus (Wassenegger et al. 1994). Subsequently, posttranscriptional silencing was shown to result in DNA methylation in a variety of contexts, including transgenes and integrated viruses (Jones et al. 1999; Aufsatz et al. 2002). Furthermore, DNA methylation was required to maintain transcriptional silencing after the RNA trigger was removed (Jones et al. 2001). These observations set the stage for investigating the connection between RNAi and chromatin modification.

Despite their significant role in transgene posttranscriptional gene silencing (PTGS), mutations in the components of the RNAi machinery had very little impact on transposon silencing (Hamilton et al. 2002; Z. Lippman et al., in prep.). One explanation was that these components are encoded by multigene families in *Arabidopsis*, which has nine *ARGONAUTE* genes and four *DICER*-like genes. (*C. elegans* is similarly endowed with RNAi genes.) It should be noted that individual genes in *Arabidopsis* can have severe developmental defects, indicating that they are not all redundant.

### Transposons, Repeats, and Silencing in Fission Yeast

Unlike in other eukaryotes, the genes encoding ARGONAUTE, DICER RNase/helicase, and RNA-dependent RNA polymerase are single copy in *S. pombe* (fission yeast). Furthermore, the budding yeast *Saccharomyces cerevisiae* has none of these genes (Avarind et al. 2000; Wood et al. 2002). Interestingly, budding yeast also lacks the histone H3 Lys-9 methyltransferase Clr4, as well as the heterochromatin protein Swi6. Homologs of these genes, as well as those required for RNAi, are found in all other eukaryotes, and *S. pombe* has a full complement of histone methylation, acetylation, and deacetylation enzymes. This makes *S. pombe* an ideal organism for investigating the link between transcriptional and posttranscriptional mechanisms of gene silencing.

*S. pombe* has relatively few retrotransposable elements and no known DNA transposons. Nonetheless, repeats are found at fission yeast centromeres, as they are in higher eukaryotes, providing a relatively simple model for heterochromatin. Each centromere comprises a central core surrounded by *dg* and *dh* repeats (also known as K repeats) in complex arrangements inverted around the central core. Individual copies of these repeats are found at two other locations in the fission yeast genome: the telomere and the silent mating-type locus (Wood et al. 2002). Until recently, the parallels between transposon-mediated silencing and silencing in yeast remained unexplored. This was in part because the repeats found at the centromere and at the mating-type locus did not appear to be transcribed, and encoded no transposase-like protein, so the possibility that they were transposons was dismissed soon after they were first discovered (Baum and Clark 1989; Grewal and Klar 1997). Nonetheless, integration of reporter genes into the centromeric repeats resulted in silencing and a form of position-effect variegation, depending on the site of integration (Allshire et al. 1994). Furthermore, silencing at the cryptic mating-type loci resembled transposon-mediated silencing in many respects.

The cryptic mating-type loci *mat2* and *mat3* are located about 20 kb proximal to *mat1*, which has an active copy of the mating-type cassette. The *mat2/mat3* locus is interrupted by a truncated copy of the centomeric repeat (*cenH*), comprising the *dg*IIa and *dh*IIa repeats from the centromere of chromosome II. When *cenH* was replaced by a reporter transgene, silencing was alleviated, but alternate expression states could still be selected and were stably maintained (Grewal and Klar 1996; Thon and Friis 1997). In addition, when *cenH* was integrated elsewhere in the genome, it could weakly confer silencing on an adjacent *ade6*[+] reporter gene (Ayoub et al. 2000).

Careful dissection of the sequences required indicated that nonoverlapping segments as small as 10% of the *cenH* region could confer silencing equally well (Ayoub et al. 2000). However, by selecting colonies with different reporter phenotypes, and following their subsequent clonal elaboration, it was demonstrated that silencing was extremely unstable, reverting at a frequency of more than 50%. Maintenance of silencing required a second sequence, REII, derived from a region adjacent to the silent mating-type locus *mat2-P*. However, this sequence alone could not confer silencing. Thus, silencing was initiated by the *cenH* homology, but maintained by REII (Ayoub et al. 2000).

The vast majority of complementation groups required for silencing at the mating-type locus also impact centromeric silencing, suggesting a common silencing mechanism (Ekwall et al. 1999). These genes encode enzymes responsible for histone Lys-9 methylation (Clr4) as well as histone deacetylation (Clr3 and Clr6). However, the use of reporter genes has allowed the isolation of a number of mutants that effect centromeric, but not mating-type, silencing. These mutants are known as *csp* mutants for *suppressor of centromeric position effect* (Ekwall et al. 1999). Given the low allele frequencies discovered, this is likely to be a large class of mutants in fission yeast.

## RNAi and Centromeric Silencing

In the course of our work on the *ARGONAUTE* gene in *Arabidopsis*, we realized that gene redundancy might underlie the differences observed in the phenotypic effects of disruption in plants and animals (see above). We therefore set out to disrupt the *AGO1*-like gene in fission yeast (Volpe et al. 2002). Initially, *ago1* gene disruption in *S. pombe* was disappointing. The cells were viable and mated normally. We examined the possibility that mating-type silencing might be affected, but no effects were noted. However, given that *csp* mutants affected centromeric and not mating-type silencing, we speculated that *argonaute* might have similar specificity. We used strains in which the *ura4*[+] reporter gene had been integrated into *cen1*, the same strains used to select *csp* mutants (Ekwall et al. 1999). Sure enough, *ago* disruption led to loss of gene silencing (Volpe et al. 2002). This was puzzling because gene silencing at the centromere was known to depend on transcriptional, rather than posttranscriptional, mechanisms.

We therefore explored whether silencing was indirect, depending on silencing of the repeats, rather than of the reporter genes themselves. Just as transposons can bring genes under their control by integrating nearby, we rationalized that integration of reporter genes near centromeric repeats could have the same effect. Transcripts from centromeric repeats could be readily detected in *argonaute* knockouts, accumulating from both strands. In addition, disruption of other genes required for RNAi had a similar effect (see below). We used nuclear run-ons to demonstrate that the two strands were differentially regulated. Transcripts corresponding to the upper strand could not be detected by nuclear run-on in wild-type cells, whereas those corresponding to the lower strand could be detected. Thus, although the lower strand is regulated posttranscriptionally, the upper strand is regulated transcriptionally (Figure 7.4). Put another way, the lower strand is transcribed all the time, but it is turned over so rapidly that it cannot be detected as long contiguous RNA species by reverse transcriptase–polymerase chain reaction (RT-PCR) or conventional northern blotting. However, an important prediction was that a short interfering RNA (siRNA) corresponding to these transcripts should be readily detected. Satisfyingly, Reinhardt and Bartel (2002) cloned siRNA from *S. pombe* and found several siRNAs corresponding to the 5′ end of both the lower-strand transcript and the upper-strand transcript. In each case, both antisense and sense siRNAs were found, indicating that they were indeed the products of dsRNA.

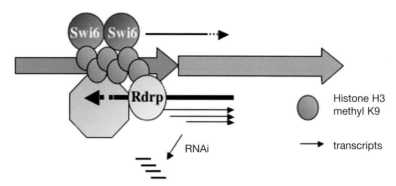

**FIGURE 7.4.** A model for recruitment of histone H3 Lys-9 methylation machinery (*orange* octagon) to centromeric repeats by RNA interference. siRNA targets RdRP to centromeric transcripts which in turn recruits H3mK9 and Swi6. These silence the complementary strand.

## The Chromatin Connection

How, then, does posttranscriptional silencing of the centromeric repeats result in transcriptional silencing of not only the transgenes integrated at *cen1*, but also the forward transcripts from the repeats on *cen3*? We explored this genetically by examining transcript levels in double mutants with *swi6*. We found that *agoswi6* double mutants accumulated much more centromeric transcript than *swi6* alone. In addition, whereas *swi6* only accumulated forward transcripts, the double mutant accumulated both forward and reverse transcripts, confirming that only the reverse strand was subject to RNAi. These results demonstrated that RNAi operated in the absence of *swi6* and was either upstream or redundant (Volpe et al. 2002). Therefore, we examined chromatin modifications known to lie upstream of *swi6*.

Swi6 is an HP-1-like protein that specifically binds to H3mK9 (Bannister et al. 2001). We found that each of the RNAi mutants had significantly reduced levels of H3mK9 associated with the centromeric repeats, assessed by chromatin immunoprecipitation. Relative to H3mK9 levels associated with the mating-type locus, each of the three RNAi mutants lost between 80% and 95% of the H3mK9 associated with repeats (Volpe et al. 2002). A similar increase in H3mK4 accompanied this change. Presumably, this accounts for the loss of transcriptional silencing observed both on the forward strand of *cen3* and at the *ura4*+ transgenes integrated at *cen1*.

How could a loss of RNAi lead to this change in nucleosome modification? We speculated that one way would be for the RNAi machinery itself to be localized at the centromeric repeats, where it could recruit the histone methyltransferase (HMT) Clr4. We tested this by examining the direct association of Dcr and Rdp with the centromere. Tagged Rdp functionally complemented the deletion strain and was tightly bound to centromeric repeats (Volpe et al. 2002). Tagged Dcr failed to bind but was only partially functional, indicating that the tagged protein may have been improperly localized. We conclude that the RNAi machinery is localized to chromatin, where it could potentially recruit the HMT machinery directly or indirectly (Figure 7.4). This leaves open the question of how the RNAi machinery itself becomes localized. One possibility is that nascent transcripts from *cen* repeats are tethered to the chromatin during elongation via the polymerase responsible for transcribing them. This would be especially true if the transcripts failed to terminate efficiently, and were very long. RNA-dependent RNA polymerase (RdRP) would then be "targeted" to these transcripts by the presence of complementary small RNA (siRNA). RNA-dependent transcription would then generate dsRNA, which would become the substrate for processing by DICER, generating further siRNA and amplifying the process.

One prediction from this model is that the siRNA complementary to the 5´ end of the reverse transcript would accumulate, rather than those corresponding to the 3´ end, and this is in fact observed. Another prediction is that Rdp binding to centromeric repeats would be abrogated in mutants that block siRNA production, as well as those that block transcription in the first place.

An excellent candidate that might regulate transcription from centromeric repeats is the highly conserved centromere-binding protein, Cenp-b. Cenp-b resembles a transposase from the Mariner superfamily of transposons, except that the DDE protein sequence motif, which is required for endonucleolytic cleavage, is missing (Plasterk et al. 1999). It has a helix-turn-helix DNA-binding domain and an acidic region found in many transcription factors. Furthermore, reduction in Cenp-b levels has a modest, but measurable, effect on H3mK9 associated with centromeric repeats (Nakagawa et al. 2002).

## Establishment or Maintenance?

Ayoub et al. (2000) and Partridge et al. (2002) demonstrated that silencing could be conferred on transgenes by repeats from the mating-type locus and the centromere, respectively, when they were integrated nearby. These observations led Hall et al. (2002) and Volpe et al. (2003) to ask whether silencing conferred by ectopic mating-type and centromeric repeats was regulated in the same way. In each case, silencing was relieved in the RNAi mutants. In addition, silencing was immediately restored upon backcrossing, as it was at the centromere (Volpe et al. 2002). Thus, silencing at centromeres as well as at ectopic loci could not be maintained without RNAi, but was efficiently reestablished on backcrossing.

Approximately 5% of the heterochromatin in *S. pombe* is found at the silent *mat2/mat3* loci. Silencing at the mating-type locus was not effected by RNAi (Volpe et al. 2002), indicating that a separate mechanism was responsible for maintaining silence in this region (Ekwall et al. 1999). When histone modification was removed either genetically or pharmacologically from the mating-type locus, the efficiency of reestablishment could be measured (Hall et al. 2002). Silencing was restored rapidly on backcrossing to wild type, but only slowly on backcrossing into an RNAi mutant background. This demonstrated that although RNAi increased the efficiency of establishment, it was not required for maintenance, in sharp contrast with the situation at the centromere. Thus, another independent mechanism is required for maintenance of silencing at the mating-type locus, as demonstrated previously by Ayoub et al. (2000).

What might this mechanism be? In *Arabidopsis*, as described above, the SWI/SNF chromatin remodeling factor, *DDM1*, is required to maintain the high concentration of K9-modified nucleosomes found associated with transposable elements and heterochromatin (Gendrel et al. 2002). Remarkably, a SWI/SNF homolog in fission yeast, *hrp3*⁺, is also required for mating-type silencing, but not centromeric silencing (Jae Yoo et al. 2002). However, only silencing of inner repeats, and not outer repeats, was tested at the centromere, so the possibility remains that it is required for silencing in both regions. Furthermore, unlike *DDM1*, *HRP3* has a chromodomain similar to those found in the CHD family of SWI/SNF remodelers. Even so, SWI/SNF ATPases are involved in maintenance of heterochromatin in both yeast and plants.

## Centromere Function

The intimate connection between heterochromatin and centromere function is well established (Bernard and Allshire 2002). In *S. pombe*, for example, along with the central core (cc), the heterochromatic repeats in the outermost region have an important role in chromosome segregation. This role depends on histone modification, because mutants in *clr4, clr6*, and *swi6* each have pronounced segregation defects resulting in frequent lagging chromosomes during anaphase. These defects are mediated by cohesin, which is recruited by Swi6 and keeps sister chromatids together long enough to orient them correctly on the spindle. Given that RNAi is required for H3mK9 and *swi6* recruitment to centromeres, it should perhaps come as no surprise that mutants in RNAi are defective in cohesin recruitment and chromosome segregation (Ekwall et al. 2002; Hall et al. 2003; Volpe et al. 2003).

The RNAi mutants have defects in chromosome segregation that are indistinguishable from *csp* mutants, and *csp8-13* mutants are also defective in RNAi, in that RNA accumulates from both strands of the centromere (Volpe et al. 2003). *csp9* is an allele of *argonaute*, having suffered a nonsense mutation that eliminates the highly conserved, carboxy-terminal Piwi domain. Thus, mutants in RNAi are defective in centromere function, and mutants in centromere function are defective in RNAi. This concordance strongly suggests that centromere function is an ancient role for RNAi and that the *csp* phenotype will provide a rich source of components in the RNAi pathway itself.

## CONCLUSIONS

RNAi is an attractive mechanism for distinguishing repetitive sequences from the remainder of the genome, because transposons and repeats are frequently associated with dsRNA. Therefore, the recruitment of the RNAi machinery to heterochromatin is likely a critical step in targeting histone modification. At least in *S. pombe*, RNAi is also required to maintain this modification at centromeres, indicating that heterochromatin is not only transcribed, but also far more dynamic than previously supposed. Other mechanisms, such as chromatin remodeling and DNA methylation, can maintain heterochromatin in other regions of the genome.

The ancient role of transposons in determining not only chromosomal organization, but also chromosome function indicates that the traditional view of transposons as parasitic agents may need to be revised. Transposon-like repeats have evolved into sophisticated molecular machines dedicated to the precise segregation of the genome into daughter cells, providing for the faithful inheritance of not only themselves, but also the remainder of the genome. The realization that centromeric repeats are transcribed from heterochromatin in at least some contexts has also changed our view of heterochromatin as a transcriptionally silent "junkyard" in higher eukaryotic genomes. McClintock's controversial vision of heterochromatin as a dynamic and essential component of the epigenetic landscape responsible for development, although flawed perhaps in many respects, has opened the doors to a new appreciation for transposons as elements controlling chromosome function as well as gene regulation (McClintock 1951).

## REFERENCES

Allfrey V.G., Faulkner R., and Mirsky A.E.A. 1964. Acetylation and methylation of histones and their possible role in regulation of RNA synthesis. *Proc. Natl. Acad. Sci.* **51:** 786–794.

Allshire R.C., Javerzat J.P., Redhead N.J., and Cranston G. 1994. Position effect variegation at fission yeast centromeres. *Cell* **76:** 157–169.

Amedeo P., Habu Y., Afsar K., Scheid O.M., and Paszkowski J. 2000. Disruption of the plant gene *MOM* releases transcriptional silencing of methylated genes. *Nature* **405:** 203–206.

The *Arabidopsis* Genome Initiative. 2000. Analysis of the genome sequence of the flowering plant *Arabidopsis thaliana*. *Nature* **408:** 796–815.

Aravin A.A., Naumova N.M., Tulin A.V., Vagin V.V., Rozovsky Y.M., and Gvozdev V.A. 2001. Double-stranded RNA-mediated silencing of genomic tandem repeats and transposable elements in the *D. melanogaster* germline. *Curr. Biol.* **11:** 1017–1027.

Aravind L., H. Watanabe, Lipman D.J., and Koonin E.V. 2000. Lineage-specific loss and divergence of functionally linked genes in eukaryotes. *Proc. Natl. Acad. Sci.* **97:** 11319–11324.

Aufsatz W., Mette M.F., Van Der Winden J., Matzke A.J., and Matzke M. 2002. RNA-directed DNA methylation in *Arabidopsis. Proc. Natl. Acad. Sci.* **99:** 16499–16506.

Ayoub N., Goldshmidt I., Lyakhovetsky R., and Cohen A. 2000. A fission yeast repression element cooperates with centromere-like sequences and defines a *mat* silent domain boundary. *Genetics* **156:** 983–994.

Bannister A.J., Zegerman P., Partridge J.F., Miska E.A., Thomas J.O., Allshire R.C., and Kouzarides T. 2001. Selective recognition of methylated lysine 9 on histone H3 by the HP1 chromo domain. *Nature* **410:** 120–124.

Barkan A. and Martienssen R.A. 1991. Inactivation of maize transposon *Mu* suppresses a mutant phenotype by activating an outward-reading promoter near the end of *Mu1. Proc. Natl. Acad. Sci.* **88:** 3502–3506.

Bartee L., Malagnac F., and Bender J. 2001. *Arabidopsis* cmt3 chromomethylase mutations block non-CG methylation and silencing of an endogenous gene. *Genes Dev.* **15:** 1753–1758.

Baum M. and Clarke L. 2000. Fission yeast homologs of human CENP-B have redundant functions affecting cell growth and chromosome segregation. *Mol. Cell. Biol.* **20:** 2852–2864.

Baylin S.B., Esteller M., Rountree M.R., Bachman K.E., Schuebel K., and Herman J.G. 2001. Aberrant patterns of DNA methylation, chromatin formation and gene expression in cancer. *Hum. Mol. Genet.* **10:** 687–692 (review).

Bazett-Jones D.P., Cote J., Landel C.C., Peterson C.L., and Workman J.L. 1999. The SWI/SNF complex creates loop domains in DNA and polynucleosome arrays and can disrupt DNA-histone contacts within these domains. *Mol. Cell. Biol.* **19:** 1470–1478.

Bender J. 1998. Cytosine methylation of repeated sequences in eukaryotes: The role of DNA pairing. *Trends Biochem. Sci.* **23:** 252–256.

Bennetzen J.L. 1996. The Mutator transposable element system of maize. *Curr. Top. Microbiol. Immunol.* **204:** 195–229.

Bercury S.D., Panavas T., Irenze K., and Walker E.L. 2001. Molecular analysis of the Doppia transposable element of maize. *Plant Mol. Biol.* **47:** 341–351

Bernard P. and Allshire R. 2002. Centromeres become unstuck without heterochromatin. *Trends Cell. Biol.* **12:** 419–424.

Bernstein E., Caudy A.A., Hammond S.M., and Hannon G.J. 2001. Role for a bidentate ribonuclease in the initiation step of RNA interference. *Nature* **409:** 363–366.

Bestor T.H. 2000. The DNA methyltransferases of mammals. *Mol. Genet.* **9:** 2395–2402.

Braunstein M., Rose A.B., Holmes S.G., Allis C.D., and Broach J.R. 1993. Transcriptional silencing in yeast is associated with reduced nucleosome acetylation. *Genes Dev.* **7:** 592–604.

Brzeski J. and Jerzmanowski A. 2003. Deficient in DNA methylation 1 (DDM1) defines a novel family of chromatin-remodeling factors. *J. Biol. Chem.* **278:** 823–828.

Bryk M., Banerjee M., Murphy M., Knudsen K.E., Garfinkel D.J., and Curcio M.J. 1997. Transcriptional silencing of Ty1 elements in the RDN1 locus of yeast. *Genes Dev.* **11:** 255–269.

Cameron E.E., Bachman K.E., Myohanen S., Herman J.G., and Baylin S.B. 1999. Synergy of demethylation and histone deacetylase inhibition in the re-expression of genes silenced in cancer. *Nat. Genet.* **21:** 103–107.

Cao X. and Jacobsen S.E. 2002. Role of the *Arabidopsis* DRM methyltransferases in de novo DNA methylation and gene silencing. *Curr. Biol.* **12:** 1138–1144.

CSHWU Consortium (Cold Spring Harbor Laboratory, Washington University Genome Sequencing Center, and PE Biosystems Arabidopsis Sequencing Consortium). 2000. The complete sequence of a heterochromatic island from a higher eukaryote. *Cell* **100:** 377–386.

Colot V., Maloisel L., and Rossignol J.L. 1996. Interchromosomal transfer of epigenetic states in *Ascobolus:* Transfer of DNA methylation is mechanistically related to homologous recombination. *Cell* **86:** 855–864.

Copenhaver G.P., Nickel K., Kuromori T., Benito M.I., Kaul S., Lin X., Bevan M., Murphy G., Harris B., Parnell L.D., McCombie W.R., Martienssen R.A., Marra M., and Preuss D. 1999. Genetic definition and sequence analysis of *Arabidopsis* centromeres. *Science* **286:** 2468–2474.

Corona D.F., Langst G., Clapier C.R., Bonte E.J., Ferrari S., Tamkun J.W., and Becker P.B. 1999.

ISWI is an ATP-dependent nucleosome remodeling factor. *Mol. Cell* **3:** 239–245.

Cui H. and Fedoroff N.V. 2002. Inducible DNA demethylation mediated by the maize Suppressor-mutator transposon-encoded TnpA protein. *Plant Cell* **14:** 2883–2899.

Dalmay T., Hamilton A., Rudd S., Angell S., and Baulcombe D.C. 2000. An RNA-dependent RNA polymerase gene in *Arabidopsis* is required for posttranscriptional gene silencing mediated by a transgene but not by a virus. *Cell* **101:** 543–553.

Deuring R., Fanti L., Armstrong J.A., Sarte M., Papoulas O., Prestel M., Daubresse G., Verardo M., Moseley S.L., Berloco M., Tsukiyama T., Wu C., Pimpinelli S., and Tamkun J.W. 2000. The ISWI chromatin-remodeling protein is required for gene expression and the maintenance of higher order chromatin structure in vivo. *Mol. Cell* **5:** 355–365.

Djikeng A., Shi H., Tschudi C., and Ullu E. 2001. RNA interference in *Trypanosoma brucei:* Cloning of small interfering RNAs provides evidence for retroposon-derived 24-26-nucleotide RNAs. *RNA* **7:** 1522–1530.

Ekwall K., Cranston G., and Allshire R. 1999. Fission yeast mutants that alleviate transcriptional silencing in centromeric flanking repeats and disrupt chromosome segregation. *Genetics* **153:** 1153–1169.

Fagard M., Boutet S., Morel J., Bellini C., and Vaucheret H. 2000. AGO1, QDE-2, and RDE-1 are related proteins required for post-transcriptional gene silencing in plants, quelling in fungi, and RNA interference in animals. *Proc. Natl. Acad. Sci.* **97:** 11650–11654.

Feng Q. and Zhang Y. 2001. The MeCP1 complex represses transcription through preferential binding, remodeling, and deacetylating methylated nucleosomes. *Genes Dev.* **15:** 827–832.

Finnegan E. and Dennis E. 1993. Isolation and identification by sequence homology of a putative cytosine methyltransferase from *Arabidopsis thaliana. Nucleic Acids Res.* **21:** 2383–2388.

Fransz P.F., Armstrong S., de Jong J.H., Parnell L.D., van Drunen C., Dean C., Zabel P., Bisseling T., and Jones G.H. 2000. Integrated cytogenetic map of chromosome arm 4S of *A. thaliana:* Structural organization of heterochromatic knob and centromere region. *Cell* **100:** 367–376.

Fuks F., Burgers W.A., Brehm A., Hughes-Davies L., and Kouzarides T. 2000. DNA methyltransferase Dnmt1 associates with histone deacetylase activity. *Nat. Genet.* **24:** 88–91.

Gendrel A.-V., Lippman Z., Yordan C., Colot V., and Martienssen R. 2002. Heterochromatic histone H3 methylation patterns depend on the *Arabidopsis* gene *DDM1. Science* **297:** 1871–1873.

Gerasimova T.I. and Corces V.G. 2001. Chromatin insulators and boundaries: Effects on transcription and nuclear organization. *Annu. Rev. Genet.* **35:** 193–208.

Gibbons R.J., McDowell T.L., Raman S., O'Rourke D.M., Garrick D., Ayyub H., and Higgs D.R. 2000. Mutations in ATRX, encoding a SWI/SNF-like protein, cause diverse changes in the pattern of DNA methylation. *Nat. Genet.* **24:** 368–371.

Guy J., Hendrich B., Holmes M., Martin J.E., and Bird A. 2001. A mouse Mecp2-null mutation causes neurological symptoms that mimic Rett syndrome. *Nat. Genet.* **27:** 322–326.

Hall I.M., Shankaranarayana G.D., Noma K., Ayoub N., Cohen A., and Grewal S.I. 2002. Establishment and maintenance of a heterochromatin domain. *Science* **297:** 2232–2237.

Hamilton A.J. and Baulcombe D.C. 1999. A species of small antisense RNA in posttranscriptional gene silencing in plants. *Science* **286:** 950–952.

Hamilton A., Voinnet O., Chappell L., and Baulcombe D. 2002. Two classes of short interfering RNA in RNA silencing. *EMBO J.* **21:** 4671–4679.

Hammond S.M., Bernstein E., Beach D., and Hannon G.J. 2000. An RNA-directed nuclease mediates post-transcriptional gene silencing in *Drosophila* cells. *Nature* **404:** 293–296.

Happel A.M., Swanson M.S., and Winston F. 1991. The *SNF2, SNF5* and *SNF6* genes are required for *Ty* transcription in *Saccharomyces cerevisiae. Genetics* **128:** 69–77.

Heard E., Rougeulle C., Arnaud D., Avner P., Allis C.D., and Spector D.L. 2001. Methylation of histone H3 at Lys-9 is an early mark on the X chromosome during X inactivation. *Cell* **107:** 727–738.

Henikoff S. and Comai L. 1998. A DNA methyltransferase homolog with a chromodomain exists in multiple polymorphic forms in *Arabidopsis. Genetics* **149:** 307–318.

Heslop-Harrison J.S. 2000. Comparative genome organization in plants: From sequence and markers to chromatin and chromosomes. *Plant Cell* **12:** 617–636.

Hirochika H., Okamoto H., and Kakutani T. 2000. Silencing of retrotransposons in *Arabidopsis* and reactivation by the *ddm1* mutation. *Plant Cell* **12:** 357–369.

Hirschhorn J.N., Brown S.A., Clark C.D., and Winston F. 1992. Evidence that SNF2/SWI2 and SNF5 activate transcription in yeast by altering chromatin structure. *Genes Dev.* **6:** 2288–2298.

Holliday R. 1996. DNA methylation in eukaryotes: 20 years on. In *Epigenetic mechanisms of gene regulation* (ed. V.E.A. Russo et al.), pp. 5–29. Cold Spring Harbor Laboratory Press, Cold Spring Harbor, New York.

Jackson J.P., Lindroth A.M., Cao X., and Jacobsen S.E. 2002. Control of CpNpG DNA methylation by the KRYPTONITE histone H3 methyltransferase. *Nature* **416:** 556–560.

Jacobsen S.E. and Meyerowitz E.M. 1997. Hypermethylated SUPERMAN epigenetic alleles in *Arabidopsis. Science* **277:** 1100–1103.

Jae Yoo E., Kyu Jang Y., Ae Lee M., Bjerling P., Bum Kim J., Ekwall K., Hyun Seong R., and Dai Park S. 2002. Hrp3, a chromodomain helicase/ATPase DNA binding protein, is required for heterochromatin silencing in fission yeast. *Biochem. Biophys. Res. Commun.* **295:** 970–974.

Jeddeloh J.A., Bender J., and Richards E.J. 1998. The DNA methylation locus *DDM1* is required for maintenance of gene silencing in *Arabidopsis. Genes Dev.* **12:** 1714–1725.

Jeddeloh J.A., Stokes T.L., and Richards E.J. 1999. Maintenance of genomic methylation requires a SWI2/SNF2-like protein. *Nat. Genet.* **22:** 94–97.

Jenuwein T. and Allis C.D. 2001. Translating the histone code. *Science* **293:** 1074–1080 (review).

Johnson L., Cao X., and Jacobsen S. 2002. Interplay between two epigenetic marks. DNA methylation and histone H3 lysine 9 methylation. *Curr. Biol.* **12:** 1360–1367.

Jones P.A. and Baylin S.B. 2002. The fundamental role of epigenetic events in cancer. *Nat. Rev. Genet.* **3:** 415–428.

Jones P.A. and Takai D. 2001. The role of DNA methylation in mammalian epigenetics. *Science* **293:** 1068–1070.

Jones L., Ratcliff F., and Baulcombe D.C. 2001. RNA-directed transcriptional gene silencing in plants can be inherited independently of the RNA trigger and requires Met1 for maintenance. *Curr. Biol.* **11:** 747–757.

Jones L., Hamilton A.J., Voinnet O., Thomas C.L., Maule A.J., and Baulcombe D.C. 1999. RNA-DNA interactions and DNA methylation in post-transcriptional gene silencing. *Plant Cell* **11:** 2291–2301.

Jorgensen R. 1990. Altered gene expression in plants due to trans interactions between homologous genes. *Trends Biotechnol.* **8:** 340–344.

Kankel M.W., Ramsey D.E., Stokes T.L., Flowers S.K., Haag J.R., Jeddeloh J.A., Riddle N.C., Verbsky M.L., and Richards E.J. 2003. *Arabidopsis* MET1 cytosine methyltransferase mutants. *Genetics* **163:** 1109–1122.

Kapitonov V.V. and Jurka J. 2001. Rolling-circle transposons in eukaryotes. *Proc. Natl. Acad. Sci.* **98:** 8714–8719.

Kouzminova E. and Selker E.U. 2001. *dim-2* encodes a DNA methyltransferase responsible for all known cytosine methylation in *Neurospora. EMBO J.* **20:** 4309–4323.

Lagos-Quintana M., Rauhut R., Lendeckel W., and Tuschl T. 2001. Identification of novel genes coding for small expressed RNAs. *Science* **294:** 853–858.

Le Q.H., Wright S., Yu Z., and Bureau T. 2000. Transposon diversity in *Arabidopsis thaliana. Proc. Natl. Acad. Sci.* **97:** 7376–7381.

Lee J.-K., Huberman J.A., and Hurwitz J. 1997. Purification and characterization of a CENP-B homologue protein that binds to the centromeric K-type repeat DNA of *Schizosaccharomyces pombe. Proc. Natl. Acad. Sci.* **94:** 8427–8432.

Lenoir A., Lavie L., Prieto J.L., Goubely C., Cote J.C., Pelissier T., and Deragon J.M. 2001. The evolutionary origin and genomic organization of SINEs in *Arabidopsis thaliana. Mol. Biol. Evol.* **18:** 2315–2322

Lipardi C., Wei Q., and Paterson B.M. 2001. RNAi as random degradative PCR: siRNA primers convert mRNA into dsRNAs that are degraded to generate new siRNAs. *Cell* **107:** 297–307.

Lindroth A.M., Cao X., Jackson J.P., Zilberman D., McCallum C.M., Henikoff S., and Jacobsen S.E. 2001. Requirement of CHROMOMETHYLASE3 for maintenance of CpXpG methylation. *Science* **292:** 2077–2080.

Litt M.D., Simpson M., Gaszner M., Allis C.D., and Felsenfeld G. 2001. Correlation between histone lysine methylation and developmental changes at the chicken beta-globin locus. *Science* **293:** 2453–2455.

Llave C., Kasschau K.D., Rector M.A., and Carrington J.C. 2002. Endogenous and silencing-associated small RNAs in plants. *Plant Cell* **14:** 1605–1619.

Lynn K., Fernandez A., Aida M., Sedbrook J., Tasaka M., Masson P., and Barton M.K. 1999. The *PINHEAD/ZWILLE* gene acts pleiotropically in *Arabidopsis* development and has overlapping functions with the *ARGONAUTE1* gene. *Development* **126:** 469–481.

Malagnac F., Bartee L., and Bender J. 2002. An *Arabidopsis* SET domain protein required for maintenance but not establishment of DNA methylation. *EMBO J.* **21:** 6842–6852.

Mann J.R. 2001. Imprinting in the germ line. *Stem Cells* **19:** 287–294.

Martienssen R. 1998a. Chromosomal imprinting in plants. *Curr. Opin. Genet. Dev.* **8:** 240–244.

———. 1998b. Transposons, DNA methylation and gene control. *Trends Genet.* **14:** 263–264.

Martienssen R. and Colot V. 2001. DNA methylation and epigenetic inheritance in plants and filamentous fungi. *Science* **293:** 1070–1074.

Martienssen R.A. and Richards E.J. 1995. DNA methylation in eukaryotes. *Curr. Opin. Genet. Dev.* **5:** 234–242.

Martienssen R., Barkan A., Taylor W.C., and Freeling M. 1990. Somatically heritable switches in the DNA modification of *Mu* transposable elements monitored with a suppressible mutant in maize. *Genes Dev.* **4:** 331–343.

Matzke M., Matzke A.J., and Kooter J.M. 2001. RNA: Guiding gene silencing. *Science* **293:** 1080–1083.

McClintock B. 1951. Chromosome organization and genic expression. *Cold Spring Harbor Symp. Quant. Biol.* **16:** 13–47.

———. 1954. Mutations in maize and chromosomal aberrations in *Neurospora*. *Carnegie Inst. Wash. Year Book* **53:** 254–260.

———. 1984. The significance of responses of the genome to challenge. *Science* **226:** 792–801.

Mette M.F., Van Der Winden J., Matzke M., and Matzke A.J. 2002. Short RNAs can identify new candidate transposable element families in *Arabidopsis*. *Plant Physiol.* **130:** 6–9.

Miura A., Yonebayashi S., Watanabe K., Toyama T., Shimada H., and Kakutani T. 2001. Mobilization of transposons by a mutation abolishing full DNA methylation in *Arabidopsis*. *Nature* **411:** 212–214.

Murfett J., Wang X.J., Hagen G., and Guilfoyle T.J. 2001. Identification of *Arabidopsis* histone deacetylase HDA6 mutants that affect transgene expression. *Plant Cell* **13:** 1047–1061.

Nakagawa H., Lee J.K., Hurwitz J., Allshire R.C., Nakayama J., Grewal S.I., Tanaka K., and Murakami Y. 2002. Fission yeast CENP-B homologs nucleate centromeric heterochromatin by promoting heterochromatin-specific histone tail modifications. *Genes Dev.* **16:** 1766–1778.

Nakayama J., Rice J.C., Strahl B.D., Allis C.D., and Grewal S.I. 2001. Role of histone H3 lysine 9 methylation in epigenetic control of heterochromatin assembly. *Science* **292:** 110–113.

Nan X., Ng H.H., Johnson C.A., Laherty C.D., Turner B.M., Eisenman R.N., and Bird A. 1998. Transcriptional repression by the methyl-CpG-binding protein MeCP2 involves a histone deacetylase complex. *Nature* **393:** 386–389.

Ng H.H., Zhang Y., Hendrich B., Johnson C.A., Turner B.M., Erdjument-Bromage H., Tempst P., Reinberg D., and Bird A. 1999. MBD2 is a transcriptional repressor belonging to the MeCP1 histone deacetylase complex. *Nat. Genet.* **23:** 58–61.

Noma K., Allis C.D., and Grewal S.I. 2001. Transitions in distinct histone H3 methylation patterns at the heterochromatin domain boundaries. *Science* **293:** 1150–1155.

Ogas J., Kaufmann S., Henderson J., and Somerville C. 1999. PICKLE is a CHD3 chromatin-remodeling factor that regulates the transition from embryonic to vegetative in *Arabidopsis*. *Proc. Natl. Acad. Sci.* **96:** 13839–13844.

Parinov S., Sevugan M., De Y., Yang W.C., Kumaran M., and Sundaresan V. 1999. Analysis of flanking sequences from dissociation insertion lines: A database for reverse genetics in *Arabidopsis*. *Plant Cell* **11:** 2263–2270.

Partridge J., Scott K., Bannister A., Kouzarides T., and Allshire R. 2002. cis-Acting DNA from fission yeast centromeres mediates histone H3 methylation and recruitment of silencing factors and cohesin to an ectopic site. *Curr. Biol.* **12:** 1652–1660.

Peters A.H., Mermoud J.E., O'Carroll D., Pagani M., Schweizer D., Brockdorff N., and Jenuwein T. 2001. Histone H3 lysine 9 methylation is an epigenetic imprint of facultative heterochromatin. *Nat. Genet.* **30:** 77–80.

Plasterk R.H. 2002. RNA silencing: The genome's immune system. *Science* **296:** 1263–1265.

Plasterk R.H., Izsvak Z., and Ivics Z. 1999. Resident aliens: The Tc1/mariner superfamily of transposable elements. *Trends Genet.* **15:** 326–332.

Rabinowicz P.D., Schutz K., Dedhia N., Yordan C., Parnell L.D., Stein L., McCombie W.R., and Martienssen R.A. 1999. Differential methylation of genes and retrotransposons facilitates shotgun sequencing of the maize genome. *Nat. Genet.* **23:** 305–308.

Rea S., Eisenhaber F., O'Carroll D., Strahl B.D., Sun Z.W., Schmid M., Opravil S., Mechtler K., Ponting C.P., Allis C.D., and Jenuwein T. 2000. Regulation of chromatin structure by site-specific histone H3 methyltransferases. *Nature* **406**: 593–599.

Reinhart B.J. and Bartel D.P. 2002. Small RNAs correspond to centromere heterochromatic repeats. *Science* **297**: 1831.

Robertson K.D., Ait-Si-Ali S., Yokochi T., Wade P.A., Jones P.L., and Wolffe A.P. 2000. DNMT1 forms a complex with Rb, E2F1 and HDAC1 and represses transcription from E2F-responsive promoters. *Nat. Genet.* **25**: 338–342.

Ronemus M.J., Galbiati M., Ticknor C., Chen J., and Dellaporta S.L. 1996. Demethylation-induced developmental pleiotropy in *Arabidopsis*. *Science* **273**: 654–657.

Round E.K., Flowers S.K., and Richards E.J. 1997. *Arabidopsis thaliana* centromere regions: Genetic map positions and repetitive DNA structure. *Genome Res.* **7**: 1045–1053.

Rountree M.R., Bachman K.E., and Baylin S.B. 2000. DNMT1 binds HDAC2 and a new co-repressor, DMAP1, to form a complex at replication foci. *Nat. Genet.* **25**: 269–277.

Rudert F., Bronner S., Garnier J.M., and Dolle P. 1995. Transcripts from opposite strands of gamma satellite DNA are differentially expressed during mouse development. *Mamm. Genome* **6**: 76–83.

Rundlett S.E., Carmen A.A., Kobayashi R., Bavykin S., Turner B.M., and Grunstein M. 1996. HDA1 and RPD3 are members of distinct yeast histone deacetylase complexes that regulate silencing and transcription. *Proc. Natl. Acad. Sci.* **93**: 14503–14508.

Scheid O.M., Probst A.V., Afsar K., and Paszkowski J. 2002. Two regulatory levels of transcriptional gene silencing in *Arabidopsis*. *Proc. Natl. Acad. Sci.* **21**: 13659–13662.

Sijen T., Fleenor J., Simmer F., Thijssen K.L., Parrish S., Timmons L., Plasterk R.H., and Fire A. 2001. On the role of RNA amplification in dsRNA-triggered gene silencing. *Cell* **107**: 465–476.

Singer T., Yordan C., and Martienssen R.A. 2001. Robertson's *Mutator* transposons in *Arabidopsis thaliana* are regulated by DNA methylation and chromatin remodeling. *Genes Dev.* **15**: 591–602.

Strahl B.D., Ohba R., Cook R.G., and Allis C.D. 1999. Methylation of histone H3 at lysine 4 is highly conserved and correlates with transcriptionally active nuclei in *Tetrahymena*. *Proc. Natl. Acad. Sci.* **96**: 14967–14972.

Sudarsanam P. and Winston F. 2000. The Swi/Snf family: Nucleosome-remodeling complexes and transcriptional control. *Trends Genet.* **16**: 345–351.

Tamaru H. and Selker E.U. 2001. A histone H3 methyltransferase controls DNA methylation in *Neurospora crassa*. *Nature* **414**: 277–283.

Tauton J., Hassig C.A., and Schreiber S.L. 1996. A mammalian histone deacetylase related to the yeast transcriptional regulator Rpd3p. *Science* **272**: 408–411.

Tompa R., McCallum C.M., Delrow J., Henikoff J.G., van Steensel B., and Henikoff S. 2002. Genome-wide profiling of DNA methylation reveals transposon targets of CHROMOMETHYLASE3. *Curr. Biol.* **12**: 65–68.

Tuschl T., Zamore P.D., Lehmann R., Bartel D.P., and Sharp P.A. 1999. Targeted mRNA degradation by double-stranded RNA in vitro. *Genes Dev.* **13**: 3191–3197.

Vaistij F.E., Jones L., and Baulcombe D.C. 2002. Spreading of RNA targeting and DNA methylation in RNA silencing requires transcription of the target gene and a putative RNA-dependent RNA polymerase. *Plant Cell* **14**: 857–867.

Verbsky M. and Richards E. 2001. Chromatin remodeling in plants. *Curr. Opin. Plant Biol.* **4**: 494–500.

Vidal M. and Gaber R.F. 1991. RPD3 encodes a second factor required to achieve maximum positive and negative transcriptional states in *Saccharomyces cerevisiae*. *Mol. Cell. Biol.* **11**: 6317–6327.

Volpe T., Kidner C., Hall I., Teng G., Grewal S., and Martienssen R. 2002. Regulation of heterochromatic silencing and histone H3 lysine 9 methylation by RNA interference. *Science* **297**: 1833–1837.

Volpe T., Schramke V., Hamilton G., White S., Teng G., Martienssen R., and Allshire R.C. 2003. RNA interference is required for normal centromere function in fission yeast. *Chromosome Res.* **11**: 137–146.

Vongs A., Kakutani T., Martienssen R.A., and Richards E.J. 1993. *Arabidopsis thaliana* DNA methylation mutants. *Science* **260**: 1926–1928.

Wassenegger M., Heimes S., Riedel L., and Sanger H.L. 1994. RNA-directed de novo methylation of genomic sequences in plants. *Cell* **76**: 567–576.

Witte C.P., Le Q.H., Bureau T., and Kumar A. 2001. Terminal-repeat retrotransposons in miniature (TRIM) are involved in restructuring plant genomes. *Proc. Natl. Acad. Sci.* **98**: 13778–13783.

Wolffe A.P. and Matzke M.A. 1999. Epigenetics: Regulation through repression. *Science* **286:** 481–486.

Wood V., Gwilliam R., Rajandream M.A., Lyne M., Lyne R., Stewart A., et al. 2002. The genome sequence of *Schizosaccharomyces pombe. Nature* **415:** 871–880.

Wright D.A. and Voytas D.F. 2002. Athila4 of *Arabidopsis* and Calypso of soybean define a lineage of endogenous plant retroviruses. *Genome Res.* **12:** 122–131.

Wu K., Tian L., Malik K., Brown D., and Miki B. 2000. Functional analysis of HD2 histone deacetylase homologues in *Arabidopsis thaliana. Plant J.* **22:** 19–27.

Wu-Scharf D., Jeong B., Zhang C., and Cerutti H. 2002. Transgene and transposon silencing in *Chlamydomonas reinhardtii* by a DEAH-box RNA helicase. *Science* **290:** 1159–1162.

Zhang Y., LeRoy G., Seelig H.P., Lane W.S., and Reinberg D. 1998. The dermatomyositis-specific autoantigen Mi2 is a component of a complex containing histone deacetylase and nucleosome remodeling activities. *Cell* **95:** 279–289.

CHAPTER 8

# The Ribonuclease III Superfamily: Forms and Functions in RNA Maturation, Decay, and Gene Silencing

Allen W. Nicholson

*Department of Chemistry, Temple University, Philadelphia, Pennsylvania 19122*

THE EMERGENCE OF RNA INTERFERENCE (RNAI) as a central eukaryotic posttranscriptional regulatory mechanism has sparked renewed interest in the nucleases that recognize and cleave double-stranded RNA (dsRNA). Given the diverse functional roles of RNAi, its associated enzymatic activities, and anticipated applications in biotechnology and biomedicine, it is of interest to revisit early studies which led to the discovery of the archetypal dsRNA-specific enzyme, ribonuclease III (RNase III). The characterization of an RNA bacteriophage (Loeb and Zinder 1961) launched pioneering studies on the cellular machinery responsible for the replication, packaging, and propagation of an RNA molecule. Infection time-course experiments using f2 bacteriophage and male *Escherichia coli* cells demonstrated the synthesis of an RNA species complementary to the infecting (genomic) RNA, whereas other experiments ruled out DNA replicative intermediates (for review, see Zinder 1980). An RNA replicase was characterized, as well as a putative replicative form (RF) of the phage RNA, which, as obtained by phenol extraction, exhibited the properties of a dsRNA (Ammann et al. 1964; Nonoyama and Ikeda 1964; Billeter et al. 1966a).

An f2 mutant was characterized that failed to synthesize functional capsid protein. This deficit caused deregulation of replicase synthesis with concomitant overproduction of the antigenomic RNA. The mutant phage-infected cells contained dsRNA species that were significantly shorter than the full-length RF (Billeter et al. 1966b; Lodish and Zinder 1966a,b), and it was speculated that the truncated dsRNAs were created by a host-cell nuclease (Zinder 1980). In fact, attempts to reconstruct the RNA replication process in *E. coli* cell-free extracts revealed an activity that could efficiently cleave phage RF as well as other added dsRNAs. These observations led to the characterization of RNase III (Robertson et al. 1967, 1968; Libonati 1968a,b).

The original preparations of RNase III were impure, and several chromatographic steps were needed to remove other activities such as RNase H1 (Crouch 1974; Robertson and Dunn 1975). Purified *E. coli* RNase III exhibits strict dsRNA specificity and is a divalent metal ion-dependent phosphodiesterase (Dunn 1976). The isolation of a point mutation in the chromosomal gene (*rnc105*) that abolished processing activity (Kindler et al.

1973) revealed that RNase III is not essential and also allowed identification of two roles for RNase III: (1) the processing of the ribosomal RNA precursor and (2) the maturation of phage T7 mRNAs (Dunn and Studier 1973; Nikolaev and Schlessinger 1973).

On the basis of these observations, it was speculated that RNase III could have an antiviral role by cleaving dsRNA phage genomes or the RF of single-stranded RNA (ssRNA) phage. However, the dsRNA phage chromosomes are kept in a sequestered intracellular environment, and for ssRNA phage, it was demonstrated that the putative RF instead was an artifact of phenol extraction, which is known to accelerate nucleic acid strand reassociation. Thus, under these conditions, the complementary ssRNAs "collapse" to form double-stranded structures sensitive to RNase III. The biologically relevant template (replicative intermediate, or RI) turned out to be an antigenomic or genomic RNA strand that carries multiple nascent strands generated by bound replicases. The nascent strands can efficiently form intramolecular structures that are resistant to RNase III. In this regard, the impact of RNase III on RNA phage replication was revealed by deliberate introduction of an RNase III cleavage site into a noncoding region of MS2 phage, which suppressed replication. Moreover, the small number of survivor genomes exhibited small insertions, deletions, or base-pair mismatches near the cleavage site that could block RNase III action (Klovins et al. 1997).

In retrospect, these original studies portended a cellular mechanism such as RNAi, and the original notion that bacterial RNase III might act as an "RNA restriction enzyme" (Robertson 1982) has come full circle with the central involvement of Dicer in RNAi. Indeed, long dsRNAs can be processed in vitro by *E. coli* RNase III, creating short duplexes that can function as short interfering RNAs (siRNAs) in transfected mammalian cells (Yang et al. 2002). This chapter discusses the structural forms, functional roles, and mechanistic features of RNase III superfamily members. The nature and diversity of RNase III substrate reactivity epitopes are presented, especially with respect to the role of substrate structure in establishing the pattern of cleavage, which, in turn, can control gene expression. Emerging features of the RNase III catalytic mechanism are presented against a backdrop of new structural information.

## OVERVIEW OF THE RNase III SUPERFAMILY

RNase III was regarded for a time as unique to the bacteria. Although dsRNA-cleaving activities had been detected in a variety of eukaryotic cells, their relationship to bacterial RNase III was not known (Nicholson 1996). For example, specific members of the mechanistically distinct ribonuclease A family also can cleave dsRNA (Libonati and Sorrentino 1992; Sorrentino and Libonati 1997). Biochemical-genetic studies on yeast RNA-processing reactions and genome-sequencing projects enabled the identification of RNase III orthologs in fungi, plants, and animals, ultimately leading to the establishment of the RNase III superfamily (Rotondo and Frendewey 1996; Mian 1997; Aravind and Koonin 2001). Figure 8.1 displays the general forms of RNase III superfamily members. The two distinguishing functional domains are the dsRNA-binding motif (dsRBM) and the catalytic (nuclease) domain. These domains are vital functional features and, as discussed below, cooperatively function to carry out dsRNA cleavage.

A classification scheme for RNase III orthologs has been proposed recently (Blasczyk et al. 2001). The Class 1 enzymes, including bacterial orthologs and the *Saccharomyces cerevisiae* ortholog Rnt1p, exhibit the simplest structure, with a single dsRBM and catalytic domain. The amino-terminal extension of Rnt1p has been shown to be important for dimer stability and optimal processing activity (Lamontagne et al. 2000). The Class 2 and 3 RNase III orthologs occur in eukaryotic organisms and exhibit larger sizes and additional domain elements (Figure 8.1). The Class 2 enzymes display a single dsRBM, but

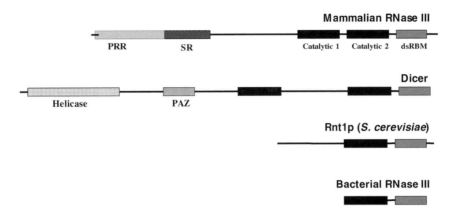

**FIGURE 8.1.** Domain structures of RNase III superfamily members. The positions of specific domains are indicated by the textured rectangles. (PRR) Proline-rich region; (SR) serine-arginine-rich region; (PAZ) "Piwi-Argo-Zwille/Pinhead" domain (Cerutti et al. 2000).

they have duplicated catalytic domain elements. The amino-terminal region exhibits variability in length and the type of conserved motifs. The Class 3 enzymes include Dicer (see also below) and exhibit a tandem duplication of the catalytic domain, with an amino-terminal RNA helicase domain (Figure 8.1) (see also Nicholson and Nicholson 2002).

## Bacterial RNase III Functions

All characterized bacterial genomes, even the genome of the "minimalist" organism *Mycoplasma genitalium* (Fraser et al. 1995), contain a gene for RNase III. This conservation implies an involvement in one or more cellular functions of fundamental importance. The most studied ortholog is *E. coli* RNase III, the primary subject of this discussion. A role for *E. coli* RNase III as a global regulator of gene expression has been discussed (Wilson et al. 2002) and is also indicated by the RNase III cleavage reactivity of transcripts of an abundant class of repetitive sequence elements in *Neisseria* sp. that flank expressed gened (De Gregorio et al. 2002). For more comprehensive reviews on *E. coli* RNase III and its substrates, see Dunn (1982), Robertson (1982), Court (1993), and Nicholson (1999). Protocols for the purification of *E. coli* RNase III and assays for substrate binding and cleavage are presented elsewhere (Amarasinghe et al. 2001).

### Maturation of rRNA and tRNA

*E. coli* RNase III cleaves the approximately 5500-nucleotide primary transcripts of the seven rRNA operons to provide the immediate precursors to the 16S, 23S, and 5S rRNAs (for review, see Srivastava and Schlessinger 1990). The full-length RNA is not normally observed because cleavage occurs during transcription. The RNase III cleavage sites are formed by the pairing of complementary sequences that flank the 5′ and 3′ ends of the 16S and 23S rRNAs (Figure 8.2A). RNase III also participates in the maturation of other structural RNAs including tRNAs. A bacteriophage T4 transcript encoding eight tRNAs is cleaved by RNase III immediately upstream of the first tRNA sequence (tRNA$^{Gln}$). This cleavage is required for correct maturation of the tRNA$^{Gln}$ species (Pragai and Apirion 1981; Gurewitz and Apirion 1983).

### Maturation of mRNA

*E. coli* RNase III site-specifically cleaves cell- and phage-encoded mRNA precursors to provide the mature fully functional species. RNase III cleaves at five intercistronic sites within the bacteriophage T7 polycistronic early mRNA precursor (Figure 8.2B), providing the

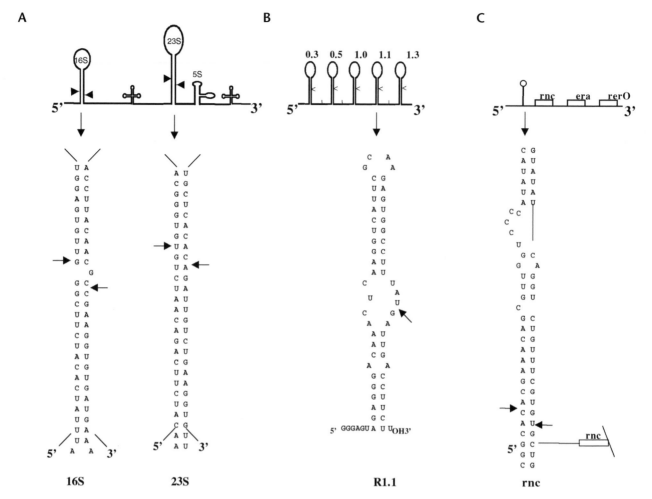

**FIGURE 8.2.** Substrates for *E. coli* RNase III. (*A*) RNase III processing sites in the *E. coli* 30S rRNA precursor. Arrows show additional cleavage sites. (*B*) Processing sites in the T7 phage polycistronic early mRNA precursor (~7100 nucleotides). The sequence displayed below corresponds to the R1.1 processing signal. (*C*) The processing site in the 5´ leader of the RNase III (*rnc*) mRNA. Only a portion of the hairpin-stem structure is shown.

mature species. RNase III action is important for optimal translation of several of these mRNAs, where cleavage creates 3´ stem-loop structures that confer protection against 3´-5´ exonuclease digestion (Dunn and Studier 1983). *E. coli* RNase III cleaves bacteriophage λ transcripts and controls the lysis-lysogeny decision (for review, see Court 1993). One regulatory component involves RNase III cleavage of the N mRNA 5´ leader, which is required for maximal synthesis of N protein. Cleavage relieves negative translational autocontrol by N, without interfering with the antiterminating activity of N on the downstream transcriptional complex (Wilson et al. 2001).

## Degradation of mRNA and antisense RNA regulation

*E. coli* RNase III action can initiate mRNA degradation. Site-specific cleavage within the 5´-untranslated region (5´UTR) of the polynucleotide phosphorylase (PNPase) mRNA promotes rapid subsequent degradation by other ribonucleases. PNPase also participates in this mechanism, thereby conferring autocontrol of production (Jarrige et al. 2001). RNase III also autoregulates its own production through site-specific cleavage within the 5´UTR of its mRNA, promoting rapid subsequent degradation (Figure 8.2C) (Bardwell et al. 1989; Matsunaga et al. 1996, 1997). Negative regulation of gene expression has also been shown to occur by RNase III cleavage within a coding sequence of a plasmid-encoded mRNA (Koraimann et al. 1993). RNase III participates in bacterial antisense RNA action. RNase III cleavage of sense-antisense RNA duplexes can negatively regulate bacterial plasmid copy number or inhibit protein production (for review, see Wagner and Simons 1994). Although in a number of instances the formation of a sense-antisense RNA complex is sufficient for inhibition, RNase III cleavage of the duplex confers irreversibility to the interaction.

## Functions of Other Bacterial RNase III Orthologs

Studies on other bacterial RNase III orthologs have revealed additional organism-specific processing pathways as well as indicating a common role in rRNA maturation. RNase III of the α-Proteobacteria group (which includes *Rhodobacter* and *Rhizobium* spp.) excises intervening sequences from the large-subunit rRNA to provide a fragmented, albeit functional, species (Evguenieva-Hackenberg and Klug 2000; Zahn et al. 2000). RNase III of *Streptomyces coelicolor* regulates antibiotic production (Aceti et al. 1998; Price et al. 1999). However, the RNA target(s) is unknown. *Bacillus subtilis* RNase III cleaves the precursor to the 4.5S RNA, which is a component of the signal recognition particle (Oguro et al. 1998). In contrast to the *E. coli* enzyme, *B. subtilis* RNase III is essential for cell viability (Herskovitz and Bechhofer 2000). As discussed below, the substrate reactivity epitopes are not conserved across bacterial species.

## Substrate Reactivity Epitopes of Bacterial RNase III Orthologs

The functional consequences of RNase III action depends on the precise manner in which the substrates are cleaved. The choice of cleavage site can dictate mRNA half-life and translational activity (see above). The specificity of cleavage of cellular substrates contrasts with the apparent random cleavage of long dsRNAs of complex or simple sequence. Reconciliation of these differing actions and determination of substrate reactivity epitopes have assumed increased importance with the ability of Dicer to site-specifically cleave small regulatory RNA precursors, as well as process virtually any dsRNA to produce precisely sized siRNA duplexes (see below).

*E. coli* RNase III substrates have been examined in some detail. The cleavage sites are contained within hairpin stem-loop structures of limited helical length (typically ≤20 bp) (e.g., see Figure 8.2). The small size necessarily limits the potential sites of cleavage. Although the standard reaction at a given target site in dsRNA involves cleavage of both strands, the presence of an internal loop can direct cleavage of a single phosphodiester (Chelladurai et al. 1993). This pattern of cleavage provides hairpins at RNA 3′ ends, which protect against 3′-5′ exonuclease action. This is seen with T7 bacteriophage mRNA precursors, which contain RNase-III-processing signals with internal loops and whose cleavage contributes to the high stability of the mature species (Dunn and Studier 1983). In addition, internal-loop-mediated single-strand cleavage may allow, in principle, a *trans*-acting antisense RNA to direct cleavage of the target RNA strand while preserving its own structure. This pattern would allow reuse of the antisense RNA. A recent study has shown that internal loop asymmetry and size, rather than specific sequence, are the primary determinants of single-site cleavage (I. Calin-Jageman and A.W. Nicholson, in prep.).

Watson-Crick (W-C) base-pair sequences also influence substrate reactivity. In particular, introduction of specific W-C base-pair sequences at defined positions relative to the cleavage site can block cleavage by inhibiting binding of RNase III (Figure 8.3A) (Zhang

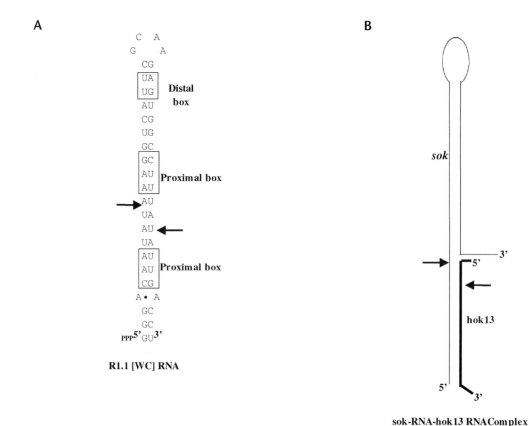

**A**

C   A
G     A
CG
UA
UG   **Distal**
AU   **box**
CG
UG
GC
GC
AU   **Proximal box**
AU
AU
UA
AU
UA
AU   **Proximal box**
AU
CG
A • A
GC
GC
ₚₚₚ5′GU3′

**R1.1 [WC] RNA**

**B**

*sok*

3′
5′

hok13

5′
3′

**sok-RNA-hok13 RNA Complex**

**FIGURE 8.3.** *E. coli* RNase III reactivity epitopes. (*A*) Structure of R1.1[WC] RNA, showing the regions sensitive to antideterminant W-C base-pair substitution (Zhang and Nicholson 1997). The proximal and distal box elements are indicated. Arrows indicate cleavage sites. (*B*) Coaxial stacking in the bacterial plasmid *hok* RNA–*sok* RNA interaction. Arrows indicate the RNase III cleavage sites formed by coaxial stacking of the two helices (Franch et al. 1999).

and Nicholson 1997). These W-C base-pair sequences have been termed "antideterminants"—analogous to antideterminant elements in tRNAs, which regulate tRNA-protein recognition (Rudinger et al. 1996)—and are proposed to participate in cleavage site determination. Thus, otherwise reactive phosphodiesters can be "masked" by antideterminant W-C base-pair sequences, with the scissile bond(s) identified, by default, through the absence of antideterminants at the relevant positions (Zhang and Nicholson 1997). As such, W-C base-pair antideterminants are not analogous to restriction enzyme sites, which provide positive signals for binding. The existence of W-C base-pair sequences does not compromise the ability of RNase III to degrade random sequence dsRNAs and may also serve to protect cellular dsRNAs with vital functions from unwanted cleavage (Zhang and Nicholson 1997).

The striking versatility of RNA in creating RNase III substrates is demonstrated by studies on the interaction of a bacterial plasmid antisense RNA with its target sequence. The plasmid-encoded *sok* antisense RNA binding to the target *hok* mRNA creates a short helix that is not recognized by RNase III. However, the helix can stack coaxially on an adjacent helix, forming a quasi-continuous helix which is recognized by RNase III (Figure 8.3B) (Franch et al. 1999). The cleavage is necessary to maintain the population of plasmid-containing cells. A similar coaxial stacking mechanism has been proposed for substrates of yeast RNase III (see below). In summary, RNA internal loops and coaxial stacking of helices reveal the flexibility of *E. coli* RNase III in recognizing diverse substrates. Although this flexibility underscores the importance of RNA structure in controlling cleavage, it also has been speculated that *E. coli* RNase III may regulate gene expression by binding to RNA without cleavage (Altuvia et al. 1987; Court 1993; Nicholson 1999). Important goals would be to identify whether such sequences exist, and, if so, how they would allow binding but prevent cleavage.

RNase III substrate reactivity epitopes are not strictly conserved among the bacteria. Thus, substrates of *B. subtilis* RNase III are not recognized in an equivalent manner by the *E. coli* enzyme (Mitra and Bechhofer 1994), and *Rhodobacter capsulatus* RNase III and *E. coli* RNase III do not recognize heterologous substrates (Conrad et al. 1998). One would anticipate that as bacterial species diverged, substrate reactivity epitopes would covary with enzyme structure, so as to maintain reactivity. It remains to be demonstrated to what extent internal loops, W-C base-pair sequence antideterminants, and coaxial stacking are recurrent recognition motifs for bacterial RNase III orthologs. As described below, substrates of eukaryotic RNase III orthologs recognize substrate features quite distinct from those seen in bacterial substrates.

## ON THE MECHANISM OF BACTERIAL RNase III

Given the conserved pairing of the dsRBM and catalytic domain among RNase III orthologs (Figure 8.1), it is expected that the mechanism of dsRNA binding and cleavage is also conserved, albeit with some variation, especially with respect to substrate recognition. Current knowledge of the mechanism of action of RNase III family members has benefited primarily from studies on the *E. coli* enzyme. Biochemical and genetic approaches have been employed to determine how the dsRBM and catalytic domain cooperate in the catalytic cycle. The structures of the dsRBM and catalytic domain of *E. coli* RNase III and *Aquifex aeolicus* RNase III, respectively, now provide fresh approaches to determining a conserved mechanism of dsRNA cleavage (Zamore 2001a).

### The Catalytic Domain: Functional and Structural Features

The *E. coli* RNase III catalytic domain requires the dsRBM to cleave the substrate in vitro under normal conditions (physiological salt concentrations and $Mg^{2+}$) (Sun et al. 2001). However, a truncated form of RNase III lacking the dsRBM is active when the salt concentration is lowered and $Mg^{2+}$ replaced by $Mn^{2+}$ (Sun et al. 2001). A similarly truncated form of *Rhodobacter* RNase III also can cleave substrate under similar conditions (Conrad et al. 2001). The low-salt enhancement of activity may be explained by a reduction in competitive ionic interactions, whereas the $Mn^{2+}$ requirement may reflect enhancement of otherwise suboptimal substrate-enzyme interactions at the active site. The *E. coli* RNase III catalytic domain is dimeric, retains strict specificity for dsRNA, and is inhibited by ethidium bromide, all of which are behaviors of the holoenzyme (Sun et al. 2001). The catalytic domain also is a key participant in substrate recognition, because the canonical cleavage sites are used by the truncated enzyme (Sun et al. 2001) (see also below).

The crystal structure of the catalytic domain of *A. aeolicus* (Aa) RNase III (Blaszczyk et al. 2001) rationalizes these biochemical properties and predicts specific features of the catalytic mechanism. The Aa-RNase III catalytic domain crystallizes as a dimer, with the subunit interface stabilized by extensive hydrophobic interactions, but relatively few hydrogen bonds or salt bridges. The extensive size of the subunit interface indicates that the catalytic domain is the main determinant of dimeric behavior of the holoenzye, previously shown by gel filtration and chemical cross-linking (Dunn 1976; March and Gonzalez 1990; Li et al. 1993).

The catalytic domain crystal structure also sheds light on a function of the approximately 11-amino-acid signature motif (HNERLE*F*LGDS) that distinguishes RNase III superfamily members. An intersubunit "ball-and-socket" motif is employed in which the side chain ("ball") of the phenylalanine in the signature sequence of one subunit (underlined in the sequence shown above) inserts into a "socket" of hydrophobic residues provided by the other subunit. The motif stabilizes, as well as positions, the two subunits relative to each other. The first characterized mutation of *E. coli* RNase III (*rnc105*) changes the glycine in the signature motif (G44) to an aspartic acid and abolishes all detectable catalytic activity (Kindler et al. 1973; Nashimoto and Uchida 1985). The inactivity may reflect a folding defect and/or disruption of the subunit interface. Attempts to overproduce the G44D RNase III mutant yielded insoluble protein (H. Li and A.W. Nicholson, unpubl.).

The Aa-RNase III catalytic domain structure also suggests how dsRNA specificity is established. The α3 helices of each subunit are aligned in an antiparallel fashion along the dimer interface, forming a 50- by 20-Å cleft (Blaszczyk et al. 2001). This cleft provides a properly sized binding site for a dsRNA (Figure 8.4). Specific side chains, perhaps also located in the α3 helix, may participate in determining whether both strands are ribose. It is also predicted that subunit association is required to create the dsRNA-binding site. Thus, an RNase III mutant unable to dimerize would be expected to be inactive for at least this reason.

## The dsRBM: Structural and Functional Features

A consideration of the structure of the dsRBM is necessary to understand its role in RNase III action. As mentioned above, the dsRBM is required for *E. coli* RNase III activity under standard assay conditions in vitro. The dsRBM may provide the necessary substrate-binding affinity under physiological conditions. Although the *E. coli* RNase III dsRBM is not strictly responsible for cleavage site selection or for conferring dsRNA specificity, it is possible that it confers an additional level of substrate specificity and possible regulation of activity (see below). The dsRBM is seen in many other dsRNA-binding proteins with diverse functions (St. Johnston et al. 1992; Fierro-Monti and Mathews 2000). The solution structure of the *E. coli* RNase III dsRBM as determined by nuclear magnetic reso-

A          B

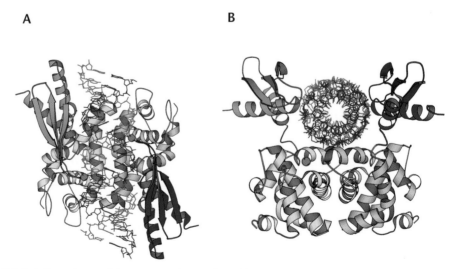

**FIGURE 8.4.** Two views of the proposed complex of bacterial RNase III bound to a dsRNA. The catalytic domains are shown in yellow and light blue, and the dsRBMs are shown in green and purple. (Reprinted, with permission, from Blaszczyk et al. 2001; ©Elsevier Science.)

nance (NMR) (Kharrat et al. 1995) exhibits an αβββα supersecondary structure, with the two α helices positioned on the same face of a three-stranded, antiparallel β sheet. This fold is maintained among all characterized dsRBMs and, as discussed below, recognizes only dsRNA.

A crystallographic analysis of a protein-RNA cocrystal provided atomic-level information on the dsRBM-dsRNA interaction (Ryter and Schultz 1998). The dsRBM was from the *Xenopus* RNA-binding protein XlrbpA and was crystallized in the presence of a self-complementary, ten-nucleotide RNA. In the unit cell, two duplexes coaxially stack to form a quasi-continuous helix. The dsRBM-dsRNA interaction spans 16 bp, and dsRBM contacts are made on one face of the helix, involving two minor grooves and the interposed major groove. Three specific regions of the dsRBM that contact RNA include the α1 helix (region 1), the loop connecting the β1 and β2 strands (region 2), and the amino terminus of α2 (region 3). The majority of the protein-RNA contacts are hydrogen bonds—either direct or water-mediated—many of which involve ribose 2′-OH groups on both RNA strands. Ribose recognition of both strands provides a basis for the dsRNA specificity (Ryter and Schultz 1998), and the preponderance of hydrogen bonds explains the relative salt insensitivity of the dsRBM-dsRNA interaction (Bevilacqua and Cech 1996). Finally, the paucity of observed dsRBM-nucleic acid base interactions is consistent with the lack of demonstrable base-pair sequence specificity for all dsRBMs studied to date.

The interactions seen in the cocrystal are anticipated to be generally similar to the RNase III dsRBM-dsRNA complex, and no major conformational changes in the dsRBM are expected upon binding to dsRNA. Using the cocrystal structural data, a model of an RNase III holoenzyme-dsRNA complex was developed using molecular modeling techniques, wherein the dsRBM was "joined" to the Aa-RNase III catalytic domain, which was also bound to dsRNA (Figure 8.4) (Blaszczyk et al. 2001). The holoenzyme engages a 23-bp duplex, in which the two composite active sites (see below) can engage in separate cleavage reactions to release a 9-bp duplex with two-nucleotide 3′ overhangs (Blaszczyk et al. 2001). The complex exhibits a twofold symmetry, which is consistent with footprinting and interference analyses (Li and Nicholson 1996). Random binding of holoenzyme to long dsRNAs is expected to provide a range of product sizes (~12–15 bp) under exhaustive digestion conditions, which is the experimental observation (Robertson and Dunn 1975).

## On the Active Site Structure and Divalent Metal Ion Involvement

*E. coli* RNase III is a phosphodiesterase and employs water as a nucleophile to create 5′-phosphate, 3′-hydroxyl product termini. That the chemical step is an irreversible hydrolytic reaction was revealed by the observation that only a single water oxygen is incorporated into the product phosphomonoester (Campbell et al. 2002). The 2′-OH group adjacent to the scissile phosphodiester is not required for cleavage, as 2′-deoxyribose- or 2′-fluororibose-containing substrates are reactive (Nicholson 1992; A. Harmouch and A.W. Nicholson, unpubl.). RNase III can also cleave a phosphorothiodiester linkage of the Rp configuration (Nicholson et al. 1988). The reactivity of the Sp isomer is not known.

*E. coli* RNase III phosphodiesterase activity requires a divalent metal ion, with $Mg^{2+}$ as the preferred species. However, $Mn^{2+}$, $Co^{2+}$, and $Ni^{2+}$ also support catalysis (Dunn 1982; Li et al. 1993; Amarasinghe et al. 2001; Campbell et al. 2002). The ability of different divalent metal ions to support catalysis was exploited in a kinetic study showing that the rate of the chemical step was dependent on the pKa of the hydrated metal ion (Campbell et al. 2002). This dependence indicates direct participation of one or more metal ions in catalysis, perhaps in generating the nucleophilic hydroxide anion, which may be a metal-bound species (Campbell et al. 2002).

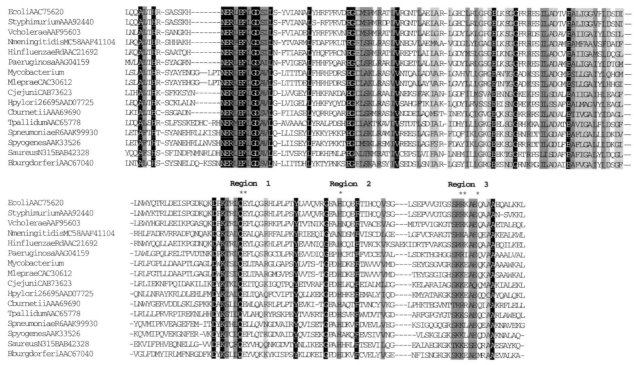

**FIGURE 8.5.** Sequence alignment of bacterial RNase III orthologs. Alignment was carried out by the CLUSTALW program. Regions 1, 2, and 3 refer to sites of dsRBM interaction with dsRNA, as shown by Ryter and Schultz (1998).

The structure of the Aa-RNase III catalytic domain has provided a first look at the structure of the active site and how the divalent metal ion, as well as conserved residues of the signature sequence, may function in catalysis (Blaszczyk et al. 2001). The catalytic domain dimer contains two clusters of six acidic side chains, symmetrically positioned at each end of the intersubunit cleft. The acidic residues are highly conserved (Figure 8.5), and several of them interact with the bound metal ($Mg^{2+}$ or $Mn^{2+}$) (Figure 8.6). The Aa-RNase III acidic cluster includes E37, E40, and D44, all of which are contained within the signature motif, and E64, D107, and E110. E37 forms a hydrogen bond with the amide of E64 of the other subunit and may not directly participate in catalysis. The E64 side chain spans the interface and is a defining feature of the composite nature of the proposed active site (Figure 8.6) (Blaszczyk et al. 2001).

The divalent metal ion interacts with the E40, D107, and E110 side chains. The E40 and D107 interactions are monodentate, whereas E110 engages in a bidentate interaction (Figure 8.6). Mutation of E117 of *E. coli* (Ec) RNase III (equivalent to E110 of Aa-RNase III) to alanine, lysine, or glutamine abolishes cleavage (Li and Nicholson 1996; Dasgupta et al. 1998), which can be explained by the loss of metal ion binding (Blaszczyk et al. 2001; Sun and Nicholson 2001). The divalent metal ion also has a significant effect on protein stability. In the absence of a divalent metal ion, the Aa-RNase III catalytic domain exhibits greater thermal fluctuation in structure, and basic side chains from other parts of the polypeptide associate with the acidic cluster and serve to reduce the negative charge density. The divalent metal ion displaces the basic groups, with a concomitant stabilization of structure (Blaszczyk et al. 2001).

The Aa-RNase III D44 side chain forms a hydrogen bond with a water molecule, which is bound to the metal ion (Figure 8.6). This interaction suggests that D44 may act as a general base that deprotonates the metal-bound water, providing the reaction nucleophile. Mutation of Ec-RNase III D45 (equivalent to the Aa-RNase III D44) to alanine abolishes activity in vivo (Blaszczyk et al. 2001) and in vitro (G. Li and A.W. Nicholson,

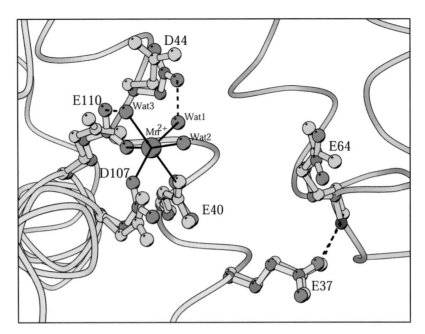

**FIGURE 8.6.** Structure of proposed active site of *Aquifex aeolicus* RNase III. The bound $Mn^{2+}$ is shown along with the interacting side chains. (Reprinted, with permission, from Blaszczyk et al. 2001; ©Elsevier Science.)

unpubl.). An examination of the pH dependence of cleavage rate under single-turnover conditions indicates the involvement of two acidic groups in catalysis (Campbell et al. 2002). It is not known what the groups are, but they could include one or more of the aforementioned carboxyl side chains. In this regard, mutation of Ec-RNase III E38 to valine or E65 to alanine inhibits activity in vivo (Blaszczyk et al. 2001).

### Dual hydrolytic chemistries at the active site?

Cleavage of dsRNA by RNase III creates two-nucleotide 3´ overhangs. In the A-form helix, the scissile phosphodiesters are positioned on the same face of the double helix, across the minor groove. It has been proposed that the composite active site employs dual chemistries to cleave each phosphodiester (Blaszczyk et al. 2001). Specifically, two groupings of residues are thought to be responsible for cleavage of each phosphodiester. One grouping includes D44 and E110, whereas the other group includes E37 and E64 (Blaszczyk et al. 2001). Given the proximity of the scissile phosphodiesters, an alternative possibility would be that a single chemistry is used to cleave both bonds. Thus, a random-order initial cleavage event may be followed by local repositioning and cleavage of the second phosphodiester. In this regard, mutation of *E. coli* RNase III E117—a residue that is proposed to participate in one of the dual mechanisms—abolishes all detectable cleavage of dsRNA (Li et al. 1996; DasGupta et al. 1998). The two mechanistic schemes may be tested by creating specific mutations expected to block only one of the two proposed cleavage chemistries.

### On the inhibitory action of $Mn^{2+}$

The original characterization of *E. coli* RNase III revealed that $Mn^{2+}$ inhibits as well as supports activity (Robertson et al. 1968). Experimental evidence indicates that the inhibitory site is on the enzyme, perhaps near the active site. Thus, although mutation of E117 to aspartic acid strongly reduces catalytic activity, as would be expected for a role providing a precisely positioned catalytic metal ion, this mutation also abolishes $Mn^{2+}$ inhibition (Sun and Nicholson 2001). These authors proposed that E117 participates in the inhibitory site as well as the active site and that the two sites are physically close (Sun and Nicholson 2001). Because each putative active site of the Aa-RNase III catalytic domain only shows a single bound $Mn^{2+}$, which is most likely the catalytic species, it is possible that substrate binding is required to create the inhibitory $Mn^{2+}$ site. The inhibitory mechanism remains to be determined, but because cleavage reactions carried out under single-turnover conditions are not inhibited by $Mn^{2+}$, product release could be the affected step (Sun and Nicholson 2001). It is not believed that $Mn^{2+}$ is a physiologically relevant regulator of RNase III activity. What is intriguing, however, is the similarity of inhibition to that observed with *E. coli* RNase H1. For the latter enzyme, it has been shown that the $Mn^{2+}$ inhibitory site is adjacent to the catalytic site (Goedken and Marqusee 2001) and has prompted the proposal of a mechanistic relatedness between RNase H and RNase III (Sun et al. 2001) (see also below).

## A Summary Mechanism of Action for Bacterial RNase III

Although precise details of the catalytic mechanism remain to be determined, an overall kinetic pathway for RNase III has been described. Assays in the steady state using purified enzyme and small model substrates revealed that *E. coli* RNase III exhibits Michaelis-Menten behavior (Li et al. 1993). A single-turnover kinetic analysis of *E. coli* RNase III showed that the steady-state rate of substrate cleavage is significantly slower than the rate of the chemical step (Campbell et al. 2002). Thus, the steady-state rate is determined by a step(s) subsequent to cleavage, which could include product release. Finally, the apparent dissociation constant ($K_D$) for the E-S complex (~5 nM) is comparable in magnitude to the $K_m$ value, indicating that substrate dissociation is rapid relative to the chemical step (Campbell et al. 2002). The same study followed the fluorescence of the single tryptophan residue to demonstrate an enzyme conformational change associated with substrate binding.

The substrate binding and cleavage steps are not obligatorily linked. Thus, mutations in the catalytic domain that block hydrolysis do not necessarily inhibit binding (Li and Nicholson 1996), and intercalating agents such as ethidium bromide can uncouple substrate binding and cleavage (Calin-Jageman and Nicholson 2001). Ethidium bromide inhibits RNase III by site-specifically binding to substrate, creating an unreactive complex that can compete with unbound substrate for enzyme binding. This provides apparent competitive inhibitory kinetics, and it has been proposed that ethidium inhibits a conformational change in the enzyme-substrate complex required to reach the transition state. In this regard, a stopped-flow kinetic assay using tryptophan fluorescence as the signal revealed a conformational change associated with the chemical step (Campbell et al. 2002). What remains to be shown is how the dsRBM and catalytic domains cooperate under catalytic conditions to perform cleavage. Although the dsRBM at the minimum serves to provide binding energy, it is not clear whether both dsRBMs in the holoenzyme are required for optimal function and how substrate reactivity epitopes that are recognized by the weakly binding catalytic domain (Sun and Nicholson, 2001) are sufficient to inhibit substrate binding by the holoenzyme (Zhang and Nicholson 1997). The analysis of mutant heterodimers of RNase III (Conrad et al. 2002; W. Meng and A.W. Nicholson, unpubl.) should provide insight into these questions.

## Yeast RNase III Orthologs: Substrates and Functions

Studies on the nucleases Rnt1p of *S. cerevisiae* and Pac1p of *Schizosaccharomyces pombe* have provided important information on the substrate reactivity epitopes recognized by

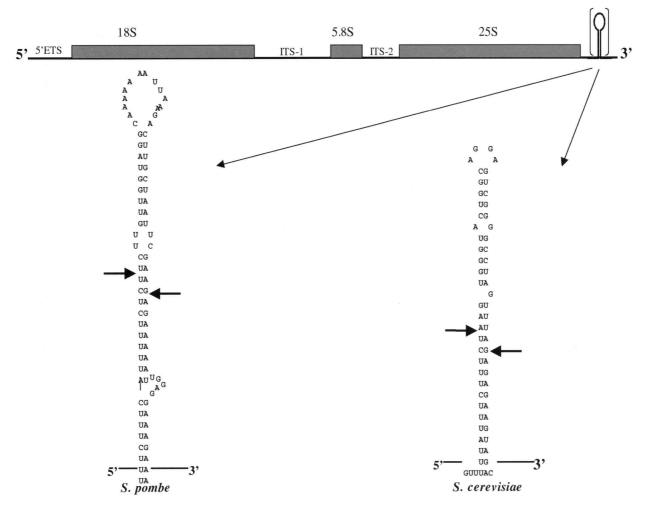

**FIGURE 8.7.** Preribosomal RNA cleavage sites for *S. pombe* Pac1p and *S. cerevisiae* Rnt1p. Shown at the top is a linear diagram of the unprocessed primary transcript. The position of the processing signal is indicated by the lollipop, with the brackets indicating the differing positions between the two organisms. Arrows indicate the cleavage sites in the hairpin sequences.

eukaryotic RNase III orthologs, and how alternative processing pathways involving RNase III may regulate gene expression. A review of yeast RNase III orthologs is also provided elsewhere (Lamontagne et al. 2001), as are purification protocols and activity assays (Lamontagne and Abou Elela 2001; Rotondo and Frendewey 2001).

Rnt1p and Pac1p are involved in stable RNA maturation. A common role is initiation of maturation of the 35S rRNA precursor through cleavage within a stem-loop structure downstream from the 3′ end of the 25S rRNA (Figure 8.7) (Abou Elela et al. 1996; Rotondo et al. 1997; Kufel et al. 1999). Cleavage at this site promotes subsequent processing steps and has been described as a "quality control" mechanism wherein only the completely synthesized precursor is allowed to undergo maturation (Lee et al. 1995; Kufel et al. 1999). Although both 3′-external transcribed sequence (3′ETS) processing signals present themselves as canonical hairpins (Figure 8.7), Rnt1p and Pac1p recognize different features (see also below), and neither enzyme is a functional substitute in the heterologous system (Rotondo et al. 1997). Pac1p cleavage of the 3′ETS site is stimulated by a multisubunit protein complex named RAC (ribosome assembly chaperone) (Lalev and Nazar 2001; Spasov et al. 2002). The RAC complex does not possess nuclease activity, but instead binds to the 3′ETS, allowing Pac1p to carry out additional cleavage at the immediate 3′ end of the 25S rRNA (Spasov et al. 2002). Whether RAC enhancement of processing involves Pac1p interaction with the complex is not known.

Rnt1p and Pac1p also cleave precursors to small nuclear RNAs (snRNAs) and small nucleolar RNAs (snoRNAs) (Chanfreau et al. 1997, 1998a,b; Abou Elela and Ares 1998; Qu et al. 1999; Zhou et al. 1999). The endonucleolytic step provides entry sites for 5′-3′ exonucleases (e.g., Rat1p) or 3′-5′ exonucleases (e.g., the exosome), which create the mature 5′ and 3′ termini. Studies on Rnt1p processing of these precursors as well as the rRNA precursor have allowed characterization of substrate reactivity epitopes for Rnt1p. A primary reactivity epitope is a conserved tetraloop (A/U)GNN motif (Figure 8.4). Rnt1p recognizes features of this motif (see below), and selects the scissile phosphodiester(s) by its distance (14–16 bp) from the motif. At least several substrates appear to use a strategy whereby coaxial stacking of helices juxtaposes an AGNN tetraloop-containing helix with a helical stem that contains the scissile bond(s) (Chanfreau et al. 1998a; Qu et al. 1999). The proposed coaxial stacking is similar to that seen with the bacterial *hok/sok* RNA interaction (see above).

The (A/U)GNN tetraloop provides a specialized motif for recognition by Rnt1p. NMR studies on small hairpins containing the tetraloop (Lebars et al. 2001; Wu et al. 2001) show that the (A/U)G dinucleotide step exhibits a 5′ base stack, with the invariant G in a *syn* conformation. The NN dinucleotide step is engaged in a 3′ base stack, which creates a sharp turn between the second and third tetraloop nucleotides. The structure provides a solvent-exposed G residue and an adjacent nonbridging phosphodiester oxygen, which may provide recognition sites for Rnt1p. In this regard, a biochemical analysis showed a direct interaction of the Rnt1p dsRBM with the tetraloop (Nagel and Ares 2000). A structural study of a dsRBM of *Drosophila* Staufen protein bound to a 12-bp RNA stem-loop structure also reveals an interaction of the dsRBM α1 helix with an all-C tetraloop (Ramos et al. 2000). However, as discussed below, the tetraloop motif is not the only reactivity epitope for Rnt1p.

Rnt1p participates in an alternative RNA maturation pathway with the potential for gene regulation. Although many snoRNAs are transcribed as monocistronic or polycistronic species, the *S. cerevisiae* U18 snoRNA is encoded in the intron of the mRNA for elongation factor 1β (EF-1β). U18 snoRNA undergoes maturation by either of two pathways, one of which involves Rnt1p (Villa et al. 1998, 2000; Giorgi et al. 2001). The primary pathway initiates with the splicing of the pre-mRNA and release of the intron lariat. The lariat 2′-5′ phosphodiester is cleaved by the debranching phosphodiesterase, Dbr1p (Ooi et al. 1998), and the linearized intron undergoes 5′-3′ and 3′-5′ exonucleolytic trimming to provide the mature snoRNA. The secondary pathway begins with Rnt1p cleavage within the pre-mRNA intron, releasing a U18 snoRNA precursor which is trimmed to the mature species. The secondary pathway necessarily suppresses the splicing and therefore maturation of the EF-1β mRNA.

The alternative maturation pathways reveal how a eukaryotic RNase III ortholog can recognize the substrate by a second type of reactivity epitope. Rnt1p cleavage of the U18 snoRNA precursor requires the fibrillarin homolog Nop1p, which binds to the U18 snoRNA sequence within the unspliced mRNA to form an snoRNP complex. Although other proteins are also present in this complex, Rnt1p recognizes the scissile phosphodiesters through an interaction with Nop1p, rather than with a tetraloop motif (Figure 8.8) (Giorgi et al. 2001). It would be of interest to determine what portion of Rnt1p (e.g., the amino-terminal domain or the dsRBM) recognizes Nop1p.

A gene regulatory capacity may be provided by the alternative maturation pathways. When the snoRNP-intron complex is allowed to form, Rnt1p cleavage would suppress production of EF-1β1. The primary regulatory signal is provided by Nop1p binding to the unprocessed U18 sequence (Villa et al. 1998, 2000; Giorgi et al. 2001). It would be of interest to examine the interrelationship between EF-1β and Nop1p levels and transla-

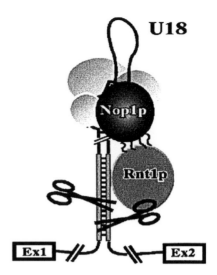

**FIGURE 8.8.** Structure of the Nop1p-U18 snoRNA complex recognized by *S. cerevisiae* Rnt1p. The scissors indicate the two cleavage events performed by Rnt1p on the U18 snoRNA precursor. Ex1 and Ex2 refer to exons 1 and 2 in the EF-1β mRNA. (Reprinted, with permission, from Giorgi et al. 2001; ©Oxford University Press.)

tion rates. The potential for Rnt1p regulation of U5 snRNA production by alternative pathways also has been proposed (Chanfreau et al. 1997).

Rnt1p also recognizes Gar1p, a component of the H/ACA snoRNA pseudo-uridylase complex (Tremblay et al. 2002). The functional importance of this interaction is revealed by its disruption in vivo, which prevents nucleolar localization of the H/ACA snoRNA-associated proteins, and a concomitant inhibition of pre-rRNA processing and pseudo-uridylation (Tremblay et al. 2002).

## Higher Eukaryotic RNase III Ortholog Structures and Functions

Eukaryotic genomes encode multiple RNase III orthologs. The genes are not necessarily physically linked, and there is growing evidence that the proteins have specialized functional roles. At least two mammalian genomes appear to encode only two RNase III orthologs (see below), whereas the *Arabidopsis thaliana* genome encodes at least nine proteins identifiable as RNase III orthologs (Z. Xie and J. Carrington, pers. comm.). Given the conserved pairing of catalytic domain and dsRBM, it is reasonable to expect that the same catalytic mechanism is used for all orthologs (see also above). However, intriguing mechanistic variations on the theme are possible with the occurrence of additional domains.

### Mammalian RNase III

The mouse and human genomes encode two RNase III orthologs, one of which is Dicer (see below) and the other named (for simplicity) RNase III. Each protein is encoded by a single-copy gene, each of which is located on separate chromosomes at sites that are syntenic between the two organisms (Figure 8.9). The RNase III polypeptides of human and mouse share high sequence similarity and exhibit a single carboxy-terminal dsRBM with tandem catalytic domains (Figure 8.1). Assuming that the polypeptide dimerizes, the holoenzyme may be expected to contain up to four active sites. Because both of the catalytic domains appear to retain all conserved residues proposed to be important for catalytic function, it has been proposed (Blasczcyk et al. 2001) that this enzyme would generate products of dsRNA cleavage similar to those of the *E. coli* enzyme. The amino-terminal portion of mammalian RNase III contains a PRR and an SR domain. In other proteins, these domains have been shown to engage in protein-protein interactions, with the SR domain in particular involved in RNA maturation and transport processes (Graveley 2000).

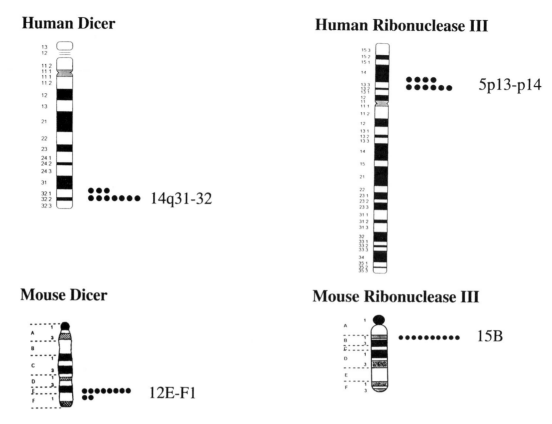

**FIGURE 8.8.** Chromosomal locations of the mouse and human Dicer and RNase III genes. The genes were mapped by fluorescent in situ hybridization (FISH) using probes specific for each gene (Fortin et al. 2002; Nicholson and Nicholson 2002).

The mouse and human RNase III transcripts are broadly expressed (Wu et al. 2000; Fortin et al. 2002), suggesting involvement in one or more fundamental cellular processes. Human RNase III has been implicated in rRNA maturation. Thus, a reduction in human RNase III levels by antisense-oligodeoxynucleotide-directed mRNA ablation leads to accumulation of an 5.8S rRNA precursor that has failed to undergo cleavage within the internal transcribed spacer 2 (ITS-2) (Wu et al. 2000). However, it is not known whether RNase III directly cleaves within the ITS-2 or is involved in other RNA-processing events, which, in turn, influence 5.8S rRNA maturation. A role in stable RNA maturation is also indicated by the observation that human RNase III localizes to the nucleolus in a cell-cycle-dependent manner (Wu et al. 2000). The localization may depend on protein-protein interactions mediated by one or both of the amino-terminal domains and/or one or more covalent modifications. As it is anticipated that mammalian RNase III has multiple roles in RNA metabolic pathways, it is clear that there is much still to be understood about the function of this enzyme and its substrates.

## DICER STRUCTURE AND FUNCTION IN RNAi AND MATURATION OF SMALL REGULATORY RNAs

The term "RNA interference" (RNAi) was originally used to describe antisense RNA inhibition of gene expression. RNAi assumed a more specialized definition when it was shown that dsRNA molecules are potent, homology-dependent inhibitors of gene expression (Fire et al. 1998). The RNAi mechanism involves dsRNA processing, and it is now established that the RNase III ortholog Dicer is an essential participant in this event (Bernstein et al. 2001). Involvement of such an enzyme had also been suggested by bioinformatic-phylogenetic approaches (Aravind et al. 2000; Cerutti et al. 2000) or to explain how small dsRNAs are generated during posttranscriptional gene silencing (Bass 2000).

Dicer may be most accurately regarded as an RNA maturation nuclease rather than a degradative nuclease. Thus, in one role, Dicer carries out the maturation of small regulatory RNAs, whereas in a second role, Dicer cleaves virtually any dsRNA to provide small, precisely sized duplexes that are incorporated into a complex that is responsible for homology-dependent RNA degradation (see below). Presented below is a brief discussion of Dicer structure and function. More extensive discussions of Dicer are presented elsewhere in this volume and in other reviews (Bernstein et al. 2001; Sharp 2001; Zamore 2001b).

## Dicer Production of siRNAs

The endogenous functions of RNAi include inhibition of viral gene expression and suppression of retroposon movement. In each situation, the RNAi machinery recognizes "aberrant" RNAs as double-stranded species. A required early step in the RNAi pathway is the production of small (~21 bp) duplex RNAs, termed short interfering RNAs (siRNAs). The development of an in vitro system that recapitulates key steps in the RNAi pathway revealed that the siRNAs are direct products of cleavage of the input dsRNA (Tuschl et al. 1999; Parrish et al. 2000; Yang et al. 2000; Zamore et al. 2000). G. Hannon and coworkers showed that siRNAs are the result of Dicer action (Bernstein et al. 2001). siRNAs are recruited into a complex, termed RISC (RNA-induced silencing complex), which directs degradation of homologous RNA targets (Hammond et al. 2000). *Drosophila* RISC contains an Argonaute protein family member (Argonaute2), which is also implicated in RNAi in other organisms (Hammond et al. 2001). The mechanism of target RNA cleavage is unknown, but it does not appear to involve Dicer.

## Dicer Maturation of Small Regulatory RNAs

Dicer is also an important participant in developmental processes through its processing of precursors to small regulatory RNAs. An initial indication of Dicer involvement in development was provided by the effect of a mutation (*Caf*) in one of the *A. thaliana* Dicer genes which causes defects in flower development and unregulated proliferation of floral meristem tissue (Jacobsen et al. 1999). Other studies established an essential role for Dicer in germ-line development in *Caenorhabditis elegans* (Grishok et al. 2001; Knight and Bass 2001).

The small regulatory RNAs created by Dicer action include small temporal RNAs (stRNAs), such as the *let-7* microRNA of *C. elegans* (Grishok et al. 2001; Hutvagner et al. 2001; Ketting et al. 2001). Site-specific cleavage of precursors to these RNAs provides the mature functional species (Figure 8.10). Purified recombinant human Dicer can cleave a precursor to *let-7* microRNA in vitro to create a species of the correct size for the mature RNA (Provost et al. 2002). The mature *let-7* microRNA binds to a complementary site in the 3´UTR of specific mRNAs, thereby down-regulating translation by an unknown mechanism. The number of small regulatory RNAs is expected to be large, and they are most likely produced by Dicer cleavage of polycistronic as well as monocistronic precursors (Lau et al. 2001). Compared to the formation of siRNAs, the cleavage of precursors

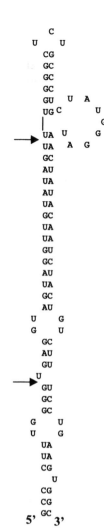

**FIGURE 8.10.** Structure of the precursor of human *let-7* microRNA (Hutvagner et al. 2001). Arrows indicate the sites of cleavage by Dicer. The cleavages provide the mature 21-nucleotide species (Provost et al. 2002).

to small regulatory RNAs is asymmetric. There is a clear involvement of substrate reactivity epitopes that have yet to be defined. It may be anticipated that internal loops, specific sequences, and coaxial stacking seen in bacterial and yeast RNase III substrates also participate in cleavage site selection by Dicer.

## On the Mechanism and Structure of Dicer

Biochemical studies on Dicer are still in their infancy, and a wealth of intriguing enzymological questions wait to be answered. Dicer is similar to bacterial RNase III in that it can cleave dsRNA in an apparent sequence-nonspecific manner. The siRNA products possess two-nucleotide 3′ overhangs, with 5′ phosphate, 3′-hydroxyl termini. On the basis of common protein domain features, the type of product termini, and the divalent metal ion requirement, it is expected that the chemistry of phosphodiester hydrolysis is the same as that used by bacterial RNase III. The sequence nonspecificity would be a requirement of the RNAi pathway to allow recognition and cleavage of virtually any dsRNA. The 21-bp size of the siRNAs is sufficient to provide the requisite specificity for a unique target sequence within the eukaryotic cytoplasm. The structural features of siRNAs important for function have been analyzed. siRNAs with two-nucleotide 3′ over-hangs were most efficient in RNAi action in *Drosophila* embryo lysates (Elbashir et al. 2001). There is not a strict ribose requirement, however, as substitution of terminal residues with 2′-deoxyribose or 2′-*O*-methylribose residues does not affect function (Elbashir et al. 2001).

The structure of the Dicer polypeptide suggests specific functional behaviors. The existence of two catalytic domains, coupled with the observation that the dimer interface of the bacterial holoenzyme involves extensive catalytic domain contacts, predicts a dimeric structure for the holoenzyme. A double-stranded ribonuclease purified from *Dictyostelium*, which is likely to be Dicer, exhibits a native gel electrophoretic mobility of about 450 kD (Novotny et al. 2001; Martens et al. 2002). Assuming the absence of unusual structural features and/or associated polypeptides, the size is consistent with a homodimeric structure. However, another scenario is possible, where intramolecular association of the two catalytic domains would provide a dsRNA-binding cleft, thus obviating the need for intermolecular association.

Crystallographic data and molecular modeling suggest how two bacterial RNase III catalytic domain homodimers, bound to adjacent positions on a dsRNA, can provide precisely sized products (Blaszcyk et al. 2001). It has been proposed that one of the two catalytic domains in Dicer is disabled through mutation of a conserved residue (Blaszcyk et al. 2001). If so, then it can be argued that the selective inactivation would allow production of 21-bp dsRNAs instead of approximately 11-bp species. Again, as mentioned above, the longer product size ensures that the siRNA-containing RISC complex would have the requisite specificity for the RNA targets.

The biochemical properties of purified recombinant human Dicer have been described (Provost et al. 2002; Zhang et al. 2002). Human Dicer requires a divalent metal ion ($Mg^{2+}$, $Mn^{2+}$, or $Co^{2+}$) for activity, with cleavage of dsRNA directly providing the mature approximately 21–23-bp siRNA species (Provost et al. 2002; Zhang et al. 2002). Prior treatment with proteases enhances Dicer catalytic activity for reasons that are presently unclear (Zhang et al. 2002). Similar to other RNase III orthologs, Dicer can bind dsRNA without concomitant cleavage in the absence of divalent metal ion, or, alternatively, at lowered temperature or elevated salt concentrations (Provost et al. 2002; Zhang et al. 2002).

Human Dicer does not require ATP to cleave dsRNA (Provost et al. 2002; Zhang et al. 2002), which is consistent with the ATP independence of *E. coli* RNase III, which nonetheless can bind ATP (Chen et al. 1990). Dicer activity in mammalian cell extracts also is not stimulated by ATP (Zhang et al. 2002). In contrast, Dicer in insect or worm cell lysates is enhanced by ATP, which may reflect postcleavage steps in the presence of RNAi pathway components specific to these organisms (Zhang et al. 2002).

The function of the amino-terminal helicase domain of Dicer has not been defined. Purified human Dicer lacks a demonstrable ATPase activity characteristic of helicases (Zhang et al. 2002). The introduction of a K70A mutation in the conserved P loop of the ATPase/helicase domain of human Dicer, which is expected to inactivate nucleotide binding, has no effect on dsRNA cleavage (Zhang et al. 2002). It is possible that the helicase domain functions in a more complete RNAi system, perhaps catalyzing enzyme translocation along substrate, or in siRNA transfer to the RISC complex, which may include an RNA unwinding step. The helicase domain also could participate in microRNA maturation by allowing efficient separation of the product strand from its complementary sequence.

Immunofluorescence analysis has localized Dicer to the endoplasmic reticulum (ER), with no evident nuclear staining (Provost et al. 2002). The ER location would place Dicer in proximity to the RISC complex, which is associated with the translational machinery (Hammond et al. 2000). The apparent lack of nuclear localization suggests that the microRNA precursors may be processed in the cytoplasm.

## ON THE ORIGIN OF RNase III

Although RNase III orthologs are present in bacteria and eukaryotes, none have been detected in any of the characterized Archaea. Moreover, the site of cleavage of the archaeal rRNA precursors is carried out by an endonuclease that is mechanistically distinct from RNase III, and which is also involved in tRNA intron excision (Thompson and Daniels 1990; Kleman-Leyer et al. 1997; Dennis et al. 1998; Li et al. 1998). It has been speculated that RNase III may have had an origin in the bacterial lineage and entered into an early eukaryote by horizontal transmission (Aravind et al. 2000). A report of the similarity between yeast mitochondrial ribosomal protein L15 and RNase III (Aravind and Koonin 2001) suggests that horizontal transmission may have occurred via a protomito-chondrial endosymbiont and that RNase III and mitochondrial ribosomal protein L15 diverged from a common ancestor. The archetypal RNase III also may only have contained the catalytic domain. Dimeric behavior may have been an original necessary feature, because the putative dsRNA-binding domain is created by dimerization, and the proposed composite active site employs residues from each subunit. The subsequent fusion of a dsRBM to the catalytic domain could have conferred additional specificity of action and possible regulation.

The RNase III catalytic domain does not exhibit any apparent sequence similarity with phosphodiesterases outside of its own superfamily, and the crystal structure of the Aa-RNase III catalytic domain reveals a hitherto undescribed all-alpha fold (Blaszczyk et al. 2001). There are intriguing biochemical similarities between *E. coli* RNase III and *E. coli* RNase H1. *E. coli* RNase H1 is a monomer of approximately 17.5 kD, which is similar in size to the *E. coli* RNase III catalytic domain. The two enzymes employ a cluster of acidic side chains to catalyze cleavage, and $Mn^{2+}$ inhibits as well as supports catalysis for each enzyme (Goedken and Marqusee 2001; Sun and Nicholson 2001). Whether the same catalytic mechanism is in fact used awaits further mechanistic studies of RNase III. A similar mechanism employed by differently folded proteins of different sequences would suggest convergent evolution of the active site.

## SUMMARY AND PROSPECTS

From its initial appearance as a biochemical novelty in bacterial cell extracts, RNase III has now been shown to occupy a central position in mediating dsRNA-dependent processes, including RNA maturation, decay, and gene silencing. For bacterial RNase III orthologs, an emerging theme is the occurrence of organism-specific RNase-III-dependent pathways in addition to a common role in ribosomal RNA maturation. Substrate reactivity epitopes are diverse and nonconserved and are devised such that RNase III family members can cleave a range of dsRNA sequences while maintaining high specificity toward cellular processing signals. The involvement of additional factors in controlling RNase-III-dependent processing pathways has been established in studies on yeast RNase III orthologs.

In higher eukaryotes, the potential exists for a regulated interplay between the Dicer-dependent RNAi pathway and that of the dsRNA-mediated antiviral interferon response. Exploration of this area will be important with respect to applying RNAi in functional genomic analyses (Ashrafi et al. 2003) and in developing new approaches for disease therapy (Kapadia et al. 2003).

## ACKNOWLEDGMENTS

The author thanks members of the laboratory for their interest and support of research on RNase III. The author is especially grateful to Irina Calin-Jageman and Rhonda H. Nicholson for help with the figures and a careful reading of the manuscript. Research in the author's laboratory is supported by grants from the National Institutes of Health.

## REFERENCES

Abou Elela S. and Ares M. 1998. Depletion of yeast RNase III blocks correct U2 3' end formation and results in polyadenylated but functional U2 snRNA. *EMBO J.* **17**: 3738–3746.

Abou Elela S., Igel H., and Ares M. 1996. RNase III cleaves eukaryotic preribosomal RNA at a U3 snoRNP-dependent site. *Cell* **85**: 115–124.

Aceti D.J. and Champness W.C. 1998. Transcriptional regulation of *Streptomyces coelicolor* pathway-specific antibiotic regulators by the *absA* and *absB* loci. *J. Bacteriol.* **180**: 3100–3106.

Altuvia S., Locker-Giladi H., Koby S., Ben-Nun O., and Oppenheim A.B. 1987. RNase III stimulates the translation of the cIII gene of bacteriophage lambda. *Proc. Natl. Acad. Sci.* **84**: 6511–6515.

Amarasinghe A.K., Calin-Jageman I., Harmouch A., Sun W., and Nicholson A.W. 2001. *Escherichia coli* ribonuclease III: Affinity purification of hexahistidine-tagged enzyme and assays for substrate binding and cleavage. *Methods Enzymol.* **342**: 143–158.

Ammann J., Delius H., and Hofschneider P.H. 1964. Isolation and properties of an intact phage-specific replicative form of RNA phage M12. *J. Mol. Biol.* **10**: 557–561.

Aravind L. and Koonin E.V. 2001. A natural classification of ribonucleases. *Methods Enzymol.* **341**: 3–28.

Aravind L., Makarova K.S., and Koonin E.V. 2000. Holliday junction resolvases and related nucleases: Identification of new families, phyletic distribution and evolutionary trajectories. *Nucleic Acids Res.* **28**: 3417–3432.

Ashrafi K., Chang F.Y., Watts J.L., Fraser A.G., Kamath R.S., Ahringer J., and Ruvkun G. 2003. Genome-wide RNAi analysis of *Caenorhabditis elegans* fat regulatory genes. *Nature* **421**: 268–272.

Bardwell J.C.A., Regnier P., Chen S.M., Nakamura Y., Grunberg-Manago M., and Court D.L. 1989. Autoregulation of RNase III operon by mRNA processing. *EMBO J.* **8**: 3401–3407.

Bass B.L. 2000. Double-stranded RNA as a template for gene silencing. *Cell* **101**: 235–238.

Bernstein E., Denli A.M., and Hannon G.J. 2001. The rest is silence. *RNA* **7**: 1509–1521.

Bernstein E., Caudy A.A., Hammond S.M., and Hannon G. J. 2001. Role for a bidentate ribonuclease in the initiation step of RNA interference. *Nature* **409**: 363–366.

Bevilacqua P.C. and Cech T.R. 1996. Minor-groove recognition of double-stranded RNA by the double-stranded RNA-binding domain from the RNA-activated protein kinase PKR. *Biochemistry* **35**: 9983–9994.

Billeter M.A., Weissmann C., and Warner R.C. 1966a. Replication of viral ribonucleic acid. IX. Properties of double-stranded RNA from *Escherichia coli* infected with bacteriophage MS2. *J. Mol. Biol.* **17**: 145–173.

Billeter M.A., Libonati M., Vinuela E., and Weissmann C. 1966b. Replication of viral ribonucleic acid. X. Turnover of virus-specific double-stranded ribonucleic acid during replication of phage MS2 in *Escherichia coli*. *J. Biol. Chem.* **241**: 4750–4757.

Blaszczyk J., Tropea J.E., Bubunenko M., Routzahn K.M., Waugh D.S., Court D.L., and Ji X. 2001. Crystallographic and modeling studies of RNase III suggest a mechanism for double-stranded RNA cleavage. *Structure* **9**: 1225–1236.

Calin-Jageman I. and Nicholson A.W. 2001. Ethidium-dependent uncoupling of substrate binding and cleavage by *Escherichia coli* ribonuclease III. *Nucleic Acids Res.* **29**: 1915–1925.

Campbell F.E., Cassano A.G., Anderson V.E., and Harris M.E. 2002. Pre-steady-state and stopped-flow fluorescence analysis of *Escherichia coli* ribonuclease III. Insights into mechanism and conformational changes associated with binding and catalysis. *J. Mol. Biol.* **317**: 21–40.

Cerutti L., Mian N., and Bateman A. 2000. Domains in gene silencing and cell differentiation proteins: The novel PAZ domain and redefinition of the Piwi domain. *Trends Biochem. Sci.* **25**: 481–482.

Chanfreau G., Abou Elela S., Ares M., and Guthrie C. 1997. Alternative 3′-end processing of U5 snRNA by RNase III. *Genes Dev.* **11:** 2741–2751.

Chanfreau G., Legrain P., and Jacquier A. 1998a. Yeast RNase III as a key processing enzyme in small nucleolar RNAs metabolism. *J. Mol. Biol.* **284:** 975–988.

Chanfreau G., Rotondo G., Legrain P., and Jacquier A. 1998b. Processing of a dicistronic small nuclear RNA precursor by the RNA endonuclease Rnt1p. *EMBO J.* **17:** 3726–3737.

Chelladurai B., Li H., Zhang K., and Nicholson A.W. 1993. Mutational analysis of a ribonuclease III processing signal. *Biochemistry* **32:** 7549–7558.

Chen S.M., Takiff H.E., Barber A.M., Dubois A.C., Bardwell J.C., and Court D.L. 1990. Expression and characterization of RNase III and Era proteins. Products of the *rnc* operon of *Escherichia coli*. *J. Biol. Chem.* **265:** 2888–2895.

Conrad C., Evguenieva-Hackenberg E., and Klug G. 2001. Both N-terminal catalytic and C-terminal RNA binding domains contribute to substrate specificity and cleavage site selection of RNase III. *FEBS Lett.* **509:** 53–58.

Conrad C., Rauhut R., and Klug G. 1998. Different cleavage specificities of RNases III from *Rhodobacter capsulatus* and *Escherichia coli*. *Nucleic Acids Res.* **26:** 4446–4453.

Conrad C., Schmitt J.G., Evguenieva-Hackenberg E., and Klug G. 2002. One functional subunit is sufficient for catalytic activity and substrate specificity of *Escherichia coli* endoribonuclease III artificial heterodimers. *FEBS Lett.* **518:** 93–96.

Court D. 1993. RNA processing and degradation by RNase III. In *Control of messenger RNA stability* (ed. J.G. Belasco and G. Brawerman), pp. 71–116. Academic Press, New York.

Crouch R.J. 1974. Ribonuclease III does not degrade deoxyribonucleic acid-ribonucleic acid hybrids. *J. Biol. Chem.* **249:** 1314–1316.

Dasgupta S., Fernandez L., Kameyama L., Inada T., Nakamura Y., Pappas A., and Court D.L. 1998. Genetic uncoupling of the dsRNA-binding and RNA cleavage activities of the *Escherichia coli* endoribonuclease RNase III—The effect of dsRNA binding on gene expression. *Mol. Microbiol.* **28:** 629–640.

De Gregorio E., Abrescia C., Carlomagno M.S., and DiNocera P.P. 2002. The abundant class of nemis repeats provide RNA substrates for ribonuclease III in *Neisseriae*. *Biochim. Biophys. Acta* **1576:** 39–44.

Dennis P.P., Ziesche S., and Mylvaganam S. 1998. Transcription analysis of two disparate rRNA operons in the halophilic archaeon *Haloarcula marismortui*. *J. Bacteriol.* **180:** 4804–4813.

Dunn J.J. 1976. RNase III cleavage of single-stranded RNA: Effect of ionic strength on the fidelity of cleavage. *J. Biol. Chem.* **251:** 3807–3814.

———. 1982. Ribonuclease III. In *The enzymes* (ed. P.D. Boyer), pp. 485–499. Academic Press, New York.

Dunn J.J. and Studier F.W. 1973. T7 early RNAs and *Escherichia coli* ribosomal RNAs are cut from large precursor RNAs *in vivo* by ribonuclease III. *Proc. Natl. Acad. Sci.* **70:** 3296–3300.

———. 1983. Complete nucleotide sequence of bacteriophage T7 and the locations of T7 genetic elements. *J. Mol. Biol.* **166:** 477–535.

Elbashir S.M., Martinez J., Patkaniowska A., Lendeckel W., and Tuschl T. 2001. Functional anatomy of siRNAs for mediating efficient RNAi in *Drosophila melanogaster* embryo lysate. *EMBO J.* **20:** 6877–6888.

Evguenieva-Hackenberg E. and Klug G. 2000. RNase III processing of intervening sequences found in helix 9 of 23S rRNA in the alpha subclass of Proteobacteria. *J. Bacteriol.* **182:** 4719–4729.

Fiero-Monti I. and Mathews M.B. 2000. Proteins binding to duplexed RNA: One motif, multiple functions. *Trends Biochem. Sci.* **25:** 241–246.

Fire A., Xu S., Montgomery M.K., Kostas S.A., Driver S.E., and Mello C.C. 1998. Potent and specific genetic interference by double-stranded RNA in *Caenorhabditis elegans*. *Nature* **391:** 806–811.

Fortin K.R., Nicholson R.H., and Nicholson A.W. 2002. Mouse ribonuclease III cDNA structure, expression analysis, and chromosomal location. *BMC Genomics* **3:** 26.

Franch T., Thisted T., and Gerdes K. 1999. Ribonuclease III processing of coaxially stacked RNA helices. *J. Biol. Chem.* **274:** 26572–26578.

Fraser C.M., Gocayne J.D., White O., Adams M.D., Clayton R.A. Fleishmann R.D., Bult C.J., Kerlavage A.R., Sutton G., Kelly J.M., et al. 1995. The minimal gene complement of *Mycoplasma genitalium*. *Science* **270:** 397–403.

Giorgi C., Fatica A., Nagel R., and Bozzoni I. 2001. Release of U18 snoRNA from its host intron requires interaction of Nop1p with the Rnt1p endonuclease. *EMBO J.* **20:** 6856–6865.

Goedken E.R. and Marqusee S. 2001. Co-crystal of *Escherichia coli* RNase HI with $Mn^{2+}$ ions reveals two divalent metals bound in the active site. *J. Biol. Chem.* **276:** 7266–7271.

Graveley B.R. 2000. Sorting out the complexity of SR protein functions. *RNA* **6:** 1197–1211.

Grishok A., Pasquinelli A.E., Conte D., Li N., Parrish S., Ha I., Baillie D.L., Fire A., Ruvkun G., and Mello C.C. 2001. Genes and mechanisms related to RNA interference regulate expression of the small temporal RNAs that control *C. elegans* temporal timing. *Cell* **106:** 23–34.

Gurevitz M. and Apirion D. 1983. Interplay among processing and degradative enzymes and a precursor ribonucleic acid in the selective maturation and maintenance of ribonucleic acid molecules. *Biochemistry* **22:** 4000–4005.

Hammond S.M., Bernstein E., Beach D., and Hannon G.J. 2000. An RNA-directed nuclease mediates post-transcriptional gene silencing in *Drosophila* cells. *Nature* **404:** 293–298.

Hammond S.M., Boettcher S., Caudy A.A., Kobayashi R., and Hannon G.J. 2001. Argonaute2, a link between genetic and biochemical analyses of RNAi. *Science* **293:** 1146–1150.

Herskovitz M.A. and Bechhofer D.H. 2000. Endoribonuclease RNase III is essential in *Bacillus subtilis*. *Mol. Microbiol.* **38:** 1027–1033.

Hutvagner G., McLachlan J., Pasquinelli A.E., Balint E., Tuschl T., and Zamore P.D. 2001. A cellular function for the RNA-interference enzyme Dicer in the maturation of the *let-7* small temporal RNA. *Science* **293:** 834–838.

Jacobsen S.E., Running M.P., and Meyerowitz E.M. 1999. Disruption of an RNA helicase/RNase III gene in *Arabidopsis* causes unregulated cell division in floral meristems. *Development* **126:** 5231–5243.

Jarrige A.-C., Mathy N., and Portier C. 2001. PNPase autocontrols its expression by degrading a double-stranded structure in the *pnp* mRNA leader. *EMBO J.* **20:** 6845–6855.

Kapadia S.B., Brideau-Andersen A., and Chisari F.V. 2003. Interference of hepatitis C virus RNA replication by short interfering RNAs. *Proc. Natl. Acad. Sci.* [epub ahead of print].

Ketting R.F., Fischer S.E.J., Bernstein E., Sijen T., Hannon G.J., and Plasterk R.H.A. 2001. Dicer functions in RNA interference and in synthesis of small RNA involved in developmental timing in *C. elegans*. *Genes Dev.* **15:** 2654–2659.

Kharrat A., Macias M.J., Gibson T.J., Nilges M., and Pastore A. 1995. Structure of the dsRNA-binding domain of *E. coli* RNase III. *EMBO J.* **14:** 3572–3584.

Kindler P., Keil T.U., and Hofschneider P.H. 1973. Isolation and characterization of an RNase III deficient mutant of *Escherichia coli*. *Mol. Gen. Genet.* **126:** 53–69.

Kleman-Leyer K., Armbruster D.W., and Daniels C.J. 1997. Properties of *H. volcanii* tRNA intron endonuclease reveal a relationship between the archaeal and eucaryal tRNA intron processing systems. *Cell* **89:** 839–847.

Klovins J., van Duin J., and Olsthoorn R.C.L. 1997. Rescue of the RNA phage genome from RNase III cleavage. *Nucleic Acids Res.* **25:** 4201–4208.

Knight S.W. and Bass B.L. 2001. A role for the RNase III enzyme DCR-1 in RNA interference and germ line development in *Caenorhabditis elegans*. *Science* **293:** 2269–2271.

Koraimann G., Schroller C., Graus H., Angerer D., Teferle K., and Hogenauer G. 1993. Expression of gene 19 of the conjugative plasmid R1 is controlled by RNase III. *Mol. Microbiol.* **9:** 717–727.

Kufel J., Dichtl B., and Tollervey D. 1999. Yeast Rnt1p is required for cleavage of the pre-ribosomal RNA in the 3′-ETS but not the 5′-ETS. *RNA* **5:** 909–917.

Lalev A. and Nazar R.N. 2001. A chaperone for ribosome biogenesis. *J. Biol. Chem.* **276:** 16655–16659.

Lamontagne B. and Abou Elela S. 2001. Purification and characterization of *Saccharomyces cerevisiae* Rnt1p nuclease. *Methods Enzymol.* **342:** 159–167.

Lamontagne B., Tremblay A., and Abou Elela S. 2000. The N-terminal domain that distinguishes yeast from bacterial RNase III contains a dimerization signal required for efficient double-stranded RNA cleavage. *Mol. Cell. Biol.* **20:** 1104–1115.

Lamontagne B., Larose S., Boulanger J., and Abou Elela S. 2001. The RNase III family: A conserved structure and expanding functions in eukaryotic dsRNA metabolism. *Curr. Issues Mol. Biol.* **3:** 71–78.

Lau N.C., Lim L.P., Weinstein E.G., and Bartel D.P. 2001. An abundant class of tiny RNAs with probable regulatory roles in *Caenorhabditis elegans*. *Science* **294:** 858–862.

Lebars I., Lamontagne B., Yoshizawa S., Abouelela S., and Fourmy D. 2001. Solution structure of conserved AGNN tetraloops: Insights into Rnt1p RNA processing. *EMBO J.* **20:** 7250–7258.

Lee Y., Melekhovets Y.F., and Nazar R.N. 1995. Termination as a factor in "quality control" during ribosome biogenesis. *J. Biol. Chem.* **270:** 28003–28005.

Li H. and Nicholson A.W. 1996. Defining the enzyme binding domain of a ribonuclease III processing signal. Ethylation interference and hydroxyl radical footprinting using catalytically inactive RNase III mutants. *EMBO J.* **15:** 1421–1433.

Li H., Trotta C.R., and Abelson J. 1998. Crystal structure and evolution of a transfer RNA splicing enzyme. *Science* **280:** 279–284.

Li H., Chelladurai B.S., Zhang K., and Nicholson A.W. 1993. Ribonuclease III cleavage of a bacteriophage T7 processing signal. Divalent cation specificity, and specific anion effects. *Nucleic Acids Res.* **21:** 1919–1925.

Libonati M. 1968a. Isolation and partial purification of ribonuclease III of *Escherichia coli*. *Boll. Soc. Ital. Biol. Sper.* **44:** 786–788.

———. 1968b. Some properties of ribonuclease III of *Escherichia coli*. *Boll. Soc. Ital. Biol. Sper.* **44:** 789–792.

Libonati M. and Sorrentino S. 1992. Revisiting the action of bovine ribonuclease A and pancreatic-type ribonucleases on double-stranded RNA. *Mol. Cell. Biochem.* **117:** 139–151.

Lodish H.F. and Zinder N.D. 1966a. Mutants of the bacteriophage f2. VIII. Control mechanisms for phage-specific syntheses. *J. Mol. Biol.* **19:** 333–348.

———. 1966b. Replication of the RNA of bacteriophage f2. *Science* **152:** 372–378.

Loeb T. and Zinder N.D. 1961. A bacteriophage containing RNA. *Proc. Natl. Acad. Sci.* **47:** 282–289.

March P.E. and Gonzalez M.A. 1990. Characterization of the biochemical properties of recombinant ribonuclease III. *Nucleic Acids Res.* **18:** 3293–3298.

Martens H., Novotny J., Oberstrass J., Steck T.L., Postlethwaite P., and Nellen W. 2002. RNAi in *Dictyostelium:* The role of RNA-directed RNA polymerases and double-stranded RNase. *Mol. Biol. Cell* **13:** 445–453.

Matsunaga J., Simons E.L., and Simons R.W. 1996. *E. coli* RNase III autoregulation: Structure and function of *rncO*, the posttranscriptional "operator." *RNA* **2:** 1228–1240.

———. 1997. *Escherichia coli* RNase III (*rnc*) autoregulation occurs independently of *rnc* gene translation. *Mol. Microbiol.* **26:** 1125–1135.

Mian I.S. 1997. Comparative sequence analysis of ribonucleases HII, III, II, PH and D. *Nucleic Acids Res.* **25:** 3187–3195.

Mitra S. and Bechhofer D.H. 1994. Substrate specificity of an RNase III-like activity from *Bacillus subtilis*. *J. Biol. Chem.* **269:** 31450–31456.

Nagel R. and Ares M. 2000. Substrate recognition by a eukaryotic RNase III: The double-stranded RNA-binding domain of Rnt1p selectively binds RNA containing a 5′-AGNN-3′ tetraloop. *RNA* **6:** 1142–1156.

Nashimoto H. and Uchida H. 1985. DNA sequencing of the *Escherichia coli* ribonuclease III gene and its mutations. *Mol. Gen. Genet.* **210:** 25–29.

Nicholson A.W. 1992. Accurate enzymatic cleavage *in vitro* of a 2′-deoxyribose-substituted ribonuclease III processing signal. *Biochim. Biophys. Acta* **1129:** 318–322.

———. 1996. Structure, reactivity, and biology of double-stranded RNA. *Prog. Nucleic Acids Res. Mol. Biol.* **52:** 1–65.

———. 1999. Function, mechanism and regulation of bacterial ribonucleases. *FEMS Microbiol. Rev.* **23:** 371–390.

Nicholson A.W. and Nicholson R.H. 2002. Molecular characterization of a mouse cDNA encoding Dicer, a ribonuclease III ortholog involved in RNA interference. *Mamm. Genome* **13:** 67–73.

Nicholson A.W., Niebling K.R., McOsker P.L., and Robertson H.D. 1988. Accurate *in vitro* cleavage by RNase III of phosphorothiate-substituted RNA processing signals in bacteriophage T7 early mRNA. *Nucleic Acids Res.* **16:** 1577–1591.

Nikolaev N., Silengo L., and Schlessinger D. 1973. Synthesis of a large precursor to ribosomal RNA in a mutant of *Escherichia coli*. *Proc. Natl. Acad. Sci.* **70:** 3361–3365.

Nonoyama M. and Ikeda Y. 1964. Ribonuclease-resistant RNA found in cells of *Escherichia coli* infected with RNA phage. *J. Mol. Biol.* **9:** 763–771.

Novotny J., Diegel S., Schirmacher H., Mohrle A., Hildebrandt M., Oberstrass J., and Nellen W. 2001. *Dictyostelium* double-stranded ribonuclease. *Methods Enzymol.* **342:** 193–212.

Oguro A., Kakeshita H., Nakamura K., Yamane K., Wang W., and Bechhofer D.H. 1998. *Bacillus subtilis* RNase III cleaves both 5′ and 3′ sites of the small cytoplasmic RNA precursor. *J. Biol. Chem.* **273:** 19542–19547.

Ooi S.L., Samarsky D.A., Fournier M.J., and Boeke J.D. 1998. Intronic snoRNA biosynthesis in *Saccharomyces cerevisiae* depends on the lariat-debranching enzyme: Intron length effects and activity of a precursor snoRNA. *RNA* **4:** 1096–1110.

Parrish S., Fleenor J., Xu S., Mello C., and Fire A. 2000. Functional anatomy of a dsRNA trigger: Differential requirement for the two trigger strands in RNA interference. *Mol. Cell* **6:** 1077–1087.

Pragai B. and Apirion D. 1981. Processing of bacteriophage T4 tRNAs. The role of RNase III. *J. Mol. Biol.* **153:** 619–630.

Price B., Adamidis T., Kong R., and Champness W. 1999. A *Streptomyces coelicolor* antibiotic regulatory gene, *absB*, encodes an RNase III homolog. *J. Bacteriol.* **181:** 6142–6151.

Provost P., Dishart D., Doucet J., Frendewey D., Samuelsson B., and Radmark O. 2002. ribonuclease activity and RNA binding of recombinant human Dicer. *EMBO J.* **21:** 5864–5974.

Qu L.-H., Henras A., Lu Y.-J., Zhou H., Zhou W.-X., Zhu Y.-Q., Zhao J., Henry Y., Caizergues-Ferrer M., and Bachellerie J.-P. 1999. Seven novel methylation guide small nucleolar RNAs are processed from a common polycistronic transcript by Rat1p and RNase III in yeast. *Mol. Cell. Biol.* **19:** 1144–1158.

Ramos A., Grunert S., Adams J., Micklem D.R., Proctor M.R., Freund S., Bycroft M., St. Johnston D., and Varani G. 2000. RNA recognition by a Staufen double-stranded RNA-binding domain. *EMBO J.* **19:** 1366–1377.

Robertson H.D. 1982. *Escherichia coli* ribonuclease III cleavage sites. *Cell* **30:** 669–672.

Robertson H.D. and Dunn J.J. 1975. Ribonucleic acid processing activity of *Escherichia coli* ribonuclease III. *J. Biol. Chem.* **250:** 3050–3056.

Robertson H.D., Webster R.E., and Zinder N.D. 1967. A nuclease specific for double-stranded RNA. *Virology* **12:** 718–719.

——. 1968. Purification and properties of ribonuclease III from *Escherichia coli*. *J. Biol. Chem.* **243:** 82–91.

Rotondo G. and Frendewey D. 2001. Pac1 ribonuclease of *Schizosaccharomyces pombe*. *Methods Enzymol.* **342:** 168–193.

Rotondo G., Huang J.Y., and Frendewey D. 1997. Substrate structure requirements of the Pac1 ribonuclease from *Schizosaccharomyces pombe*. *RNA* **3:** 1182–1193.

Rudinger J., Hillenbrandt R., Sprinzl M., and Giege R. 1996. Antideterminants present in minihelix(Sec) hinder its recognition by prokaryotic elongation factor Tu. *EMBO J.* **15:** 650–657.

Ryter J.M. and Schultz S.C. 1998. Molecular basis of double-stranded RNA-protein interactions: Structure of a dsRNA-binding domain complexed with dsRNA. *EMBO J.* **17:** 7505–7513.

St. Johnston D., Brown N.H., Gall J.G., and Jantsch M. 1992. A conserved double-stranded RNA binding domain. *Proc. Natl. Acad. Sci.* **89:** 10979–10983.

Sharp P.A. 2001. RNA interference—2001. *Genes Dev.* **15:** 485–490.

Sorrentino S. and Libonati S. 1997. Structure-function relationships in human ribonucleases: Main distinctive features of the major RNase types. *FEBS Lett.* **404:** 1–5.

Spasov K., Perdomo L.I., Evakine E., and Nazar R.N. 2002. RAC protein directs the complete removal of the 3′-external transcribed spacer by the Pac1 nuclease. *Mol. Cell* **9:** 433–437.

Srivastava A.K. and Schlessinger D. 1990. Mechanism and regulation of bacterial ribosomal RNA processing. *Annu. Rev. Microbiol.* **44:** 105–129.

Sun W. and Nicholson A.W. 2001. Mechanism of action of *Escherichia coli* ribonuclease III. Stringent chemical requirement for the glutamic acid 117 side chain, and Mn$^{2+}$ rescue of the Glu117Asp mutant. *Biochemistry* **40:** 5102–5110.

Sun W., Jun E.-J., and Nicholson A.W. 2001. Intrinsic double-stranded-RNA processing activity of *Escherichia coli* ribonuclease III lacking the dsRNA-binding domain. *Biochemistry* **40:** 14976–14984.

Thompson L.D. and Daniels C.J. 1990. Recognition of exon-intron boundaries by the *Halobacterium volcanii* tRNA intron endonuclease. *J. Biol. Chem.* **265:** 18104–18111.

Tremblay A., Lamontagne B., Catala M., Yam Y., Larose S., Good L., and Abou Elela S. 2002. A physical interaction between Gar1p and Rnt1p is required for the nuclear import of H/ACA small nucleolar RNA-associated proteins. *Mol. Cell. Biol.* **22:** 4792–4802.

Tuschl T., Zamore P.D., Lehmann R., Bartel D.P., and Sharp P.A. 1999. Targeted mRNA degradation by double-stranded RNA *in vitro*. *Genes Dev.* **13:** 3191–31977.

Villa T., Ceradini F., and Bozzoni I. 2000. Identification of a novel element required for processing of intron-encoded box C/D small nucleolar RNAs in *Saccharomyces cerevisiae*. *Mol. Cell. Biol.* **20:**

1311–1320.

Villa T., Ceradini F., Presutti C., and Bozzoni I. 1998. Processing of the intron-encoded U18 small nucleolar RNA in the yeast *Saccharomyces cerevisiae* relies on both exo- and endonucleolytic activities. *Mol. Cell. Biol.* **18:** 3376–3383.

Wilson H.R., Yu D., Peters H.K., Zhou J., and Court D.L. 2002. The global regulator RNase III modulates translation repression by the transcription elongation factor N. *EMBO J.* **21:** 4154–4161.

Yang D., Lu H., and Erickson J.W. 2000. Evidence that processed small dsRNAs may mediate sequence-specific mRNA degradation during RNAi in *Drosophila* embryos. *Curr. Biol.* **10:** 1191–1200.

Yang D., Buchholz F., Huang Z., Goga A., Chen C.-Y., Brodsky F.M., and Bishop J.M. 2002. Short RNA duplexes produced by hydrolysis with *Escherichia coli* RNase III mediate effective RNA interference in mammalian cells. *Proc. Natl. Acad. Sci.* **99:** 9942–9947.

Wagner E.G.H. and Simons R.W. 1994. Antisense RNA control in bacteria, phages and plasmids. *Annu. Rev. Biochem.* **48:** 713–742.

Wu H., Xu H., Miraglia L.J., and Crooke S.T. 2000. Human RNase III is a 160-kDa protein involved in preribosomal RNA processing. *J. Biol. Chem.* **275:** 36957–36965.

Wu H., Yang P.K., Butcher S.E., Kang S., Chanfreau G., and Feigon J. 2001. A novel family of RNA tetraloop structure forms the recognition site for *Saccharomyces cerevisiae* RNase III. *EMBO J.* **20:** 7240–7249.

Zahn K., Inui M., and Yukawa H. 2000. Divergent mechanisms of 5′ 23S rRNA IVS processing in the alpha-proteobacteria. *Nucleic Acids Res.* **28:** 4623–4633.

Zamore P.D. 2001a. Thirty-three years later, a glimpse at the ribonuclease III active site. *Mol. Cell* **8:** 1158–1160.

———. 2001b. RNA interference: Listening to the sound of silence. *Nature Struct. Biol.* **8:** 746–750.

Zamore P.D., Tuschl T., Sharp P.A., and Bartel D.P. 2000. RNAi: Double-stranded RNA directs the ATP-dependent cleavage of mRNA at 21 to 23 nucleotide intervals. *Cell* **101:** 25–33.

Zhang H., Kolb F.A., Brondani V., Billy E., and Filipowicz W. 2002. Human Dicer preferentially cleaves dsRNAs at their termini without a requirement for ATP. *EMBO J.* **21:** 5875–5885.

Zhang K. and Nicholson A.W. 1997. Regulation of ribonuclease III processing by double-helical sequence antideterminants. *Proc. Natl. Acad. Sci.* **94:** 13437–13441.

Zhou D., Frendewey D., and Lobo-Ruppert S.M. 1999. Pac1p, an RNase III homolog, is required for formation of the 3′ end of U2 snRNA in *Schizosaccharomyces pombe*. *RNA* **5:** 1083–1098.

Zinder N.D. 1980. Portraits of viruses: RNA phage. *Intervirology* **13:** 257–270.

# CHAPTER 9

# RNA-dependent RNA Polymerase in Gene Silencing

Luyun Huang, John Gledhill, and Craig E. Cameron

*Department of Biochemistry and Molecular Biology, Pennsylvania State University, University Park, Pennsylvania 16802*

HOMOLOGY-DEPENDENT GENE SILENCING (HDGS) has been observed in a variety of organisms. This phenomenon—termed cosuppression in plants (Jorgensen 1990; see also Chapter 1), quelling in fungi (Cogoni et al. 1996), and RNA interference (RNAi) in nematodes (Fire et al. 1998)—is normally triggered by transgenes, double-stranded RNAs (dsRNAs), or viruses that share a homologous sequence with an endogenous gene or another transgene. In some cases, it also occurs between two genomic loci (Coen and Carpenter 1988). The underlying mechanisms of HDGS may act at the transcriptional or posttranscriptional level (for review, see Bernstein et al. 2001a).

Evidence from numerous studies in various systems has revealed that dsRNA has a central role in HDGS at the posttranscriptional level (Guo and Kemphues 1995; Angell and Baulcombe 1997; Fire et al. 1998; Waterhouse et al. 1998; Cogoni and Macino 1999; Jakowitsch et al. 1999; Mette et al. 1999; Dalmay et al. 2000; Mette et al. 2000; Mourrain et al. 2000; Smardon et al. 2000; Stam et al. 2000). At the transcriptional level, dsRNA is linked to the methylation of genomic DNA in silenced plants (Wassenegger et al. 1994; Pelissier and Wassenegger 2000; Jones et al. 2001). Moreover, recent studies have shown that HDGS shares common components with other regulatory pathways, including "nonsense-mediated decay" (NMD) of mRNAs (Domeier et al. 2000) and the regulation of gene expression by small temporal RNAs (stRNAs), also referred to as microRNAs (miRNAs) (Reinhart et al. 2000; Grishok et al. 2001; Lagos-Quintana et al. 2001; Lau et al. 2001; Lee and Ambros 2001).

A schematic representation of RNAi-related gene-silencing pathways is shown in Figure 9.1 (Bernstein et al. 2001a). Triggers such as dsRNAs introduced to the cell or transcripts from transgenes or transposons might be recognized by some kind of adapter proteins and then processed to approximately 22-nucleotide short interfering dsRNAs (siRNAs) by Dicer, a member of the RNase III family of nucleases. Proteins in the Argonaute family and perhaps some other factors bind the siRNAs and form the RNA-induced silencing complex (RISC). The protein-protected siRNAs hybridize to RNAs with homologous sequences and direct the degradation of the corresponding target RNAs. stRNAs may use the same or similar machinery to regulate gene expression. Genomic modification may also take place in this process.

Guo and Kemphues (1995) have observed that injection of sense and antisense RNAs silenced the *par-1* gene in *Caenorhabditis elegans* equally well. dsRNA was later found to be tenfold more efficient in silencing than either of the single-stranded RNAs (ssRNAs)

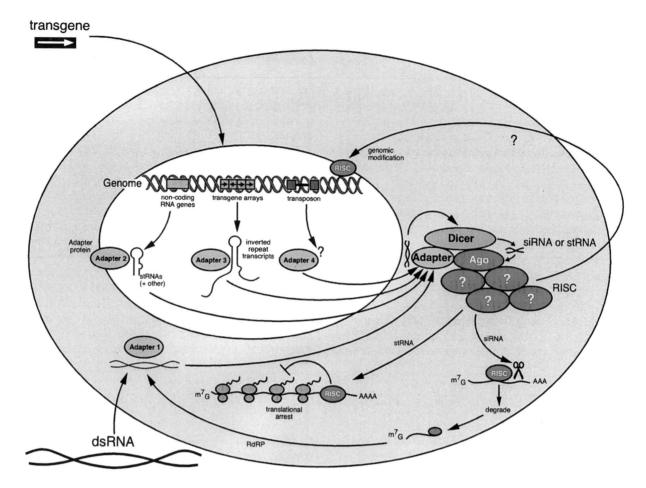

**FIGURE 9.1.** A schematic representation of RNAi-related gene silencing pathways. The silencing trigger can be generated by a number of different circumstances. A possible explanation for different types of triggers leading to different outcomes is that each may be delivered to Dicer-containing complexes by specific adapter proteins, which interact with specific Argonaute family members. The precise adapter-Argonaute interactions may determine the type of RISC complex generated and thus the silencing mode, either RNA degradation as in standard RNAi, translational suppression as for stRNAs, or perhaps also genomic modification. (Reprinted, with permission of Cambridge University Press, from Bernstein et al. 2001a.)

(Fire et al. 1998). In plants, inverted repeat arrays of transgenes that lead to transcripts with hairpin structures were much more effective triggers than direct repeat arrays (Waterhouse et al. 1998). Given our current understanding of gene silencing, it is easy to comprehend the role of dsRNA as a trigger for silencing, because this molecule is a substrate for Dicer (Bass 2000; Cerutti et al. 2000; Elbashir et al. 2001). However, it is still difficult to explain how dispersed and single-copy transgenes, sense-strand RNAs, and endogenes trigger silencing unless a cellular RNA-dependent RNA polymerase (RdRP) is involved.

Two other striking observations are that local induction of gene silencing can spread throughout the whole organism, and gene silencing is often heritable to the next generation (Palauqui and Vaucheret 1998; Voinnet et al. 1998; Grishok et al. 2000; Sonoda and Nishiguchi 2000). Systemic silencing even occurs between scion (the upper vegetative tissues) and stock (lower tissues and the root system) in plants (Palauqui et al. 1997). In addition, a few molecules of dsRNA are sufficient to silence genes in *C. elegans* and *Drosophila* (Fire et al. 1998; Kennerdell and Carthew 1998). Thus, how is the initial silencing signal spread, amplified, and maintained?

Speculation that cellular RdRPs have a critical role in posttranscriptional gene silencing (PTGS) began 9 years ago after a group of plant biologists reported the isolation and characterization of a cellular RdRP from tomato (Lindbo et al. 1993). Since this time, more and more evidence has accumulated to support this hypothesis. This chapter reviews the important discoveries that are related to the role of an RdRP in gene silencing.

## CONCEPTS AND STRATEGIES

### Tomato Cellular RdRP

RNA-dependent RNA polymerases were first observed in RNA viruses (Blumenthal and Carmichael 1979). The viral RdRP is responsible for transcription and replication of the viral genome. Interestingly, in some plants, RdRP activity can also be found in healthy tissue, for example, in Chinese cabbage (Astier-Manifacier and Cornuet 1971), cauliflower (Astier-Manifacier and Cornuet 1978), tobacco (Duda et al. 1973; Duda 1979; Takanami and Fraenkel-Conrat 1982), tomato (Boege and Sänger 1980), and cucumber (Khan et al. 1986). RdRP activity was normally demonstrated by performing RNA transcription reactions in the presence of DNA-dependent RNA polymerase inhibitors. Although the existence of the plant cellular RdRPs has been known for decades, their biological function remains unclear. Upon infection with RNA viruses, both the amount (Van der Meer et al. 1984) and specific activity of the plant RdRP increase (Astier-Manifacier and Cornuet 1971; Duda et al. 1973; Romaine and Zaitlin 1978; Takanami and Fraenkel-Conrat 1982). However, because these studies failed to purify the cellular RdRP to homogeneity, it is not possible to assign unambiguously the increased activity to the cellular RdRP alone. In addition to plants, RdRP activity has also been detected in animal cells (murine erythroleukemia cells), protozoa (*Trypanosoma brucei* extracts), and bacteria (*Halobacterium cutirubrum*) (Louis and Fitt 1972; Volloch et al. 1987; Volloch et al. 1991). However, these studies are rather preliminary. HDGS does function in mice (Wianny and Zernicka-Goetz 2000) and protozoa (Ngo et al. 1998).

In 1993, Schiebel and co-workers isolated and purified an RdRP from tomato leaves (Schiebel et al. 1993a) and characterized its catalytic properties in vitro (Schiebel et al. 1993b). Although there was no study in the literature that showed the involvement of this tomato RdRP in gene silencing, this group's reports encouraged speculation regarding a role for cellular RdRPs in silencing (Lindbo et al. 1993). Later, the same group successfully cloned the gene for this RdRP (Schiebel et al. 1998). Sequence analysis of the gene provided very intriguing information. The tomato RdRP represented a new family of enzymes with homologs of this enzyme existing in a wide variety of organisms. Genetic analysis of putative RdRP genes in several organisms has provided a direct link between this class of enzymes and gene silencing. However, the only RdRP that has been purified and characterized is that from plants. The information gathered from these studies is reviewed below and may be representative of other members of this family.

#### Purification and physical properties

The tomato RdRP was isolated from both healthy tomato leaves and viroid-infected leaves (Schiebel et al. 1993a). Viroids are subviral plant pathogens composed of a single-stranded circular RNA of 246–401 nucleotides with a compact secondary structure. Like RNA viruses, viroid infection can increase the expression of the cellular RdRP. Unlike viruses, however, viroids do not code for any protein; they depend entirely on host enzymes for their replication (Davies et al. 1974; Hall et al. 1974; Semancik et al. 1977). Hence, in protein preparations from viroid-infected leaves, the possibility for contamination of purified cellular polymerase preparations with viral enzymes is eliminated.

Elimination of contaminating cellular enzymes, such as terminal ribonucleotidyl-transferases (TNTases), especially RNA uridylyltransferase, had provided the greatest obstacle to purification (Zabel et al. 1981). It is challenging to distinguish unambiguously between RdRP-mediated transcription and TNTase-mediated addition using standard activity assays. Therefore, all TNTase contamination needed to be removed before a reliable assay could be performed. The general procedure employed is as follows:

- Healthy tomato leaves or those infected with potato spindle tuber viroid (PSTV) were homogenized, extracted, and centrifuged. Throughout the purification, the protein was kept in buffer containing 50% glycerol to maximize stability.

- DEAE-Sepharose chromatography was used to remove nucleic acids and the majority of the pigments from the tissue extracts.

- dsDNA-cellulose chromatography was used to remove most of the other proteins.

- A Mono Q column was used to separate TNTases and the desired RdRP.

- Fractions containing RdRP activity were evaluated by SDS-polyacrylamide gel electrophoresis (SDS-PAGE) and silver staining. If a single silver-stained band was not observed, then the impurities were removed by chromatography of the sample on a poly(U)-Sepharose column.

The yield of this preparation was low. From 1 kg of PSTV-infected tomato leaves, only about 50 μg of RdRP protein was isolated. However, the enzyme activity increased three- to fourfold in the infected leaves compared with RdRP isolated from the same amount of healthy leaves.

The apparent molecular mass of the RdRP determined by SDS-PAGE was 128 kD. The sedimentation coefficient of the active RdRP was about 6.6S as determined by sucrose gradient centrifugation under native conditions. Calculated from the sedimentation coefficient, the size of the RdRP was 115 kD. There is only about a 10% difference between the molecular mass values obtained under native or denaturing conditions, indicating that the RdRP is likely a globular enzyme.

RdRP activity was assayed by quantifying the incorporation of radiolabeled [5,6-$^3$H] UTP into RNA transcripts using tobacco mosaic virus (TMV) RNA as template (see Protocol 1). The apparent $K_m$ value for TMV RNA was about 1.2 μg/ml. The purified enzyme exhibited a specific activity of 500 nmoles/mg per 30 minutes, which represented a 100,000-fold enrichment from the initial tissue extract. The fact that this enzyme activity was inhibited by heparin but was insensitive to α-amanitin and actinomycin D proved that DNA-dependent RNA polymerase was not present in the RdRP preparation.

### In vitro catalytic properties

In vitro analysis of the purified protein was performed to obtain information on the biochemical properties of the RdRP (Schiebel et al. 1993b). For these studies, TMV RNA was suboptimal as a template, because heterogeneous transcripts <100 nucleotides in length were produced. This result is quite different from that observed for viral RdRPs. To obtain information on the fidelity of the reaction, short synthetic DNA and RNA molecules of defined sequence were used as templates (see Protocol 2). It was found that the tomato RdRP transcribes both short ssRNA and ssDNA molecules precisely. However, dsRNA and dsDNA were not transcribed at all. These results are consistent with previous reports for partially purified plant RdRPs from cauliflower (Astier-Manifacier and Cornuet 1978), tobacco (Duda 1979; Ikegami and Fraenkel-Conrat 1979), and cucumber (Singer et al. 1983).

Experimental analysis of the primer requirements of the RdRP revealed that the enzyme could catalyze transcription in the presence or absence of primers. Primer extension was observed with RNA or DNA dinucleotides and trinucleotides complementary to the 3′-terminal nucleotides of the template. In the absence of primers, the RdRP initiated transcription at or near the 3′-terminal nucleotide of a single-stranded template.

Although the catalytic properties of the cellular RdRP determined by this in vitro analysis may hold for the enzyme in vivo, it is important to keep in mind the possibility that the protein purified might only be the core component of a larger RdRP complex in vivo. The existence of other proteins could change the specificity and/or activity of this enzyme.

## Cloning of the gene encoding the tomato RdRP

The entire gene for the tomato RdRP was obtained from two overlapping clones (Schiebel et al. 1998). The open reading frame (ORF) encodes 1114 amino acids with a calculated molecular mass of about 127 kD. This value is in good agreement with that observed experimentally for the isolated tomato RdRP (Schiebel et al. 1993a). Unfortunately, attempts to express the protein in recombinant systems were unsuccessful. Therefore, indirect methods were employed to prove that the cloned gene (C-RdRP gene) is in fact the gene for the isolated tomato RdRP (T-RdRP).

Antisera were raised against synthetic peptides whose sequences were derived from the C-RdRP. All results with the antisera supported the consanguinity of C-RdRP and T-RdRP. In immunoblot experiments, only one protein of approximately 127 kD in the tomato leaves extract was detected by the C-RdRP-specific antibody. The changes in enzyme activity of T-RdRP fractions from each purification step were in perfect accordance with the changes in reactivity to C-RdRP-specific antibody. Finally, the C-RdRP-specific antibody inhibited T-RdRP activity.

The sequence of the C-RdRP did not share any similarity with known RNA virus replicases and lacked the "GDD" motif common to the active site of most viral RdRPs (Kamer and Argos 1984). To identify RdRP homologs in different plant species, Southern analysis and polymerase chain reaction (PCR) amplification were performed using combinations of C-RdRP-specific primers. Evidence was obtained for related nucleic acid sequences in tobacco, petunia, *Arabidopsis*, and wheat. The RdRP homolog sequences amplified from these four plants overlapped a 93-amino-acid region (Figure 9.2). Queries of existing sequence databases also suggested that cellular RdRPs might exist in other organisms, in particular, *Schizosaccharomyces pombe* and *C. elegans*.

```
              805                                                  852
Tom.     DIYFVCWDQDMIPPRQVQPMEYPPAPSIQLDHDVTIEEVEEYFTNYI
Tb.      DIYFVCWDPDLIPPRQVQSMDYTPAPTTQLDHDVTIEEVEEYFTNYI
Pt.      DIYFVCWDPDLIPPRQVQPMDYTPAPSIQLDHDVTIEEVEEYFTNYM
Ara.     DIYFVCWDQDLVPPRTSEPMDYTPEPTQILDHDVTIEEVEEYFANYI
Wh.      DIYFVSWDPDLIPTRMVAPMDYTPAPTETLDHDVMIEEVHEYFTNYI

              853                                                  897
Tom.     VNDSLGIMANAHVVFADREPDMAMSDPCKKLAELFSIAVDFPKTGV
Tb.      INDSLGIIANAHVVFADREPDMAMSDPCKQLAQLFSIAVDFPKTGV
Pt.      VNDGLGVIANAHVVFADREPNMAMSDPCIELAQLFSIAVDFPKTGV
Ara.     VNDSLGIIANAHTAFADKGPLKAFSDPCIELAKKFSTAVDFPKTGV
Wh.      VNESLGIIANAHVVFADREILKAFSTPCIKLAELFSIAVDFPKTGV
```

**FIGURE 9.2.** Comparison of the tomato RdRP (Tom.) amino acid sequence with that of tobacco (Tb.), petunia (Pt.), *Arabidopsis* (Ara.), and wheat (Wh.).

## RdRP Homolog Genes in Gene Silencing

Evidence for the involvement of the RdRP in gene silencing derives from genetic studies in a variety of organisms. Genes in *Arabidopsis* (*sgs2/sde1*), *Neurospora* (*qde-1*), *C. elegans* (*ego-1, rrf-1*), and *Dictyostelium discoideum* (*rrpA*) are essential for PTGS (Cogoni and Macino 1999; Dalmay et al. 2000; Mourrain et al. 2000; Smardon et al. 2000; Martens et al. 2002), and these genes share significant homology with the tomato RdRP (Figure 9.3). These studies are summarized below.

```
tomato    FSSSQLRDNSVWMFASRPGL---TANDIRAWMGDFSQIKNVAKYAARLGQSFGSSRET-LSVLRHEIEVIPDV
QDE-1     VVPAEEPVEQRTEFKVSQML------DWLLQLDNNTW-QPHLKLFSRIQLGLSKTYAI-MTLEPHQIRHHKTD
EGO-1     YSSSQCREYERIYQIKPPITFNPKIQAARKNLGRFETIDNIPKMMARLGQCFTQSRLSGVNLERCTYMTTYDL
RRF-1     FTDKQLDRFYKCNPTASNINFKPKIDEVRFQLGRFSEIENVPKLMARLGQCFTQSRLTGVGLGRDDYCSTYDL
SDE1      FSANQLRDRSAWFFAEDGKT---RVSDIKTWMGKFKD-KNVAKCAARMGLCFSSTYAT-VDVMPHEVDTEVPD
RrpA      NSNSQLREYSSWFVSNQIGT-----HTVKIWSGIEHV-DNVRKFFRCIGLMFSTTIPT-VTLPQNRIYRIQDI

tomato    ----KVHGTSYVFSDGIGKISGDFAHRVASKCGLQY--TPSAFQIRYGGYKGVVGVDP--------------
QDE-1     LL--SPSGTGEVMNDGVGRMSRSVAKRIRDVLGLGD--VPSAVQGRFGSAKGMWVIDVDDTGDEDWIETYPSQ
EGO-1     TGGKNLKGDEYTFSDGVGMMSYRFAQMVSEVMDFGKG-VPSCFQFRFRGMKGVISIEPLLDNLRQWSISYNIS
RRF-1     TGGRATNGSEYTFSDGVGMMSYQFAQEVSQAMQFGKA-VPSCFQIRFRGNKGVIAIEPFLDEIRKWALVNGVT
SDE1      -----IERNGYVFSDGIGTITPDLAGEVMEKLKLDVHYSPCAYQIRYAGFKGVVARWPS--------------
RrpA      ------TRNTHEFTEGCGEIGPELAKHLNENYNFRP--STCAYQVRIGGNKGVLVVNNQ--------------

tomato    ---------DSSMKLSLRKSMSKYESDNIKLD-----VLGWSKYQPCYLNRQLITTLLSTLGVKDEVLEQKQ-K
QDE-1     RKWECDFVDKHQRTLEVRSVASELKSAGLNLQ---LLPVLEDRARDKVKMRQAIGDRLINDLQRQFSEQKHAL
EGO-1     KPSD---DSSWSLNCMFRPSQIKFISKRHPRDQ--VEIVKYSSPVPVALNKPFINILDQVSEMQSLECHRRVT
RRF-1     --------SMKMAKCLFRPSQIKFQAKAISGDQ--IEMVKFSSAVLVALNKPFINILDQVSEMQSLDCHKRIT
SDE1      --------KSDGIRLALRDSMKKIFSKHTILE-----ICSWTRFQPGFLNRQIITTLLSVLGVPVEIFWDMQ-E
RrpA      --------APDPSGIYIRPSMVKFNPIDCGDEHRTLEICSVSTTSRCKLNRQVISLLSTLGTQDNVFFALQ-D

tomato    EAVDQLDAILHDSLKAQEALE------LMS--PGENTNILKAMLNCGYKPDAEPFLSMMLQTFRASKLLDLRT
QDE-1     NRPVEFRQWVYESYSSRATRVSHGRVPFLAGLPDSQEETLNFLMNSGFDPKKQKYLQDIAWDLQKRKCDTLKS
EGO-1     NRIEELLDRQMLSFAQQMVDETFCRNRLKE--LPRRVDIDYLRTTWGFTLSSEPFFRSLIKASIKFSITRQLR
RRF-1     SRIEELMDRQILSFAKQMNEETFCRNKLKE--FPRRIDIDNLRTMWGFTLSSEPFFRSLIKASIKFSITKQLC
SDE1      SMLYKLDRILVDTDVAFEVLT------AS--CAEQGNTAAIMLSAGFKPKTEPHLRGMLSSVRIAQLWGLRE
RrpA      HYLNQVAQIVNDTNASKQAIV-----------EFFP-DITEGEL------YQDPYIRRILISLYKLKMERIQQ

tomato    RSRIFIPN--GRTMMGCLDESRTLEYGQVFVQFTGAG-HGEFSDDLHPFNNSRSTNSNFILKGNVVVAKNPCL
QDE-1     KLNIRVGR--SAYIYMIADFWGVLEENEVHVGFSSK-----FRDEEESFTLLSDC--------DVLVARSPAH
EGO-1     KEQIPIPCDLGRSMLGVVDETGRLQYGQIFVQYTKN-----LALKLPPKNAARQV-----LTGTVLLTKNPCI
RRF-1     KEQIQIPSELGRSMLGVVDETGRLQYGQIFVQYTKN-----YKKKLPPRDSNNKVHGSEIVTGTVLLTKNPCI
SDE1      KSRIFVTS--GRWLMGCLDEAGILEHGQCFIQVSKPSIENCFSKHGSRFKETKKDL--EVVKGYVAIAKNPCL
RrpA      KCHIEIKD--SRMLLGVCDPTNSLPPNTVFVQLEEE-------DE-DDDDDGRKYE--KVIEGLVMVIKNPCT

tomato    HPGDIRVLKAVNVRALHHMVDCVVFPQKGKRPHPNECSGSDLDGDIYFVCWDQDMIPP-RQVQPMEYP-PAPS
QDE-1     FPSDIQRVRAVFKPELHSLKDVIIFSTKGDVPLAKKLSGGDYDGDMAWVCWDPEIVDGFVNAEMPLEPDLSRY
EGO-1     VAGDVRIFEAVDIPELHHMCDVVVFPQHGPRPHPDEMAGSDLDGDEYSIIWDQQLLLD-KNEDPYDFTSEKQK
RRF-1     VPGDVRIFEAVDIPELHHMCDVVVFPQHGPRPHPDEMAGSDLDGDEYSVIWDQELLLE-RNEEPFDFAVEKIK
SDE1      HPGDVRILEAVDVPQLHHMYDCLIFPQKGDRPHTNEASGSDLDGDLYFVAWDQKLIPPNRKSYPAMHYDAAEE
RrpA      HPGDVRYLKAVDNIRLRHLRNVLVFSTKGDVPNFKEISGSDLDGDRYFFCYDKSLIGNRSESETAYLVVETVS

tomato    IQLDHDVTIEEVEEYFTNYIVNDS---LGIIANAHVVFA-----------DREPDMAMSD---PCKKLAELFS
QDE-1     LKKDKTTFKQLMASHGTGSAAKEQTTYDMIQKSFHFALQPNFLGMCTNYKERLCYINNSVSNKPAIILSSLVG
EGO-1     ASFKEDEIDDLMREFYVKYLKLDS---VGQISNSHLHNS-----------DQYGLNARV-----CMDLAKKNC
RRF-1     VPYDREKLDVLMREFYVTYLKLDS---VGQISNSHLHNS-----------DQYGLNSRV-----CMDLAKKNC
SDE1      KSLGRAVNHQDIIDFFARNLANEQ---LGTICNAHVVHA-----------DRSEYGAMDE---ECLLLAELAA
RrpA      NNDKKANVFNDPFALSSMYSTNAERQELGKLYHSHLAIS-----------DQYGANHKY-----SIQISKECF

tomato    IAVDFPKTGVP--------A-----EIPSQLRPKEYPDFMDKPDKTS--YISERVIGKLFRKVKDK-APQAS
QDE-1     NLVDQSKQGIVFNE------ASWAQLRRELLGGALSLPDPMYKSDSWLGRGEPTHIIDYLKFSIARPAIDKEL
EGO-1     QAVDFTKSGQPPDELERKWRKDEETGEMIPPERAERVPDYHMGNDHTPM-YVSPRLCGKLFREFKA--IDDVL
RRF-1     QAVDFTKSGQPPDPLETKWRADPVTFEVIPPENPERIPDFHMGNERSPM-YVSPRLCGKLFREFQA--IDNVI
SDE1      TAVDFPKTGKI--------------VSMPFHLKPKLYPDFMGKEDYQT--YKSNKILGRLYRPVKEVYDEDAE
RrpA      KEIDYPKTGIHGT---------IPKEVNIRLQTVGYPHYMQRENSTRVYYQSKKIMGKMYDQIDQ-----LV

tomato    SIATFTRDVARRSYDADMEVDGFEDYIDEAFDYKTEYDNKLGNLMDYYGIKT-EAEILSGGIMKA-SKTFDRR
QDE-1     EAFHNAMKAAKDTEDGAHFWDPDLASYYTFFKEISDKSRSSALLFTTLKNRIGEVEKEYGRLVKNKEMRDSKD
EGO-1     KISEERDEQVEISIDETIKIDGYTEYMASAKNDLARYNAQLRSMMENYGIKT-EGEVFSGCIVDMRNRISDKD
RRF-1     KISEERDEQYNIELDETIFVTGFERYMESAQKQLSSYNGQLRSIMENYGIRS-EGEIMSGCIVEMRNRISDKD
SDE1      ASSEESTDPSAIPYDAVLEIPGFEDLIPEAWGHKCSYDGQLIGLLGQYKVQK-EEEIVTGHIWSM-PKYTSKK
RrpA      Y---IGDFLPNISLDKSNLVDGYEIYLNSAKILYSQYKLQVHSLLRHYSAES-EESIMIGFLDQG--FISDKV
```

**FIGURE 9.3.** Comparison of putative RdRP amino acid sequence from different organisms including tomato plant, *Neurospora* (QDE-1), *C. elegans* (EGO-1, RRF-1), *Arabidopsis* (SDE1), and *Dictyostelium discoideum* (RrpA). Amino acids shaded red indicate absolutely conserved residues in all sequences in the alignment. Those shaded blue and green indicate conservative substitutions and semiconservative substitutions, respectively.

## qde-1 in Neurospora crassa

Cogoni and Macino (1999) cloned the first gene of this family, *qde-1* (quelling-defective), from *Neurospora crassa*. They used the *albino-1* (*al-1*) gene, which is essential for carotenoid biosynthesis, as a visible reporter for gene silencing. A silenced *al-1* transgenic strain would display a white phenotype. Three classes of mutants, *qde-1*, *qde-2*, and *qde-3*, were identified. Random insertional mutagenesis with a tagged plasmid was introduced into an *al-1* transgenic strain to screen for genes essential for silencing. One strain of the insertional transformants showed release of gene silencing because the orange wild-type phenotype was recovered. After heterokaryon genetic analysis, this strain was assigned to the *qde-1* mutant group.

Using a plasmid-rescue method, a chromosomal DNA fragment flanking the integration site was isolated. This fragment was used to probe an *N. crassa* genomic cosmid library. The location of the *qde-1* gene was narrowed down to a 7.9-kb fragment and sequencing uncovered a long ORF (1402 amino acids). The putative QDE-1 protein has a relative molecular mass of 158 kD, which is fairly close to the size of tomato RdRP.

Analysis of the expression patterns showed that the *qde-1* mRNA had a twofold higher steady-state level in an *al-1*-silenced strain than in a wild-type strain, indicating that this gene is activated during gene silencing. QDE-1 was predicted to be an intracellular soluble protein because it does not contain a signal peptide or transmembrane domain. A BLAST search revealed that QDE-1 shares significant homology with tomato RdRP and some putative proteins in other organisms (Cogoni and Macino 1999).

## ego-1 and rrf genes in C. elegans

Unlike the *qde-1* gene in fungi isolated from a silencing-deficient mutant and subsequently found to be homologous to the tomato RdRP, the *C. elegans ego-1* gene was identified in screens for mutants defective in a pathway that signals germ-line proliferation (Qiao et al. 1995). When the *ego-1* gene was cloned and sequenced, it became clear that this gene encoded a homolog of the tomato RdRP. In addition, three other predicted products of *C. elegans* genes with unknown function, RRF-1, RRF-2, and RRF-3 (RdRP family), were also homologous to EGO-1 (Smardon et al. 2000).

Because *ego-1* is predominantly expressed in the germ line, Smardon et al. (2000) used RNAi to silence four germ-line-expressed genes and one somatic gene in *ego-1* mutants. *ego-1* mutants were defective in RNAi for three germ-line-expressed genes: *gld-1*, *mpk-1*, and *ncc-1*, but not for *lag-1*. RNAi of *unc-22*, the somatic gene, was not affected by mutation of *ego-1*.

Later, Sijen et al. (2001) discovered that another RdRP homolog, *rrf-1*, is required for RNAi in somatic cells. In *rrf-1* deletion mutants, RNAi for genes expressed in somatic tissue was lost, whereas RNAi was still functional for germ-line-expressed genes. In contrast, this study showed that *rrf-2* and *rrf-3* are not essential for RNAi in either somatic or germ-line cells. These results suggest that *ego-1* and *rrf-1* perhaps have a similar role in RNAi but function in different tissues.

## sgs2/sde1 in Arabidopsis

Two groups demonstrated independently that plant RdRP is required for gene silencing (Dalmay et al. 2000; Mourrain et al. 2000). Mourrain and co-workers isolated *sgs1* and *sgs2* (suppressor of gene silencing) mutants of *Arabidopsis* that were deficient in PTGS of a transgene. After genetic analysis, a region carrying the *sgs2* gene was identified. Computational analysis of this region revealed a sequence of 4014 bp containing one intron of 423 bp. The putative protein was 1197 amino acids and was a homolog of the tomato RdRP.

Concurrently, Dalmay et al. (2000) reported that at least four genetic loci are required for transgene-induced gene silencing in *Arabidopsis*. One mutant locus called *sde1* (silencing-defective) encodes an RdRP-related protein. The protein deduced from the cDNA is 113.7 kD and 1196 amino acids long. *sde1* and *sgs2* are, in fact, the same gene. Queries of the genome sequence of *Arabidopsis* revealed the existence of at least three additional SDE1 homologs.

Viral resistance studies performed by both groups suggested that *sgs2/sde1* is not required for virus-induced gene silencing (VIGS). *sgs2/sde1* mutants only showed enhanced susceptibility to a cucumovirus (CMV) but not to potyvirus (TuMV), tobamovirus (TVCV), the crucifer strain of tobacco mosaic virus (crTMV), tobacco rattle virus (TRV), or turnip crinkle virus (TCV). It is likely that the RdRP encoded by these RNA viruses is sufficient to amplify the dsRNA trigger to initiate and perpetuate gene silencing. However, the possibility that other RdRP genes participate in VIGS cannot be eliminated completely.

### *rrpA* in *Dictyostelium discoideum*

*D. discoideum* is a social ameba that has become a model system for cell and developmental biology (Maeda et al. 1997; Kessin 2001). RNAi was shown to be functional in this organism (Martens et al. 2002), when a construct containing inverted repeats of a β-galactosidase (β-gal) sequence was inserted into the ameba, the expression of β-gal could be reduced to undetectable levels.

Three RdRP-related genes were identified in the *Dictyostelium* genome database. RrpA and RrpB differ by only 49 amino acids (<3%) in the available sequence, whereas DosA is less conserved. The knockout strains of these genes were obtained by homologous recombination. RNAi experiments revealed that the *rrpA* gene is strictly required for RNAi, whereas the knockouts of *rrpB* and *dosA* had no obvious effect.

Intriguingly, one unique feature of RrpA is that the amino terminus of the enzyme resembles the helicase domain of K12H4.8 from *C. elegans*, a member of the Dicer gene family (Bass 2000). Members of the Dicer family usually consist of an amino-terminal helicase domain, a carboxy-terminal RNase III homolog, and a dsRNA-binding domain (Jacobsen et al. 1999; Bass 2000). However, the amino-terminal helicase domain is not found in the two Dicer homologs in *Dictyostelium* (Novotny et al. 2001; Martens et al. 2002). Martens and co-workers speculate that domain swapping has occurred between the nuclease and the polymerase, and the *Dictyostelium* Dicer homolog and RdRP are the components of the same complex. It is possible that Dicer and the RdRP interact in other systems as well.

## Degradative PCR and Transitive RNAi

All of the evidence presented thus far provide a link between the tomato RdRP homolog genes and gene silencing. However, it was not completely clear whether any of these genes really encoded an RdRP, as RdRP activity has never been recovered by expressing these genes. Moreover, organisms such as *Drosophila* perform silencing without any difficulty, yet *Drosophila* lacks an RdRP-related gene.

Late in 2001, one biochemical study in cell-free extracts of *Drosophila* embryos and one genetic study in *C. elegans* provided very compelling evidence that RdRP activity is an essential component of the silencing mechanism (Lipardi et al. 2001; Sijen et al. 2001). In the past, production of sense and antisense RNA fragments of 21–25 nucleotides, siRNAs, had been observed in silenced plants (Hamilton and Baulcombe 1999), *C. elegans* undergoing RNAi (Parrish et al. 2000), and *Drosophila* embryo extracts incubated with dsRNA (Zamore et al. 2000). The sequences of the siRNAs detected in plants and *C. elegans* correspond to the specific gene undergoing PTGS within the organism. However, in the case of in vitro studies using *Drosophila* embryo extracts, formation of the siRNAs from the dsRNA did not require the presence of the corresponding mRNA. It has been shown that Dicer, a member of the RNase III family of nucleases, specifically cleaves dsRNA into siRNAs (Bernstein et al. 2001b). The siRNAs derived from dsRNA incubated in *Drosophila* extracts have 5′-monophosphate and 3′-hydroxyl groups (Elbashir et al. 2001) and can be used to direct RNAi in vitro. However, if the 3′-hydroxyl is not present, then RNAi is inhibited. This observation is consistent with the possible role of siRNAs as primers for an RdRP (Lipardi et al. 2001).

### Degradative PCR

Using a cell-free extract prepared from *Drosophila* embryos, Lipardi et al. (2001) showed that when single-stranded, double-stranded, or capped/adenylated RNAs were used as templates, radiolabeled siRNAs were incorporated specifically into full-length RNA (Figure 9.4). Neither ssDNA nor dsDNA substrates served as templates for siRNA uptake. The full-length RNAs labeled with siRNA were double-stranded. When incubated in *Drosophila* embryo extract, the siRNA-labeled dsRNA was processed to produce new siRNAs. This second generation of siRNAs was termed secondary siRNAs.

Incorporation of siRNAs into full-length dsRNA products demonstrated for the first time that siRNAs could be used as primers for an RdRP in the *Drosophila* embryo extract. Because the secondary siRNAs can continue to prime synthesis of additional dsRNA, this process was termed degradative PCR.

RNAi can also be initiated by using synthetic 21-nucleotide duplexes in both *Drosophila* embryo extracts and Schneider cells (Elbashir et al. 2001). To demonstrate further that the siRNAs serve as primers for transcription, Lipardi et al. (2001) designed an experiment using a [32]P-labeled, 21-nucleotide siRNA as the trigger and either a sense or antisense strand of GFP RNA as the target (Lipardi et al. 2001). The duplex RNA corre-

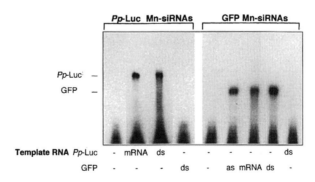

**FIGURE 9.4.** siRNAs are mRNA-specific primers for dsRNA synthesis. [32P]UTP-labeled GFP or *Pp*-Luc Mn-siRNAs, prepared from the full-length dsRNAs by the micrococcal nuclease/CIP method, were incubated in extract with the cognate and heterologous RNA templates and assayed for incorporation into larger RNAs on 1.5% agarose formaldehyde gels. *Pp*-Luc and GFP siRNAs were incorporated up to the corresponding full-size RNAs only with the homologous ssRNA or dsRNA templates. Incubation of siRNAs in the absence of template or with the heterologous template resulted in no product. Single-stranded antisense RNA (as), double-stranded RNA (ds), and capped-poly(A) RNA (mRNA) were all templates for primer incorporation. GFP (746 bp) and *Pp*-Luc (1682 bp) mark the positions of the corresponding full-length RNAs. (Reprinted, with permission of Elsevier Science, from Lipardi et al. 2001.)

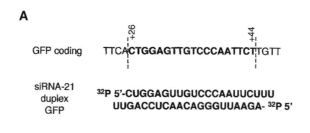

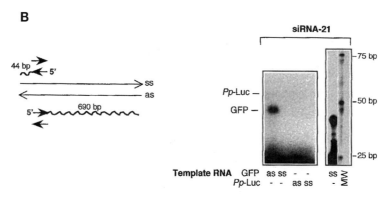

**FIGURE 9.5.** Incorporation of a synthetic 21-nucleotide duplex GFP siRNA into dsRNA. (*A*) The GFP-coding region and the sequence of the synthetic 21-nucleotide duplex GFP siRNA are indicated. The siRNA corresponds to nucleotides 26–44 in the GFP-coding region with two additional uridine residues on the 3′ end. The 5′ end of each strand of the siRNA was labeled with [$\gamma$-$^{32}$P]ATP. (*B left*) Expected product lengths for incorporation of each strand of synthetic 21-nucleotide GFP siRNA using full-length GFP sense and antisense RNA templates are shown: 690 bp with the antisense strand template and 44 bp with the sense strand template. (*Right*) $^{32}$P-labeled siRNA incorporation into full-length GFP dsRNA occurs only with the GFP antisense RNA template and gives the expected 44-bp product with the sense-strand template. *Pp*-Luc templates give no product. GFP and *Pp*-Luc mark the position of the full-length RNAs and MW denotes the 25-bp ladder. (Reprinted, with permission of Elsevier Science, from Lipardi et al. 2001.)

sponded to nucleotides 26–44 in the GFP sequence with two additional uridine residues on the 3′ end of each strand (Figure 9.5). As expected, when antisense GFP RNA was used as the template, nearly full-length dsRNA (690 nucleotides) was produced. A 44-nucleotide product was obtained for the sense-strand template.

Although a tomato RdRP homolog has not been identified in *Drosophila* through sequence comparison, the studies presented above strongly support the presence of an RdRP. It is very likely that the RdRP in *Drosophila* is not a tomato homolog because the biochemical properties appear to be unique (i.e., long RNAs are produced and DNA does not serve as a template). Protocols 3, 4, and 5 might be useful for carrying out similar biochemical studies in *Drosophilia*.

### Transitive RNAi

If siRNAs serve as primers for the RdRP to produce new dsRNAs, then, because of the orientation of a RdRP reaction, the secondary siRNAs might contain sequences from the primary target RNA upstream of the initial trigger region while lacking downstream sequences. Moreover, the secondary siRNAs would be expected to induce a secondary RNAi reaction, so the targets of the secondary, tertiary, (and so on) siRNAs should not require homology with the initial trigger RNA. This hypothesis was verified by Sijen et al. (2001) in *C. elegans* by using an RNase protection assay to monitor the siRNA population.

Initially, the worms were fed with bacteria engineered to express high levels of dsRNA homologous to a specific *C. elegans* gene (Timmons and Fire 1998; Fraser et al. 2000). The dsRNA trigger would induce gene silencing and, in turn, produce siRNAs. [32]P-labeled ssRNA probes were constructed and hybridized to denatured cellular RNA. Any unhybridized probe was degraded by treatment with ssRNA-specific ribonucleases. To distinguish siRNAs clearly from degraded mRNA, probes were constructed from the sense-strand sequence.

Figure 9.6 shows the results for two target genes: the muscle-specific gene *unc-22* and the germ-line-specific gene *pos-1*. In each case, the siRNA corresponding to the primary target sequence gave rise to the strongest signals. Moreover, signals were also detected upstream of the trigger region (closer to the 5′ end of the target mRNA). Additional analy-

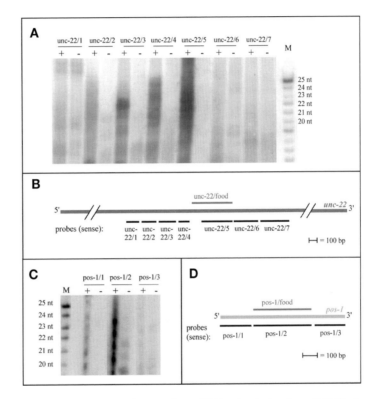

**FIGURE 9.6.** Biochemical detection of secondary siRNAs. Analysis of small RNAs from wild-type *C. elegans* grown on *E. coli* expressing dsRNA segments of *unc-22-* or *pos-1*-specific probes (all sense polarity). (*A*) Products of RNase protection assay (*right:* protected fragments of probe resolved on polyacrylamide-urea gel; *left:* detail of 16–30-nucleotide portion of gel). Feeding on *unc-22* dsRNA yielded siRNAs from the dsRNA segment comprising the food, but also produced siRNAs mapping upstream of this region. Lanes indicated with a + designate RNA from animals fed *unc-22*. To determine levels of probe-derived background, negative controls (–) were carried out by performing RNase protections with yeast tRNA as input RNA. A similar background in the siRNA size range was observed in RNase protection on RNA from animals grown on induced bacteria containing the feeding vector L4440 with no insert (data not shown). RNase protection assays have also been carried out using RNA from IPTG-induced *E. coli* producing *unc-22* dsRNA; these showed some level of probe protection but no protected fragments in the siRNA size range. Labels above the lanes indicated probes. M indicates [32]P-labeled 25-nucleotide RNA oligonucleotide marker. (*B*) Map of *unc-22* mRNA with positions of probes and bacterially produced dsRNA. (*C*) Secondary siRNAs are also produced upon feeding with *E. coli* producing *pos-1* dsRNA. Since *pos-1* is a germ-line-specific gene, RNA was isolated from egg preparations. (+) *C. elegans* populations fed with *E. coli* producing *pos-1* dsRNA; (–) equivalent RNA preparations from animals grown on *E. coli* containing the empty L4440 vector. (*D*) Map of *pos-1* mRNA with positions of probes and bacterially produced dsRNA. (Reprinted, with permission of Elsevier Science, from Sijen et al. 2001.)

sis proved that the secondary siRNAs were capable of triggering RNAi and degrading homologous mRNA sequences as well.

Using a "transitive RNAi" assay, Sijen et al. (2001) were able to distinguish between targeting by the initial dsRNA trigger and secondary siRNAs. Basically, this type of experiment employs a strain with two populations of target RNAs. The primary target RNA has a segment that is homologous to the dsRNA trigger. The secondary target, on the other hand, has no homology with the dsRNA trigger, but it has a segment that is identical to the primary target.

In the experiment shown in Figure 9.7, both primary and secondary targets are RNAs carrying *gfp*. The primary target encodes a GFP-LacZ fusion protein that is targeted to the nucleus (NLS-GFP-LacZ); the secondary target encodes a GFP that is targeted to the mitochondrion and lacks the β-gal domain (MtGFP). Injection of dsRNA segments from *lacZ* into the line carrying both transgenes produced a transitive RNAi effect. Besides the silencing of the nuclear GFP-LacZ, the expression of mitochondrial GFP was also reduced. Two triggers were used to test how the siRNAs work as a function of distance from the reporter gene. The trigger *ds-lacZU* was located just 3′ to the *gfp::lacZ* junction and produced a more significant effect than the other trigger tested, *ds-lacZL*, which is located further downstream. This "transitive RNAi" effect was illustrated nicely in a diagram presented by Nishikura (2001) (Figure 9.8).

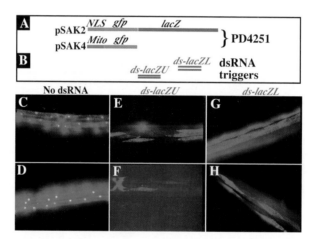

**FIGURE 9.7.** Assays for transitive RNAi using a distinct *gfp* transgene. (*A,B*) Transgenic line used for this assay (PD4251) carries two different *gfp* reporter constructs. pSAK2 produces nucleus-localized GFP fused at the carboxyl terminus to additional sequences encoding *E. coli* β-galactosidase (*lacZ*). pSAK4 produces mitochondrially localized GFP with no additional sequences at the carboxyl terminus. PD4251 animals express both nuclear and mitochondrial GFP forms in all cells of the body musculature. Young adult progeny of adult animals injected with specific dsRNA segments were examined to determine the level of interference with nucleus- and mitochondria-targeted *gfp*s. (*C,D*) Mock-injected control animals with both GFP isoforms expressed in each muscle cell. (*E,F*) Progeny of animals injected with *ds-lacZU*. This injection produced a strong transitive RNAi effect, interfering in a majority of cells not only with the nuclear targeted *gfp::lacZ* transgene, but also with the mitochondrial targeted *gfp*. (A bright "X" shape in *F* shows vulval muscles fortuitously included in the photo; these cells are generally nonresponsive to parentally injected dsRNA.) (*G,H*) Progeny of animals injected with *ds-lacZL*. This segment had only a modest effect on the expression of mitochondrially targeted *gfp*, so that the majority of cells continue to produce GFP in mitochondria but not nuclei. (*F,H*) Representative of the strongest transitive RNAi response in each population; (*E,G*) representative of the weakest effect. As negative controls, PD4251 animals injected with a variety of unrelated dsRNA segments (*unc-22A, unc-22B, lin-26IVS3*) showed no evident decrease in either nuclear or mitochondrial GFP. Animals injected with *gfp* dsRNA show near-complete (98%) loss of both nuclear and mitochondrial GFP. (Reprinted, with permission of Elsevier Science, from Sijen et al. 2001.)

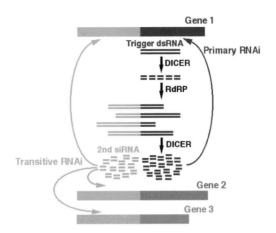

**FIGURE 9.8.** Transitive RNAi by secondary siRNA. Secondary siRNAs, generated from the dsRNA extended (*orange*) upstream of the primary target region (*blue*) by RdRP, promote transitive RNAi against their sequence homologous gene family members. (Reprinted, with permission of Elsevier Science, from Nishikura 2001.)

Similar results were observed when the primary target was an *unc-22::gfp* fusion transgene and the secondary target was an endogenous gene, *unc-22*. Injection of *ds-gfp* RNA into animals carrying the *unc-22::gfp* transgene produced the twitching phenotype, which is characteristic of loss of *unc-22* expression. In a third experiment, both targets were endogenous genes. In-frame deletion alleles of *unc-22* and *unc-52* produce proteins that lose a fraction of the coding region but retain full wild-type function (Kiff et al. 1988; Fire et al. 1991; Rogalski et al. 1993; Mullen et al. 1999). When dsRNA triggers that correspond to the deleted regions were introduced into heterozygous animals carrying both wild-type and mutant alleles, a strong transitive RNAi effect was found. These experiments showed that transitive RNAi works on both foreign and cellular genes.

## A Working Model for Gene Silencing

A unified model for gene silencing is shown in Figure 9.9 (Sijen et al. 2001). In this model, the introduction of a dsRNA trigger prompts cleavage by Dicer. The siRNAs, protected in a nuclease-resistant protein complex (RISC), find their target RNA sequences in cellular mRNAs and prime the synthesis of dsRNA by the RdRP. These RdRP-synthesized dsRNA products become new substrates for Dicer and are subsequently degraded to secondary siRNAs. The stable nature of the siRNAs facilitates rounds of signal amplification and degradation, which makes PTGS a persistent phenomenon in the affected system.

### Strand preferences

By using this "primer model," it is possible to explain and/or clarify previous observations. Parrish et al. (2000) reported that some chemical modifications on the antisense strand of the triggers showed a more dramatic reduction in RNAi than the same modifications on the sense strand. This group proposed that the antisense strand interacts directly with the target RNA while the sense strand participates less directly. This observation is now easier to understand. If the antisense strand of the siRNAs derived from the trigger is used as the primer for the first round of synthesis by the RdRP, then chemical modification of this strand could have a significant impact on production of secondary siRNA. The modified sense strand would not cause the same problem.

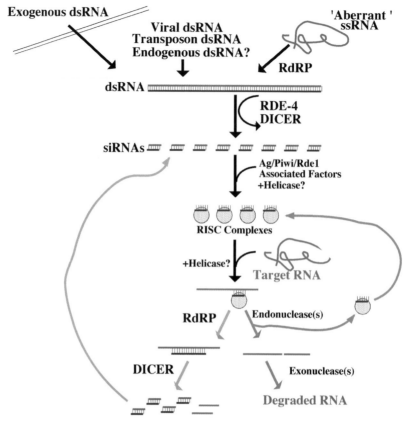

**FIGURE 9.9.** A working model for RNA interference. Two different aspects of the model enhance the potency of the RNAi reaction. Reuse of RNA-loaded RISC complexes (*magenta arrows*) should provide the reaction with a catalytic component, whereas physical amplification by the RdRP (*orange arrows*) provides a physical amplification of the initial trigger RNA. (Reprinted, with permission of Elsevier Science, from Sijen et al. 2001.)

## RNAi triggered by short antisense RNAs

A very recent study in *C. elegans* also supports the primer model (Tijsterman et al. 2002). These authors found that injection of 25-nucleotide dsRNAs only triggered a low level of RNAi, whereas the corresponding antisense-strand RNA was very efficient. Very short (15 and 18 nucleotides) antisense-strand RNAs were ineffective, but longer (22–40 nucleotides) antisense-strand RNAs were fully active in triggering RNAi (Tijsterman et al. 2002). A reasonable hypothesis is that these antisense-strand RNAs serve as primers for dsRNA synthesis by the RdRP. Consistent with this possibility is the finding that modification of the 3′ end of antisense-strand RNA causes a substantial reduction in the efficacy of the antisense-strand RNA.

## Origin of siRNAs in silenced tomato

Han and co-workers investigated the origin of siRNAs in a silenced tomato line containing a truncated ripening-specific polygalacturonase transgene (Han and Grierson 2002). They discovered that the siRNAs are synthesized from transgene transcripts before the transcription of the endogenous polygalacturonase gene occurs. Interestingly, the siRNAs are preferentially produced from the 3′ end of the transgene and cause the preferential degradation of the target polygalacturonase mRNA in the corresponding region. This result is in good agreement with the catalytic properties of the tomato RdRP determined in vitro: Unprimed transcription by the tomato RdRP is initiated at the 3′ terminus of the template and transcription products are no more than 100 nucleotides long (Schiebel et al. 1993b). If the siRNAs were produced from inverted repeats of the transgene and the RdRP was not involved, then there should not be any preference for a specific region of the target reflected in the siRNA population.

## Is There an Evolutionary Relationship between Cellular and Viral RdRPs?

RNA viruses can be divided into two categories on the basis of the nature of their genome: single-stranded and double-stranded. Viruses containing ssRNA genomes can be further divided on the basis of the polarity of the genome. Genomes of "positive" polarity are functional mRNAs, whereas those of "negative" polarity are the reverse-complement of the coding strand. Transcription and replication of RNA virus genomes require a virus-encoded RdRP. Although the viral RdRP can be produced by translation of the viral genome in the case of positive-stranded RNA viruses, it must be incorporated into the virions of negative- and double-stranded RNA viruses. In some cases, the virion-associated RdRP has evolved to function as an organizational component of the capsid (Reinisch et al. 2000; Butcher et al. 2001). Variability between viral RdRPs also derives from the variable requirement for additional enzymatic activities. For example, viral mRNAs that are translated by a cap-dependent mechanism often require guanylyltransferase and methyltransferase activities. These activities are often associated with the polymerase polypeptide (Egloff et al. 2002). Consequently, the size (50–200 kD) and sequence complexity of viral RdRPs can vary on the basis of the functional complexity of the enzyme.

Despite the sequence diversity of this class of enzymes, X-ray crystallographic studies of nonstructural and structural RdRPs have demonstrated not only a close relationship between all viral RdRPs, but also a close relationship with other classes of nucleic acid polymerases (Hansen et al. 1997; Bressanelli et al. 1999; Lesburg et al. 1999; Butcher et al. 2001; Ng et al. 2002). Shown in Figure 9.10 is the first structure for a viral RdRP, the

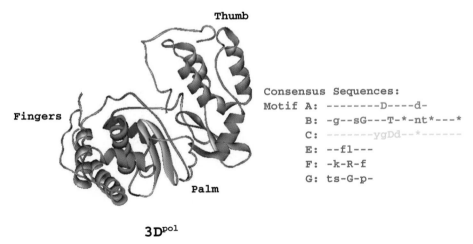

```
Consensus Sequences:
Motif A:  --------D----d-
      B:  -g--sG---T-*-nt*---*
      C:  -------ygDd--*------
      E:  --fl---
      F:  -k-R-f
      G:  ts-G-p-
```

**3D^pol**

**FIGURE 9.10.** Crystal structure of poliovirus polymerase (3D^pol) and the consensus sequences of conserved motifs in viral RdRPs that employ the canonical palm subdomain. The 3D^pol molecule is light blue. Structural motif designations are according to Hansen et al. (1997) and are colored as follows: (*red*) motif A; (*green*) motif B; (*gold*) motif C; (*dark blue*) motif D; (*purple*) motif E. The consensus sequences of the structural motifs were derived from sequence alignment of RdRPs of RNA viruses employing the canonical palm subdomain. Uppercase residues are absolutely conserved residues; lowercase residues are partly conserved. Asterisks indicate the presence of Ile, Leu, Val, or Met. The lengths of the various motifs are defined by the structure; therefore, dashes indicate residues of the structural motif for which no sequence conservation is observed.

enzyme from poliovirus (PV) that is often referred to as 3D$^{pol}$ (Hansen et al. 1997). This enzyme exhibits an overall topology similar to that described for other polymerases. The enzyme can be compared to a cupped, right hand with "fingers," "palm," and "thumb" subdomains. The fingers and thumb subdomains are involved primarily in interactions with nucleic acid and vary significantly from one class of nucleic acid polymerase to another. In contrast, the palm subdomain is very conserved, and the corresponding domains of all nucleic acid polymerases superimpose quite well (Hansen et al. 1997; Bressanelli et al. 1999; Lesburg et al. 1999; Butcher et al. 2001; Ng et al. 2002).

The conservation of the palm subdomain likely reflects its critical role in catalysis. The palm subdomain contains five conserved structural motifs (A–E) that, in most cases, correspond to conserved sequence motifs (Figure 9.10). The one exception is motif D, which is thought to impart structural stability to the palm subdomain and is therefore highly variable. The other motifs have more conserved functional roles, for example, interactions with the nascent polynucleotide chain (motif E), incoming nucleotide (motifs A and B), and divalent cations required for catalysis (motifs A and C) (Gohara et al. 2000). Finally, two additional motifs have been discovered recently. Motif F (Figure 9.10) is a part of the fingers domain and is involved in nucleotide binding (Hansen et al. 1997; Bressanelli et al. 1999; Lesburg et al. 1999; Gohara et al. 2000; Butcher et al. 2001; Ng et al. 2002). Motif G has no known function but is present in all animal virus RdRPs (A. Gorbalenya, pers. comm.).

Unique features of the viral RdRP also exist relative to the other classes of nucleic acid polymerases. In particular, there is an amino-terminal extension of the fingers, referred to as the "fingertips," that interacts with the thumb subdomain. This interaction results in a completely encircled active site (Figure 9.11). This observation was made first for the RdRP from hepatitis C virus (HCV NS5B) (Figure 9.11) (Ago et al. 1999; Bressanelli et al. 1999; Lesburg et al. 1999), but it has also been suggested to exist in poliovirus polymerase (Figure 9.11) (D. Gohara and C. Cameron, unpubl.) and related enzymes (Hansen et al. 1997; Bressanelli et al. 1999; Lesburg et al. 1999; Butcher et al. 2001; Ng et al. 2002). One possible consequence of this interaction between the fingers and thumb is the formation of a more rigid nucleic-acid-binding pocket and a corresponding increase in the relative stability of this class of polymerases with the nascent chain and/or template relative to other nucleic acid polymerases (Arnold and Cameron 2000).

Viral RdRPs can be distinguished in vitro on the basis of their mechanism of initiation. Some enzymes are strictly primer-dependent (e.g., poliovirus 3D$^{pol}$), whereas others initiate RNA synthesis de novo (e.g., HCV NS5B). Comparison of a poliovirus polymerase

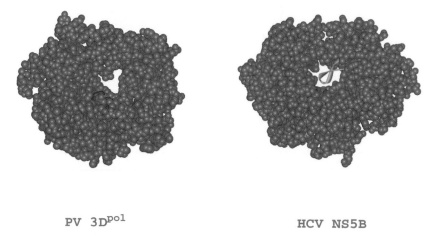

PV 3D$^{pol}$                    HCV NS5B

**FIGURE 9.11.** Comparison of polymerases from poliovirus (PV 3D$^{pol}$) and hepatitis C virus (HCV NS5B). Motif C (GDD motif) in the active site is colored in red for both polymerases. In HCV NS5B, the β-loop consisting of residues 443–454 is colored in yellow.

model to the structure of HCV NS5B provides a possible explanation for this difference in the mechanism of initiation. As shown in Figure 9.11, the HCV polymerase has a loop protruding from the thumb into the nucleic-acid-binding pocket that is absent in the poliovirus polymerase. The presence of this loop likely precludes binding of dsRNA (Ago et al. 1999; Bressanelli et al. 1999; Lesburg et al. 1999). Interestingly, truncation of this loop permits this polymerase derivative to use a primer-dependent mechanism for initiation (Hong et al. 2001). The presence of this loop appears to correspond quite well to the ability of an RdRP to initiate RNA synthesis de novo (Laurila et al. 2002).

Together, structural studies of viral RdRPs combined with computational studies of the corresponding sequences have uncovered structural/sequence motifs that illuminate conserved signatures of this class of nucleic acid polymerases and provide a basis set for evaluation of the evolutionary relationships of these enzymes. In this context, where do the cellular RdRPs fit? Analysis of the most conserved elements of the cellular RdRP reveals no obvious evolutionary relationship of this enzyme family to the viral RdRPs (A. Gorbalenya, pers. comm.). Therefore, the cellular RdRP likely represents a new structural organization for polymerases that supports an alternative mechanism of phosphoryl transfer (i.e., phosphodiester bond formation).

## RdRP ACTIVITY OF *QDE-1*

Very recently, Makeyev and Bamford (2002) demonstrated that the *QDE-1* gene from *Neurospora crassa* encodes a functional RdRP, establishing a direct link between cellular RdRPs and PTGS. The *QDE-1* gene was modified to encode a protein (Qde-1p) with a hexa-histidine tag on its carboxyl terminus, expressed in *Saccharomyces cerevisiae*, and Qde-1p was purified from the soluble fraction of an *S. cerevisiae* extract by using nickel-affinity chromatography. In addition to full-length Qde-1p, a derivative containing a deletion of the first 376 amino acids was produced. Both proteins exhibited RdRP activity in vitro. This activity was dependent on the terminal aspartate residue of the conserved GXDXGD motif found in this class of enzymes (see Figure 9.3).

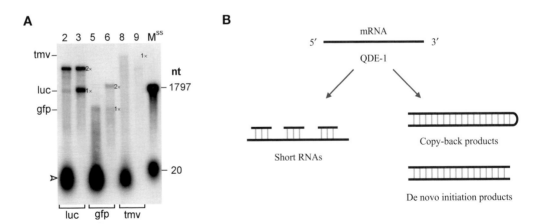

**FIGURE 9.12.** *QDE-1* generates two types of reaction products: (*A*) Formaldehyde-containing 1.5% agarose gel of *QDE-1* RdRP reactions: 90 μg/ml luciferase mRNA (lanes *2* and *3*), 80 μg/ml GFP mRNA (lanes *5* and *6*), or 100 μg/ml TMV genomic RNA (lanes *8* and *9*) was used as template. Reactions contained [α-$^{32}$P]UTP and either 10 μg/ml *QDE-1* (lanes *2*, *5*, and *8*) or 20 μg/ml φ6Pol (lanes *3*, *6*, and *9*). M$^{ss}$ indicates $^{32}$P-labeled ssRNA markers (20 and 1797 nucleotides). (Reprinted, with permission, from Makeyev et al. 2002.) (*B*) Model of action for *QDE-1* RdRP activity. Short RNAs are produced by *QDE-1* via internal de novo initiation. Full-length products are produced by either de novo initiation or a copy-back mechanism.

Purified Qde-1p used a number of ssRNA templates, producing two distinct types of products (Figure 9.12). The first class of products was similar to that observed for the better-characterized viral RdRPs, i.e., full-length monomeric and dimeric species produced by de novo initiation or a copy-back mechanism, respectively. The second class of products, however, was unique to the cellular RdRP; these RNA products ranged in length from 9 to 21 nucleotides. These short RNA products accumulated to a much higher level than any other reaction product. Rather than originating from the template 3′ end, these short RNAs originated from template regions evenly distributed along the length of the input RNA. Qde-1p did not use blunt-ended dsRNA as template and, under the assay condition employed, did not catalyze primer extension very efficiently.

Together, these observations suggest that Qde-1p not only is capable of producing products required for transitive silencing, but may also be involved in producing short RNAs capable of directing Dicer to target RNA. Curiously, Qde-1p appears to lack specificity for RNA templates in vitro, suggesting the possibility that this protein must be targeted/recruited to appropriate templates in vivo. Targeting may be a reflection of secondary and/or tertiary structures present in target RNAs or even something as simple as the intracellular concentration of the target RNA. Alternatively, target RNAs may be recognized by protein factors, which, in turn, recruit the RdRP to these RNAs. The ability to isolate active enzyme will clearly permit many of these remaining questions to be addressed experimentally.

## SUMMARY

Since 1993, there has been an exponential increase in our understanding of PTGS. Although there has been some debate on the involvement of a cellular RdRP in PTGS, data in numerous systems support the existence and requirement of this enzyme for this process. In most organisms that undergo silencing, it is likely that the RdRP is homologous to that isolated from tomato. However, some organisms may have unique enzymes. For example, although RdRP activity has been demonstrated biochemically in *Drosophila* embryo extracts, a gene encoding a tomato RdRP homolog has not been identified. The function of the RdRP appears to be production of dsRNA primed by individual strands of silencing RNAs. In doing so, this enzyme is capable of both amplifying and maintaining the PTGS response.

Finally, it is important to note that RdRP activity has very recently been demonstrated directly by expression of the RdRP gene from *N. crassa* in *S. cerevisiae*. This advance will now permit a more rigorous biochemical evaluation of the function of this enzyme. It is likely that this 128-kD protein encodes multiple functions because very processive viral RdRPs exist that are in the 50–60-kD range. Moreover, the apparent lack of template specificity for this enzyme suggests the requirement of other elements/factors for proper biological function. Given the clear involvement of the RdRP in PTGS, this enzyme and its corresponding complexes will likely become a higher priority for investigation in the future.

# TECHNIQUES

**Protocol 1:** An Assay for RdRP Activity

**Protocol 2:** Primed Transcription of Short Synthetic Templates

**Protocol 3:** Preparation and Fractionation of *Drosophila* Embryo Extracts

**Protocol 4:** Preparation of siRNAs from *Drosophila* Embryo Extracts

**Protocol 5:** Assay for siRNA-primed RNA Synthesis

## PROTOCOL 1: AN ASSAY FOR RdRP ACTIVITY

This protocol provides a quick way to assay for RdRP activity in crude samples (Schiebel et al. 1993a). Following paper chromatography, transcription products remain at the origin. Therefore, RdRP activity can be monitored by measuring the time-dependent increase in radioactivity present at the origin.

### Procedure

**▼ CAUTION**
*See Appendix for appropriate handling of materials marked with <!>.*

---

**MATERIALS**

**REAGENTS**

ATP, CTP, and GTP (10 mM each)

Dithioerythritol (1 M) <!>

Magnesium acetate (0.1 M) <!>

Paper chromatography solvent (2 M sodium acetate [pH 5.2]/ethanol [1:1 v/v]) <!>

RdRP preparation

Tobacco mosaic virus (TMV) RNA

Tris acetate (1 M, pH 7.8)

    Prepare Tris acetate by adjusting the pH of Tris base <!> with glacial acetic acid. <!>

[5,6-$^3$H]UTP <!>

**EQUIPMENT**

Whatman No. 3MM strips (2 x 10 cm)

---

1. Set up a 45-μl reaction containing:

   50 mM Tris acetate (pH 7.8)

   10 mM magnesium acetate

   2 mM dithioerythritol

   0.3 mM ATP, 0.3 mM CTP, and 0.3 mM GTP

   0.02 mM [5,6-$^3$H]UTP (5 Ci/mmole)

   1 μg of TMV RNA

   > A negative control reaction should be set up without RdRP. A positive control reaction can be set up using an active viral RdRP (Arnold and Cameron 2000).

2. Initiate the reaction by adding 5 μl of RdRP preparation.

3. Incubate the reaction for 30 minutes at 37°C.

4. Spot 45 μl of the reaction on a Whatman No. 3MM paper strip (2 x 10 cm).

5. Separate the reaction products by chromatography in 2 M sodium acetate (pH 5.2)/ethanol (1:1 v/v).

6. Quantify the radioactivity of the RNA products remaining at the origin by liquid scintillation counting.

## PROTOCOL 2: PRIMED TRANSCRIPTION OF SHORT SYNTHETIC TEMPLATES

For short templates with known sequence, products of RdRP-catalyzed primer extension can be analyzed by denaturing polyacrylamide gel electrophoresis. The following protocol was adopted from the work of Schiebel et al. (1993b). Alternative protocols are also available in the literature (Arnold and Cameron 2000; Xie et al. 2001).

### Procedure

▼ CAUTION
*See Appendix for appropriate handling of materials marked with <!>.*

---

**MATERIALS**

**REAGENTS**

[α-$^{32}$P]ATP <!>

Bovine serum albumin (BSA), acetylated

Dithiothreitol (DTT) (1 M) <!>

Magnesium acetate (100 mM) <!>

Magnesium chloride (100 mM) <!>

Polynucleotide kinase

RdRP solution

Ribodinucleoside monophosphate primer

Ribonucleoside triphosphates (10 mM of each)

Template solution
    See Step 5.

Tris-HCl (1 M, pH 7.8 and pH 8.0) <!>

**EQUIPMENT**

Denaturing sequencing gel equipment

Water bath (boiling)

---

1. Set up a 10-μl reaction containing:

   200 pmoles of ribodinucleoside monophosphate primer

   200 pmoles (600 μCi) of [α-$^{32}$P]ATP

   32 units of polynucleotide kinase

   50 mM Tris-HCl (pH 8.0)

   5 mM DTT

   10 mM MgCl$_2$

   5 μg/ml acetylated BSA

   Incubate the reaction for 1 hour at 37°C.

2. Heat the reaction for 1 minute at 100°C to inactivate the kinase. Chill mixture on ice.

3. Prepare a 40-μl reaction containing:

0.3 mM ATP, 0.3 mM CTP, 0.3 mM GTP, and 0.3 mM UTP

10 mM magnesium acetate

50 mM Tris-HCl (pH 7.8)

10 mM DTT

15 milliunits of RdRP

4. Add 5 μl of template solution (0.01–0.04 $OD_{260}$ unit) and 5 μl of the phosphorylation mixture (from Step 3). Incubate the reaction for 1 hour at 37°C.

> One unit of enzyme activity was defined as the amount of enzyme that catalyzes the incorporation of 1 nmole of uridine 5´-monophosphate into high-molecular-weight RNA products within 30 minutes.

5. Analyze the reaction products on a denaturing sequencing gel.

## PROTOCOL 3: PREPARATION AND FRACTIONATION OF *DROSOPHILA* EMBRYO EXTRACTS

These extracts are used for in vitro characterization of RNAi. A cell-free system for RNAi can be prepared using *Drosophila* embryos (Tuschl et al. 1999). In this system, RNAi is sequence-specific and is promoted by dsRNA. This extract can be fractionated to produce a more robust system (Hammond et al. 2000).

## Procedure

**▼ CAUTION**

*See Appendix for appropriate handling of materials marked with <!>.*

---

**MATERIALS**

**REAGENTS**

Bleach (50% v/v) <!>

Hypotonic buffer

1 mM $MgCl_2$ <!>

300 mM potassium acetate

Lysis buffer

5 mM dithiothreitol <!>

30 mM HEPES-KOH (pH 7.4) <!>

2 mM magnesium acetate <!>

1 mg/ml Pefabloc SC (Boehringer Mannheim)

100 mM potassium acetate

Oregon R fly embryos

See Step 1.

**EQUIPMENT**

Potter-Elvehjem tissue grinder (Kontes), chilled in ice water

Ultracentrifuge

---

1. Collect 0–2-hour-old embryos from Oregon R flies on yeasted molasses agar at 25°C.

2. Dechorionate the embryos for 4–5 minutes in 50% (v/v) bleach.

3. Wash the embryos with $H_2O$. Blot them dry, and transfer them to a chilled Potter-Elvehjem tissue grinder.

4. Lyse the embryos at 4°C in lysis buffer with ten strokes, a stroke being one cycle up and down of a tight-fitting Potter-type Teflon pestle. To ensure that no air is introduced into the homogenate, use slow and controlled strokes. Use 1 ml of buffer per gram of damp embryos.

5. Centrifuge the lysate at 14,500*g* for 25 minutes at 4°C.

6. Centrifuge the crude extracts at 200,000*g* for 3 hours at 4°C.

7. Extract the pellet, which contains the ribosomes, in hypotonic buffer.

8. Centrifuge at 100,000*g* for 1 hour at 4°C.

9. Store aliquots of the P100 extracts at –80°C.

> This material is good forever at –80°C.

## PROTOCOL 4: PREPARATION OF siRNAs FROM *DROSOPHILA* EMBRYO EXTRACTS

Lipardi et al. (2001) have shown that siRNAs produced upon the addition of dsRNA to *Drosophila* embryo extract can be enriched in a micrococcus-nuclease-resistant fraction. After proteinase K treatment and dephosphorylation with calf intestinal phosphatase, these siRNAs mediate efficient RNAi in vitro.

### Procedure

### MATERIALS

#### REAGENTS

α-Amanitin <!>

ATP

$CaCl_2$ <!>

Calf intestinal alkaline phosphatase

Creatine kinase

Creatine phosphate

Dithiothreitol (DTT) (5 mM) <!>

*Drosophila* embryo S100 extract

dsRNA, unlabeled or [$^{32}$P]UTP uniformly labeled <!>

EDTA (0.5 M)

EGTA (0.5 M)

Endonuclease buffer 3 (New England Biolabs)

Ethanol <!>

Glycogen

HEPES (pH 7.4)

Magnesium acetate <!>

Micrococcal nuclease (Roche)

Phenol:chloroform:isoamyl alcohol (25:24:1) <!>

Potassium acetate

Proteinase K <!>

NaCl

SDS <!>

Tris-HCl (pH 7.5) <!>

#### EQUIPMENT

Water bath preset to 65°C

1. Add 0.5–10 µg of unlabeled dsRNA or 1 µg of [$^{32}$P]UTP uniformly labeled dsRNA to 100 µl of S100 extract.

2. To the materials from Step 1, add the following to a final concentration of:

   100 mM potassium acetate

   30 mM HEPES (pH 7.4)

   2 mM magnesium acetate

   5 mM DTT

   10 mM creatine phosphate

   1 µg/ml creatine kinase

   500 µM ATP

   500 µM α-amanitin

3. Incubate the reaction for 30 minutes at 25°C.

4. Add micrococcal nuclease to a final concentration of 60 units/50 µl of reaction in the presence of 2 mM CaCl$_2$. Incubate for 30 minutes at 37°C.

5. Repeat Step 4.

6. Stop the reaction by the addition of EGTA to a final concentration of 10 mM.

7. Digest the siRNA preparation with 100 µg/ml proteinase K for 15 minutes at 65°C in 100 mM Tris-HCl (pH 7.5), 12.5 mM EDTA, 150 mM NaCl, and 1% SDS.

8. Extract the siRNAs with phenol and precipitate them with ethanol using 40 µg of glycogen as carrier.

9. Dissolve the precipitated material in 50 µl of H$_2$O.

10. Treat 10 µl of siRNAs with 20 units of calf intestinal alkaline phosphatase in NEB endonuclease buffer 3 in a final volume of 50 µl for 1 hour at 37°C.

11. Extract the sample with phenol and precipitate with ethanol.

12. Dissolve the precipitated material in 10 µl of H$_2$O for further use.

## PROTOCOL 5: ASSAY FOR siRNA-PRIMED RNA SYNTHESIS

siRNAs prepared using Protocol 4 can be used by the RdRP as primers for specific cellular mRNAs, forming dsRNA products capable of inducing transitive RNAi (Lipardi et al. 2001).

## Procedure

**▼ CAUTION**
*See Appendix for appropriate handling of materials marked with <!>.*

**MATERIALS**

**REAGENTS**

α-Amanitin <!>

Amino acids

Creatine kinase

Creatine phosphate

Dithiothreitol (DTT) <!>

*Drosophila* embryo S100 extract

Isopropanol <!>

Lysis buffer

> 100 mM potassium acetate
> 30 mM HEPES-KOH (pH 7.4) <!>
> 2 mM magnesium acetate <!>
> 5 mM DTT <!>
> 1 mg/ml Pefabloc SC (Boehringer Mannheim)

Phenol:chloroform:isoamyl alcohol (25:24:1) <!>

2x PK buffer

> 200 mM Tris-HCl (pH 7.5) <!>
> 25 mM EDTA
> 300 mM NaCl
> SDS (2% w/v) <!>

Potassium acetate

Proteinase K (E.M. Merck; dissolved in water) <!>

Ribonucleoside triphosphates

RNAsin (Promega)

RNA template

$[^{32}P]$UTP-labeled siRNAs <!>

**EQUIPMENT**

Formaldehyde-agarose or DNA sequencing gels and equipment

Water baths preset to 25°C and 65°C

1. Set up reactions containing:

   50% (v/v) S100 extract

   10% lysis buffer

   100 mM potassium acetate (final)

   500 μM α-amanitin

   2.5 x $10^5$ cpm of $[^{32}P]$UTP-labeled siRNAs

   3 μg of RNA template

   10 mM creatine phosphate

   10 μg/ml creatine kinase

   100 μM GTP, 100 μM UTP, 100 μM CTP

   500 μM ATP

   5 mM DTT

   0.1 unit/μl RNAsin

   100 μM of each amino acid

   Set up negative control reactions without template RNA or with a heterologous template.

2. Incubate the reactions for 60 minutes at 25°C or various times for time course reactions.

3. Stop the primer extension reaction with 40 volumes of 2x PK buffer.

4. Add proteinase K to a final concentration of 465 μg/ml.

5. Incubate the digest for 15 minutes at 65°C.

6. Extract the RNA products with phenol:chloroform:isoamyl alcohol (25:24:1), and precipitate them with an equal volume of isopropanol.

7. Analyze the labeled RNA products on agarose formaldehyde gels or DNA sequencing gels.

## ACKNOWLEDGMENTS

We thank our collaborator, Dr. Eleanor Maine, for critical evaluation of the manuscript. We also thank Dr. Bruce Paterson for providing detailed information for Protocol 3.

## REFERENCES

Ago H., Adachi T., Yoshida A., Yamamoto M., Habuka N., Yatsunami K., and Miyano M. 1999. Crystal structure of the RNA-dependent RNA polymerase of hepatitis C virus. *Struct. Fold Des.* **7:** 1417–1426.

Angell S.M. and Baulcombe D.C. 1997. Consistent gene silencing in transgenic plants expressing a replicating potato virus X RNA. *EMBO J.* **16:** 3675–3684.

Arnold J.J. and Cameron C.E. 2000. Poliovirus RNA-dependent RNA polymerase (3D[pol]). Assembly of stable, elongation-competent complexes by using a symmetrical primer-template substrate (sym/sub). *J. Biol. Chem.* **275:** 5329–5336.

Astier-Manifacier S. and Cornuet P. 1971. RNA-dependent RNA polymerase in Chinese cabbage. *Biochim. Biophys. Acta* **232:** 484–493.

———. 1978. Purfication and molecular weight of an RNA-dependant RNA polymerase from *Brassicae oleracea* var. Botrytis. *C.R. Acad. Sci. Hebd. Seances. Acad. Sci. D* **287:** 1043–1046.

Bass B.L. 2000. Double-stranded RNA as a template for gene silencing. *Cell* **101:** 235–238.

Bernstein E., Denli A.M., and Hannon G.J. 2001a. The rest is silence. *RNA* **7:** 1509–1521.

Bernstein E., Caudy A.A., Hammond S.M., and Hannon G.J. 2001b. Role for a bidentate ribonuclease in the initiation step of RNA interference. *Nature* **409:** 363–366.

Blumenthal T. and Carmichael G.G. 1979. RNA replication: Function and structure of Qbeta-replicase. *Annu. Rev. Biochem.* **48:** 525–548.

Boege F. and Sänger H.L. 1980. RNA-dependent RNA polymerase from healthy tomato leaf tissue. *FEBS Lett.* **121:** 91–96.

Bressanelli S., Tomei L., Roussel A., Incitti I., Vitale R.L., Mathieu M., De Francesco R., and Rey F.A. 1999. Crystal structure of the RNA-dependent RNA polymerase of hepatitis C virus. *Proc. Natl. Acad. Sci.* **96:** 13034–13039.

Butcher S.J., Grimes J.M., Makeyev E.V., Bamford D.H., and Stuart D.I. 2001. A mechanism for initiating RNA-dependent RNA polymerization. *Nature* **410:** 235–240.

Cerutti L., Mian N., and Bateman A. 2000. Domains in gene silencing and cell differentiation proteins: The novel PAZ domain and redefinition of the Piwi domain. *Trends Biochem. Sci.* **25:** 481–482.

Coen E.S. and Carpenter R. 1988. A semi-dominant allele, *niv-525*, acts in trans to inhibit expression of its wild-type homologue in *Antirrhinum majus. EMBO J.* **7:** 877–883.

Cogoni C. and Macino G. 1999. Gene silencing in *Neurospora crassa* requires a protein homologous to RNA-dependent RNA polymerase. *Nature* **399:** 166–169.

Cogoni C., Irelan J.T., Schumacher M., Schmidhauser T.J., Selker E.U., and Macino G. 1996. Transgene silencing of the *al-1* gene in vegetative cells of *Neurospora* is mediated by a cytoplasmic effector and does not depend on DNA-DNA interactions or DNA methylation. *EMBO J.* **15:** 3153–3163.

Dalmay T., Hamilton A., Rudd S., Angell S., and Baulcombe D.C. 2000. An RNA-dependent RNA polymerase gene in *Arabidopsis* is required for posttranscriptional gene silencing mediated by a transgene but not by a virus. *Cell* **101:** 543–553.

Davies J.W., Kaesberg P., and Diener T.O. 1974. Potato spindle tuber viroid. XII. An investigation of viroid RNA as a messenger for protein synthesis. *Virology* **61:** 281–286.

Domeier M.E., Morse D.P., Knight S.W., Portereiko M., Bass B.L., and Mango S.E. 2000. A link between RNA interference and nonsense-mediated decay in *Caenorhabditis elegans*. *Science* **289**: 1928–1931.

Duda C.T. 1979. Synthesis of double-stranded RNA. II. Partial purification and characterization of an RNA-dependent RNA polymerase in healthy tobacco leaves. *Virology* **92**: 180–189.

Duda C.T., Zaitlin M., and Siegel A. 1973. In vitro synthesis of double-stranded RNA by an enzyme system isolated from tobacco leaves. *Biochim. Biophys. Acta* **319**: 62–71.

Egloff M.P., Benarroch D., Selisko B., Romette J.L., and Canard B. 2002. An RNA cap (nucleoside-2′-O-)-methyltransferase in the flavivirus RNA polymerase NS5: Crystal structure and functional characterization. *EMBO J.* **21**: 2757–2768.

Elbashir S.M., Lendeckel W., and Tuschl T. 2001. RNA interference is mediated by 21- and 22-nucleotide RNAs. *Genes Dev.* **15**: 188–200.

Fire A., Albertson D., Harrison S.W., and Moerman D.G. 1991. Production of antisense RNA leads to effective and specific inhibition of gene expression in *C. elegans* muscle. *Development* **113**: 503–514.

Fire A., Xu S., Montgomery M.K., Kostas S.A., Driver S.E., and Mello C.C. 1998. Potent and specific genetic interference by double-stranded RNA in *Caenorhabditis elegans*. *Nature* **391**: 806–811.

Fraser A.G., Kamath R.S., Zipperlen P., Martinez-Campos M., Sohrmann M., and Ahringer J. 2000. Functional genomic analysis of *C. elegans* chromosome I by systematic RNA interference. *Nature* **408**: 325–330.

Gohara D.W., Crotty S., Arnold J.J., Yoder J.D., Andino R., and Cameron C.E. 2000. Poliovirus RNA-dependent RNA polymerase (3Dpol): Structural, biochemical, and biological analysis of conserved structural motifs A and B. *J. Biol. Chem.* **275**: 25523–25532.

Grishok A., Tabara H., and Mello C.C. 2000. Genetic requirements for inheritance of RNAi in *C. elegans*. *Science* **287**: 2494–2497.

Grishok A., Pasquinelli A.E., Conte D., Li N., Parrish S., Ha I., Baillie D.L., Fire A., Ruvkun G., and Mello C.C. 2001. Genes and mechanisms related to RNA interference regulate expression of the small temporal RNAs that control *C. elegans* developmental timing. *Cell* **106**: 23–34.

Guo S. and Kemphues K.J. 1995. *par-1*, a gene required for establishing polarity in *C. elegans* embryos, encodes a putative Ser/Thr kinase that is asymmetrically distributed. *Cell* **81**: 611–620.

Hall T.C., Wepprich R.K., Davies J.W., Weathers L.G., and Semancik J.S. 1974. Functional distinctions between the ribonucleic acids from citrus exocortis viroid and plant viruses: Cell-free translation and aminoacylation reactions. *Virology* **61**: 486–492.

Hamilton A.J. and Baulcombe D.C. 1999. A species of small antisense RNA in posttranscriptional gene silencing in plants. *Science* **286**: 950–952.

Hammond S.M., Bernstein E., Beach D., and Hannon G.J. 2000. An RNA-directed nuclease mediates post-transcriptional gene silencing in *Drosophila* cells. *Nature* **404**: 293–296.

Han Y. and Grierson D. 2002. Relationship between small antisense RNAs and aberrant RNAs associated with sense transgene mediated gene silencing in tomato. *Plant J.* **29**: 509–519.

Hansen J.L., Long A.M., and Schultz S.C. 1997. Structure of the RNA-dependent RNA polymerase of poliovirus. *Structure* **5**: 1109–1122.

Hong Z., Cameron C.E., Walker M.P., Castro C., Yao N., Lau J.Y., and Zhong W. 2001. A novel mechanism to ensure terminal initiation by hepatitis C virus NS5B polymerase. *Virology* **285**: 6–11.

Ikegami M. and Fraenkel-Conrat H. 1979. Characterization of the RNA-dependent RNA polymerase of tobacco leaves. *J. Biol. Chem.* **254**: 149–154.

Jacobsen S.E., Running M.P., and Meyerowitz E.M. 1999. Disruption of an RNA helicase/RNAse III gene in *Arabidopsis* causes unregulated cell division in floral meristems. *Development* **126**: 5231–5243.

Jakowitsch J., Papp I., Moscone E.A., van der Winden J., Matzke M., and Matzke A.J. 1999. Molecular and cytogenetic characterization of a transgene locus that induces silencing and methylation of homologous promoters in trans. *Plant J.* **17**: 131–140.

Jones L., Ratcliff F., and Baulcombe D.C. 2001. RNA-directed transcriptional gene silencing in plants can be inherited independently of the RNA trigger and requires Met1 for maintenance. *Curr. Biol.* **11**: 747–757.

Jorgensen R. 1990. Altered gene expression in plants due to trans interactions between homologous genes. *Trends Biotechnol.* **8**: 340–344.

Kamer G. and Argos P. 1984. Primary structural comparison of RNA-dependent polymerases from plant, animal and bacterial viruses. *Nucleic Acids Res.* **12:** 7269–7282.

Kennerdell J.R. and Carthew R.W. 1998. Use of dsRNA-mediated genetic interference to demonstrate that frizzled and frizzled 2 act in the wingless pathway. *Cell* **95:** 1017–1026.

Kessin R.H. 2001. Dictyostelium: *Evolution cell biology, and the development of multicellularity.*, Cambridge University Press, New York.

Khan Z.A., Hiriyanna K.T., Chavez F., and Fraenkel-Conrat H. 1986. RNA-directed RNA polymerases from healthy and from virus-infected cucumber. *Proc. Natl. Acad. Sci.* **83:** 2383–2386.

Kiff J.E., Moerman D.G., Schriefer L.A., and Waterston R.H. 1988. Transposon-induced deletions in *unc-22* of *C. elegans* associated with almost normal gene activity. *Nature* **331:** 631–633.

Lagos-Quintana M., Rauhut R., Lendeckel W., and Tuschl T. 2001. Identification of novel genes coding for small expressed RNAs. *Science* **294:** 853–858.

Lau N.C., Lim L.P., Weinstein E.G., and Bartel D.P. 2001. An abundant class of tiny RNAs with probable regulatory roles in *Caenorhabditis elegans*. *Science* **294:** 858–862.

Laurila M.R., Makeyev E.V., and Bamford D.H. 2002. Bacteriophage phi 6 RNA-dependent RNA polymerase: Molecular details of initiating nucleic acid synthesis without primer. *J. Biol. Chem.* **277:** 17117–17124.

Lee R.C. and Ambros V. 2001. An extensive class of small RNAs in *Caenorhabditis elegans*. *Science* **294:** 862–864.

Lesburg C.A., Cable M.B., Ferrari E., Hong Z., Mannarino A.F., and Weber P.C. 1999. Crystal structure of the RNA-dependent RNA polymerase from hepatitis C virus reveals a fully encircled active site. *Nat. Struct. Biol.* **6:** 937–943.

Lindbo J.A., Silva-Rosales L., Proebsting W.M., and Dougherty W.G. 1993. Induction of a highly specific antiviral state in transgenic plants: Implications for regulation of gene expression and virus resistance. *Plant Cell* **5:** 1749–1759.

Lipardi C., Wei Q., and Paterson B.M. 2001. RNAi as random degradative PCR: siRNA primers convert mRNA into dsRNAs that are degraded to generate new siRNAs. *Cell* **107:** 297–307.

Louis B.G. and Fitt P.S. 1972. Purification and properties of the ribonucleic acid-dependent ribonucleic acid polymerase from *Halobacterium cutirubrum*. *Biochem. J.* **128:** 755–762.

Maeda Y., Inouye K., and Takeuchi I. 1997. Dictyostelium—*A model system for cell and developmental biology*. Universal Academy Press, Tokyo, Japan.

Makeyev E.V. and Bamford D.H. 2002. Cellular RNA-dependent RNA polymerase involved in posttranscriptional gene silencing has two distinct activity modes. *Mol. Cell* **10:** 1417–1427.

Martens H., Novotny J., Oberstrass J., Steck T.L., Postlethwait P., and Nellen W. 2002. RNAi in *Dictyostelium:* The role of RNA-directed RNA polymerases and double-stranded RNase. *Mol. Biol. Cell* **13:** 445–453.

Mette M.F., van der Winden J., Matzke M.A., and Matzke A.J. 1999. Production of aberrant promoter transcripts contributes to methylation and silencing of unlinked homologous promoters in trans. *EMBO J.* **18:** 241–248.

Mette M.F., Aufsatz W., van der Winden J., Matzke M.A., and Matzke A.J. 2000. Transcriptional silencing and promoter methylation triggered by double-stranded RNA. *EMBO J.* **19:** 5194–5201.

Mourrain P., Beclin C., Elmayan T., Feuerbach F., Godon C., Morel J.B., Jouette D., Lacombe A.M., Nikic S., Picault N., Remoue K., Sanial M., Vo T.A., and Vaucheret H. 2000. *Arabidopsis* SGS2 and SGS3 genes are required for posttranscriptional gene silencing and natural virus resistance. *Cell* **101:** 533–542.

Mullen G.P., Rogalski T.M., Bush J.A., Gorji P.R., and Moerman D.G. 1999. Complex patterns of alternative splicing mediate the spatial and temporal distribution of perlecan/UNC-52 in *Caenorhabditis elegans*. *Mol. Biol. Cell* **10:** 3205–3221.

Ng K.K., Cherney M.M., Vazquez A.L., Machin A., Alonso J.M., Parra F., and James M.N. 2002. Crystal structures of active and inactive conformations of a caliciviral RNA-dependent RNA polymerase. *J. Biol. Chem.* **277:** 1381–1387.

Ngo H., Tschudi C., Gull K., and Ullu E. 1998. Double-stranded RNA induces mRNA degradation in *Trypanosoma brucei*. *Proc. Natl. Acad. Sci.* **95:** 14687–14692.

Nishikura K. 2001. A short primer on RNAi: RNA-directed RNA polymerase acts as a key catalyst. *Cell* **107:** 415–418.

Novotny J., Diegel S., Schirmacher H., Mohrle A., Hildebrandt M., Oberstrass J., and Nellen W. 2001. *Dictyostelium* double-stranded ribonuclease. *Methods Enzymol.* **342:** 193–212.

Palauqui J.C. and Vaucheret H. 1998. Transgenes are dispensable for the RNA degradation step of cosuppression. *Proc. Natl. Acad. Sci.* **95:** 9675–9680.

Palauqui J.C., Elmayan T., Pollien J.M., and Vaucheret H. 1997. Systemic acquired silencing: Transgene-specific post-transcriptional silencing is transmitted by grafting from silenced stocks to non-silenced scions. *EMBO J.* **16:** 4738–4745.

Parrish S., Fleenor J., Xu S., Mello C., and Fire A. 2000. Functional anatomy of a dsRNA trigger. Differential requirement for the two trigger strands in RNA interference. *Mol. Cell* **6:** 1077–1087.

Pelissier T. and Wassenegger M. 2000. A DNA target of 30 bp is sufficient for RNA-directed DNA methylation. *RNA* **6:** 55–65.

Qiao L., Lissemore J.L., Shu P., Smardon A., Gelber M.B., and Maine E.M. 1995. Enhancers of *glp-1*, a gene required for cell-signaling in *Caenorhabditis elegans*, define a set of genes required for germline development. *Genetics* **141:** 551–569.

Reinhart B.J., Slack F.J., Basson M., Pasquinelli A.E., Bettinger J.C., Rougvie A.E., Horvitz H.R., and Ruvkun G. 2000. The 21-nucleotide *let-7* RNA regulates developmental timing in *Caenorhabditis elegans*. *Nature* **403:** 901–906.

Reinisch K.M., Nibert M.L., and Harrison S.C. 2000. Structure of the reovirus core at 3.6 Å resolution. *Nature* **404:** 960967.

Rogalski T.M., Williams B.D., Mullen G.P., and Moerman D.G. 1993. Products of the *unc-52* gene in *Caenorhabditis elegans* are homologous to the core protein of the mammalian basement membrane heparan sulfate proteoglycan. *Genes Dev.* **7:** 1471–1484.

Romaine C.P. and Zaitlin M. 1978. RNA-dependent RNA polymerases in uninfected and tobacco mosaic virus-infected tabacco leaves: Viral induced stimulation of a host polymerase activity. *Virology* **86:** 241–253.

Schiebel W., Haas B., Marinkovic S., Klanner A., and Sanger H.L. 1993a. RNA-directed RNA polymerase from tomato leaves. I. Purification and physical properties. *J. Biol. Chem.* **268:** 11851–11857.

———. 1993b. RNA-directed RNA polymerase from tomato leaves. II. Catalytic in vitro properties. *J. Biol. Chem.* **268:** 11858–11867.

Schiebel W., Pelissier T., Riedel L., Thalmeir S., Schiebel R., Kempe D., Lottspeich F., Sanger H.L., and Wassenegger M. 1998. Isolation of an RNA-directed RNA polymerase-specific cDNA clone from tomato. *Plant Cell* **10:** 2087–2101.

Semancik J.S., Conejero V., and Gerhart J. 1977. Citrus exocortis viroid: Survey of protein synthesis in *Xenopus laevis* oocytes following addition of viroid RNA. *Virology* **80:** 218–221.

Sijen T., Fleenor J., Simmer F., Thijssen K.L., Parrish S., Timmons L., Plasterk R.H., and Fire A. 2001. On the role of RNA amplification in dsRNA-triggered gene silencing. *Cell* **107:** 465–476.

Singer B., Kusmierek J.T., and Fraenkel-Conrat H. 1983. In vitro discrimination of replicases acting on carcinogen-modified polynucleotide templates. *Proc. Natl. Acad. Sci.* **80:** 969–972.

Smardon A., Spoerke J.M., Stacey S.C., Klein M.E., Mackin N., and Maine E.M. 2000. EGO-1 is related to RNA-directed RNA polymerase and functions in germ-line development and RNA interference in *C. elegans*. *Curr. Biol.* **10:** 169–178.

Sonoda S. and Nishiguchi M. 2000. Graft transmission of post-transcriptional gene silencing: Target specificity for RNA degradation is transmissible between silenced and non-silenced plants, but not between silenced plants. *Plant J.* **21:** 1–8.

Stam M., de Bruin R., van Blokland R., van der Hoorn R.A., Mol J.N., and Kooter J.M. 2000. Distinct features of post-transcriptional gene silencing by antisense transgenes in single copy and inverted T-DNA repeat loci. *Plant J.* **21:** 27–42.

Takanami Y. and Fraenkel-Conrat H. 1982. Comparative studies on ribonucleic acid dependent RNA polymerases in cucumber mosaic virus infected cucumber and tobacco and uninfected tobacco plants. *Biochemistry* **21:** 3161–3167.

Tijsterman M., Ketting R.F., Okihara K.L., Sijen T., and Plasterk R.H. 2002. RNA helicase MUT-14-dependent gene silencing triggered in *C. elegans* by short antisense RNAs. *Science* **295:** 694–697.

Timmons L. and Fire A. 1998. Specific interference by ingested dsRNA. *Nature* **395:** 854.

Tuschl T., Zamore P.D., Lehmann R., Bartel D.P., and Sharp P.A. 1999. Targeted mRNA degradation by double-stranded RNA in vitro. *Genes Dev.* **13:** 3191–3197.

van der Meer J., Dorssers L., van Kammen A., and Zabel P. 1984. The RNA-dependent RNA polymerase of cowpea is not involved in cowpea mosaic virus RNA replication: Immunological evidence. *Virology* **132:** 413–425.

Voinnet O., Vain P., Angell S., and Baulcombe D.C. 1998. Systemic spread of sequence-specific transgene RNA degradation in plants is initiated by localized introduction of ectopic promoter-less DNA. *Cell* **95:** 177–187.

Volloch V., Schweitzer B., and Rits S. 1987. Synthesis of globin RNA in enucleated differentiating murine erythroleukemia cells. *J. Cell Biol.* **105:** 137–143.

Volloch V., Schweitzer B., Zhang X., and Rits S. 1991. Identification of negative-strand complements to cytochrome oxidase subunit III RNA in *Trypanosoma brucei. Proc. Natl. Acad. Sci.* **88:** 10671–10675.

Wassenegger M., Heimes S., Riedel L., and Sanger H.L. 1994. RNA-directed de novo methylation of genomic sequences in plants. *Cell* **76:** 567–576.

Waterhouse P.M., Graham M.W., and Wang M.B. 1998. Virus resistance and gene silencing in plants can be induced by simultaneous expression of sense and antisense RNA. *Proc. Natl. Acad. Sci.* **95:** 13959–13964.

Wianny F. and Zernicka-Goetz M. 2000. Specific interference with gene function by double-stranded RNA in early mouse development. *Nat. Cell. Biol.* **2:** 70–75.

Xie Z., Fan B., Chen C., and Chen Z. 2001. An important role of an inducible RNA-dependent RNA polymerase in plant antiviral defense. *Proc. Natl. Acad. Sci.* **98:** 6516–6521.

Zabel P., Dorssers L., Wernars K., and Van Kammen A. 1981. Terminal uridylyl transferase of *Vigna unguiculata:* Purification and characterization of an enzyme catalyzing the addition of a single UMP residue to the 3´-end of an RNA primer. *Nucleic Acids Res.* **9:** 2433–2453.

Zamore P.D., Tuschl T., Sharp P.A., and Bartel D.P. 2000. RNAi: Double-stranded RNA directs the ATP-dependent cleavage of mRNA at 21 to 23 nucleotide intervals. *Cell* **101:** 25–33.

# CHAPTER 10

# Structure and Function of Heterochromatin: Implications for Epigenetic Gene Silencing and Genome Organization

Ira M. Hall* and Shiv I.S. Grewal*[†]

*Watson School of Biological Sciences, Cold Spring Harbor Laboratory, Cold Spring Harbor, New York 11724; [†]Laboratory of Molecular Cell Biology, National Cancer Institute, National Institutes of Health, Bethesda, Maryland 20892

THE IMPORTANT ROLE OF HIGHER-ORDER CHROMATIN STRUCTURE in the regulation of eukaryotic genomes is demonstrated by a few simple observations: First, chromosomes are condensed up to 10,000-fold from their extended length and undergo phases of relaxation, condensation, and segregation within the cell cycle; second, chromatin domains of distinct transcriptional activity are differentially established and stably maintained within the cellular lineages of an organism; and third, repetitive sequences make up a sizeable portion of eukaryotic genomes and pose a considerable threat to genomic integrity by way of illegitimate recombination and transposition. Considering that the processes of transcription, recombination, and segregation are sensitive to changes in chromatin structure, the regulation of higher-order chromatin structure is likely a preeminent factor in the evolution and diversification of large genomes.

Broadly speaking, chromatin is packaged into "active" or "silent" structural configurations that can determine the behavior of the underlying DNA. The genome can be cytologically divided into euchromatin and heterochromatin. Euchromatin is typically generich and composed of unique sequences; it has high rates of meiotic recombination and a relatively "relaxed" chromatin structure. In contrast, heterochromatin is relatively devoid of genes and composed mostly of repetitive sequences. The chromatin in such regions typically replicates late in S phase and is relatively inaccessible and highly condensed, in some cases visibly so. Recombination rates are very low, and reporter genes inserted into heterochromatin are transcriptionally repressed in a region-specific sequence-independent manner.

A distinguishing feature of heterochromatic gene silencing is that it is transmitted to daughter cells in a clonal, metastable fashion, implying that specialized chromatin structures are recapitulated during chromosome replication and inherited through cell division (Grewal 2000). Such inheritance is "epigenetic," defined for our purposes as mitotically and/or meiotically heritable changes in gene function that cannot be explained by changes in DNA sequence (Russo et al. 1996). This chapter examines the recent advances in our understanding of the initiation and propagation of heterochromatic structures. Our discussion here is restricted to cases of epigenetic gene silencing that are chromosomal in origin and function, and we draw from studies on an assortment of metazoan systems and the fission yeast *Schizosaccharomyces pombe*. Studies of chromatin structure and gene silencing in the budding yeast *Saccharomyces cerevisiae* have contributed greatly to our current understanding of the subject and are reviewed comprehensively elsewhere (Grunstein 1998; Moazed 2001).

## PROPERTIES OF HETEROCHROMATIN

The discovery of heterochromatin dates to the early part of the 20th century, when Ernest Heitz made the observation that some of the nuclear material remained condensed and deeply staining as the cell made the transition from metaphase to interphase (Heitz 1928). He termed these portions the "heterochromatin." The concept of heterochromatin has evolved considerably since its original discovery and has come to be defined by a constellation of genetic, molecular, and biochemical properties as discussed below.

### Condensation

Visible heterochromatic structures generally correspond to the large tracts of pericentric heterochromatin that flank the site of microtubule attachment at the centromeres of most eukaryotic organisms and to the telomeric heterochromatin that protects chromosome ends (Henikoff et al. 2000). In addition, highly condensed heterochromatic regions are visible at interstitial locations along the chromosome arms in various organisms, the most notable cases being the prominent "knobs" found in plants (McClintock 1929; Bennetzen 2000) and the G-bands found in mammals (Craig and Bickmore 1993). In flies, multiple rounds of DNA replication in the salivary gland cells allow direct visualization of polytene chromosomes, revealing distinct patterns of condensed heterochromatic segments and loosely organized euchromatic segments (Figure 10.1). Perhaps the most dramatic case of cytological condensation is that of the female inactive X chromosome found in most mammalian species (Lyon 1972; Riggs and Porter 1996; Lee and Jaenisch 1997). To regulate gene dosage, an entire X chromosome is assembled into a highly condensed, transcriptionally silent structure known as the Barr body. Although heterochromatic regions in the more simple eukaryotes are comparably small, similar condensed structures can be observed in fission yeast by electron microscopy (Kniola et al. 2001). Chromatin condensation can also be molecularly inferred by testing the general accessibility of chromatin to enzymatic probes such as nucleases, DNA methylases, and restriction enzymes. Heterochromatic DNA is relatively protected from these enzymes, and nucleosomes within heterochromatin are more densely and evenly distributed than within euchromatin (Flick et al. 1986; Gottschling 1992; Singh and Klar 1992; Wines et al. 1996; Sun et al. 2001).

### Transcriptional Repression

Studies of the phenomenon of position effect variegation (PEV) in *Drosophila melanogaster* have demonstrated that reporter genes placed within or adjacent to heterochromatin are

A

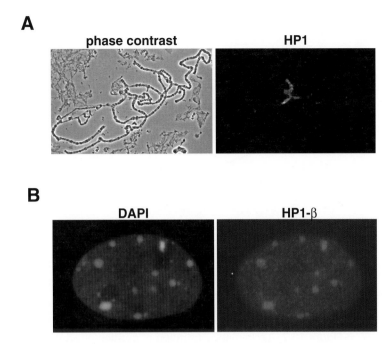

B

**Figure 10.1.** Heterochromatin. (*A*) Polytene chromosomes stained with HP1. Note intense staining at the chromocenter (a cluster of the centromeres) and at discrete foci along the chromosome arms. (*B*) Mouse cells stained with DAPI (DNA) and HP1-β, one of the three HP1 homologs in mouse. The foci visible by DAPI staining correspond to the highly condensed pericentric heterochromatin, which are also the sites of concentrated HP1-β staining. The only densely staining DAPI spot that does not correspond to HP1-β staining is the X chromosome, which uses alternative factors. (Photographs courtesy of S.C. Elgin [*A*], and E. Heard [*B*].)

subject to transcriptional repression (Muller 1930; Weiler and Wakimoto 1995). In PEV, founder cells pass on alternate transcriptionally active ("on") and silent ("off") states of a reporter gene to their descendants, resulting in a variegated phenotype (Figure 10.2) (Eissenberg et al. 1995). Importantly, the repressive influence of heterochromatin is regional and does not depend on particular sequences within the reporter gene. A key observation was that when the endogenous *white* gene was placed adjacent to pericentric heterochromatin via chromosomal rearrangement, the chromosomal region including the marker gene was physically condensed in the cells in which the marker gene was "off," but not in the cells in which it was "on" (Zhimulev et al. 1986). Thus, the condensed state of heterochromatin is able stochastically to spread across and encompass the marker gene, and this structurally condensed appearance correlates with the transcriptionally inactive state of the marker. The spread of condensed chromatin structure across the entire inactive X chromosome in mammals renders genes on that chromosome transcriptionally inactive in a similar fashion (Avner and Heard 2001). In fission yeast, the insertion of reporter genes within or adjacent to heterochromatic regions results in clonally inherited gene-silencing phenomena that exhibit the characteristic properties of classical PEV (Thon and Klar 1992; Allshire et al. 1994; Grewal and Klar 1996, 1997). Mysteriously, those few genes found within heterochromatic regions appear to require such an environment for their proper expression and exhibit PEV when placed within euchromatin (see Lu et al. 2000).

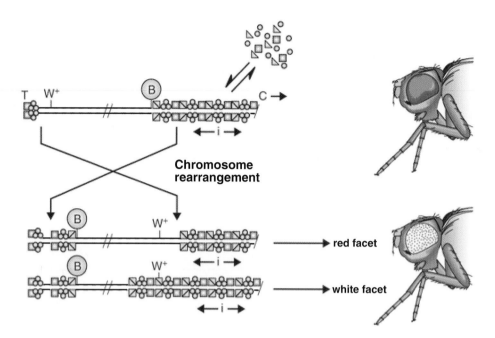

**Figure 10.2.** A schematic illustration of position-effect variegation of the *white* gene in the X chromosome inversion *ln(1)w^{m4}* of *Drosophila*. The *white* locus (*w^+*), located in the distal euchromatin of the wild-type X chromosome, provides a function essential for normal red pigmentation of the fly's eye. The inversion shown is a result of chromosomal breaks (X-ray-induced) that occurred adjacent to the *white* locus and within the pericentric heterochromatin; this inversion places the *white* gene within 25 kb of the heterochromatic breakpoint. This abnormal juxtaposition gives rise to flies with mottled (variegated) eyes composed of both fully pigmented red facets (*white* active) and less pigmented white facets (*white* inactive). It is suggested that in the normal chromosome, the assembly of heterochromatin-specific proteins (represented by colored geometric symbols) extends cooperatively from initiation sites "i" until a barrier "B" is reached; in the absence of the barrier, the *w^+* gene may be so packaged (and silenced), the extent of spreading being the result of competition for the heterochromatic components. The model shown is not intended to imply a requirement for strict continuous linear spreading, as a gene closer to the breakpoint may escape silencing in a case where the more distal gene is silenced. (C) Centromere; (T) telomere. (Reprinted, with permission, from Grewal and Elgin 2002 ©Elsevier Science. Figure courtesy of J.C. Eissenberg.)

## Suppression of Recombination

In addition to transcriptional silencing, heterochromatin also exerts a repressive influence on recombination. The factors that are required for heterochromatin formation are also important for recombination suppression, and, at the silent mating-type region of fission yeast, heterochromatic gene silencing is inseparable from recombination suppression (Klar and Bonaduce 1991; Thon et al. 1994; Grewal and Klar 1997). Recombination rates fall dramatically in the vicinity of centromeres (Nakaseko et al. 1986; Choo 1998), and, in maize, which contains extensive tracts of noncoding repetitive sequences in intergenic regions, recombination is almost entirely limited to intragenic regions (SanMiguel et al. 1996). The suppression of recombination within nongenic repetitive sequences is likely to underlie the so-called recombination value paradox, which stems from the observation that organisms with genomes of vastly different DNA content have genetic maps of similar length in centiMorgans (cM) (Thuriaux 1977). Further suggesting that recombination between repetitive elements is suppressed, meiotic recombination between transposable elements at nonallelic sites in *S. cerevisiae* occurs at a reduced frequency (Kupiec and Petes 1988), whereas recombination between ectopically placed genes is relatively common (Goldman and Lichten 1996). The packaging of large repetitive regions of the genome into heterochromatin might help ensure that they do not participate in illegitimate crossovers that can lead to inversions, translocations, insertions, or deletions. As such, the ability of heterochromatin to suppress recombination is thought to be crucial for the maintenance of genomic integrity (Grewal and Klar 1997; Davis et al. 2000; Gartenberg 2000; Guarente 2000).

## Chromosome Segregation

Heterochromatic structures also have the important property of facilitating chromosome segregation. This is most obvious and well-described within the centric and pericentric regions of chromosomes, but can also occur at distal blocks of heterochromatin (Platero et al. 1999). Heterochromatin formation and chromosome segregation require many of the same factors (Allshire et al. 1995; Kellum and Alberts 1995; Grewal et al. 1998; Peters et al. 2001). Centromere function appears to be facilitated by the condensed structural properties of heterochromatin, which may help to orient and assemble the kinetochore (Pidoux and Allshire 2000; Sullivan et al. 2001). Studies in fission yeast have directly implicated components of heterochromatin in recruiting cohesin to centromeres as well as to heterochromatic locations along the chromosome arms, such as the silent mating-type interval (Bernard et al. 2001; Nonaka et al. 2001; Partridge et al. 2002; Hall et al. 2003). Moreover, cohesin levels at the 20-kb silent mating-type interval strictly mirror the distribution of heterochromatin components (Figure 10.3) (K. Noma and S. Grewal, unpubl.). Centric heterochromatin in *Drosophila* can facilitate the achiasmate (nonexchange) segregation of chromosomes during meiosis in a dosage-dependent manner, in part due to pairing interactions (Dernburg et al. 1996; Karpen et al. 1996), and interstitial heterochromatic knobs can lead to the preferential meiotic segregation of chromosomes in maize (Rhoades and Vilkomerson 1942). The extent to which segregation function is an intrinsic property of heterochromatin or an acquired function is not clear; however, in cases where heterochromatin is known to be involved in segregation, this function is inseparable from its other properties of transcriptional silencing and recombination suppression.

## Sequence Composition

The DNA sequence found within heterochromatin is generally repetitive and gene-poor. The silent nature of these repetitive genomic regions is widely believed to reflect the need for organisms to prohibit the transcription and proliferation of "useless" or "parasitic" DNA elements (Bestor 1998a; Matzke and Matzke 1998; Selker 1999). The origin of other common, seemingly unrelated classes of repeats such as satellite, knob, and subtelomeric repeats is unclear. Repeats can be amplified or deleted by transposition, replication errors, DNA-repair processes, illegitimate mitotic recombination, and/or unequal crossover events during meiotic recombination. The repeat unit size within such regions can vary widely, and in some cases, different repeat unit lengths and symmetries coexist within the same sequence in a hierarchical fashion. The centromeric heterochromatin and heterochromatic islands of higher eukaryotes are characterized by long tracts of tandem satellite repeats punctuated by simple insertions and/or complex arrays of transposable elements and their truncated derivatives (Le et al. 1995; CSHL/WUGSC/PEB *Arabidopsis* Sequencing Consortium 2000). In fission yeast, *dg* and *dh* centromeric repeats surround the unique central core region and are assembled into heterochromatic structures (Nakaseko et al. 1986; Chikashige et al. 1989; Clarke and Baum 1990; Takahashi et al. 1992). Although repeat sequences are important for centromere function, there is little or no conservation between the centromeric repeats of different species, supporting

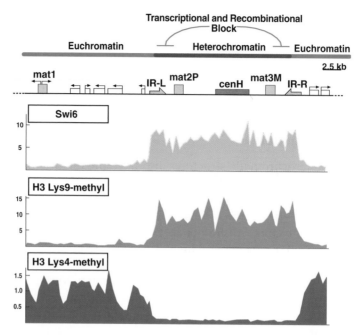

**Figure 10.3.** Distribution of Swi6 and distinct site-specific histone H3 methylation patterns marking the euchromatic and heterochromatic domains at the fission yeast mating-type region. A physical map of the mating-type region with *mat1*, *mat2*, and *mat3* loci is shown. The IR-L and IR-R inverted repeats flanking the *mat2-mat3* interval are shown as *orange/yellow* half arrows. The *gray* box labeled *cenH* represents sequences sharing homology with the centromeric repeats. Open boxes indicate the location of open reading frames; arrows indicate the direction of transcription. The graphs below represent results from high-resolution mapping of levels of Swi6, H3 Lys-9 methylation, and H3 Lys-4 methylation levels. H3 Lys-9 methylation and Swi6 are specifically enriched throughout the 20-kb heterochromatic interval that displays transcriptional silencing and suppression of recombination. In contrast, H3 Lys-4 methylation is enriched in transcriptionally poised euchromatic regions containing genes. IR-L and IR-R inverted repeats act as boundaries of the heterochromatin domain and prevent spreading of repressive chromatin complexes into neighboring genes. (Reprinted, with permission, from Grewal and Elgin 2002 ©Elsevier Science.)

an epigenetic model for centromere specification (Karpen and Allshire 1997; Richards and Dawe 1998). In a variety of systems, repetitive elements of diverse origin appear to attract the silencing machinery preferentially (Dorer and Henikoff 1994; Sabl and Henikoff 1996; Hsieh and Fire 2000), and their emerging role in chromosome segregation and gene regulation suggests that otherwise "parasitic" DNA elements may sometimes contribute to important biological processes (Henikoff 2000).

## MODIFIERS OF POSITION-EFFECT VARIEGATION

The transcriptional repression of reporter genes subjected to PEV has been exploited to identify mutations that either suppress (*Suppressor of variegation* [*Su(var)*]) or enhance (*Enhancer of variegation* [*E(var)*]) variegation. Extensive genetic screens in *Drosophila* and fission yeast have identified a large number of modifiers of PEV, and characterization of these modifiers has provided important clues about the composition and properties of heterochromatin. Most of these modifiers fall into two general classes: histones and non-histone chromosomal proteins, and enzymes involved in the posttranslational modification of histones. Demonstrating the importance of dosage, one of the earliest recognized suppressors of PEV in flies was an additional copy of the heterochromatic Y chromosome (Spofford 1976). The presence of an extra Y chromosome is believed to titrate limiting structural proteins, leading to the reduction of silencing at variegating breakpoints.

## Heterochromatin Protein 1

The most well-studied chromosomal proteins involved in silent chromatin assembly and epigenetic inheritance are members of the heterochromatin protein 1 (HP1) family, which exert dosage-dependent effects on PEV (Eissenberg and Elgin 2000; Nakayama et al. 2000). The HP1 protein was originally found in a screen of monoclonal antibodies as a protein primarily localized to the pericentric heterochromatin of *Drosophila* chromosomes (James and Elgin 1986; James et al. 1989), and it is encoded by the *Su(var)2-5* gene that is required for PEV (Eissenberg et al. 1990, 1992). HP1 is a highly conserved protein, with homologs found from fission yeast (Swi6, Chp2, and Chp1) to humans (HP1$^{\alpha}$, HP1$^{\beta}$, and HP1$^{\gamma}$) (Eissenberg and Elgin 2000; Thon and Verhein-Hansen 2000), although interestingly not in the budding yeast *S. cerevisae*. These proteins contain an amino-terminal chromodomain, a short variable hinge region, and, with the exception of Chp1, a chromoshadow domain (Paro and Hogness 1991; Lorentz et al. 1994; Aasland and Stewart 1995). The functional conservation of HP1 proteins is illustrated by the fact that the chromodomain of *S. pombe* Swi6 can be complemented by that of mouse HP1 (Wang et al. 2000) and that human HP1 can promote silencing in flies (Ma et al. 2001).

Several factors that interact with HP1/Swi6 have been identified. These include proteins involved in DNA replication, gene regulation, and chromatin assembly (Eissenberg and Elgin 2000; Nakayama et al. 2001a; Li et al. 2002). The self-association of HP1 family members through their chromoshadow domain is thought to mediate chromatin compaction and heterochromatic spreading (Brasher et al. 2000; Cowieson et al. 2000; Wang et al. 2000). The chromodomain, which has an important role in the targeting of Swi6/HP1 to silenced chromosomal domains (Platero et al. 1995; Wang et al. 2000), is a specific interaction motif for the histone H3 tail methylated at Lys-9 (Bannister et al. 2001; Jacobs et al. 2001; Lachner et al. 2001; Nakayama et al. 2001b). Structural studies of the HP1 chromodomain bound to histone H3 peptide-methylated Lys-9 reveal that the H3 tail inserts as a β-strand, completing the β-sandwich architecture of the chromodomain (Jacobs and Khorasanizadeh 2002; Nielsen et al. 2002).

## Histones and Histone Modifications

The fundamental unit of chromatin is the nucleosome, which consists of 147 bp of genomic DNA wrapped twice around the highly conserved histone octamer containing two molecules each of the four core histones: H2A, H2B, H3, and H4 (van Holde 1989). Studies from several different systems have provided valuable knowledge concerning the role of histones and histone-modifying factors in heterochromatin formation. Mutation of the histone tails in *S. cerevisiae* or a partial deletion of the histone gene cluster in *Drosophila* results in suppression of PEV (Moore et al. 1983; Grunstein 1998). There is a growing appreciation for the information encoded in the covalent modification of histones, which specify the interactions between nucleosomes and nonhistone chromosomal proteins (Jenuwein and Allis 2001; Zhang and Reinberg 2001). Histones H3 and H4 are largely hypoacetylated in the heterochromatic chromosomal regions of organisms as diverse as yeast, *Drosophila*, and mammals (Braunstein et al. 1993; Turner 1993). Genetic and biochemical studies suggest that histone tails and their acetylation status have crucial roles in the formation of heterochromatin (see Grunstein 1997; Grewal and Elgin 2002). Inhibition of histone deacetylation by treatment with trichostatin A or butyrate, specific inhibitors of histone deacetylases (HDAC), is correlated with loss of heterochromatin-mediated silencing (Mottus et al. 1980; Ekwall et al. 1997; Grewal et al. 1998). In *S. pombe*, the HDACs (Clr3 and Clr6) were identified in an unbiased screen for silencing-defective mutants (Ekwall and Ruusala 1994; Thon et al. 1994; Grewal et al. 1998), and specific mutations in *Drosophila* HDAC1 protein suppress PEV (Mottus et al. 2000). Moreover, the Sir2 family of NAD-dependent HDACs modify PEV in *S. cerevisiae* (Hoppe et al. 2002; Rusche et al. 2002), *Drosophila* (Neuman et al. 2002; Rosenberg and Parkhurst 2002), and *S. pombe* (G. Shankaranarayana et al., unpubl.).

Another key connection between histone modifications and heterochromatin assembly was established by recent findings that the evolutionarily conserved human SUV39H1 and mouse Suv39h1—mammalian homologs of *Drososophila Su(var)3-9*—encode a methyltransferase specific for Lys-9 of histone H3 (H3 Lys-9) (Rea et al. 2000). This methyltransferase (HMTase), originally identified as suppressor of variegation in *Drosophila*, has been found in all organisms examined that contain HP1. SUV39H1 and its *S. pombe* homolog Clr4 contain an amino-terminal chromodomain and a carboxy-terminal SET domain (Ivanova et al. 1998). Although only the SET domain and the surrounding cysteine-rich regions of SUV39H1/Clr4 are required for their HMTase activity in vitro, both the chromodomain and SET domain are required for histone methylation in vivo (Nakayama et al. 2001b), indicating that the chromodomain might be important for the targeting of these enzymes to silenced chromosomal domains.

The methylation of H3 Lys-9 is a conserved mark for heterochromatin assembly (Boggs et al. 2002; Peters et al. 2002) and is required for HP1 localization in diverse species (Nakayama et al. 2001b; Peters et al. 2001; Schotta et al. 2002). Moreover, histone-modifying enzymes such as HDACs and HMTases are believed to cooperate with each other to establish a specific histone modification pattern that is recognized by the HP1/Swi6 chromodomain (Grewal and Elgin 2002). To this end, it has been reported that in addition to methylation of H3 Lys-9, the deacetylation of H3 Lys-14 is also required for localization of Swi6 protein at heterochromatic loci in *S. pombe* (Nakayama et al. 2001b). Furthermore, the histone deacetylase HDAC1 and methyltransferase *Su(var)3-9* in *Drosophila* associate in vivo (Czermin et al. 2001), and these enzymes cooperate with each other to methylate preacetylated histones (Czermin et al. 2001; Nishioka et al. 2002; Schultz et al. 2002). These discoveries lend considerable support to the "histone code" hypothesis for chromatin regulation (Strahl and Allis 2000), whereby distinct combinations of histone modifications form a code that is read by other proteins to specify downstream regulatory events.

In many organisms, DNA methylation is used as a marker for epigenetic gene silencing (Bestor 1998b; Martienssen and Colot 2001). However, some organisms, such as *Drosophila* and *S. pombe*, have very low or undetectable levels of methylated DNA and yet still form large and stable regions of heterochromatin. Most likely, DNA methylation and histone modification are complementary and, to some extent, redundant mechanisms to mark silent loci. Emerging evidence points to a feedback loop between cytosine methylation and H3 Lys-9 methylation, such that one promotes the maintenance of the other (Tamaru and Selker 2001; Johnson et al. 2002; Malagnac et al. 2002; Soppe et al. 2002).

## HETEROCHROMATIN-LIKE STRUCTURES IN EPIGENETIC CONTROL OF EUCHROMATIN

It is necessary to recognize that silent chromatin states exist on vastly different scales and in different genomic contexts. At the most extreme, condensed heterochromatic structure can encompass an approximately 9-Mb maize centromere or the entire mammalian X chromosome, whereas repressive effects can also be observed upon the modification of specific nucleosomes within gene promoters (Choo 1997; Nielsen et al. 2001). Although it seems clear that silent chromatin regions of different sizes and chromosomal contexts will have varying structural, and thus phenotypic, properties, epigenetically inactive regions of the genome appear fundamentally similar in their behavior. Our discussion thus far has been limited to repetitive, gene-poor regions of the genome such as pericentric heterochromatin and interstitial knobs; but similar processes, and in many cases identical factors, govern epigenetic gene silencing within euchromatic regions.

## HP1 in Control of "Euchromatin"

Studies on the distribution of HP1 have observed a banded pattern across the small fourth chromosome of *Drosophila* and at a small number of euchromatic sites dispersed throughout the genome (see Figure 10.1) (James et al. 1989; Sun et al. 2000; Li et al. 2002). This indicates that although HP1 is primarily concentrated at pericentric heterochromatin, specific locations along the chromosome arms are also under its control. Recent studies have shown that H3 Lys-9 methylation and HP1 are recruited to specific promoters for gene silencing, directly implicating integral components of heterochromatin in the regulated silencing of euchromatic genes (Nielsen et al. 2001; Schultz et al. 2002; Tachibana et al. 2002).

## Heritable Gene Repression in Development

The maintenance of heritable transcriptional states is essential for the development of multicellular organisms, and accumulating evidence indicates that misregulation of such processes contributes to cellular transformation and cancer progression (Warnecke and Bestor 2000; Jacobs and van Lohuizen 2002; Varambally et al. 2002). During the early development of the *Drosophila* embryo, patterns of homeotic gene expression are established by the combinatorial action of transcription factors within segments. These proteins quickly degrade, and the Polycomb (Pc-G) and trithorax (Tr-G) group proteins act in conjunction with DNA sequences termed polycomb response elements (PRE) to maintain lineage-specific "off" or "on" transcriptional states throughout subsequent development (Ringrose and Paro 2001). The Pc-G proteins function to maintain the transcriptionally repressed "off" state, and the founding member of this family, Polycomb, contains a chromodomain similar to HP1 (Messmer et al. 1992). The *Drosophila* Pc-G complex ESC-E(Z) and its human counterpart contain methyltransferase activity for histone H3 Lys-27 (Cao et al. 2002; Muller et al. 2002) and H3 Lys-9 (Czermin et al. 2002; Kuzmichev et al. 2002). These marks colocalize with Polycomb-binding sites (Czermin et al. 2002), and methylation of Lys-27 facilitates the binding of the Polycomb protein to histone H3 (Cao et al. 2002). The transcriptional repression conferred by Pc-G complexes shares many characteristics with heterochromatin, including altered chromatin structure, mitotic heritability, and the ability to spread in *cis* (Paro et al. 1998; Pirrotta 1998). Thus, the pathways leading to gene silencing during development appear to follow rules similar to those for the formation of heterochromatin by HP1.

## Epialleles

The phenomenon of "epialleles" has been observed in fungi (Grewal and Klar 1996) (see below), insects (Cavalli and Paro 1998), plants (Finnegan 2002), and mammals (Rakyan et al. 2001). Different epialleles of a gene differ in their phenotypic expression but not their DNA sequence, and by definition, they are meiotically heritable and metastable.

Although the reasons are not clear, certain genes or gene structures appear to be more susceptible to this phenomenon than others (Jacobsen 1999). Perhaps the most dramatic example of epialleles comes from the plant species *Linaria vulgaris* (Cubas et al. 1999), in which a naturally occurring variant described more than 250 years ago by Linnaeus has an altered flower morphology displaying radial symmetry, rather than bilateral symmetry. Genetic experiments demonstrated that a single locus containing a developmental gene was responsible for the phenotype, but DNA sequencing of that locus revealed no mutations. Further characterization showed that the mutant phenotype was caused by epigenetic gene silencing of this gene, as displayed by extensive DNA methylation and lack of transcription. Interestingly, this gene appears to be epigenetically regulated within the developing plant, providing a plausible mechanism for occasional conversion to the "off" epiallelic state.

The phenomenon of epialleles demonstrates that alternative states of gene function, as determined by alternative chromatin configurations, can be inherited through meiosis in a manner similar to that of conventional alleles determined by DNA sequence changes (Grewal and Klar 1996) (see below). Although the frequency of epialleles in natural populations remains almost entirely unexamined, it may be significant that the first morphological mutant to be characterized is a naturally occurring epigenetic variant that has persisted alongside its "wild-type" relatives.

## Neocentromeres

Not only can euchromatic genes be silenced in a manner resembling heterochromatin formation, but under certain circumstances, sizeable regions of euchromatin can be assembled into a novel centromere, a so-called "neocentromere" (Amor and Choo 2002). Loss of the natural centromere by chromosomal rearrangement can create an unstable acentric chromosome. In humans, neocentromeres can form at novel positions along the arms of these chromosomes; these genomic sites do not differ in sequence from normal relatives and do not contain an elevated density of centromere-like sequences such as α-satellite repeats (Barry et al. 2000). In *Drosophila*, however, there is some evidence that centromere proximity and the local volume of repetitive DNA aid in neocentromere formation (Maggert and Karpen 2001). Neocentromeres closely resemble normal centromeres in structure and function and are able to recruit kinetochore components as well as heterochromatic factors such as *Su(var)3-9* and HP1 (Aagaard et al. 2000; Saffery et al. 2000). The existence of neocentromeres further supports an epigenetic mechanism for centromere specification (Steiner and Clarke 1994; Murphy and Karpen 1998) and demonstrates that heterochromatic structure can be an acquired characteristic.

## MECHANISMS FOR THE INITIATION OF HETEROCHROMATIN

A large body of work indicates that there are distinct requirements for the establishment and maintenance of heterochromatin. Whereas much progress has been made in defining the pathways by which histone modification patterns lead to the formation and inheritance of heterochromatic structures, the manner in which these histone modifications are targeted in the first place remains enigmatic. In general, the major targets for heterochromatin formation are repeated sequences such as transposons and satellite repeats. Additional genomic loci may be the target of heterochromatin formation during certain periods of development or the cell cycle, or for the purpose of parental imprinting. There are at least three ways to confer specificity for chromosomal targeting: protein-DNA, RNA-DNA, or DNA-DNA interactions.

## DNA-binding Proteins

Heterochromatin can be nucleated directly by specific *cis*-acting DNA sequences. In this mechanism, sequences are recognized by DNA-binding proteins that are then able to recruit more general silencing factors such as histone-modifying enzymes or structural chromatin components. This method is probably most prominent in the regulated silencing of specific genes during development or the cell cycle and would evolve in a manner similar to that of other forms of recruitment-governed transcriptional control (Ptashne and Gann 1997). This form of recruitment may also operate at repetitive centromeric regions, where proteins such as CENP-B localize (Earnshaw and Cooke 1989). CENP-B proteins bind specific sequences within *S. pombe* centromeres and contribute to the establishment of heterochromatin-specific histone modification patterns at these regions (Murakami et al. 1996; Baum and Clarke 2000; Nakagawa et al. 2002).

## Repeat-induced Heterochromatin Formation

Transposons comprise a major fraction of heterochromatic sequences, and substantial evidence suggests that eukaryotic genomes have evolved mechanisms to suppress the mobility of transposons by epigenetic silencing mechanisms, including chromatin modification (Bestor 1998a,b) and RNA interference (RNAi) (Tabara et al. 1999). The ability of transposons to preferentially attract the silencing machinery may underlie their ability to affect the expression of nearby genes (McClintock 1956; Martienssen et al. 1989; Whitelaw and Martin 2001). It is not clear what aspects of transposon identity are recognized by the silencing machinery, but it has been suggested that their repetitive nature might be a key factor. Dispersed repeated sequences of any composition are subject to epigenetic silencing in filamentous fungi (Rossignol and Faugeron 1994; Selker 2002), and, in plants and animals, increased copies of transgenes can lead to homology-dependent transcriptional and posttranscriptional silencing, termed cosuppression (Jorgensen et al. 1996; Matzke and Jorgensen 1996; Birchler et al. 2000) (see also Chapter 1). Tandem repeats of transgenes can also be potent triggers for silencing, particularly when they are in an inverted orientation (Dorer and Henikoff 1994; Sabl and Henikoff 1996; Garrick et al. 1998; Luff et al. 1999).

The mechanisms by which repeated sequences per se act as triggers for heterochromatin formation are poorly understood. Most models hinge upon the detection of repeats by RNA intermediates (see below) or by the physical pairing of homologous DNA sequences (Henikoff 1998; Hsieh and Fire 2000; Matzke et al. 2001). Supporting a role for pairing mechanisms in the transfer of epigenetic information, transcriptional silencing and DNA methylation can be transferred between homologous genes by a recombination-like mechanism in *Ascobolus* (Colot et al. 1996). Polycomb-mediated silencing is subject to mitotic pairing effects in *Drosophila* (Pirrotta 1999), and meiotic transvection effects involving silencing and DNA methylation have been reported in the mouse (Rassoulzadegan et al. 2002). In such a model, paired DNA intermediates form unique structures that are recognized by chromatin-modifying factors.

## RNA-directed Silencing

Accumulating evidence points to an important role for noncoding RNAs in the targeting of chromatin modifications (Park and Kuroda 2001). In mammals, the noncoding Xist RNA initiates H3 Lys-9 methylation and heterochromatin formation from the site of its synthesis (Jaenisch et al. 1998; Heard et al. 2001), and silencing requires the antisense Tsix transcript (Lee 2000; Avner and Heard 2001). In plants, the production of RNA from a viral vector (Wassenegger et al. 1994; Angell and Baulcombe 1997; Vance and Vaucheret 2001), a virus-derived transgene (Dalmay et al. 2000), or a transgene that produces self-complementary transcripts (Mette et al. 1999; Smith et al. 2000) can trigger the transcriptional silencing and chromatin modification of homologous sequences. Double-stranded RNA (dsRNA) is a potential intermediate in all of these cases, and similar to dsRNA-mediated posttranscriptional silencing (Hamilton and Baulcombe 1999; Hannon 2002), transcriptional silencing triggered by dsRNA is accompanied by the production of short interfering RNAs (siRNAs) (Mette et al. 2000). Furthermore, components of the RNAi machinery are required for cosuppression in *Drosophila* (Pal-Bhadra et al. 2002), for heterochromatin formation and the targeting of H3 Lys-9 methylation in *S. pombe* (Hall et al. 2002; Volpe et al. 2002), and for programmed DNA elimination in *Tetrahymena* (Mochizuki et al. 2002), which is also specified by H3 Lys-9 methylation (Taverna et al. 2002). More recently, a member of the Argonaute family involved in RNAi was shown to be required for transcriptional silencing and the targeting of chromatin modification to silent epialleles in *Arabidopsis* (Zilberman et al. 2003). The exact mechanisms by which RNAs target silent histone modifications to genomic loci are not clear; however, all known examples suggest homology recognition between RNA and DNA (Matzke et al. 2001).

A possible link between RNA and DNA-based detection mechanisms is illustrated by the requirement for an RNA-dependent RNA polymerase for "meiotic silencing by unpaired DNA" in *Neurospora* (Aramayo and Metzenberg 1996), in which sequences with no homologous counterpart are silenced during meiosis (Aramayo and Metzenberg 1996). Given the paucity of potential targeting mechanisms, it is likely that RNA sensing and DNA pairing will both be found to operate and, perhaps in some cases, cooperate in the silencing of repeated sequences.

## CASE STUDY: HETEROCHROMATIN FORMATION IN FISSION YEAST

Silencing at centromeres, telomeres, and the silent mating-type region of fission yeast is linked to the formation of heterochromatin structures, which globally inhibit both transcription and recombination throughout large chromosomal domains (Grewal 2000). As described above, several *trans*-acting factors involved in assembly of heterochromatin formation in fission yeast are conserved in higher eukayotic species (Grewal and Elgin 2002), and recent studies strongly suggest that similar mechanisms of heterochromatin assembly might operate in fission yeast and mammals (Heard et al. 2001; Nakayama et al. 2001b; Noma et al. 2001; Peters et al. 2001). This makes *S. pombe* the model system of choice for exploring heterochromatin function in a genetically tractable eukaryote.

## Centromeres and Telomeres

Fission yeast centromeres are large and complex structures ranging in size from 35 kb to 110 kb. Electron micrographic analysis reveals that the domain structure, including the localization patterns of centromere-associated proteins, is conserved from fission yeast to humans (Kniola et al. 2001). Large inverted repeat structures (*imr* and *otr*) surround the central core (*cnt*) domain, which is the site of kinetochore formation and microtubule attachment. The *imr* regions are composed of clusters of tRNA genes and relatively nonconserved sequence (Wood et al. 2002). In contrast, the *otr* region is composed of tandem copies of the *dg* and *dh* centromeric repeats, which although highly variable in copy number among the three centromeres, are remarkably conserved in sequence and are arranged symmetrically about the central core domain in an inverted orientation (Nakaseko et al. 1986; Chikashige et al. 1989). The formation of heterochromatin in fission yeast is tightly linked to centromere function. Mutations in factors including *clr1-clr4, clr6, swi6,* and *rik1*—originally identified for their role in silencing at the mating-type region—affect both centromeric silencing and chromosome segregation (Allshire et al. 1995; Grewal et al. 1998). In addition, a number of genetic loci have been identified that have a specific role in centromeric silencing (Ekwall et al. 1999). Distinct *trans*-acting factors interact with the central core domain and the flanking repeat sequences: The essential centromeric protein Mis6 and the centromere-specific histone H3 variant Cnp1 (CENP-A) primarily interact with the central core domain (Saitoh et al. 1997), whereas Swi6 and Chp1 are confined predominantly to the outer centromeric repeats. The localization of Swi6 and Chp1 depends on H3 Lys-9 methylation by Clr4 (Partridge et al. 2000; Nakayama et al. 2001b).

Marker genes placed adjacent to telomeric regions are repressed in a metastable epigenetic manner (Nimmo et al. 1994). Telomere-specific silencing factors, such as Taz1 and Rap1, are required for the silencing of reporter genes (Cooper et al. 1997; Nimmo et al. 1998; Kanoh and Ishikawa 2001). Swi6 is also found at telomeres, and, as at the centromeres and mating-type region, its localization is dependent on Clr4 and Rik1 (Ekwall et al. 1996). Interestingly, telomere-specific silencing factors are also required for unique functions such as the maintenance of telomere length and the clustering of telomeres at the spindle pole body during meiosis (Cooper et al. 1998; Nimmo et al. 1998).

## Heterochromatin at the Silent *mat* Locus Regulates Silencing, Recombination, and Mating-type Switching

The mating-type (*mat*) region of *S. pombe* is composed of three linked loci: *mat1, mat2,* and *mat3*. The *mat2* and *mat3* loci, which serve as donors of *P* and *M* genetic information for the active *mat1* locus during mating-type switching, are located within a 20-kb silent chromosomal domain (Figure 10.3) (Grewal 2000). In addition to transcriptional silencing, the entire *mat2/mat3* interval exhibits severe recombination suppression (Egel et al. 1989). Despite the dramatic suppression of interchromosomal recombination within this region, the intrachromosomal recombination event that switches mating types occurs efficiently. Switching-competent cells switch to the opposite mating type in about 72–90% of cell divisions, and donor choice is determined by chromosomal position rather than the sequence information present at the donor loci (Thon and Klar 1993; Grewal and Klar 1997). This high degree of heterologous mating-type switching, termed directionality, implies a nonrandom choice of donors during each switch such that *mat2* is the preferred donor in M cells and *mat3* is the preferred donor in P cells.

Epigenetic gene silencing, suppression of recombination, and the nonrandom choice of donors are governed by the heterochromatin domain within the *mat2/mat3* interval and require the same *trans*-acting factors (Klar and Bonaduce 1991; Thon and Klar 1992; Ekwall and Ruusala 1994; Thon et al. 1994; Grewal et al. 1998) and *cis*-acting sequences (Grewal and Klar 1997). High-resolution mapping of heterochromatin complexes at the *mat* locus has revealed that the entire 20-kb silent *mat2/mat3* interval is marked by high levels of H3 Lys-9 methylation and Swi6 protein (Figure 10.3) (Noma et al. 2001). Furthermore, mutations that adversely affect these properties of the mating-type region also disrupt levels of H3 Lys-9 methylation and Swi6 protein (Nakayama et al. 2000, 2001b; Hall et al. 2002). In addition to silencing and recombination suppression, heterochromatin at the *mat2/mat3* region is believed to preferentially bring one donor close to *mat1* in a cell-type-specific manner, controlling the directionality of switching. Similar mechanisms may operate in the specification of site-specific recombination events that regulate such diverse processes as programmed DNA elimination in *Tetrahymena* and V(D)J recombination in mammals.

## Centromeric Repeats in Heterochromatin Formation

The factors that define specific chromosomal domains as sites of heterochromatin assembly are poorly understood, but it has been suggested that repeats and transposons are preferential targets (see above). In this regard, our previous work has demonstrated that regional silencing at the *mat* region is mediated by a 4.3-kb sequence (*cenH*) that is 96% homologous to the *dg* and *dh* centromeric repeats (Grewal and Klar 1997) and that may be the relic of an ancient transposition. The *cenH* repeat element operates in the same pathway as the *clr1-4* and *swi6* genes, indicating a cooperative interaction between *cenH* and histone-modifying activities (Grewal and Klar 1997). Replacement of *cenH* with a marker gene results in variegation of the marker and results in a defect specifically in the establishment of heterochromatin. This gives rise to two functional epigenetic states: The "on" state in which the marker gene is transcribed, meiotic recombination is allowed, and directionality of mating-type switching is impaired; and the "off" state, in which the marker is transcriptionally silent, meiotic recombination is suppressed, and switching efficiency is close to wild-type (Grewal and Klar 1996; Thon and Friis 1997).

The respective epigenetic states in *cenH* deletion cells are remarkably stable through mitosis and meiosis, and furthermore are inherited in *cis*. Unlike "on" cells, the *mat* region of "off" cells is marked by histone H3 hypoacetylation, H3 Lys-9 methylation, and high levels of Swi6/HP1. Histone H3 Lys-9 methylation and Swi6 proteins cosegregate with the "off" epiallele through meiosis, demonstrating that at the silent mating-type region of fission yeast, the "gene" is composed of the DNA sequence plus an associated macromolecular complex of H3 Lys-9-methylated histones and Swi6 protein (Nakayama et al. 2000; Hall et al. 2002).

Further evidence for the contribution of *cenH* to the targeting of heterochromatin formation is provided by the demonstration that *cenH* can confer epigenetic repression on an adjacent reporter gene when inserted at an ectopic, otherwise euchromatic, site (Ayoub et al. 2000). This repression requires the same *trans*-acting chromatin-modifying factors as heterochromatin formation within the endogenous mating-type region and is associated with recruitment of H3 Lys-9 methylation and Swi6 (Hall et al. 2002). Independently, examination of a centromere-derived repeat similar to *cenH* demonstrated that it can also direct silent chromatin assembly repeat at an ectopic location (Partridge et al. 2002).

## RNAi and Heterochromatin Assembly in *S. pombe*

Our recent work has led to the discovery that an RNAi-related process is involved in the targeting of histone modification and Swi6/HP1 to the centromeres (Volpe et al. 2002) and the mating-type region (Hall et al. 2002) and that targeting to these regions is mediated by their respective centromeric repeat sequences. The fission yeast genome contains a single homolog for three of the proteins known to be involved in RNAi: (1) an Argonaute family member (*ago1*), (2) an RNase-III-like enzyme similar to Dicer (*dcr1*), and (3) an RNA-dependent RNA polymerase (*rdp1*). Targeted deletion of these genes causes derepression of *ura4*[+] reporter genes inserted within centromeric repeats and concomitant loss of Swi6 and H3 Lys-9 methylation from the *ura4*[+] reporter, as well as from surrounding centromeric regions (Volpe et al. 2002). The extent to which heterochromatin is compromised in the RNAi mutants is apparent from the severe mitotic (Provost et al. 2002) and meiotic segregation defects and disrupted centromeric cohesion observed in RNAi deletion strains (Hall et al. 2003).

Directly implicating an RNA-like process in centromeric silencing, transcripts from the *dg* and *dh* centromeric repeats accumulate in the mutant strains, and "small heterochromatic" RNAs resembling *Dicer* cleavage products have been isolated that correspond to

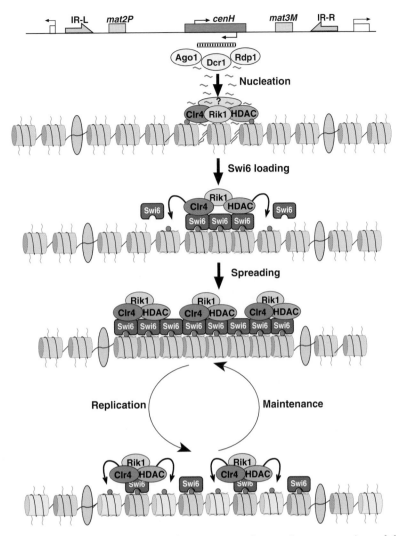

**Figure 10.4.** Model for heterochromatin formation at the mating-type region of fission yeast. Transcription from *cenH* results in RNAs (*large red lines*) that yield small heterochromatic RNAs (*small red lines*) in a process requiring Dcr1, Ago1, and Rdp1. These small RNAs specify the initiation of heterochromatin by interacting with homologous DNA. Histone modification is initiated in a process requiring an HDAC(s), the HMtase Clr4, the β-propeller domain protein Rik1, and potentially an unidentified protein (*question mark*). Swi6 recognizes deacetylated, H3 Lys-9-methylated histones (*red lollipop*) and binds to chromatin. Chromatin-bound Swi6 recruits HDAC(s) and Clr4 to modify adjacent nucleosomes, which are subsequently bound by additional molecules of Swi6. Such a spreading process is repeated until the flanking IR-L and IR-R boundaries (*orange/yellow ovals*) are encountered or until Swi6 is limiting. Upon chromosome replication, parental histones H3/H4 (*darker nucleosomes*) are randomly distributed to daughter chromatids, and newly assembled histones H3/H4 (*lighter nucleosomes*) acquire parental histone modification patterns by the same mechanism that allows spreading in *cis*: Swi6 recruits HDAC(s) and Clr4 to modify adjacent nucleosomes, which become bound by additional molecules of Swi6.

these sequences (Reinhart and Bartel 2002). Although transcripts from centromeric repeats do not accumulate in wild-type strains, nuclear run-on assays demonstrate that the forward strand of centromeric repeats is transcribed at low levels and is subject to posttranscriptional control by the RNAi components and that transcription from the reverse strand is repressed. Loss of centromeric silencing leads to transcription of the reverse strand as well, providing a potential dsRNA trigger for the reestablishment of heterochromatic silencing (Volpe et al. 2002). The essential role of *rdp1* in this process suggests that a posttranscriptional step involving synthesis and/or amplification of dsRNA from a single-stranded template is required, for which there is precedent (Lipardi et al. 2001; Tang et al. 2003).

The capacity of *cenH* to confer epigenetic repression on a reporter gene at an ectopic site is strictly dependent on the RNAi machinery (Hall et al. 2002), suggesting that a similar RNAi-mediated process may operate in heterochromatin formation at the *mat* region. Further evidence comes from the characteristics of *cenH* at its endogenous location. Unlike flanking sequences within the silent *mat* interval, the *cenH* repeat is a hot spot for H3 Lys-9 methylation; *cenH* is able to recruit H3 Lys-9 methylation in a Swi6-independent manner, and H3 Lys-9 methylation becomes restricted to *cenH* in the absence of the spreading activity conferred by Swi6 (Hall et al. 2002). The sequences from *cenH* that are transcribed correspond to the minimal region of *cenH* that confers silencing activity (Ayoub et al. 2000) and to some of the isolated small heterochromatic RNAs (Reinhart and Bartel 2002).

Despite the prominent role that *cenH* has in silencing at the mating-type region and the strict requirement for the RNAi machinery in *cenH*-mediated repression at an ectopic site, the endogenous *mat* region does not display obvious defects when introduced into RNAi mutant backgrounds. This apparent paradox was resolved by the observation that the RNAi machinery is required exclusively for the establishment of heterochromatin at the *mat* region, but not for its subsequent inheritance. RNAi mutants are not able to establish heterochromatin efficiently after treatment with the histone deacetylase inhibitor Trichostatin-A, which has previously been shown to erase the epigenetic imprint governing silencing at the *mat* region (Grewal et al. 1998). Furthermore, genetic analysis demonstrated that RNAi mutant strains are able to maintain silencing when the *mat* region is introduced from a wild-type parental background, but establish silencing in a stochastic, inefficient manner if the *mat* region is derived from a *clr4* mutant parent (Hall et al. 2002). This shows that the RNAi factors can operate to nucleate heterochromatin at a *mat* region devoid of H3 Lys-9 methylation and Swi6 protein, but they are dispensable for the clonal propagation of a *mat* region that is preassembled into the heterochromatic state.

It is not immediately clear why the RNAi machinery is strictly required for heterochromatin at the centromeres, whereas it is only required for the establishment of heterochromatin at the mating-type region. One possibility is that silencing at the *mat* region is more tightly regulated and more redundantly organized. Whereas heterochromatic structure is quickly and efficiently reestablished at the *mat* region in wild-type cells (Hall et al. 2002), reestablishment at centromeres in wild-type cells is relatively stochastic and inefficient. Treatment with Trichostatin-A (Ekwall et al. 1997) or propagation in RNAi mutant backgrounds (I.M. Hall and S.I. Grewal, unpubl.) can give rise to epigenetically "on" centromeres that are heritable through mitosis and meiosis. Another possibility is that the unique nature of centromere function, or of the factors found at centromeres, necessitates more frequent initiation events.

## Boundaries of Heterochromatin

In their natural chromosomal context, heterochromatin domains are frequently juxtaposed with regions of active, even essential, genes. Specialized DNA elements, termed boundary elements, have been suggested to mark the borders between adjacent chromatin domains and to act as barriers to contain heterochromatin within defined genomic regions (Bell and Felsenfeld 1999; Labrador and Corces 2002).

Whereas Swi6 and H3 Lys-9 methylation is localized throughout the 20-kb silent domain, H3 Lys-4 methylation is strictly localized to the surrounding euchromatic regions where active genes reside. Two 2.1-kb inverted repeats (IR-L and IR-R) flank the 20-kb silent domain and mark a sharp transition in distinct histone modification patterns (Figure 10.3) (Noma et al. 2001). The IR-L and IR-R repeats share perfect identity with each other in *S. pombe* and are conserved in a closely related species (Singh and Klar 2002). Deletion of these elements results in the spread of H3 Lys-9 methylation and Swi6 protein into surrounding euchromatin (Noma et al. 2001) and stochastic silencing of reporter genes placed well outside of the boundaries (Thon et al. 2002), indicating that the IR-L and IR-R boundary elements serve to contain heterochromatin within the *mat2/mat3* interval. Similar sharp transitions in histone modification patterns have been observed at the boundary elements of the chicken β-globin locus (Litt et al. 2001), suggesting that chromatin boundaries may function by preventing the spread of histone modification along the chromosome.

Given the complex nature of eukaryotic genomes and the observed ability of heterochromatic complexes to spread in *cis*, boundary elements are likely to serve an essential role in the organization of chromatin domains.

## Establishment and Maintenance of a Heterochromatin Domain

The above observations allow for a model of heterochromatin formation in which at least three phases are required: initiation, spreading, and maintenance.

At the centromeres and mating-type region of fission yeast, H3 Lys-9 methylation and heterochromatic silencing are targeted through a pathway involving repeated elements and the RNAi machinery. The exact mechanism by which this occurs is currently not known, but it appears to involve transcription from the centromeric repeats, processing by the RNAi machinery, and a sequence-specific interaction between RNA and DNA. A possible scenario is that an RNA-DNA complex serves as a substrate for the binding of a histone-modifying protein or a protein that can recruit one. In this respect, it may be noteworthy that a chromodomain motif has been shown to interact with RNA (Akhtar et al. 2000) and that Clr4 has a chromodomain that is required for its H3 Lys-9 methyltransferase activity in vivo. The β-propeller domain protein Rik1 is also strictly required for recruitment of H3 Lys-9 methylation to the *mat* region (Nakayama et al. 2001b), and although its mechanistic contribution is still poorly described, it may function in RNA-directed heterochromatin formation.

Once initiation sites have been marked by silent histone modifications, Swi6 is recruited to these foci and becomes bound to chromatin. Swi6 then recruits H3 Lys-9 methylation and histone deacetylation to adjacent nucleosomes through direct interactions with histone-modifying enzymes. This allows silent histone modifications and chromatin-bound Swi6 to spread in *cis* beyond the original nucleation site(s) and explains the dosage-critical role of Swi6/HP1 in epigenetic silencing (Eissenberg et al. 1992; Nakayama et al. 2000; Hall et al. 2002). Lending mechanistic support to this model, Swi6/HP1 and Clr4/Su(var)3-9 are known to associate with each other in *Drosophila* (Aagaard et al. 1999) and in *S. pombe* (G. Xiao and S.I. Grewal, unpubl.).

Swi6 molecules bound to nearby nucleosomes dimerize through the chromoshadow domain, potentially mediating condensation and higher-order folding. H3 Lys-9 methylation, hypoacetylation, and chromatin-bound Swi6 continue to spread in *cis* until some physical boundary is encountered (see above) (Litt et al. 2001; Noma et al. 2001), or until Swi6 itself becomes limiting. It has recently been suggested that Swi6/HP1 proteins require an RNA component to become localized to visible foci of pericentric heterochromatin in mammalian cells (see Figure 10.1) (Maison et al. 2002), and this RNA-binding activity appears to reside within the hinge region linking the chromodomain with the chromoshadow domain (Muchardt et al. 2002). The identity of the RNA component is not clear, but it appears to operate in the formation of a higher-order structure composed of H3 Lys-9-methylated tails and Swi6/HP1. Supporting a role for RNAs in the higher-order association of heterochromatic sequences, the RNAi machinery is required for the clustering of telomeres in *S. pombe* (Hall et al. 2003). RNAs may act in conjunction with Swi6/HP1 as a "glue" to promote the folding or clustering of heterochromatic regions into higher-order structures that may serve to concentrate heterochromatic portions of the genome into specific nuclear compartments (Gasser 2001) or to promote long-range regulatory interactions (Zaman et al. 2002).

The model outlined above for the establishment of heterochromatin is similar to mammalian X-chromosome inactivation, where an H3 Lys-9 methylation hot spot and Xist-noncoding RNA cooperate to initiate the formation of a heterochromatic structure that spreads in *cis* to encompass the entire chromosome (Heard et al 2001). Similar mechanisms could also account for parental imprinting in mammals, where H3 Lys-9 methylation, noncoding RNAs, and nucleation centers have been proposed to establish silenced genomic regions (Ferguson-Smith and Surani 2001). Finally, such mechanisms may have a pivotal role in the lineage-specific establishment of silenced domains during development, where active chromatin must be quickly and efficiently silenced to allow for cellular differentiation. In this regard, it may be noteworthy that some developmental genes in plants are subject to posttranscriptional silencing by the RNAi machinery (Llave et al 2002) and that the small RNAs produced by such a process are potential triggers for heterochromatin formation.

Once established, heterochromatin domains must be reproduced through chromosome replication and cell division. At the crux of any model for the inheritance of chromatin structure must be the segregation of some component other than DNA. Biochemical experiments in mammalian cells indicate that histones H3 and H4 remain associated with DNA and are randomly distributed to daughter chromatids in a replication-coupled manner (Jackson and Chalkley 1985). Supporting a link between DNA replication and the inheritance of silent chromatin states, DNA polymerase α is required for silencing at the *S. pombe* mating-type region and interacts with Swi6 (Ahmed et al. 2001; Nakayama et al. 2001a), and in budding yeast, chromatin assembly factor 1 (CAF1) and proliferating cell nuclear antigen (PCNA) cooperate to maintain epigenetic gene silencing (Zhang et al. 2000). Implying that replication-coupled chromatin assembly is involved in HP1-mediated silencing, HP1 and CAF1 interact directly in mammalian cells (Murzina et al. 1999), the origin recognition complex (ORC) has a role in silencing (Bell et al. 1993), and HP1 and ORC interact in *Drosophila* (Pak et al. 1997).

The dilution of parental histones by chromosome replication makes necessary a mechanism to reinforce histone modification patterns on the newly replicated daughter chromatids. Given the requirement for Swi6 in the maintenance of H3 Lys-9 methylation (Hall et al. 2002), we propose that this is accomplished through the same mechanism as spreading in *cis*; Swi6 associates with parental histones and recruits histone-modifying activities to adjacent, newly loaded histones. In this model, an "epigenetic loop" between histone modification and structural proteins such as Swi6/HP1 underlies the propagation of heterochromatic complexes.

## CONCLUDING REMARKS

The regulation of higher-order chromatin assembly imposes structural organization on the genetic information of eukaryotes and is likely a major force in the origin and evolution of genes, genomes, and organisms. In particular, the packaging of genomic regions into heterochromatin exerts epigenetic control over important biological processes. The ability of chromatin states to maintain clonal patterns of gene expression contributes to the development of diverse cellular lineages within a single organism and to the stable response of cells and organisms to environmental signals. Moreover, heterochromatin has a crucial role in maintenance of genomic integrity. The genomes of most eukaryotes have been extensively colonized by "parasitic" repetitive elements such as transposons and retroviruses. The containment of such repetitive regions within silent chromosomal domains renders these elements transcriptionally silent and recombinationally inert, prohibiting potentially mutagenic transposition events and genomic rearrangements. Heterochromatin also facilitates chromosome segregation through its effect on kinetochore formation, sister chromatid cohesion, and meiotic pairing. The maintenance of genomic integrity is essential for the accurate transmission of genetic information through cell division and sexual reproduction, and thus is crucial to the identity of cells and species. Importantly, all of the above processes contribute to the abnormal transcriptional profiles and aneuploidy frequently observed in cancer cells and to the chromosomal rearrangements that underlie many heritable human disorders.

We expect that future investigations into the inheritance of chromatin structure will have profound implications for diverse fields of biology, including gene regulation, genome organization, phenotypic variation, and evolution.

## ACKNOWLEDGMENTS

We thank Edith Heard, Sarah C. Elgin, Joel C. Eissenberg, and Ken-ichi Noma for providing material for the figures. We also thank Joshua Chang Mell, Elizabeth Murchison, and Sabrina Nuñez for critical reading of the manuscript. I.M.H. is an Arnold and Mabel Beckman Fellow. Work in the author's laboratory was supported by the Ellison Medical Foundation and by a National Institutes of Health grant to S.G.

## REFERENCES

Aagaard L., Schmid M., Warburton P., and Jenuwein T. 2000. Mitotic phosphorylation of SUV39H1, a novel component of active centromeres, coincides with transient accumulation at mammalian centromeres. *J. Cell Sci.* **113:** 817–829.

Aagaard L., Laible G., Selenko P., Schmid M., Dorn R., Schotta G., Kuhfittig S., Wolf A., Lebersorger A., Singh P.B., Reuter G., and Jenuwein T. 1999. Functional mammalian homologues of the *Drosophila* PEV-modifier Su(var)3-9 encode centromere-associated proteins which complex with the heterochromatin component M31. *EMBO J.* **18:** 1923–1938.

Aasland R. and Stewart A.F. 1995. The chromo shadow domain, a second chromo domain in heterochromatin-binding protein 1, HP1. *Nucleic Acids Res.* **23:** 3168–3174.

Ahmed S., Saini S., Arora S., and Singh J. 2001. Chromodomain protein Swi6-mediated role of DNA polymerase-α in establishment of silencing in fission yeast. *J. Biol. Chem.* **276:** 47814–47821.

Akhtar A., Zink D., and Becker P.B. 2000. Chromodomains are protein-RNA interaction modules. *Nature* **407:** 405–409.

Allshire R.C., Javerzat J.P., Redhead N.J., and Cranston G. 1994. Position effect variegation at fission yeast centromeres. *Cell* **76:** 157–169.

Allshire R.C., Nimmo E.R., Ekwall K., Javerzat J.P., and Cranston G. 1995. Mutations derepressing silent centromeric domains in fission yeast disrupt chromosome segregation. *Genes Dev.* **9:** 218–233.

Amor D.J. and Choo K.H. 2002. Neocentromeres: Role in human disease, evolution, and centromere study. *Am. J. Hum. Genet.* **71:** 695–714.

Angell S.M. and Baulcombe D.C. 1997. Consistent gene silencing in transgenic plants expressing a replicating potato virus X RNA. *EMBO J.* **16:** 3675–3684.

*Arabidopsis* Sequencing Consortium: Cold Spring Harbor Laboratory, Washington University Genome Sequencing Center, and PE Biosystems. 2000. The complete sequence of a heterochromatic island from a higher eukaryote. *Cell* **100:** 377–386.

Aramayo R. and Metzenberg R.L. 1996. Meiotic transvection in fungi. *Cell* **86:** 103–113.

Avner P. and Heard E. 2001. X-chromosome inactivation: Counting, choice and initiation. *Nat. Rev. Genet.* **2:** 59–67.

Ayoub N., Goldshmidt I., Lyakhovetsky R., and Cohen A. 2000. A fission yeast repression element cooperates with centromere-like sequences and defines a *mat* silent domain boundary. *Genetics* **156:** 983–994.

Bannister A.J., Zegerman P., Partridge J.F., Miska E.A., Thomas J.O., Allshire R.C., and Kouzarides T. 2001. Selective recognition of methylated lysine 9 on histone H3 by the HP1 chromo domain. *Nature* **410:** 120–124.

Barry A.E., Bateman M., Howman E.V., Cancilla M.R., Tainton K.M., Irvine D.V., Saffery R., and Choo K.H. 2000. The 10q25 neocentromere and its inactive progenitor have identical primary nucleotide sequence: Further evidence for epigenetic modification. *Genome Res.* **10:** 832–838.

Baum M. and Clarke L. 2000. Fission yeast homologs of human CENP-B have redundant functions affecting cell growth and chromosome segregation. *Mol. Cell. Biol.* **20:** 2852–2864.

Bell A.C. and Felsenfeld G. 1999. Stopped at the border: Boundaries and insulators. *Curr. Opin. Genet. Dev.* **9:** 191–198.

Bell S.P., Kobayashi R., and Stillman B. 1993. Yeast origin recognition complex functions in transcription silencing and DNA replication. *Science* **262:** 1844–1849.

Bennetzen J.L. 2000. The many hues of plant heterochromatin. *Genome Biol.* **1:** REVIEWS107.

Bernard P., Maure J.F., Partridge J.F., Genier S., Javerzat J.P., and Allshire R.C. 2001. Requirement of heterochromatin for cohesion at centromeres. *Science* **11:** 2539–2542.

Bestor T.H. 1998a. The host defence function of genomic methylation patterns. *Novartis Found. Symp.* **214:** 187–195; discussion: 195–189, 228–132.

———. 1998b. Gene silencing. Methylation meets acetylation. *Nature* **393:** 311–312.

Birchler J.A., Bhadra M.P., and Bhadra U. 2000. Making noise about silence: Repression of repeated genes in animals. *Curr. Opin. Genet. Dev.* **10:** 211–216.

Boggs B.A., Cheung P., Heard E., Spector D.L., Chinault A.C., and Allis C.D. 2002. Differentially methylated forms of histone H3 show unique association patterns with inactive human X chromosomes. *Nat. Genet.* **30:** 73–76.

Brasher S.V., Smith B.O., Fogh R.H., Nietlispach D., Thiru A., Nielsen P.R., Broadhurst R.W., Ball L.J., Murzina N.V., and Laue E.D. 2000. The structure of mouse HP1 suggests a unique mode of single peptide recognition by the shadow chromo domain dimer. *EMBO J.* **19:** 1587–1597.

Braunstein M., Rose A.B., Holmes S.G., Allis C.D., and Broach J.R. 1993. Transcriptional silencing in yeast is associated with reduced nucleosome acetylation. *Genes Dev.* **7:** 592–604.

Cao R., Wang L., Wang H., Xia L., Erdjument-Bromage H., Tempst P., Jones R.S., and Zhang Y. 2002. Role of histone H3 lysine 27 methylation in Polycomb-group silencing. *Science* **298:** 1039–1043.

Cavalli G. and Paro R. 1998. The *Drosophila* Fab-7 chromosomal element conveys epigenetic inheritance during mitosis and meiosis. *Cell* **93:** 505–518.

Chikashige Y., Kinoshita N., Nakaseko Y., Matsumoto T., Murakami S., Niwa O., and Yanagida M. 1989. Composite motifs and repeat symmetry in *S. pombe* centromeres: Direct analysis by integration of *Not*I restriction sites. *Cell* **57:** 739–751.

Choo K.H. 1997. Centromere DNA dynamics: Latent centromeres and neocentromere formation. *Am. J. Hum. Genet.* **61:** 1225–1233.

———. 1998. Why is the centromere so cold? *Genome Res.* **8:** 81–82.

Clarke L. and Baum M.P. 1990. Functional analysis of a centromere from fission yeast: A role for centromere-specific repeated DNA sequences. *Mol. Cell. Biol.* **10:** 1863–1872.

Colot V., Maloisel L., and Rossignol J.L. 1996. Interchromosomal transfer of epigenetic states in *Ascobolus:* Transfer of DNA methylation is mechanistically related to homologous recombination. *Cell* **86:** 855–864.

Cooper J.P., Watanabe Y., and Nurse P. 1998. Fission yeast Taz1 protein is required for meiotic telomere clustering and recombination. *Nature* **392:** 828–831.

Cooper J.P., Nimmo E.R., Allshire R.C., and Cech T.R. 1997. Regulation of telomere length and function by a Myb-domain protein in fission yeast. *Nature* **385:** 744–747.

Cowieson N.P., Partridge J.F., Allshire R.C., and McLaughlin P.J. 2000. Dimerisation of a chromo shadow domain and distinctions from the chromodomain as revealed by structural analysis. *Curr. Biol.* **10:** 517–525.

Craig J.M. and Bickmore W.A. 1993. Chromosome bands—Flavours to savour. *BioEssays* **15:** 349–354.

Cubas P., Vincent C., and Coen E. 1999. An epigenetic mutation responsible for natural variation in floral symmetry. *Nature* **401:** 157–161.

Czermin B., Melfi R., McCabe D., Seitz V., Imhof A., and Pirrotta V. 2002. *Drosophila* Enhancer of Zeste/ESC complexes have a histone H3 methyltransferase activity that marks chromosomal Polycomb sites. *Cell* **111:** 185–196.

Czermin B., Schotta G., Hulsmann B.B., Brehm A., Becker P.B., Reuter G., and Imhof A. 2001. Physical and functional association of SU(VAR)3-9 and HDAC1 in *Drosophila*. *EMBO Rep.* **2:** 915–919.

Dalmay T., Hamilton A., Rudd S., Angell S., and Baulcombe D.C. 2000. An RNA-dependent RNA polymerase gene in *Arabidopsis* is required for posttranscriptional gene silencing mediated by a transgene but not by a virus. *Cell* **101:** 543–553.

Davis E.S., Shafer B.K., and Strathern J.N. 2000. The *Saccharomyces cerevisiae RDN1* locus is sequestered from interchromosomal meiotic ectopic recombination in a SIR2-dependent manner. *Genetics* **155:** 1019–1032.

Dernburg A.F., Sedat J.W., and Hawley R.S. 1996. Direct evidence of a role for heterochromatin in meiotic chromsome segregation. *Cell* **86:** 135–146.

Dorer D.R. and Henikoff S. 1994. Expansions of transgene repeats cause heterochromatin formation and gene silencing in *Drosophila*. *Cell* **77:** 993–1002.

Earnshaw W.C. and Cooke C.A. 1989. Proteins of the inner and outer centromere of mitotic chromosomes. *Genome* **31:** 541–552.

Egel R., Willer M., and Nielsen O. 1989. Unblocking of meiotic crossing over between the silent mating-type cassettes of fission yeast, conditioned by the recessive, pleiotropic mutant *rik1*. *Curr Genet.* **15:** 407–410.

Eissenberg J.C. and Elgin S.C. 2000. The HP1 protein family: Getting a grip on chromatin. *Curr. Opin. Genet. Dev.* **10:** 204–210.

Eissenberg J.C., Elgin S.C., and Paro R. 1995. Epigenetic regulation in *Drosophila:* A conspiracy of silence. In *Chromatin structure and gene expression* (ed. S.C. Elgin), pp. 147–169. Oxford University Press, England.

Eissenberg J.C., Morris G.D., Reuter G., and Hartnett T. 1992. The heterochromatin-associated protein HP-1 is an essential protein in *Drosophila* with dosage-dependent effects on position-effect variegation. *Genetics* **131:** 345–352.

Eissenberg J.C., James T.C., Foster-Hartnett D.M., Hartnett T., Ngan V., and Elgin S.C. 1990. Mutation in a heterochromatin-specific chromosomal protein is associated with suppression of position-effect variegation in *Drosophila melanogaster. Proc. Natl. Acad. Sci.* **87:** 9923–9927.

Ekwall K. and Ruusala T. 1994. Mutations in *rik1, clr2, clr3* and *clr4* genes asymmetrically derepress the silent mating-type loci in fission yeast. *Genetics* **136:** 53–64.

Ekwall K., Cranston G., and Allshire R.C. 1999. Fission yeast mutants that alleviate transcriptional silencing in centromeric flanking repeats and disrupt chromosome segregation. *Genetics* **153:** 1153–1169.

Ekwall K., Olsson T., Turner B.M., Cranston G., and Allshire R.C. 1997. Transient inhibition of histone deacetylation alters the structural and functional imprint at fission yeast centromeres. *Cell* **91:** 1021–1032.

Ekwall K., Nimmo E.R., Javerzat J.P., Borgstrom B., Egel R., Cranston G., and Allshire R. 1996. Mutations in the fission yeast silencing factors *clr4+* and *rik1+* disrupt the localisation of the chromo domain protein Swi6p and impair centromere function. *J. Cell. Sci.* **109:** 2637–2648.

Ferguson-Smith A.C. and Surani M.A. 2001. Imprinting and the epigenetic asymmetry between parental genomes. *Science* **293:** 1086–1089.

Finnegan E.J. 2002. Epialleles—A source of random variation in times of stress. *Curr. Opin. Plant Biol.* **5:** 101–106.

Flick J.T., Eissenberg J.C., and Elgin S.C. 1986. Micrococcal nuclease as a DNA structural probe: Its recognition sequences, their genomic distribution and correlation with DNA structure determinants. *J. Mol. Biol.* **190:** 619–633.

Garrick D., Fiering S., Martin D.I., and Whitelaw E. 1998. Repeat-induced gene silencing in mammals. *Nat. Genet.* **18:** 56–59.

Gartenberg M.R. 2000. The Sir proteins of *Saccharomyces cerevisiae:* Mediators of transcriptional silencing and much more. *Curr. Opin. Microbiol.* **3:** 132–137.

Gasser S.M. 2001. Positions of potential: Nuclear organization and gene expression. *Cell* **104:** 639–642.

Goldman A.S. and Lichten M. 1996. The efficiency of meiotic recombination between dispersed sequences in *Saccharomyces cerevisiae* depends upon their chromosomal location. *Genetics* **144:** 43–55.

Gottschling D.E. 1992. Telomere-proximal DNA in *Saccharomyces cerevisiae* is refractory to methyltransferase activity in vivo. *Proc. Natl. Acad. Sci.* **89:** 4062–4065.

Grewal S.I. 2000. Transcriptional silencing in fission yeast. *J. Cell. Physiol.* **184:** 311–318.

Grewal S.I. and Elgin S.C. 2002. Heterochromatin: New possibilities for inheritance of structure. *Curr. Opin. Genet. Dev.* **12:** 178–187.

Grewal S.I. and Klar A.J. 1996. Chromosomal inheritance of epigenetic states in fission yeast during mitosis and meiosis. *Cell* **86:** 95–101.

———. 1997. A recombinationally repressed region between *mat2* and *mat3* loci shares homology to centromeric repeats and regulates directionality of mating-type switching in fission yeast. *Genetics* **146:** 1221–1238.

Grewal S.I., Bonaduce M.J., and Klar A.J. 1998. Histone deacetylase homologs regulate epigenetic inheritance of transcriptional silencing and chromosome segregation in fission yeast. *Genetics* **150:** 563–576.

Grunstein M. 1997. Histone acetylation in chromatin structure and transcription. *Nature* **389:** 349–352.

———. 1998. Yeast heterochromatin: Regulation of its assembly and inheritance by histones. *Cell* **93:** 325–328.

Guarente L. 2000. Sir2 links chromatin silencing, metabolism, and aging. *Genes Dev.* **14:** 1021–1026.

Hall I.M., Noma K., and Grewal S.I. 2003. RNA interference machinery regulates chromosome dynamics during mitosis and meiosis in fission yeast. *Proc. Natl. Acad. Sci.* **100:** 193–198.

Hall I.M., Shankaranarayana G., Noma K., Ayoub N., Cohen A., and Grewal S.I. 2002. Establishment and maintenance of a heterochromatic domain. *Science* **297:** 2232–2237.

Hamilton A.J. and Baulcombe D.C. 1999. A species of small antisense RNA in posttranscriptional gene silencing in plants. *Science* **286:** 950–952.

Hannon G.J. 2002. RNA interference. *Nature* **418:** 244–251.

Heard E., Rougeulle C., Arnaud D., Avner P., Allis C.D., and Spector D.L. 2001. Methylation of histone H3 at Lys-9 is an early mark on the X chromosome during X inactivation. *Cell* **107:** 727–738.

Heitz E. 1928. Das heterochromatin der Moose. *Jehrb. Wiss. Botanik* **69:** 762–818.

Henikoff S. 1998. Conspiracy of silence among repeated transgenes. *BioEssays* **20:** 532–535.

———. 2000. Heterochromatin function in complex genomes. *Biochim. Biophys. Acta* **1470:** 1–8.

Henikoff S., Eissenberg J.C., Hilliker A.J., Schmidt E.R., and Wallrath L.L. 2000. Reaching for new heitz. *Genetica* **109:** 7–8.

Hoppe G.J., Tanny J.C., Rudner A.D., Gerber S.A., Danaie S., Gygi S.P., and Moazed D. 2002. Steps in assembly of silent chromatin in yeast: Sir3-independent binding of a Sir2/Sir4 complex to silencers and role for Sir2-dependent deacetylation. *Mol. Cell. Biol.* **22:** 4167–4180.

Hsieh J. and Fire A. 2000. Recognition and silencing of repeated DNA. *Annu. Rev. Genet.* **34:** 187–204.

Ivanova A.V., Bonaduce M.J., Ivanov S.V., and Klar A.J. 1998. The chromo and SET domains of the Clr4 protein are essential for silencing in fission yeast. *Nat. Genet.* **19:** 192–195.

Jackson V. and Chalkley R. 1985. Histone segregation on replicating chromatin. *Biochemistry* **24:** 6930–6938.

Jacobs J.J. and van Lohuizen M. 2002. Polycomb repression: From cellular memory to cellular proliferation and cancer. *Biochim. Biophys. Acta* **1602:** 151–161.

Jacobs S.A. and Khorasanizadeh S. 2002. Structure of HP1 chromodomain bound to a lysine 9-methylated histone H3 tail. *Science* **295:** 2080–2083.

Jacobs S.A., Taverna S.D., Zhang Y., Briggs S.D., Li J., Eissenberg J.C., Allis C.D., and Khorasanizadeh S. 2001. Specificity of the HP1 chromo domain for the methylated N-terminus of histone H3. *EMBO J.* **20:** 5232–5241.

Jacobsen S.E. 1999. Gene silencing: Maintaining methylation patterns. *Curr. Biol.* **9:** 617–619.

Jaenisch R., Beard C., Lee J., Marahrens Y., and Panning B. 1998. Mammalian X chromosome inactivation. *Novartis Found. Symp.* **214:** 200–232.

James T.C. and Elgin S.C. 1986. Identification of a nonhistone chromosomal protein associated with heterochromatin in *Drosophila melanogaster* and its gene. *Mol. Cell. Biol.* **6:** 3862–3872.

James T.C., Eissenberg J.C., Craig C., Dietrich V., Hobson A., and Elgin S.C. 1989. Distribution patterns of HP1, a heterochromatin-associated nonhistone chromosomal protein of *Drosophila. Eur. J. Cell. Biol.* **50:** 170–180.

Jenuwein T. and Allis C.D. 2001. Translating the histone code. *Science* **293:** 1074–1080.

Johnson L., Cao X., and Jacobsen S. 2002. Interplay between two epigenetic marks. DNA methylation and histone H3 lysine 9 methylation. *Curr. Biol.* **12:** 1360.

Jorgensen R.A., Cluster P.D., English J., Que Q., and Napoli C.A. 1996. Chalcone synthase cosuppression phenotypes in petunia flowers: Comparison of sense vs. antisense constructs and single-copy vs. complex T-DNA sequences. *Plant Mol. Biol.* **31:** 957–973.

Kanoh J. and Ishikawa F. 2001. spRap1 and spRif1, recruited to telomeres by Taz1, are essential for telomere function in fission yeast. *Curr. Biol.* **11:** 1624–1630.

Karpen G.H. and Allshire R.C. 1997. The case for epigenetic effects on centromere identity and function. *Trends Genet.* **13:** 489–496.

Karpen G.H., Le M.H., and Le H. 1996. Centric heterochromatin and the efficiency of achiasmate disjunction in *Drosophila* female meiosis. *Science* **273:** 118–122.

Kellum R. and Alberts B.M. 1995. Heterochromatin protein 1 is required for correct chromosome segregation in *Drosophila* embryos. *J. Cell Sci.* **108:** 1419–1431.

Klar A.J. and Bonaduce M.J. 1991. *swi6*, a gene required for mating-type switching, prohibits meiotic recombination in the *mat2-mat3* "cold spot" of fission yeast. *Genetics* **129:** 1033–1042.

Kniola B., O'Toole E., McIntosh J.R., Mellone B., Allshire R., Mengarelli S., Hultenby K., and Ekwall K. 2001. The domain structure of centromeres is conserved from fission yeast to humans. *Mol. Biol. Cell.* **12:** 2767–2775.

Kupiec M. and Petes T.D. 1988. Allelic and ectopic recombination between Ty elements in yeast. *Genetics* **119:** 549–559.

Kuzmichev A., Nishioka K., Erdjument-Bromage H., Tempst P., and Reinberg D. 2002. Histone methyltransferase activity associated with a human multiprotein complex containing the Enhancer of Zeste protein. *Genes Dev.* **16:** 2893–2905.

Labrador M. and Corces V.G. 2002. Setting the boundaries of chromatin domains and nuclear organization. *Cell* **111:** 151–154.

Lachner M., O'Carroll D., Rea S., Mechtler K., and Jenuwein T. 2001. Methylation of histone H3 lysine 9 creates a binding site for HP1 proteins. *Nature* **410:** 116–120.

Le M.H., Duricka D., and Karpen G.H. 1995. Islands of complex DNA are widespread in *Drosophila* centric heterochromatin. *Genetics* **141:** 283–303.

Lee J.T. 2000. Disruption of imprinted X inactivation by parent-of-origin effects at Tsix. *Cell* **103:** 17–27.

Lee J.T. and Jaenisch R. 1997. The (epi)genetic control of mammalian X-chromosome inactivation. *Curr. Opin. Genet. Dev.* **7:** 274–280.

Li Y., Kirschmann D.A., and Wallrath L.L. 2002. Does heterochromatin protein 1 always follow code? *Proc. Natl. Acad. Sci.* **4:** 16462–16469.

Lipardi C., Wei Q., and Paterson B.M. 2001. RNAi as random degradative PCR: siRNA primers convert mRNA into dsRNAs that are degraded to generate new siRNAs. *Cell* **107:** 297–307.

Litt M.D., Simpson M., Gaszner M., Allis C.D., and Felsenfeld G. 2001. Correlation between histone lysine methylation and developmental changes at the chicken beta-globin locus. *Science* **293:** 2453–2455.

Llave C., Xie Z., Kasschau K.D., and Carrington J.C. 2002. Cleavage of Scarecrow-like mRNA targets directed by a class of *Arabidopsis* mRNA. *Science* **297:** 2053–2056.

Lorentz A., Ostermann K., Fleck O., and Schmidt H. 1994. Switching gene *swi6*, involved in repression of silent mating-type loci in fission yeast, encodes a homologue of chromatin-associated proteins from *Drosophila* and mammals. *Gene* **143:** 139–143.

Lu B.Y., Emtage P.C., Duyf B.J., Hilliker A.J., and Eissenberg J.C. 2000. Heterochromatin protein 1 is required for the normal expression of two heterochromatin genes in *Drosophila. Genetics* **155:** 699–708.

Luff B., Pawlowski L., and Bender J. 1999. An inverted repeat triggers cytosine methylation of identical sequences in *Arabidopsis. Mol. Cell* **3:** 505–511.

Lyon M.F. 1972. X-chromosome inactivation and developmental patterns in mammals. *Biol. Rev.*

*Camb. Philos. Soc.* **47:** 1–35.

Ma J., Hwang K.K., Worman H.J., Courvalin J.C., and Eissenberg J.C. 2001. Expression and functional analysis of three isoforms of human heterochromatin-associated protein HP1 in *Drosophila*. *Chromosoma* **109:** 536–544.

Maggert K.A. and Karpen G.H. 2001. The activation of a neocentromere in *Drosophila* requires proximity to an endogenous centromere. *Genetics* **158:** 1615–1628.

Maison C., Bailly D., Peters A.H., Quivy J.P., Roche D., Taddei A., Lachner M., Jenuwein T., and Almouzni G. 2002. Higher-order structure in pericentric heterochromatin involves a distinct pattern of histone modification and an RNA component. *Nat. Genet.* **19:** 19.

Malagnac F., Bartee L., and Bender J. 2002. An *Arabidopsis* SET domain protein required for maintenance but not establishment of DNA methylation. *EMBO J.* **21:** 6842–6852.

Martienssen R.A. and Colot V. 2001. DNA methylation and epigenetic inheritance in plants and filamentous fungi. *Science* **293:** 1070–1074.

Martienssen R.A., Barkan A., Freeling M., and Taylor W.C. 1989. Molecular cloning of a maize gene involved in photosynthetic membrane organization that is regulated by Robertson's Mutator. *EMBO J.* **8:** 1633–1639.

Matzke M.A. and Jorgensen R.A. 1996. From plants to mammals. *Science* **271:** 1347–1348.

Matzke M.A. and Matzke A.J. 1998. Epigenetic silencing of plant transgenes as a consequence of diverse cellular defence responses. *Cell. Mol. Life Sci.* **54:** 94–103.

Matzke M.A., Matzke A.J., and Kooter J.M. 2001. RNA: Guiding gene silencing. *Science* **293:** 1080–1083.

McClintock B. 1929. Chromosome morphology in *Zea mays*. *Science* **69:** 629.

———. 1956. Controlling elements and the gene. *Cold Spring Harbor Symp. Quant. Biol.* **21:** 197–216.

Messmer S., Franke A., and Paro R. 1992. Analysis of the functional role of the Polycomb chromo domain in *Drosophila melanogaster*. *Genes Dev.* **6:** 1241–1254.

Mette M.F., van der Winden J., Matzke M.A., and Matzke A.J. 1999. Production of aberrant promoter transcripts contributes to methylation and silencing of unlinked homologous promoters in trans. *EMBO J.* **18:** 241–248.

Mette M.F., Aufsatz W., van der Winden J., Matzke M.A., and Matzke A.J. 2000. Transcriptional silencing and promoter methylation triggered by double-stranded RNA. *EMBO J.* **19:** 5194–5201.

Moazed D. 2001. Common themes in mechanisms of gene silencing. *Mol. Cell* **8:** 489–498.

Mochizuki K., Fine N.A., Fujisawa T., and Gorovsky M.A. 2002. Analysis of a piwi-related gene implicates small RNAs in genome rearrangement in *Tetrahymena*. *Cell* **110:** 689–699.

Moore G.D., Sinclair D.A.R., and Grigliatti T.A. 1983. Histone gene multiplication and position-effect variegation in *Drosophila melanogaster*. *Genetics* **105:** 327.

Mottus R., Reeves R., and Grigliatti T.A. 1980. Butyrate suppression of position-effect variegation in *Drosophila melanogaster*. *Mol. Gen. Genet.* **178:** 465–469.

Mottus R., Sobel R.E., and Grigliatti T.A. 2000. Mutational analysis of a histone deacetylase in *Drosophila melanogaster:* Missense mutations suppress gene silencing associated with position effect variegation. *Genetics* **154:** 657–668.

Muchardt C., Guilleme M., Seeler J.S., Trouche D., Dejean A., and Yaniv M. 2002. Coordinated methyl and RNA binding is required for heterochromatin localization of mammalian HP1-α. *EMBO Rep.* **13:** 13.

Muller H.J. 1930. Types of visible variations induced by X-rays in *Drosophila*. *J. Genet.* **22:** 299–334.

Muller J., Hart C.M., Francis N.J., Vargas M.L., Sengupta A., Wild B., Miller E.L., O'Connor M.B., Kingston R.E., and Simon J.A. 2002. Histone methyltransferase activity of a *Drosophila* polycomb group repressor complex. *Cell* **111:** 197–208.

Murakami Y., Huberman J.A., and Hurwitz J. 1996. Identification, purification, and molecular cloning of autonomously replicating sequence-binding protein 1 from fission yeast *Schizosaccharomyces pombe*. *Proc. Natl. Acad. Sci.* **93:** 502–507.

Murphy T.D. and Karpen G.H. 1998. Centromeres take flight: Alpha satellite and the quest for the human centromere. *Cell* **93:** 317–320.

Murzina N., Verreault A., Laue E., and Stillman B. 1999. Heterochromatin dynamics in mouse cells: Interaction between chromatin assembly factor 1 and HP1 proteins. *Mol. Cell* **4:** 529–540.

Nakagawa H., Lee J., Hurwitz J., Allshire R.C., Nakayama J., Grewal S.I., Tanaka K., and Murakami Y. 2002. Fission yeast CENP-B homologs nucleate centromeric heterochromatin by promoting heterochromatin-specific histone tail modifications. *Genes Dev.* **16:** 1766–1778.

Nakaseko Y., Adachi Y., Funahashi S., Niwa O., and Yanagida M. 1986. Chromosome walking shows a highly homologous repetitive sequence present in all the centromere regions of fission yeast. *EMBO J.* **5:** 1011–1021.

Nakayama J., Klar A.J., and Grewal S.I. 2000. A chromodomain protein, Swi6, performs imprinting functions in fission yeast during mitosis and meiosis. *Cell* **101:** 307–317.

Nakayama J., Allshire R.C., Klar A.J., and Grewal S.I. 2001a. A role for DNA polymerase-α in epigenetic control of transcriptional silencing in fission yeast. *EMBO J.* **20:** 2857–2866.

Nakayama J., Rice J.C., Strahl B.D., Allis C.D., and Grewal S.I.S. 2001b. Role of histone H3 lysine 9 methylation in epigenetic control of heterochromatin assembly. *Science* **292:** 110–113.

Neuman B.L., Lundblad J.R., Chen Y., and Smolik S.M. 2002. A *Drosophila* homologue of *sir2* modifies position-effect variegation but does not affect life span. *Genetics* **162:** 1675–1685.

Nielsen P.R., Nietlispach D., Mott H.R., Callaghan J., Bannister A., Kouzarides T., Murzin A.G., Murzina N.V., and Laue E.D. 2002. Structure of the HP1 chromodomain bound to histone H3 methylated at lysine 9. *Nature* **416:** 103–107.

Nielsen S.J., Schneider R., Bauer U.M., Bannister A.J., Morrison A., O'Carroll D., Firestein R., Cleary M., Jenuwein T., Herrera R.E., and Kouzarides T. 2001. Rb targets histone H3 methylation and HP1 to promoters. *Nature* **412:** 561–565.

Nimmo E.R., Cranston G., and Allshire R.C. 1994. Telomere-associated chromosome breakage in fission yeast results in variegated expression of adjacent genes. *EMBO J.* **13:** 3801–3811.

Nimmo E.R., Pidoux A.L., Perry P.E., and Allshire R.C. 1998. Defective meiosis in telomere-silencing mutants of *Schizosaccharomyces pombe*. *Nature* **392:** 825–828.

Nishioka K., Chuikov S., Sarma K., Erdjument-Bromage H., Allis C.D., Tempst P., and Reinberg D. 2002. Set9, a novel histone H3 methyltransferase that facilitates transcription by precluding histone tail modifications required for heterochromatin formation. *Genes Dev.* **16:** 479–489.

Noma K., Allis C.D., and Grewal S.I. 2001. Transitions in distinct histone H3 methylation patterns at the heterochromatin domain boundaries. *Science* **293:** 1150–1155.

Nonaka N., Kitajima T., Yokobayashi S., Xiao G., Yamamoto M., Grewal S.I., and Watanabe Y. 2001. Recruitment of cohesin to heterochromatic regions by Swi6/HP1 in fission yeast. *Nat. Cell Biol.* **1:** 89–93.

Pak D.T., Pflumm M., Chesnokov I., Huang D.W., Kellum R., Marr J., Romanowski P., and Botchan M.R. 1997. Association of the origin recognition complex with heterochromatin and HP1 in higher eukaryotes. *Cell* **91:** 311–323.

Pal-Bhadra M., Bhadra U., and Birchler J.A. 2002. RNAi related mechanisms affect both transcriptional and posttranscriptional transgene silencing in *Drosophila*. *Mol. Cell* **9:** 315–327.

Park Y. and Kuroda M.I. 2001. Epigenetic aspects of X chromosome dosage compensation. *Science* **293:** 1083–1085.

Paro R. and Hogness D.S. 1991. The Polycomb protein shares a homologous domain with a heterochromatin-associated protein of *Drosophila*. *Proc. Natl. Acad. Sci.* **88:** 263–267.

Paro R., Strutt H., and Cavalli G. 1998. Heritable chromatin states induced by the Polycomb and trithorax group genes. *Novartis Found. Symp.* **214:** 51–61.

Partridge J.F., Borgstrom B., and Allshire R.C. 2000. Distinct protein interaction domains and protein spreading in a complex centromere. *Genes Dev.* **14:** 783–791.

Partridge J., Scott K., Bannister A., Kouzarides T., and Allshire R. 2002. *cis*-Acting DNA from fission yeast centromeres mediates histone H3 methylation and recruitment of silencing factors and cohesin to an ectopic site. *Curr. Biol.* **12:** 1652.

Peters A.H., Mermoud J.E., O'Carroll D., Pagani M., Schweizer D., Brockdorff N., and Jenuwein T. 2002. Histone H3 lysine 9 methylation is an epigenetic imprint of facultative heterochromatin. *Nat. Genet.* **30:** 77–80.

Peters A.H., O'Carroll D., Scherthan H., Mechtler K., Sauer S., Schofer C., Weipoltshammer K., Pagani M., Lachner M., Kohlmaier A., Opravil S., Doyle M., Sibilia M., and Jenuwein T. 2001. Loss of the suv39h histone methyltransferases impairs mammalian heterochromatin and genome stability. *Cell* **107:** 323–337.

Pidoux A.L. and Allshire R.C. 2000. Centromeres: Getting a grip of chromosomes. *Curr. Opin. Cell. Biol.* **12:** 308–319.

Pirrotta V. 1998. Polycombing the genome: PcG, trxG, and chromatin silencing. *Cell* **93:** 333–336.

———. 1999. Transvection and chromosomal trans-interaction effects. *Biochim. Biophys. Acta* **1424:** M1–8.

Platero J.S., Hartnett T., and Eissenberg J.C. 1995. Functional analysis of the chromo domain of

HP1. *EMBO J.* **14:** 3977–3986.

Platero J.S., Ahmad K., and Henikoff S. 1999. A distal heterochromatic block displays centromeric activity when detached from a natural centromere. *Mol. Cell* **4:** 995–1004.

Provost P., Silverstein R.A., Dishart D., Walfriddson J., Djupedal I., Kniola B., Wright A., Samuelsson B., Radmark O., and Ekwall K. 2002. Dicer is required for chromosome segregation and gene silencing in fission yeast cells. *Proc. Natl. Acad. Sci.* **99:** 16648–16653.

Ptashne M. and Gann A. 1997. Transcriptional activation by recruitment. *Nature* **386:** 569–577.

Rakyan V.K., Preis J., Morgan H.D., and Whitelaw E. 2001. The marks, mechanisms and memory of epigenetic states in mammals. *Biochem. J.* **356:** 1–10.

Rassoulzadegan M., Magliano M., and Cuzin F. 2002. Transvection effects involving DNA methylation during meiosis in the mouse. *EMBO J.* **21:** 440–450.

Rea S., Eisenhaber F., O'Carroll D., Strahl B.D., Sun Z.W., Schmid M., Opravil S., Mechtler K., Ponting C.P., Allis C.D., and Jenuwein T. 2000. Regulation of chromatin structure by site-specific histone H3 methyltransferases. *Nature* **406:** 593–599.

Reinhart B.J. and Bartel D.P. 2002. Small RNAs correspond to centromere heterochromatic repeats. *Science* **297:** 1831.

Rhoades M.M. and Vilkomerson H. 1942. On the anaphase movement of chromosomes. *Proc. Natl. Acad. Sci.* **28:** 433–436.

Richards E.J. and Dawe R.K. 1998. Plant centromeres: Structure and control. *Curr. Opin. Plant Biol.* **1:** 130–135.

Riggs A.D. and Porter T.N. 1996. X-chromosome inactivation and epigenetic mechanisms. In *Epigenetic mechanisms of gene regulation* (ed. V.E.A. Russo et al.), pp. 231–248. Cold Spring Harbor Laboratory Press, Cold Spring Harbor, New York.

Ringrose L. and Paro R. 2001. Remembering silence. *BioEssays* **23:** 566–570.

Rosenberg M.I. and Parkhurst S.M. 2002. *Drosophila* Sir2 is required for heterochromatic silencing and by euchromatic Hairy/E(Spl) bHLH repressors in segmentation and sex determination. *Cell* **109:** 447–458.

Rossignol J.L. and Faugeron G. 1994. Gene inactivation triggered by recognition between DNA repeats. *Experientia* **50:** 307–317.

Rusche L.N., Kirchmaier A.L., and Rine J. 2002. Ordered nucleation and spreading of silenced chromatin in *Saccharomyces cerevisiae. Mol. Biol. Cell* **13:** 2207–2222.

Russo V.E.A., Martienssen R.A., and Riggs A.D. 1996. *Epigenetic mechanisms in gene regulation.* Cold Spring Harbor Laboratory Press, Cold Spring Harbor, New York.

Sabl J.F. and Henikoff S. 1996. Copy number and orientation determine the susceptibility of a gene to silencing by nearby heterochromatin in *Drosophila. Genetics* **142:** 447–458.

Saffery R., Irvine D.V., Griffiths B., Kalitsis P., Wordeman L., and Choo K.H. 2000. Human centromeres and neocentromeres show identical distribution patterns of >20 functionally important kinetochore-associated proteins. *Hum. Mol. Genet.* **9:** 175–185.

Saitoh S., Takahashi K., and Yanagida M. 1997. Mis6, a fission yeast inner centromere protein, acts during G1/S and forms specialized chromatin required for equal segregation. *Cell* **90:** 131–143.

SanMiguel P., Tikhonov A., Jin Y.K., Motchoulskaia N., Zakharov D., Melake-Berhan A., Springer P.S., Edwards K.J., Lee M., Avramova Z., and Bennetzen J.L. 1996. Nested retrotransposons in the intergenic regions of the maize genome. *Science* **274:** 765–768.

Schotta G., Ebert A., Krauss V., Fischer A., Hoffmann J., Rea S., Jenuwein T., Dorn R., and Reuter G. 2002. Central role of *Drosophila* SU(VAR)3-9 in histone H3-K9 methylation and heterochromatic gene silencing. *EMBO J.* **21:** 1121–1131.

Schultz D.C., Ayyanathan K., Negorev D., Maul G.G., and Rauscher F.J., 3rd. 2002. SETDB1: A novel KAP-1-associated histone H3, lysine 9-specific methyltransferase that contributes to HP1-mediated silencing of euchromatic genes by KRAB zinc-finger proteins. *Genes Dev.* **16:** 919–932.

Selker E.U. 1999. Gene silencing: Repeats that count. *Cell* **97:** 157–160.

———. 2002. Repeat-induced gene silencing in fungi. *Adv. Genet.* **46:** 439–450.

Singh G. and Klar A.J. 2002. The 2.1-kb inverted repeat DNA sequences flank the *mat2,3* silent region in two species of *Schizosaccharomyces* and are involved in epigenetic silencing in *Schizosaccharomyces pombe. Genetics* **162:** 591–602.

Singh J. and Klar A.J. 1992. Active genes in budding yeast display enhanced in vivo accessibility to foreign DNA methylases: A novel in vivo probe for chromatin structure of yeast. *Genes Dev.* **6:** 186–196.

Smith N.A., Singh S.P., Wang M.B., Stoutjesdijk P.A., Green A.G., and Waterhouse P.M. 2000. Total silencing by intron-spliced hairpin RNAs. *Nature* **407:** 319–320.

Soppe W.J., Jasencakova Z., Houben A., Kakutani T., Meister A., Huang M.S., Jacobsen S.E., Schubert I., and Fransz P.F. 2002. DNA methylation controls histone H3 lysine 9 methylation and heterochromatin assembly in *Arabidopsis. EMBO J.* **23:** 6549–6559.

Spofford J.B. 1976. Position-effect variegation in *Drosophila.* In *The genetics and biology of* Drosophila (ed. M. Ashburner and E. Novitski), vol. 1c, p. 955. Academic Press, New York.

Steiner N.C. and Clarke L. 1994. A novel epigenetic effect can alter centromere function in fission yeast. *Cell* **79:** 865–874.

Strahl B.D. and Allis C.D. 2000. The language of covalent histone modifications. *Nature* **403:** 41–45.

Sullivan B.A., Blower M.D., and Karpen G.H. 2001. Determining centromere identity: Cyclical stories and forking paths. *Nat. Rev. Genet.* **2:** 584–596.

Sun F.L., Cuaycong M.H., and Elgin S.C. 2001. Long-range nucleosome ordering is associated with gene silencing in *Drosophila melanogaster* pericentric heterochromatin. *Mol. Cell. Biol.* **21:** 2867–2879.

Sun F.L., Cuaycong M.H., Craig C.A., Wallrath L.L., Locke J., and Elgin S.C. 2000. The fourth chromosome of *Drosophila melanogaster:* Interspersed euchromatic and heterochromatic domains. *Proc. Natl. Acad. Sci.* **97:** 5340–5345.

Tabara H., Sarkissian M., Kelly W.G., Fleenor J., Grishok A., Timmons L., Fire A., and Mello C.C. 1999. The *rde-1* gene, RNA interference, and transposon silencing in *C. elegans. Cell* **99:** 123–132.

Tachibana M., Sugimoto K., Nozaki M., Ueda J., Ohta T., Ohki M., Fukuda M., Takeda N., Niida H., Kato H., and Shinkai Y. 2002. G9a histone methyltransferase plays a dominant role in euchromatic histone H3 lysine 9 methylation and is essential for early embryogenesis. *Genes Dev.* **16:** 1779–1791.

Takahashi K., Murakami S., Chikashige Y., Funabiki H., Niwa O., and Yanagida M. 1992. A low copy number central sequence with strict symmetry and unusual chromatin structure in fission yeast centromere. *Mol. Biol. Cell* **3:** 819–835.

Tamaru H. and Selker E.U. 2001. A histone H3 methyltransferase controls DNA methylation in *Neurospora crassa. Nature* **414:** 277–283.

Tang G., Reinhart B.J., Bartel D.P., and Zamore P.D. 2003. A biochemical famework for RNA silencing in plants. *Genes Dev.* **17:** 49–63.

Taverna S.D., Coyne R.S., and Allis C.D. 2002. Methylation of histone H3 at lysine 9 targets programmed DNA elimination in *Tetrahymena. Cell* **110:** 701–711.

Thon G. and Friis T. 1997. Epigenetic inheritance of transcriptional silencing and switching competence in fission yeast. *Genetics* **145:** 685–696.

Thon G. and Klar A.J. 1992. The *clr1* locus regulates the expression of the cryptic mating-type loci of fission yeast. *Genetics* **131:** 287–296.

———. 1993. Directionality of fission yeast mating-type interconversion is controlled by the location of the donor loci. *Genetics* **134:** 1045–1054.

Thon G. and Verhein-Hansen J. 2000. Four chromo-domain proteins of *Schizosaccharomyces pombe* differentially repress transcription at various chromosomal locations. *Genetics* **155:** 551–568.

Thon G., Cohen A., and Klar A.J. 1994. Three additional linkage groups that repress transcription and meiotic recombination in the mating-type region of *Schizosaccharomyces pombe. Genetics* **138:** 29–38.

Thon G., Bjerling P., Bunner C.M., and Verhein-Hansen J. 2002. Expression-state boundaries in the mating-type region of fission yeast. *Genetics* **161:** 611–622.

Thuriaux P. 1977. Is recombination confined to structural genes on the eukaryotic genome? *Nature* **268:** 460–462.

Turner B.M. 1993. Decoding the nucleosome. *Cell* **75:** 5–8.

van Holde K.E. 1989. *Chromatin,* p. 497. Springer-Verlag, New York.

Vance V. and Vaucheret H. 2001. RNA silencing in plants—Defense and counterdefense. *Science* **292:** 2277–2280.

Varambally S., Dhanasekaran S.M., Zhou M., Barrette T.R., Kumar-Sinha C., Sanda M.G., Ghosh D., Pienta K.J., Sewalt R.G., Otte A.P., Rubin M.A., and Chinnaiyan A.M. 2002. The polycomb group protein EZH2 is involved in progression of prostate cancer. *Nature* **419:** 624–629.

Volpe T.A., Kidner C., Hall I.M., Teng G., Grewal S.I., and Martienssen R.A. 2002. Regulation of heterochromatic silencing and histone H3 lysine-9 methylation by RNAi. *Science* **297:** 1833–1837.

Wang G., Ma A., Chow C.M., Horsley D., Brown N.R., Cowell I.G., and Singh P.B. 2000. Conservation of heterochromatin protein 1 function. *Mol. Cell. Biol.* **20**: 6970–6983.

Warnecke P.M. and Bestor T.H. 2000. Cytosine methylation and human cancer. *Curr. Opin. Oncol.* **12**: 68–73.

Wassenegger M., Heimes S., Riedel L., and Sanger H.L. 1994. RNA-directed de novo methylation of genomic sequences in plants. *Cell* **76**: 567–576.

Weiler K.S. and Wakimoto B.T. 1995. Heterochromatin and gene expression in *Drosophila. Annu. Rev. Genet.* **29**: 577–605.

Whitelaw E. and Martin D.I. 2001. Retrotransposons as epigenetic mediators of phenotypic variation in mammals. *Nat. Genet.* **27**: 361–365.

Wines D.R., Talbert P.B., Clark D.V., and Henikoff S. 1996. Introduction of a DNA methyltransferase into *Drosophila* to probe chromatin structure in vivo. *Chromosoma* **104**: 332–340.

Wood V., Gwilliam R., Rajandream M.A., Lyne M., Lyne R., Stewart A., Sgouros J., Peat N., Hayles J., Baker S., Basham D., Bowman S., Brooks K., Brown D., Brown S., Chillingworth T., Churcher C., Collins M., Connor R., Cronin A., Davis P., Feltwell T., Fraser A., Gentles S., Goble A., Hamlin N., Harris D., Hidalgo J., Hodgson G., Holroyd S., Hornsby T., Howarth S., Huckle E.J., Hunt S., Jagels K., James K., Jones L., Jones M., Leather S., McDonald S., McLean J., Mooney P., Moule S., Mungall K., Murphy L., Niblett D., Odell C., Oliver K., O'Neil S., Pearson D., Quail M.A., Rabbinowitsch E., Rutherford K., Rutter S., Saunders D., Seeger K., Sharp S., Skelton J., Simmonds M., Squares R., Squares S., Stevens K., Taylor K., Taylor R.G., Tivey A., Walsh S., Warren T., Whitehead S., Woodward J., Volckaert G., Aert R., Robben J., Grymonprez B., Weltjens I., Vanstreels E., Rieger M., Schafer M., Muller-Auer S., Gabel C., Fuchs M., Fritzc C., Holzer E., Moestl D., Hilbert H., Borzym K., Langer I., Beck A., Lehrach H., Reinhardt R., Pohl T.M., Eger P., Zimmermann W., Wedler H., Wambutt R., Purnelle B., Goffeau A., Cadieu E., Dreano S., Gloux S., Lelaure V., Mottier S., Galibert F., Aves S.J., Xiang Z., Hunt C., Moore K., Hurst S.M., Lucas M., Rochet M., Gaillardin C., Tallada V.A., Garzon A., Thode G., Daga R.R., Cruzado L., Jimenez J., Sanchez M., del Rey F., Benito J., Dominguez A., Revuelta J.L., Moreno S., Armstrong J., Forsburg S.L., Cerrutti L., Lowe T., McCombie W.R., Paulsen I., Potashkin J., Shpakovski G.V., Ussery D., Barrell B.G. and Nurse P. 2002. The genome sequence of *Schizosaccharomyces pombe. Nature* **415**: 871–880.

Zaman Z., Heid C., and Ptashne M. 2002. Telomere looping permits repression "at a distance" in yeast. *Curr. Biol.* **12**: 930–933.

Zhang Y. and Reinberg D. 2001. Transcription regulation by histone methylation: Interplay between different covalent modifications of the core histone tails. *Genes Dev.* **15**: 2343–2360.

Zhang Z., Shibahara K., and Stillman B. 2000. PCNA connects DNA replication to epigenetic inheritance in yeast. *Nature* **408**: 221–225.

Zhimulev I.F., Belyaeva E.S., Formina O.V., Protopopov M.O., and Bolshakov V.N. 1986. Cytogenetic and molecular aspects of position effect variegation in *Drosophila melanogaster*. I. Morphology and genetic activity of the 2AB region in chromosome rearrangement T(1;2)dor-var7. *Chromosoma* **94**: 492–504.

Zilberman D., Cao X., and Jacobsen S.E. 2003. ARGONAUTE4 control of locus-specific siRNA acumulation and DNA and histone methylation. *Science* **299**: 716–719.

# CHAPTER 11

# Total Interference: Genome-wide RNAi Screens in *C. elegans*

Andrew G. Fraser and Julie Ahringer

*Wellcome Trust/Cancer Research UK Institute and Department of Genetics, University of Cambridge, Cambridge CB2 1QR, United Kingdom*

I N 1998, THE COMPLETE SEQUENCE OF THE *CAENORHABDITIS ELEGANS* GENOME was published (*C. elegans* Sequencing Consortium 1998) and the phenomenon of RNAi (RNA interference) was discovered (Fire et al. 1998). These achievements together allow unprecedented comprehensive approaches to understanding gene function in the worm. This chapter describes how RNAi-mediated interference has been used to carry out genome-wide analyses of gene function in *C. elegans* thus far and discusses the prospects for the future.

*C. elegans* is without doubt one of the easiest animal model organisms to maintain and use (Wood 1988). It can grow as a self-fertilizing hermaphrodite with a short life cycle (~3 days), is maintained in the lab on a diet of *Escherichia coli*, and can be frozen easily. In addition to its low cost of culturing and ease of use, *C. elegans* has certain unusual aspects of its development that make it very attractive as a simple animal to study complex metazoan processes. Unlike most other model organisms, the development and physiology of *C. elegans* are understood at a single-cell resolution: Every adult hermaphrodite worm contains 959 somatic cells that have arisen in an essentially invariant manner (for review, see Wood 1988). This means that there is an exquisite sensitivity in examining mutant phenotypes in the worm because a single misplaced or additional cell is an informative and detectable phenotype. Furthermore, even complex tissues such as the neuromuscular system can be understood at a high resolution. *C. elegans* has approximately 300 neurons, and its network of neuron-neuron and neuron-muscle contacts has been entirely mapped using electron microscopy (Ward et al. 1975; Albertson and Thomson 1976; White et al. 1976, 1983; Chalfie et al. 1985), making it the best physically understood neuronal system in any animal.

Despite having this invariant pattern of cell division and migration, signaling still has a vital role in *C. elegans* development, and experiments in the worm have played a key part in elucidating many of the major conserved signaling pathways, including the RTK-*ras*-*raf*-MAPK (for review, see Sternberg and Han 1998) and insulin-PI3-kinase-Akt pathways (for review, see Gems and Partridge 2001), among others. Worms have also proven to be an excellent model organism for unraveling other aspects of complex metazoan biology; for example, analyses of genes involved in programmed cell death in nematodes revolutionized the field of apoptosis (for review, see Metzstein et al. 1998).

Historically, gene function in the worm has been analyzed primarily by classical forward genetics (Jorgensen and Mango 2002), in which mutant animals are isolated following treatment with a mutagen (e.g., ethylmethane sulfonate), and the mutated gene is subsequently isolated by positional cloning. This is a relatively slow process for understanding gene function. In the last 30 years, only about 5% of sequenced worm genes have been assigned a biological function by forward genetics. Although the speed of positional cloning has been greatly accelerated by single-nucleotide (SNP) mapping (Wicks et al. 2001), forward genetics remains a slow and laborious way of connecting biological function with gene sequence.

The sequencing of the complete *C. elegans* genome (*C. elegans* Sequencing Consortium 1998) has opened up entirely new challenges and possibilities for understanding gene function. Rather than examining a single gene, it became possible to analyze the biology of all predicted genes in the genome. This goal has been pursued in different ways by several groups, including work in the lab of Stuart Kim on microarray analysis of gene expression (Kim et al. 2001), the attempt to delineate a comprehensive map of protein-protein interactions by the lab of Marc Vidal (Walhout et al. 2000a,b; http://vidal.dfci.harvard.edu/interactome.html), and the analysis of in situ expression patterns of genes by the Kohara lab (Tabara et al. 1996; http://nematode.lab.nig.ac.jp/). In addition to these approaches, functions for the great majority of predicted genes in the worm have been assessed using RNAi (Fraser et al. 2000; Gonczy et al. 2000; Piano et al. 2000; Maeda et al. 2001; Kamath et al. 2003). This has added greatly to the understanding of gene function in *C. elegans* and has provided a firm platform for future genome-wide functional screens. This chapter describes the RNAi-based screens carried out thus far in the worm, along with a description of the techniques involved, and assesses how genome-wide RNAi screens may be used in the future.

## CONCEPTS AND STRATEGIES

### Methods for Delivery of dsRNA in *C. elegans*

We do not discuss the complexities of RNAi as a process, but concentrate solely on the methodologies and the applications for functional genomics in the worm. For worm biologists, RNAi has always been a relatively easy technique. Double-stranded RNA (dsRNA) introduced into the animal results in the specific and effective targeting of any gene of complementary sequence in both the affected animal and its progeny (Fire et al. 1998). dsRNA can be introduced in one of three different ways: injection (Fire et al. 1998), soaking (Tabara et al. 1998), or feeding (Timmons and Fire 1998) (see also Chapter 4). The three methods vary in their ease of use, reproducibility, and scalability, and it is thus important to choose the best method for the desired application.

#### Injection of dsRNA

An adult hermaphrodite is injected with a high concentration (~1 mg/ml) of dsRNA and her progeny are assayed for a mutant phenotype (Fire et al. 1998). This method reliably delivers a high one-off dose of dsRNA into the mother, inhibiting target gene expression in her tissues, including the germ line, through which a dose is also delivered to the progeny. This method is very effective for studying gene function in the embryo and is probably the method of choice for such studies due to the high penetrance of phenotype produced. However, RNAi by injection appears to work less well for genes expressed during larval development, possibly because the dsRNA is introduced into the animal only once at the beginning of development. The process of injection is learned quickly (~1 week of practice is usually sufficient), and it normally takes only 30 minutes to inject 10–30 hermaphrodites for an experiment. However, because each dsRNA must be injected manually, RNAi by injection is not ideally suited to high-throughput RNAi analyses. Furthermore, RNAi by injection is clearly not useful for any experiment that requires a large affected population, such as microarray or biochemical experiments.

### Introduction of dsRNA by soaking

Worms of any developmental stage are soaked in a solution of high-concentration dsRNA (1–5 mg/ml) for 24 hours, and the worms and their progeny are then scored for phenotypic changes (Tabara et al. 1998; for example, the screen done by Maeda et al. 2001). This method has good efficacy and can easily be scaled up to treat a large number of worms with one dsRNA or to test a large number of dsRNAs in parallel in a 96-well format. The drawback of this method is cost and time; a large amount of dsRNA must be synthesized and purified for each gene tested.

### Feeding dsRNA to worms

Both the techniques of RNAi by injection and RNAi by soaking rely on the in vitro synthesis of dsRNA, a process that is ultimately quite costly. The alternative technique, RNAi by feeding, uses *E. coli* to express the desired dsRNA. The dsRNA-expressing bacteria are fed to worms of any developmental stage, usually continuously, and the ingested dsRNA spreads throughout the animal (Timmons and Fire 1998; Kamath et al. 2001; Timmons et al. 2001). Compared to injection or soaking, RNAi by feeding has a similar rate for detection of a phenotype, but with a variable penetrance. However, possibly because dsRNA is being continually ingested, it appears to be more effective than RNAi by injection for looking at postembryonic phenotypes (Kamath et al. 2001). This method is inexpensive and can be scaled up easily. Not only is it relatively easy to analyze many genes in a short period of time, but a large number of worms can also be treated for biochemical experiments. The principal drawback of RNAi by feeding is that the gene of interest must first be cloned into a vector for dsRNA expression in bacteria. Once made, however, the bacterial strain may be reused an unlimited number of times at minimal cost. Bacterial strains for 86% of *C. elegans* genes are already available (Fraser et al. 2000; Kamath et al. 2003; http://www.hgmp.mrc.ac.uk/Biology/descriptions/Celegans.html), and thus, a great majority of genes can be studied using RNAi by feeding using publicly available reagents.

In summary, all three standard protocols for RNAi in *C. elegans* are effective, but they have key differences—the ideal method is dependent on the exact experiment being carried out.

- To target a maternally expressed gene in a small number of animals, either injection or soaking is most likely to give a reproducible and strong loss-of-function phenotype.

- For high-throughput RNAi screens, or to target a large number of animals, then RNAi by feeding or soaking is better.

These differences are important factors both in the design of experiments and in the interpretation of data—all RNAi experiments are not equal, and this is an important consideration when comparing different data sets.

## RNAi as a Tool to Analyze the Function of Large Sets of Genes

Whatever the chosen method of RNAi, it is a very powerful technique to investigate gene function in the worm. Furthermore, the ease with which RNAi can be carried out makes it ideal for analyzing the functions of large numbers of genes. In this section, we illustrate how high-throughput RNAi experiments can be an invaluable way to assess the biological relevance of sets of genes selected on the basis of in silico or in vitro analyses or, alternatively, as a tool to systematically study all genes in the genome.

The most frequent (and most basic) in silico use of a *C. elegans* sequence has been to mine the genome to find the worm homolog(s) of genes identified in other organisms, especially mammals. RNAi makes it easy to identify the loss-of-function phenotype of the worm gene, and this kind of experiment has proven to be an excellent way to understand the functions of hitherto elusive mammalian genes. In some cases, the studied genes are members of a gene family, and, using RNAi, each gene in the family can be inhibited individually to determine whether they have unique roles or RNAi can be performed in combination to determine whether family members have shared functions. For example, RNAi was used to show that two of the three *C. elegans nanos* homologs function redundantly to maintain germ cell viability (Subramaniam and Seydoux 1999) and that two of 20 $G\alpha$ subunits function redundantly to control spindle positioning (Gotta and Ahringer 2001). This approach can be extended to determine the RNAi phenotypes of all genes thought to be involved in a particular biochemical process. For example, Candido and co-workers identified a set of *C. elegans* genes involved in ubiquitination and used RNAi to target each individually, and in some cases, in combination (Jones et al. 2002). By examining the loss-of-function phenotypes of all the members of a gene family, it is possible to draw conclusions not just about the single genes, but about the ways in which gene functions overlap and how they have evolved and diverged.

RNAi can also be particularly useful for investigating the biological relevance of genes identified in in vitro screens, for example, yeast two-hybrid interaction screens or microarray analyses of gene expression. Since the speed of gathering data through such screens is becoming ever greater, the ability to investigate the function of the identified genes through RNAi is a vital addition to functional genomics. An excellent example of this use of RNAi is the recent analysis of the DNA repair machinery in *C. elegans* (Boulton et al. 2002). Using the yeast two-hybrid system, a map of the physical interactions between known components of the DNA repair machinery were identified and novel interactors were found. Using RNAi, many of the novel interacting genes were demonstrated to have a role in DNA damage sensing and repair, experiments that would have been very difficult without such an efficient means of analyzing gene function.

The above examples show that RNAi can provide a rapid and easy way to assess the biological relevance of sets of genes selected on the basis of in silico or in vitro analyses. A highly complementary approach is to use RNAi to screen all of the genes in the genome for those involved in a particular process. This approach assumes no prior knowledge, but is instead an unbiased and comprehensive way of examining gene function. Two groups have carried out this kind of RNAi screen thus far. The Hyman group used RNAi by injection to identify genes with essential roles in early embryogenesis (Gonczy et al. 2000). They screened approximately 90% of the genes on chromosome III and identified 133 (~6%) to be essential. Furthermore, these authors characterized the RNAi phenotypes in exquisite detail and, in many cases, pinpointed the precise defect resulting from RNAi of the target gene. This led to the identification of new genes with roles in cytokinesis, spindle assembly, and other basic cellular processes. The second systematic screen has been carried out by the Ahringer lab and used RNAi by feeding to identify genes with grossly detectable loss-of-function phenotypes, ranging from lethality or sterility through to defects in body shape or animal movement (Fraser et al. 2000; Kamath et al. 2003). More than 86% of the genes were analyzed in this manner, and 1722 genes (~10.3% of those screened) were found to have a detectable RNAi phenotype. This data set has provided a starting point for more detailed analyses of the genes identified; for example, Zipperlen et al. (2001) analyzed the RNAi phenotypes of the embryonic lethal genes identified on chromosome I using time-lapse video microscopy. The major advantage of these two large data sets is that they have been compiled in an essentially unbiased manner, with no preselection on the types or properties of genes analyzed.

The availability of this RNAi feeding library (http://www.hgmp.mrc.ac.uk/Biology/descriptions/Celegans.html) will allow any investigator to carry out global RNAi screens for genes involved in the process of interest and should greatly accelerate research in many fields. Thus far, no other system is easily amenable to whole-animal genome-wide functional screening. In addition, comprehensive data sets generated from a large number of screens will provide a unique platform for functional genomics studies such as exploring the relationship between the organismal function of a gene and its sequence or chromosomal location. These data will also provide numerous starting points for more directed research, for example, determining the physical interactions between genes with similar RNAi phenotypes or examining in greater detail what roles genes with particular RNAi phenotypes (e.g., sterile or uncoordinated movement) have in development.

## Limitations and Differences between RNAi and Forward Genetics

RNAi-based screens are a rapid way to add knowledge of gene function in *C. elegans*. However, such screens clearly differ from the classical genetic screens that have been the standard approach to understanding gene function in the worm for more than 30 years. RNAi-based screens have both clear advantages and substantial disadvantages.

One of the biggest advantages of RNAi-based screens is their incredible speed at connecting function with predicted gene sequence. Although classical genetic screens can rapidly identify mutant animals, the positional cloning of the mutated gene is still very time-consuming—moving from mutant phenotype to affected gene can take several months or longer. With RNAi, the sequences of the targeted genes are already known, and thus any observed loss-of-function phenotype is automatically linked to the predicted gene.

RNAi-based screens have other benefits in addition to their efficiency. In an RNAi-based screen, each gene is targeted once and once only, which may allow a more comprehensive identification of all the molecular components of a particular pathway or process than in a classical forward genetic screen where saturation may be harder to achieve. In addition, a range of gene inhibition is often seen in an RNAi-affected population, from weak to strong loss-of-function phenotypes. This can allow relevant genes with lethal null phenotypes that were missed in a classical forward genetic screen to be identified in an RNAi screen. Finally, RNAi targets both zygotic and maternal messages and thus can be used to investigate the maternal effect of genes that are required for zygotic development. For example, RNAi experiments demonstrated a requirement for *ncc-1* (a *cdc-2* homolog) to complete mitosis in early embryogenesis. Null *ncc-1* mutants are viable as embryos due to maternal contribution, but they develop into sterile adults, making it difficult to determine the maternal role for the gene using a conventional mutant (Boxem et al. 1999).

Although RNAi-based screens have certain advantages over classical forward genetics, there are also key shortfalls when compared to traditional mutagenesis. First, there are alleles that cannot be replicated with RNAi, most obviously dominant gain-of-function alleles. Because these comprise some of the most informative mutations isolated, this is a key advantage of forward genetics. In addition, the isolation of multiple alleles of the same gene can tell us much about the domain structure and function of the encoded protein. RNAi results only in the reduction in levels of wild-type protein and thus provides no such information. Thus, although it takes longer to identify the affected gene using forward genetics, the amount of informative data gleaned once this is done may be greater than obtained with RNAi.

Another advantage of classical forward genetics is the fact that certain tissues and certain genes appear to be refractory to RNAi. In particular, genes expressed in neurons and sperm appear not to be as sensitive to RNAi as are genes expressed in other tissues (Timmons and Fire 1998; Fraser et al. 2000; Kamath et al. 2001; Timmons et al. 2001), and thus screens to identify genes that affect neurons or sperm are better carried out at present through classical mutagenesis and positional cloning strategies. However, this should change in the future with the development of *C. elegans* strains that are supersensitive to RNAi. Indeed, one such strain has already been isolated. Mutations in the RNA-dependent RNA polymerase gene *rrf-3* result in a general increase in sensitivity to RNAi over wild-type worms (Simmer et al. 2002). In many tissues, including neurons, the penetrance and detectability of mutant phenotypes are greatly increased. This strain will undoubtedly prove to be very useful in future RNAi analyses.

Finally, like any interference technique, a negative result with RNAi is not informative, because low (even undetectable) levels of protein may persist and be sufficient for wild-type function. This is not true with a genetic null, and genetic lesions still remain the gold standard for defining null phenotypes. Therefore, there is no hard-and-fast rule about which type of screen is "better": classical forward genetics or RNAi. Indeed, when used in a genome-wide manner, RNAi is essentially a forward genetics screening tool. Each has its strengths and weaknesses, but fortunately, these are highly complementary approaches.

## Considerations When Designing an RNAi-based Screen

Before embarking on any screen, whether using RNAi or classical forward genetics, it is essential to optimize the assay for detection of mutants. These assays in worms can range from the crude (e.g., lethality) to the subtle (e.g., the ability of a worm to detect a light touch on the head with an eyelash!), and from observing a phenotype directly to using a molecular marker such as activation of a green fluorescent protein (GFP) reporter to detect an effect. Furthermore, there is now an excellent worm sorter that can greatly enhance the sensitivity and speed of phenotypic analysis (Union Biometrica Incorporated; http://www.unionbiometrica.com).

In addition to developing a sensitive assay for the desired mutant phenotype, however, it is also absolutely critical in an RNAi screen for the investigator to optimize the RNAi conditions for the assay. The key variables for RNAi depend on the method used to deliver the dsRNA, and the developmental stages of both the treated and affected worms must be considered. Outlined below are the critical issues and options for conducting a genome-wide screen, concentrating on RNAi by feeding. For detailed protocols for RNAi by feeding, see Kamath and Ahringer (2003) and Chapter 4.

### What stage of worms is to be assayed?

If the assay is to be conducted in embryos, the mothers must be fed the dsRNA-expressing bacteria (usually from the early L4 stage until adulthood) and her progeny assayed. This method will inhibit both maternal and zygotic gene activities. If the assay is to be conducted on larvae or adults, then the investigator must decide whether to inhibit maternal as well as zygotic gene expression. Many genes have roles both in the early embryo (often through maternally contributed mRNA and protein) and later in postembryonic development. If a gene has an essential role in the embryo and RNAi efficiently inhibits gene expression there, then the affected progeny will die in embryogenesis, making it impossible to observe any later role of the gene. There are two possible ways to overcome this problem. First, the gene can be targeted directly in larvae by exposing them (and not their mothers) to dsRNA using either soaking or feeding, leaving maternal and embryonic gene expression unaffected (e.g., see Solari and Ahringer 2000). Note, however, that for many genes, the maternal gene product is sufficient for postembryonic as well as embryonic development, and thus inhibiting only postembryonic gene expression may not induce a mutant phenotype. An alternative is to expose adult hermaphrodites to dsRNA for a shorter time before assaying progeny, to inhibit maternal gene function only partially.

There are also other strategies to generate weak rather than strong loss-of-function phenotypes using RNAi by soaking or feeding. With soaking, the concentration of the dsRNA in solution can be reduced until the effect is attenuated. With feeding, the expression of dsRNA in the bacterial strain is induced by the addition of IPTG. Varying the concentration of IPTG provides a simple, regulable method for reducing the strength of the RNAi effect to generate only partially penetrant or weak loss-of-function phenotypes; 1 $\mu$M IPTG instead of 1 mM IPTG is a good starting point (Kamath et al. 2001). This latter method is also useful for assaying the postembryonic role of genes required for embryogenesis.

### Is a synchronized population needed?

A synchronized population can be obtained in one of two ways. If maternal gene expression is to be inhibited, adult hermaphrodites are exposed to dsRNA for a period of time and then are individually moved to fresh plates to lay eggs for a set period of time (e.g., 24 hours) before being removed. Phenotypes are scored in the progeny (eggs) laid in this time window. However, it is very time consuming to put single hermaphrodites onto plates and later remove them, so this method should be considered only if it is essential to the screen. Alternatively and more simply, worms can by synchronized by collecting eggs by hypochlorite treatment and then allowing them to hatch in buffer containing no food (Wood 1988). These arrest at the beginning of the L1 stage and will grow synchronously when fed. Having a synchronized population can be useful for scoring some assays, but it is not always necessary.

### Is there sufficient food for the worms?

Providing an adequate amount of bacteria for the worms is merely a matter of choosing the best plate format for the number of worms required in each RNAi experiment. If a 24-hour egg lay from one hermaphrodite will be collected in each well, then 12-well agar plates will contain enough bacteria for feeding. If several adult hermaphrodites will be allowed to lay eggs indefinitely, then 6-well agar plates or individual plates should be used so that there will be enough bacteria to feed the progeny. The amount of bacteria and worms required must be determined empirically in pilot assays.

Once a set of candidate genes is identified in a screen, varying parameters, such as time and temperature, can affect the strength and penetrance of the observed phenotype. Different genes can have stronger or weaker RNAi phenotypes depending on the temperature at which the RNAi and subsequent development are carried out. However, no single temperature is best overall. For many genes, feeding an early L4 hermaphrodite for 72 hours at 15°C and then assaying progeny gives results similar to feeding for 36 hours at 25°C, but some genes will have a stronger RNAi phenotype at one temperature.

The length of time hermaphrodites are exposed to the dsRNA before assaying progeny will also affect the results. For example, it is sometimes possible to increase the strength and penetrance of a phenotype either by prolonged feeding of adult hermaphrodites or by waiting 48–72 hours after injection of dsRNA before assaying progeny. In this case, it may be necessary to mate the treated hermaphrodites with males to prolong egg laying.

Many variables can thus affect the efficacy of RNAi-based screens. By assessing the ability to detect previously identified genes known to be positive in the assay, the screening conditions can be optimized. This type of analysis sets out the hit-and-miss rate for the screen under the chosen conditions. In some cases, optimizing the hit rate may need to be balanced by the time taken to conduct the screen. For example, it may be acceptable to use screening conditions that are substantially faster even if this results in a small decrease in the hit rate. If no positive controls are yet known, then it can be useful to carry out a few small pilot screens, varying the RNAi parameters before embarking on a genome-wide screen.

In addition to using the positive control genes for optimizing conditions for a screen beforehand, an informative measure of the efficacy of a genome-wide RNAi screen is determining the hit rate of these genes during the actual screen. Throughput is not always easily compatible with sensitivity, and many of the genes detected in a pilot screen might be missed during the genome-wide screen. The "known gene hit rate" is thus an invaluable marker for the quality of data in any RNAi-based screen.

## RNAi SCREENS: THE FUTURE

The initial RNAi-based screens either have been very broad-brush screens to identify genes with an easily detectable phenotype or were focused on embryonic development (Fraser et al. 2000; Gonczy et al. 2000; Piano et al. 2000; Maeda et al. 2001; Kamath et al. 2003). However, such screens are merely the tip of the iceberg in terms of the biology that can be investigated using RNAi. Using the RNAi feeding library, high-throughput functional screens can be conducted for genes involved in a wide range of biological processes. For example, it has recently been used to study aging (Dillin et al. 2002; Lee et al. 2003), fat metabolism (Ashrafi et al. 2003), and genome stability (Pothof et al. 2003). Not only should this have a large impact on understanding metazoan biology, but combining data from a large number of screens will give an unprecedented data set for exploring the relationships between gene sequence and biological function. In addition, it should prove to be a very useful resource of phenotypic data that can be used in a variety of ways, including identifying candidate genes to accelerate positional cloning.

One particularly powerful use of RNAi-based genome-wide screens is to identify genes with synthetic phenotypes, i.e., loss-of-function phenotypes that result from the mutation of two or more partially redundant genes. This type of screen is particularly difficult using forward genetics if the synthetic phenotype is nonviable (i.e., lethal or sterile). RNAi screens for genes with synthetic phenotypes may prove to be very effective for identifying pathway components that have hitherto been hard to find by classical genetics.

## CONCLUSION

RNAi has already revolutionized the analysis of gene function in *C. elegans* both through facilitating the characterization of single genes and, increasingly, in the analysis of the functions of large numbers of genes. Now that there are publically available tools for targeting the great majority of all predicted worm genes in high-throughput screens, it seems certain that RNAi-based experiments will have a major role in our understanding of worm biology.

## ACKNOWLEDGMENTS

A.G.F. was supported by a U.S. Army Breast Cancer Research Fellowship and J.A. by a Wellcome Trust Senior Research Fellowship (No. 054523).

## REFERENCES

Albertson D.G. and Thomson J.N. 1976. The pharynx of *Caenorhabditis elegans. Philos. Trans. R. Soc. Lond. B Biol. Sci.* **275**: 299–325.

Ashrafi K., Chang F., Watts J., Fraser A., Kamath R.S., Ahringer J., and Ruvkun G. 2003. Genome-wide RNAi analysis of *C. elegans* fat regulatory genes. *Nature* **412**: 268–272.

Boulton S.J., Gartner A., Reboul J., Vaglio P., Dyson N., Hill D.E., and Vidal M. 2002. Combined functional genomic maps of the *C. elegans* DNA damage response. *Science* **295**: 127–131.

Boxem M., Srinivasan D.G., and van den Heuvel S. 1999. The *Caenorhabditis elegans* gene *ncc-1* encodes a cdc2-related kinase required for M phase in meiotic and mitotic cell divisions, but not for S phase. *Development* **126**: 2227–2239.

*C. elegans* Sequencing Consortium. 1998. Genome sequence of the nematode *C. elegans:* A platform for investigating biology. *Science* **282**: 2012–2018.

Chalfie M., Sulston J.E., White J.G., Southgate E., Thomson J.N., and Brenner S. 1985. The neural circuit for touch sensitivity in *Caenorhabditis elegans. J. Neurosci.* **5**: 956–964.

Dillin A., Hsu A.-L., Arantes-Oliveira N., Lehrer-Graiwer J., Hsin H., Fraser A.G., Kamath R.S., Ahringer J., and Kenyon C. 2002. Lifelong rates of behavior and aging specified by respiratory chain activity during development. *Science* **298**: 2398–2401.

Fire A., Xu S., Montgomery M.K., Kostas S.A., Driver S.E., and Mello C.C. 1998. Potent and specific genetic interference by double-stranded RNA in *Caenorhabditis elegans. Nature* **391**: 806–811.

Fraser A.G., Kamath R.S., Zipperlen P., Martinez-Campos M., Sohrmann M., and Ahringer J. 2000. Functional genomic analysis of *C. elegans* chromosome I by systematic RNA interference. *Nature* **408**: 325–330.

Gems D. and Partridge L. 2001. Insulin/IGF signalling and ageing: Seeing the bigger picture. *Curr. Opin. Genet. Dev.* **11**: 287–292.

Gonczy P., Echeverri G., Oegema K., Coulson A., Jones S.J., Copley R.R., Duperon J., Oegema J., Brehm M., Cassin E., Hannak E., Kirkham M., Pichler S., Flohrs K., Goessen A., Leidel S., Alleaume A.M., Martin C., Ozlu N., Bork P., and Hyman A.A. 2000. Functional genomic analysis of cell division in *C. elegans* using RNAi of genes on chromosome III. *Nature* **408**: 331–336.

Gotta M. and Ahringer J. 2001. Distinct roles for Gα and Gβγ in regulating spindle position and orientation in *Caenorhabditis elegans* embryos. *Nat. Cell. Biol.* **3**: 297–300.

Jones D., Crowe E., Stevens T.A., and Candido E.P. 2002. Functional and phylogenetic analysis of the ubiquitylation system in *Caenorhabditis elegans:* Ubiquitin-conjugating enzymes, ubiquitin-activating enzymes, and ubiquitin-like proteins. *Genome Biol.* **3**: RESEARCH0002.

Jorgensen E.M. and Mango S.E. 2002. The art and design of genetic screens: *Caenorhabditis elegans. Nat. Rev. Genet.* **3**: 356–369.

Kamath R.S. and Ahringer J. 2003. Genome wide RNAi screening in *C. elegans. Methods* (in press).

Kamath R.S., Martinez-Campos M., Zipperlen P., Fraser A.G., and Ahringer J. 2001. Effectiveness of specific RNA-mediated interference through ingested double-stranded RNA in *C. elegans. Genome Biol.* **2**: RESEARCH0002.

Kamath R.S., Fraser A.G., Dong Y., Poulin G., Durbin R., Gotta M., Kanapin A., Le Bot N., Moreno S., Sohrmann M., Welchman D.P., Zipperlen P., and Ahringer J. 2003. Systematic functional analysis of the *Caenorhabditis elegans* genome using RNAi. *Nature* **16**: 220–221.

Kim S.K., Lund J., Kiraly M., Duke K., Jiang M., Stuart J.M., Eizinger A., Wylie B.N., and Davidson G.S. 2001. A gene expression map for *Caenorhabditis elegans. Science* **293**: 2087–2092.

Lee S.S., Lee R.Y.N., Fraser A.G., Kamath R.S., Ahringer J., and Ruvkun G. 2003. A comprehensive RNAi screen for *C. elegans* longevity genes. *Nat. Genet.* **33**: 40–48.

Maeda I., Kohara Y., Yamamoto M., and Sugimoto A. 2001. Large-scale analysis of gene function in *Caenorhabditis elegans* by high-throughput RNAi. *Curr. Biol.* **11**: 171–176.

Metzstein M.M., Stanfield G.M., and Horvitz H.R. 1998. Genetics of programmed cell death in *C. elegans:* Past, present and future. *Trends Genet.* **14:** 410–416.

Piano F., Schetter A.J., Mangone M., Stein L., and Kemphues K.J. 2000. RNAi analysis of genes expressed in the ovary of *Caenorhabditis elegans. Curr. Biol.* **10:** 1619–1622.

Pothof J., van Haaften G., Thijssen K., Kamath R.S., Fraser A.G., Ahringer J., Plasterk R.H.A., and Tijsterman M. 2003. Identification of genes that protect the *C. elegans* genome against mutations by genome-wide RNAi. *Genes Dev.* **17:** 443–448.

Simmer F., Tijsterman M., Parrish S., Koushika S.P., Nonet M.L., Fire A., Ahringer J., and Plasterk R.H.A. 2002. Loss of the putative RNA directed RNA polymerase *rrf-3* makes *C. elegans* hypersensitive to RNAi. *Curr. Biol.* **12:** 1317–1319.

Solari F. and Ahringer J. 2000. NURD-complex genes antagonise Ras-induced vulval development in *Caenorhabditis elegans. Curr. Biol.* **10:** 223–226.

Sternberg P.W. and Han M. 1998. Genetics of RAS signaling in *C. elegans. Trends Genet.* **14:** 466–472.

Subramaniam K. and Seydoux G. 1999. *nos-1* and *nos-2*, two genes related to *Drosophila nanos*, regulate primordial germ cell development and survival in *Caenorhabditis elegans. Development* **126:** 4861–4871.

Tabara H., Grishok A., and Mello C.C. 1998. RNAi in *C. elegans:* Soaking in the genome sequence. *Science* **282:** 430–431.

Tabara H., Motohashi T., and Kohara Y. 1996. A multi-well version of in situ hybridization on whole mount embryos of *Caenorhabditis elegans. Nucleic Acids Res.* **24:** 2119–2124.

Timmons L. and Fire A. 1998. Specific interference by ingested dsRNA. *Nature* **395:** 854.

Timmons L., Court D.L., and Fire A. 2001. Ingestion of bacterially expressed dsRNAs can produce specific and potent genetic interference in *Caenorhabditis elegans. Gene* **263:** 103–112.

Walhout A.J., Boulton S.J., and Vidal M. 2000a. Yeast two-hybrid systems and protein interaction mapping projects for yeast and worm. *Yeast* **17:** 88–94.

Walhout A.J., Sordella R., Lu X., Hartley J.L., Temple G.F., Brasch M.A., Thierry-Mieg N., and Vidal M. 2000b. Protein interaction mapping in *C. elegans* using proteins involved in vulval development. *Science* **287:** 116–122.

Ward S., Thomson N., White J.G., and Brenner S. 1975. Electron microscopical reconstruction of the anterior sensory anatomy of the nematode *Caenorhabditis elegans.* ?2UU. *J. Comp. Neurol.* **160:** 313–337.

White J.G., Southgate E., Thomson J.N., and Brenner S. 1976. The structure of the ventral nerve cord of *Caenorhabditis elegans. Philos. Trans. R. Soc. Lond. B Biol. Sci.* **275:** 327–348.

———. 1983. Factors that determine connectivity in the nervous system of *Caenorhabditis elegans. Cold Spring Harbor Symp. Quant. Biol.* **48:** 633–640.

Wicks S.R., Yeh R.T., Gish W.R., Waterston R.H., and Plasterk R.H. 2001. Rapid gene mapping in *Caenorhabditis elegans* using a high density polymorphism map. *Nat. Genet.* **28:** 160–164.

Wood W.B., ed. 1988. *The nematode* Caenorhabditis elegans. Monograph 17. Cold Spring Harbor Laboratory Press, Cold Spring Harbor, New York.

Zipperlen P., Fraser A.G., Kamath R.S., Martinez-Campos M., and Ahringer J. 2001. Roles for 147 embryonic lethal genes on *C. elegans* chromosome I identified by RNA interference and video microscopy. *EMBO J.* **20:** 3984–3992.

# CHAPTER 12

# PTGS Approaches to Large-scale Functional Genomics in Plants

Tessa M. Burch-Smith,* Jennifer L. Miller,* and
Savithramma P. Dinesh-Kumar

*Department of Molecular, Cellular, and Developmental Biology, Yale University,
New Haven, Connecticut 06520-8104*

WITH THE COMPLETE GENOME SEQUENCES of a number of important eukaryotic organisms now available, the present goal is to determine the function of all of these predicted genes. Thus, there is a need for large-scale functional assays that can address gene function on the whole-organism level. Currently, posttranscriptional gene silencing (PTGS) is the best technique available to undertake such analyses. This homology-dependent mechanism results in the knockdown of gene function through the degradation of mRNA transcripts, and the resulting PTGS phenotype mimics the loss-of-function phenotype (Baulcombe 1999; Vaucheret et al. 2001; Waterhouse et al. 2001).

PTGS, also known as RNA interference (RNAi) (Fire et al. 1998), has been used successfully for whole-chromosome studies in the nematode *Caenorhabditis elegans*. In one *C. elegans* study, all of the open reading frames (ORFs) on chromosome I were silenced, thus increasing the number of genes with known functions fivefold; however, the function of 86% of the genes is still not yet determined (Fraser et al. 2000). Chromosome III was also subjected to such analyses with the focus on genes involved in cell division (Gonczy et al. 2000). These investigators targeted 96% of the chromosome III genes and were able to assign a function in cell division to 6% of them. These studies confirm the feasibility of large-scale genomic analysis using PTGS.

Two completed plant genome sequences are currently available: the model dicotyledonous organism *Arabidopsis thaliana* and the model monocotyledonous organism *Oryza sativa* (rice). We need strategies to study these organisms at the whole-genome level to further our understanding of plant biology and eventually to improve agricultural crops. One can envision being able to develop crops, for example, with greater disease resistance, better cold, drought, salt, and heavy metal tolerance, or increased nutritional value.

## CONCEPTS AND STRATEGIES

### Forward Genetics

Traditionally, plant gene studies began by identifying an interesting mutant and then isolating the gene responsible for the phenotype. These "forward" genetic approaches relied on chemical mutagens such as ethylmethylsulfonate (EMS) or physical mutagens such as

*Both authors contributed equally to this chapter.

ionizing radiation. More recently, insertional mutagenesis using transposons and the *Agrobacterium tumefaciens*-derived T-DNA has been used to create large, saturated mutant populations (http://signal.salk.edu/cgi-bin/tdnaexpress, http://www.arabidopsis.org/info/2010_projects/comp_proj/AFGC/index.html) (Parinov et al. 1999; Meissner et al. 2000). These approaches have successfully produced collections that researchers routinely use to identify mutations in their genes of interest (http://signal.salk.edu/cgi-bin/tdnaexpress). However, there are inherent difficulties with approaches that rely on the random introduction of mutations into a plant gene:

- Plant genomes are large; for example, rice has 430 Mbp (Feng et al. 2002), and the relatively small genome of *Arabidopsis* has 125 Mbp (*Arabidopsis* Genome Initiative 2000). Thus, the chances of an insertion into a specific gene are quite low. For instance, in tomato, it is estimated that a population of 200,000–300,000 individuals carrying two to three insertions of the *Ds* transposon would be necessary for a near-saturated mutagenesis assuming that there are 40,000 genes in the tomato genome (Emmanuel and Levy 2002). Furthermore, integrations of transposons and T-DNA are not entirely random because some areas of the genome, hot spots, are particularly favored for insertion, whereas others, cold spots, are less likely to be hit. Thus, the chance of an insertion into a particular coding sequence that will result in an analyzable phenotype is further reduced. The large number of plants that must be screened to identify lines carrying insertions or mutations in a gene not only requires a large growing area, but is time consuming as well. These approaches are limited to species like *Arabidopsis* that are small and require little space for growth.

- Many mutations will not result in a knockout phenotype, and thus will be routinely missed in screening. This is particularly true for genes with redundant functions. In addition, there is no practical way to recover mutations caused by chemical and physical mutagenesis if there is no phenotype.

- All of these techniques require the generation of individual plant lines after performing the mutagenesis. Mutations that result in embryonic or early developmental lethality will never be isolated by such approaches. Thus, the genes causing such mutations will not be identified on the basis of their function.

- Insertional mutagenesis requires the generation of transgenic plants carrying a T-DNA or transposon from a heterologous system. Rapid, efficient, and reliable systems for generating transgenic plants are limited to a few species, making it difficult to utilize this technique in many species.

## Reverse Genetics

PTGS is a sequence-specific approach to reducing expression of a given gene, allowing analysis of the gene's function (Baulcombe 1999; Sharp 2001; Waterhouse et al. 2001). In contrast to traditional genetics, PTGS is a "reverse" genetics approach, because the sequence of the gene is known before its function is determined. PTGS was first described in petunia where workers trying to introduce transgenes found that, in many cases, multiple insertions caused the transgenes to be silent (Napoli et al. 1990; van der Krol et al. 1990). This observation was the first of several that eventually led to the recognition of double-stranded RNA (dsRNA) as a potent mediator of gene silencing not only in plants, but also in other model systems including *C. elegans*, *Drosophila*, and *Neurospora* (Vaucheret et al. 2001).

Many different vectors have been used to introduce dsRNA into hosts. Currently, hairpin RNA (hpRNA) constructs are considered the best design for achieving consistently high levels of silencing (Fire et al. 1998; Waterhouse et al. 1998). In the silencing construct, an intron is placed between the sense and antisense coding sequences and is spliced during transcript processing, producing dsRNA. Viruses can also be used to introduce a homologous sequence to silence an endogenous gene by PTGS, a phenomenon termed virus-induced gene silencing (VIGS) (Lindbo et al. 1993; Baulcombe 1999). Subsequent work has led to the development of several viral vectors that are now routinely used to mediate gene silencing. Some of these vectors are based on tobacco mosaic virus (TMV) (Kumagai et al. 1995), potato virus X (PVX) (Ruiz et al. 1998), and tobacco rattle virus (TRV) (Ratcliff et al. 2001; Liu et al. 2002b). These are positive-sense RNA viruses and so replicate via a dsRNA intermediate. During replication, dsRNA corresponding to the gene of interest is also formed and causes silencing of the endogenous gene. VIGS vectors based on DNA viruses have also been developed (Atkinson et al. 1998; Kjemtrup et al. 1998; Turnage et al. 2002). PTGS technology has many advantages over "forward" genetics approaches:

- PTGS is a targeted approach to determine gene function. Using a partial gene sequence in a silencing vector, the function of a gene can be determined. This bypasses the need for the laborious identification and isolation of transgenic knockout plants from large collections of mutants.

- PTGS can be used to selectively silence a single gene belonging to a gene family or to silence many members of a family simultaneously. This can be achieved through careful selection of the sequence to be used for targeting silencing. If a sequence that is highly conserved between members of a family is used, then it is likely that multiple genes will be silenced.

- PTGS can be used to determine the function of a gene whose knockout phenotype is embryonic or early developmental death. If a gene has an essential function, then its absence due to silencing will also result in death of the organism. However, the process of dying might be quite enlightening, for example, by determining if it is due to severe disruption of an essential process such as DNA replication or if it is due to the activation of apoptosis. In addition, because of the variation in levels of silencing, the phenotypes produced may mimic weaker mutations instead of null mutations.

- PTGS can easily be adapted to large-scale, high-throughput systems as has already been demonstrated in *C. elegans* (Fraser et al. 2000; Gonczy et al. 2000). Progress toward this goal is being made in plant systems (Wesley et al. 2001; Liu et al. 2002a).

## PTGS Systems

### dsRNA-mediated PTGS

The first reports of dsRNA mediating PTGS were made simultaneously in plants (Waterhouse et al. 1998) and *C. elegans* (Fire et al. 1998). Previously, sense or antisense strands had been used to mediate PTGS, most often with modest effects on gene expression. Progeny generated from crossing transgenic plants containing a gene fragment in a sense orientation with plants carrying the same fragment in an antisense orientation had the endogenous gene silenced five- to tenfold more than each transgenic parental line alone (Waterhouse et al. 1998). Considerable research has been conducted to determine the most efficient silencing construct. The smallest gene fragment used to silence a stably integrated transgene was 23 bp, whereas a 33-bp fragment was able to silence an endogenous gene (Thomas et al. 2001).

The most effective dsRNA construct designed transcribes hpRNA (Smith et al. 2000). In this type of construct, two copies of the target sequence are inserted as an inverted repeat. There is a further enhancement of silencing when an intron separates the sense and antisense arms of the hairpin construct: intron-containing hairpin RNA (ihpRNA)

(Smith et al. 2000). These researchers also found that a functional spacer intron, which is spliced from the transcript, is even more effective at mediating silencing than non-functional introns (Smith et al. 2000; Wesley et al. 2001). The reason for the spliceable intron generating higher levels and frequencies of silencing is not clear, but it is hypothesized that intron excision may bring the sense and antisense arms into closer proximity with better alignment, favoring the formation of an RNA duplex. It is also possible that there is a transient increase in dsRNA in the nucleus as the construct is retained to undergo splicing mediated by the spliceosome. Finally, removal of the intron loop from the silencing construct may generate a smaller loop that is less sensitive to nuclease activity (Smith et al. 2000).

The ihpRNA technology has been used to generate a generic vector into which a sequence from any gene of interest can be inserted to achieve efficient gene silencing (Wesley et al. 2001). This vector, pHANNIBAL (Figure 12.1), contains two polylinker sites into which the polymerase chain reaction (PCR) fragments amplified from the gene of interest can be inserted. A PCR product can be inserted into the *Xho*I.*Eco*RI.*Kpn*I polylinker in the sense orientation and into the *Cla*I.*Hin*dIII.*Bam*HI.*Xba*I polylinker in the antisense orientation. The cloning can be carried out using two separate PCR products, each possessing the appropriate single restriction sites as introduced by the primers. Alternatively, a single PCR product generated using primers carrying two restriction sites can also be used (Figure 12.1). This vector makes it possible to clone a large number of target sequences into the same vector using the same combination of restriction enzymes. The advantages of such a system immediately become obvious as the possibility of cloning large batches of fragments simultaneously presents itself. ihpRNA constructs carried by pHANNIBAL are efficient at mediating silencing. Of the plants expressing dsRNA from an ihpRNA construct, 90% showed silencing, whereas only 58%, 13%, and 12% were silenced using hpRNA and sense and antisense constructs, respectively (Wesley et al. 2001). pHANNIBAL-derived ihpRNA has been used to silence various genes involved in metabolic and developmental processes in a number of species, including *Arabidopsis*, rice, cotton, and tobacco. A drawback to cloning large numbers of products into pHANNIBAL is the multiple steps required to introduce the silencing sequences into the vector in the correct orientations by restriction digest.

pHANNIBAL-derived ihpRNA has been successfully used to silence genes in monocotyledonous plants and thus may also become an important tool for large-scale genomic analysis. The *dihydroflavonol-4-reductase* and *A1* and *Ant18* genes in maize and barley, respectively, and *Mlo* (Schweizer et al. 2000), *SGT1*, and *Rar1* genes in barley (Azevedo et al. 2002) have been successfully silenced. *GUS* has also been silenced in transgenic wheat lines (Schweizer et al. 2000).

Recently, a high-throughput vector called pHELLSGATE that mediates PTGS via ihpRNA has been developed based on pHANNIBAL. pHELLSGATE is a high-copy plasmid that uses the Gateway Cloning Technology from Invitrogen Life Technologies. The Gateway™ system is based on site-specific recombination as carried out by bacteriophage λ. pHELLSGATE contains intron 2 from the *Flaveria Pdk* gene between the recombination cassettes (Figure 12.1). A single PCR product carrying *attB1* and *attB2* sites is generated. The *attB* sites recombine with the *attP* sites of pHELLSGATE in a sequence-specific manner, *attB1* directed at *attP1*, and *attB2* at *attP2*. This introduces the target sequence of the gene of interest into the vector in the sense direction at one site and in the antisense direction at the other. The presence of the *ccdB* gene in the *attP* recombination cassette, as well as the chloramphenicol resistance gene in the vector, facilitates the selection of positive recombinants on chloramphenicol-containing media as the *ccdB* gene product is lethal in common laboratory strains of *Escherichia coli* such as DH5α and DH10B. Thus, the difficult and laborious steps of restriction and ligation are avoided in construction of

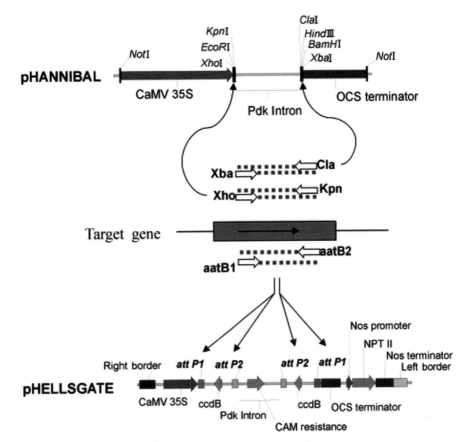

**FIGURE 12.1.** Maps and cloning strategies for pHANNIBAL and pHELLSGATE. PCR products from the target gene are cloned conventionally into the polylinkers of pHANNIBAL. Restriction sites added by the primers ensure the correct orientation of the resulting sense and antisense arms. The *attB1* and *attB2* sequences on a single PCR product facilitate the recombination of one sense-orientated and one antisense-orientated molecule into each molecule of pHELLSGATE when incubated with BP clonase. The complete sequences and annotations for pHANNIBAL and pHELLSGATE have been depositied at EMBL (Acc Nos: AJ311872 and AJ311874). (Reprinted, with permission, from Wesley et al. 2001.)

the silencing vector. After recombination using 200-bp and 400-bp PCR products and transformation into *E. coli*, 96% of colonies tested were the desired recombinants (Wesley et al. 2001). pHELLSGATE thus allows efficient and rapid construction of ihpRNA constructs.

pHELLSGATE has other features that make it attractive as a high-throughput vector. It contains a high-copy-number origin of replication, making it easy to handle in a manner that would be required for large-scale cloning. In addition, pHELLSGATE is a binary vector (see Stable Transformation later in this chapter) and thus can be directly used for *Agrobacterium*-mediated transformation of plants. Although to date there are no published reports on the efficiency of silencing using pHELLSGATE, one would expect results similar to those obtained for pHANNIBAL.

The few simple steps required to produce the ihpRNA constructs using pHELLSGATE suggest that the production of such recombinants could possibly be automated, allowing the analysis of hundreds, if not thousands, of genes (Wesley et al. 2001). The sequences used for silencing in ihpRNA constructs may range in size from as short as 98 bp to up to 800 bp. Thus, either expressed sequence tag (EST) libraries or oligonucleotides synthesized by computer predictions of gene sequence could be used as a source of targets for silencing. The automation of ihpRNA construction would undoubtedly lead us closer to high-throughput gene function analysis via PTGS.

One drawback to this system is that if PTGS in the whole organism is desired, then stably transformed plants carrying these constructs must be generated. This requires a large expenditure of effort when working with some species and takes considerable time in all species (see Stable Transformation later in this chapter). In some cases, as with many important agricultural crops such as cereals, the advantages to having a gene silenced in every cell of a plant may override these concerns. However, it is possible to express ihpRNA transiently and determine the role of some genes. A dsRNA system was used to silence *HvSGT1*, an essential gene in *R*-gene-mediated disease resistance, in a single-cell assay and in barley leaves (Azevedo et al. 2002), and a *GUS* transgene in wheat (Schweizer et al. 2000). DNA vectors encoding ihpRNA were bombarded into the desired tissues. Thus, transient dsRNA techniques may also be used to assess gene function, in addition to using transgenic plants. Other types of delivery could be adopted, such as Agroinfiltration, to develop an efficient tool for analysis in crops.

## Virus-induced Gene Silencing

VIGS uses a virus to deliver a sequence from a gene of interest into a host plant. The virus carrying the fragment of the gene of interest must be capable of replication if dsRNA is to be produced. Typically, VIGS does not involve the generation of transgenic plants that stably express the silencing construct. Instead, one or two leaves are either inoculated with *Agrobacterium* strains carrying the VIGS vector possessing the gene fragment or bombarded with DNA encoding the viral sequence. The virus then replicates and spreads throughout the plant, mediating silencing. Transient VIGS is not heritable from one generation to the next.

Many vectors have been developed for VIGS. A TMV-based vector was the earliest of these vectors (Kumagai et al. 1995). In this vector, the tomato mosaic virus (ToMV) coat protein (CP) gene replaces the TMV-U1 CP gene. The TMV-U1 CP subgenomic promoter is used to express a partial sequence of the gene to be silenced (Figure 12.2A). This arrangement avoids the possible deletion of the insert that could occur if the TMV coat protein was expressed from an additional copy of its own subgenomic promoter. In vitro transcripts were used to inoculate TMV host species. This vector carried 648 bp of sequence and was used successfully to silence the *phytoene desaturase* (*PDS*) gene in

**FIGURE 12.2.** Different VIGS vectors used for successful silencing in plants. (*A*) TMV-based VIGS vector. TMV-U1 coat protein (CP) subgenomic promoter (pCP) drives transcription of the gene fragment of interest cloned into the multiple cloning site (MCS). SP6 promoter drives in vitro transcription. This vector was described in Kumagai et al. (1995). (*B*) TGMV-based VIGS vector. The gene fragment of interest can be inserted in the MCS of either TGMV A or TGMV B. The *AR1* promoter drives transcription of the gene fragment of interest in TGMV A, whereas in TGMV B, it is cotranscribed with BR1. The TGMV A vector was described in Kjemtrup et al. (1998), and the TGMV B vector was described in Peele et al. (2001). (*C*) PVX-based VIGS vector. CP subgenomic promoter (pCP) is duplicated and drives transcription of the gene fragment of interest cloned into the MCS. T7 promoter drives in vitro transcription. This vector was described in Ruiz et al. (1998). (*D*) PVX-based VIGS vector. Same as *C*, except that the CaMV 35S promoter drives transcription of PVX in plant cells and this is a T-DNA vector. This vector was described in Ratcliff et al. (2001). (*E*) TRV-based VIGS vector. A single copy of the CaMV 35S promoter drives transcription of pBINTRA6 and pTV00. The gene fragment of interest is cloned into the MCS of pTV00. These vectors were described in Ratcliff et al. (2001). (*F*) TRV-based improved VIGS vector. The gene fragment of interest is cloned into the MCS of pTRV2 or into pTRV2-GATEWAY cassette. Transcription of pTRV1 and pTRV2 is driven by the duplicated CaMV 35S promoter. pTRV1 and pTRV2 are described in Liu et al. (2002b). The pTRV-GATEWAY vector was described in Liu et al. (2002a). (*G*) TMV-SVISS-based VIGS vector. The gene fragment of interest is cloned into the middle of the TMV satellite genome. Replication and movement proteins are provided by the TMV helper virus. This vector was described in Gossele et al. (2002). (*H*) BSMV-based VIGS vector. The gene fragment of interest can be inserted in either a sense or antisense direction and is transcribed by the γb subgenomic promoter. This vector was described in Holzberg et al. (2002).

*Nicotiana benthamiana* (Kumagai et al. 1995). The next VIGS vector developed was DNA-virus-based and derived from the tomato golden mosaic virus (TGMV) (Kjemtrup et al. 1998). TGMV, like many geminiviruses, possesses a bipartite single-stranded DNA (ssDNA) genome that replicates in the nucleus of host cells. One plasmid, TGMV A, carries the genes required for replication and expression, whereas another plasmid, TGMV B, encodes the genes required for nuclear shuttling and cell-to-cell movement (Figure 12.2B) (Hull 2002). The fragment of the gene to be silenced replaces the coat protein gene in TGMV A. The TGMV coat protein is not required for viral replication or cell-to-

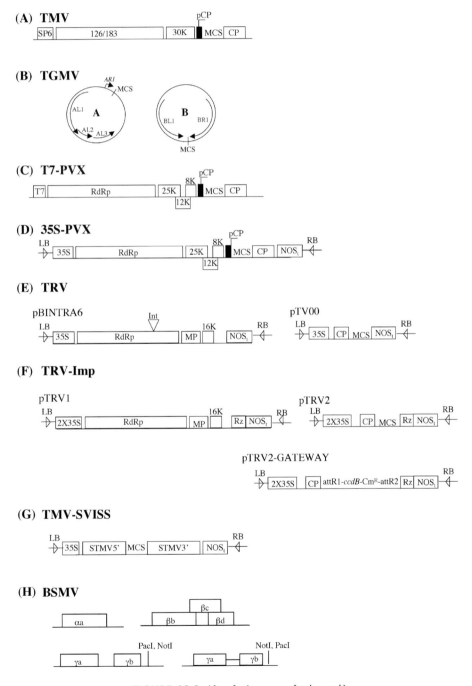

**FIGURE 12.2.** (*See facing page for legend.*)

cell movement (Kjemtrup et al. 1998). Both plasmids are then introduced into plant cells by particle bombardment. The recombinant TGMV replicates and spreads systemically, inducing silencing of the target gene. Kjemtrup et al. (1998) used fragments of varying sizes encoding different polypeptides to silence the *su* allele of magnesium chelatase (*Ch*), an endogenous gene required for chlorophyll biosynthesis. These authors also demonstrated that silencing of a stably integrated transgene, *luciferase*, could be mediated by this vector. The fragments used in their experiments ranged in size from 403 bp to 786 bp and were equally as effective when oriented in the sense or antisense direction in the vector. TGMV B could also effectively carry the fragment of the gene of interest (Peele et al. 2001). With this modified vector, less than 100 bp of homologous sequence can be used for sufficient silencing. In addition to *su*, they also silenced *proliferating cell nuclear antigen* (*PCNA*), demonstrating that although TGMV is excluded from the meristem, it effects silencing in the growing points of the plant (Peele et al. 2001). These vectors have not been adapted to be utilized in a high-throughput system.

PVX-based vectors have also been developed. PVX, like TMV, is a positive-sense ssRNA virus. The vector is a single plasmid containing a cDNA of the PVX genome with the gene of interest located between the triple gene block and the coat protein (Figure 12.2, C and D). Expression of the insert is driven by the duplicated coat protein promoter (Ruiz et al. 1998). Transcription of the entire construct is under the control of the cauliflower mosaic virus (CaMV) 35S promoter. This vector has the advantage of being more stable than the TMV-derived VIGS vector and is therefore more amenable to manipulation by researchers. The PVX vector has been used to silence a number of endogenous genes, as well as transgenes, with a variety of functions (Table 12.1). For example, a PVX VIGS vector was used to silence genes *NACK1* and *NACK2*, which are involved in cell division in *N. benthamiana* (Figure 12.3) (Nishihama et al. 2002). Other genes silenced include the *green fluorescent protein* (*GFP*) transgene, *cellulose synthase* (*CesA*), and the small subunit of *Rubisco* (*RbcS*) (Table 12.1). The silencing construct can be introduced into the host either by infection of transcripts (Ruiz et al. 1998; Burton et al. 2000) or through the use of Agroinfiltration (Thomas et al. 2001). Gene fragments from 377 bp to 670 bp have been used successfully to mediate silencing.

There are disadvantages to the above-described viral vectors. Some cause chlorosis and disease symptoms, which complicate analyses of the VIGS phenotype, especially if silencing is being used for plant defense or metabolic pathway studies. Some of the viruses are unable to enter every cell, and therefore will not result in whole-organism silencing. The latest VIGS vectors developed are TRV-based. TRV has advantages over other vectors, because the virus causes only mild disease symptoms (Ratcliff et al. 2001; Liu et al. 2002b) and is able to infect meristematic tissue (Ratcliff et al. 2001) and flowers (Y. Liu and S.P. Dinesh-Kumar, unpubl.). TRV has a bipartite positive-sense ssRNA genome. RNA 1 encodes the movement protein, the 134-kD and 194-kD replicases, and a 16-kD protein. RNA 2 encodes the coat protein (CP), a 29.4-kD protein, and a 32.8-kD protein. Ratcliff et al. (2001) developed the first TRV VIGS vector. They cloned RNA 1 and RNA 2 cDNAs separately between the left and right borders of T-DNA and under the control of a CaMV 35S promoter and the nopaline synthase terminator (NOSt) (Figure 12.2E). A multiple cloning site was placed after the CP on RNA 2, replacing the 29.4-kD and 32.8-kD proteins. Intron 3 of the *Arabidopsis nitrate reductase* gene was included in the replicase open reading frame of RNA 1 for stability in *E. coli*. This VIGS vector resulted in mild mosaic symptoms at 5 days postinfection (dpi), but there were no symptoms by 14 dpi (Ratcliff et al. 2001). These authors used this vector to silence *GFP, NFL, RbcS*, and *PDS* in *N. benthamiana*, and compared silencing efficiency and duration to the PVX vector. The TRV vector induced an equal and at times stronger silencing phenotype with a greater persistence (Ratcliff et al. 2001).

**TABLE 12.1.** Examples of genes successfully silenced using PTGS

| Method | | Plant | Gene Silenced | References |
|---|---|---|---|---|
| dsRNA | | *Arabidopsis* | AG, CLV3, AP1, PAN | Chuang and Meyerowitz (2000) |
| | | | CBL | Levin et al. (2000) |
| | | | EIN2, FLC1, CHS | Wesley et al. (2001) |
| | | | FAD2 | Smith et al. (2000); Wesley et al. (2001) |
| | | Barley | A1, Mlo | Schweizer et al. (2000) |
| | | | Rar1, SGT1 | Azevedo et al. (2002) |
| | | Cotton | FAD2, SAD | Wesley et al. (2001) |
| | | Maize | A1 | Schweizer et al. (2000) |
| | | *N. sylvestris* | CHN48 | Schob et al. (1997) |
| | | Tobacco | PPO, GUS | Wesley et al. (2001) |
| | | | PVY | Waterhouse et al. (1998) |
| | | Rice | GUS | Waterhouse et al. (1998); Wesley et al. (2001) |
| Amplicon | PVX | *Arabidopsis* | GFP | Dalmay et al. (2000) |
| | | *N. benthamiana* | PDS, RbcS | Angell and Baulcombe (1999) |
| | | Tobacco | GUS, PDS, PVX*, RbcS | Angell and Baulcombe (1997, 1999) |
| | | Tomato | DWARF | Angell and Baulcombe (1999) |
| | TYDV | Petunia | ChsA | Atkinson et al. (1998) |
| VIGS | BSMV | Barley | PDS | Holzberg et al. (2002) |
| | CbLCV | *Arabidopsis* | H42, GFP, PDS | Turnage et al. (2002) |
| | TMV | *N. benthamiana* | PDS | Kumagai et al. (1995) |
| | PVX | *N. benthamiana* | CesA | Burton et al. (2000) |
| | | | CDPK2 | Romeis et al. (2001) |
| | | | FtsH | Saitoh and Terauchi (2002) |
| | | | GFP | Anadalakshmi et al. (1998); Ruiz et al. (1998); Ratcliff et al. (2001) |
| | | | GUS | Anadalakshmi et al. (1998) |
| | | | N, NPK1, SGT1 | Jin et al. (2002) |
| | | | NACK1, NACK2 | Nishihama et al. (2002) |
| | | | PDS | Ruiz et al. (1998); Ratcliff et al. (2001) |
| | | | RbcS | Ratcliff et al. (2001) |
| | TGMV | | Ch, Lux | Kjemtrup et al. (1998) |
| | TRV | *N. benthamiana* | CSN3, CSN8, SKP1 | Liu et al. (2002c) |
| | | | EDS1 | Liu et al. (2002a); Peart et al. (2002a) |
| | | | GFP | Ratcliff et al. (2001) |
| | | | N | Liu et al. (2002a); Peart et al. (2002a) |
| | | | NFL | Ratcliff et al. (2001) |
| | | | NPK1 | Jin et al. (2002) |
| | | | PDS | Ratcliff et al. (2001); Liu et al. (2002a) |
| | | | RbcS | Ratcliff et al. (2001) |
| | | | SGT1 | Liu et al. (2002c); Peart et al. (2002b) |
| | | Tomato | CTR1, CTR2, PDS, RbcS | Liu et al. (2002a) |
| | SVISS | Tobacco | PARP, RNAPII | Gossele et al. (2002) |

A second more stable and improved TRV vector has been developed (Liu et al. 2002b). The CaMV 35S promoter was duplicated for both RNA 1 and RNA 2, driving high transcription rates (Figure 12.2F). A self-cleaving ribozyme site was placed before the NOSt to create a precise 3′ end to the RNAs. This vector is stable in *E. coli* without the addition of an intron (Liu et al. 2002b). *PDS, Rar1, NPR1/NIM1, EDS1, SKP1, SGT1*, and the *CSN3* and *CSN8* subunits of the COP9 signalosome were silenced in *N. benthamiana* using this vector (Liu et al. 2002b,c). An important advancement toward utilizing VIGS technology in model plant systems is the development of a TRV vector to silence endogenous genes in tomato (Liu et al. 2002a)—*PDS, CTR1, CTR2*, and *RbcS* (Figure 12.4) (Liu et al. 2002a). This TRV vector was adapted for high-throughput cloning (Figure 12.2F), as the tradi-

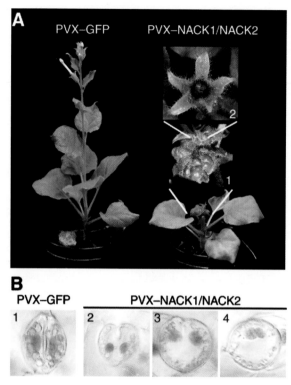

**FIGURE 12.3.** Phenotypes generated by suppression of *NACK1* and *NACK2*. (*A*) Leaves of 1-month-old plants of *N. benthamiana* were inoculated with PVX-*NACK1/NACK2* or PVX-*GFP* RNA. The gross morphology of typical plants, 30 dpi, is shown. (*Insets*) Magnified views of the shoot apex (*part 1*) and an aborted flower (*part 2*) of a PVX-*NACK1/NACK2* RNA-inoculated plant. Necrotic petals (*brown*) and small *green* carpels are visible. (*B*) Nomarski images of orcein-stained guard cells on the second highest leaves of a PVX-GFP-infected plant (*part 1*) and a PVX-*NACK1/NACK2* RNA-infected plant (*parts 2–4*), 30 dpi. These are the same plants as shown in *A*. (Reprinted, with permission, from Nishihama et al. 2002 ©Elsevier Science.)

tional cloning methods used with the original vector would have been impractical for genome-wide studies (Liu et al. 2002a).

Recently, a modified TMV-VIGS vector system has been described (Gossele et al. 2002). The satellite virus-induced silencing system (SVISS), as it is called, recruits the satellite tobacco mosaic virus (STMV) and uses it to effect silencing of endogenous genes. To date, this is the only VIGS system utilizing a satellite virus. In the SVISS, the target gene fragment is inserted into the satellite virus, whereas replication and movement of proteins are provided by the TMV helper virus (Figure 12.2G). Satellite viruses, with their small genomes, replicate efficiently, and their RNAs accumulate to high levels in plant cells. SVISS was used to silence 14 endogenous genes in *Nicotiana tabacum* (Table 12.1) (Gossele et al. 2002). These authors note that silencing efficiency was dependent on insert size as inserts larger than 300 bp resulted in lower levels of viral RNA accumulation and lower efficiency of silencing. This is the first report of a transient PTGS assay in tobacco, an important plant experimental system.

All of the VIGS vectors described thus far are limited to functioning in dicotyledonous plants. However, many of the most important agricultural and food crops are monocotyledonous. The availability of the rice genome sequence (Feng et al. 2002) makes it possible to use reverse genetics technology for gene function analysis in monocots. The recent development of a barley stripe mosaic virus (BSMV) VIGS vector is important because it demonstrates for the first time that monocotyledonous plants have the cellu-

**TRV** **TRV -tPDS**

**FIGURE 12.4.** Silencing of the tomato *PDS* gene. Infection of tomato plants with recombinant TRV alone (*a*) or TRV carrying the tomato *PDS* (TRV-*tPDS*) (*b*). Infection with TRV-*tPDS* silences endogenous PDS in Micro-Tom tomato plants and causes inhibition of carotenoid biosynthesis resulting in a photo-bleaching phenotype (*b*). (Reprinted, with permission, from Liu et al. 2002a.)

lar machinery required to carry out VIGS (Holzberg et al. 2002). BSMV is a tripartite positive-sense ssRNA virus with a limited host range that infects the agriculturally important barley crop (Holzberg et al. 2002; Hull 2002). In constructing the silencing vector, the target gene sequence was inserted downstream from the γb gene of the γ RNA (Figure 12.2H) (Holzberg et al. 2002). Using the *PDS* gene sequence from barley, maize, and rice, these investigators were able to silence endogenous *PDS* in barley and observe the photobleaching phenotype typical of *pds* mutants. The deletion of the βa coat protein gene enhanced the silencing phenotype as well as attenuated the symptoms of viral infection. This work was an important first step in the development of other VIGS vectors that will facilitate the rapid assessment of gene function by silencing in monocots that has become commonplace in the dicot *N. benthamiana*.

VIGS has obvious advantages over other methods used for genome analysis. It is a transient assay, eliminating the need for the generation of transgenic plants. Thus, in theory, VIGS should have widespread application to many different species, regardless of the existence of transformation systems. Because VIGS is performed on growing plants and is transient, gene knockouts that have lethal effects can be readily isolated. In addition, VIGS constructs result in higher expression of the dsRNA because of viral replication itself. VIGS has been used to study genes involved in many developmental (Figures 12.3 and 12.4 and Table 12.1) and metabolic processes in plants and has been particularly used to study disease resistance (Figure 12.5 and Table 12.1).

To date, however, only the tomato VIGS vector has been adapted for large-scale functional genomics (Liu et al. 2002a). This takes advantage of the Gateway™ Technology (Invitrogen), a system that does not require restriction enzymes or ligations for cloning (Figure 12.2F). PCR products can now be directly cloned into the TRV vector. The cloning efficiency of this system was found to be approximately 90% (Liu et al. 2002a). The use of tomato in large-scale functional genomics studies holds much promise. Although its genome has not been sequenced, there are numerous resources available for genetic analysis in this model crop species. These include approximately 32,000 unique tomato ESTs available to researchers as of April 25, 2002 (http://www.tigr.org/tdb/tgi/lgi/release_notes.html). There are also large numbers of bacterial artificial chromosome (BAC), yeast artificial chromosome (YAC), and other genomic libraries available for analysis, as well as dense restriction-fragment-length polymorphism (RFLP) maps among

Normal Light    UV Illumination

**FIGURE 12.5.** Analysis of the effect of *NbSGT1* and *NbSKP1* suppression on *N*-gene-mediated resistance to TMV. Wild-type *nn* (*A*) and transgenic *NN* (*B–E*) *N. benthamiana* plants were first silenced for *N* (*C*), *NbSGT1* (*D*), and *NbSKP1* (*E*) at the four-leaf stage. (*A,B*) Nonsilenced controls. The upper leaves of these plants were then infected with TMV-GFP virus to monitor resistance or susceptibility responses. The spread of TMV-GFP from the inoculated leaf into the upper uninoculated leaves indicates loss of resistance to TMV. (Reprinted, with permission, from Liu et al. 2002c ©American Society of Biologists.)

many others (Mysore et al. 2001; Emmanuel and Levy 2002). In addition, there exists the possibility of utilizing *Arabidopsis* sequence data for studies in tomato. Another important consideration for the pertinence of studies in tomato is the close relationship of tomato to other agriculturally important species such as potato, pepper, tobacco, and petunia, all members of the Solanaceae family. Thus, studies in tomato will provide information that may be applied to many other crop species. The development of a high-throughput VIGS vector is quite exciting because it makes large-scale silencing possible.

The *Arabidopsis* genome has been sequenced, and VIGS in this model dicot would be a valuable tool to plant biologists. The exciting news is that a geminivirus-based vector for VIGS has been developed for use in *Arabidopsis* (Turnage et al. 2002). The cabbage leaf curl virus (CbLCV) used in this VIGS system is closely related to the TGMV used in the VIGS vector described above (Kjemtrup et al. 1998). The CbLCV-based system was successfully used to silence the endogenous genes *PDS* and *Chlorata 42* (*CH42*) as well as the *GFP* transgene. The development of this vector is an important first step to adopting large-scale VIGS for use in the most widely used plant biology system.

## Amplicon-mediated VIGS

VIGS is not limited to use in transient assays; it can also be used for silencing in transgenic plants. In this approach, a stably integrated replication-competent viral cDNA carrying a portion of the gene of interest is the source of dsRNA that results in gene silencing. This is termed an amplicon. Amplicons based on both RNA viruses, such as PVX (Angell and Baulcombe 1997) and brome mosaic virus (BMV) (Kaido et al. 1995), as well as on the DNA virus tobacco yellow dwarf virus (TYDV) (Atkinson et al. 1998), have been described. The same general mechanism probably causes PTGS, although the generation of the dsRNA intermediate differs between the RNA virus-based vectors (Kaido et al. 1995; Angell and Baulcombe 1997) and the DNA geminivirus-based vector (Atkinson et al. 1998). The description that follows here is pertinent to the PVX-amplicon system used in *N. benthamiana* (Figure 12.2D) (Angell and Baulcombe 1997). Initially, the transgene is transcribed from the CaMV 35S promoter. The first transcript produced encodes the viral RNA-dependent RNA polymerase (RdRP). Using the positive-sense transgene as template, the RdRP then synthesizes negative-sense RNA, generating the dsRNA intermediate. The negative-sense strand is then used for the transcription of various viral genes as well as the foreign gene of interest. Transcription is initiated off a number of subgenomic promoters. The RdRP synthesizes multiple negative-sense transcripts from the transgene template, effectively amplifying the transgene RNA and resulting in higher levels of sense transcript as compared to a single-copy gene. This high level of transgene expression precipitated by viral replication can obscure variations in transcription levels between transgenic lines that result from chromosome position effects. The amplicon, by producing such a high foreign gene transcript, mediates consistent silencing of the endogenous gene. Interestingly, amplicon-based VIGS occurs in the absence of viral infection symptoms, even though some infectious viral particles can be isolated from amplicon-expressing plants (Angell and Baulcombe 1999).

PVX amplicon-mediated VIGS has been used to silence a *GUS* transgene and the endogenous gene *DWARF* (Angell and Baulcombe 1997, 1999). In both instances, the silencing phenotype was confirmed by RNA blots and, in the case of *GUS*, enzyme assays. The phenotype of the *DWARF*-silenced plants mirrored that of knockout plants (Angell and Baulcombe 1999).

Amplicon-containing transgenic tobacco expressing the full complement of either replication-competent or replication-incompetent genomic RNAs of BMV was generated (Kaido et al. 1995). Protoplasts derived from the plants expressing replication-competent virus exhibited resistance when inoculated with BMV, whereas those expressing BMV RNAs incapable of replication were susceptible. The resistance was specific for BMV because cucumber mosaic virus (CMV) was able to replicate in these protoplasts. This suggests that resistance was homology-based, indicative of VIGS. The geminivirus TYDV was also used as a silencing vector (Atkinson et al. 1998). Mutations were made in the viral sequence that eliminated its ability to spread systemically as well as the development of disease symptoms. The *chalcone synthase* (*ChsA*) gene was successfully silenced in transgenic petunia plants carrying this TYDV amplicon (Atkinson et al. 1998).

The amplicon-mediated silencing system has many attractive features. First, the high level of transgene expression reduces the between-line variation in silencing that is observed in transgenic lines that direct silencing by other mechanisms. Second, the high level of transgene transcription results in consistent high-level silencing. Third, the silencing phenotype is not obscured by symptoms of viral infection. Another additional advantage of amplicon-mediated silencing is that it can be used to induce silencing in species that are not naturally infected by the viruses from which the amplicons are derived. PVX amplicons have been used to silence a *GFP* transgene in *Arabidopsis* (Figure 12.6) (Dalmay et al. 2000). However, the observed silencing was relatively weak as com-

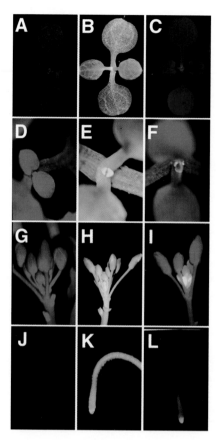

**FIGURE 12.6.** GFP fluorescence in plants carrying 35S-*GFP* and 35S-*PVX:GFP* transgenes. GFP fluorescence in plants carrying 35S-*GFP* (*B, E, H,* and *K*), 35S-*PVX:GFP* (*A, D, G,* and *J*), or a combination of both transgenes (*C, F, I,* and *L*). The images were produced under UV light in a dissecting microscope, and the red fluorescence is due to chlorophyll. GFP fluorescence appears *green, yellow,* or *blue-green* in different tissues and also depends on the photographic exposure time. Nontransformed plants look the same as the Amp243 plants (*A, D, G,* and *J*). (*A–C*) Young seedlings; (*D–F*) close up of the growing point of young seedlings; (*G–I*) immature flowers; (*J–L*) roots. There was no detectable fluorescence from the roots of the plants shown in *A, D, G,* and *J*; in the plants shown in *C, F, I,* and *L,* there was GFP fluorescence only in the growing point. (Reprinted, with permission, from Dalmay et al. 2000 ©American Society of Biologists.)

pared to silencing in the PVX host tobacco. Finally, these amplicon lines have the added advantage of conferring resistance to viruses with high similarity to the viral transgene, presumably by the same PTGS mechanism (Angell and Baulcombe 1997).

There are, however, several drawbacks to amplicon-mediated VIGS. Perhaps most important of these is the need to generate transgenic plants. One barrier to using this technique is where routine transformation protocols do not exist. Even in systems where protocols exist, such as rice and tobacco, it is still a time-consuming process. Nevertheless, in systems like *Arabidopsis*, where transformation is routine and easy, the introduction of viral transgenes can be envisioned as a viable option. Another limitation to using amplicons to initiate gene silencing is the unavailability of viruses to act as vectors for many species. For example, the potyviruses encode powerful suppressors of gene silencing (Anadalakshmi et al. 1998; Brigneti et al. 1998; Kasschau and Carrington 1998), making them unsuitable as VIGS vectors. The same is true of CMV (Beclin et al. 1998; Brigneti et al. 1998).

# TECHNIQUES

Delivery of dsRNA by Particle Bombardment

Delivery of dsDNA by Stable Transformation

Delivery of dsRNA by Agroinfiltration

Delivery of dsRNA by VIGS Methodology


## Delivery of dsRNA by Particle Bombardment

Particle bombardment can be used to transiently introduce DNA constructs that produce dsRNA (Kjemtrup et al. 1998; Schweizer et al. 2000; Peele et al. 2001). The DNA or dsRNA is coated onto tungsten or gold particles that are then forced across plant cell walls using a particle gun. The presence of dsRNA in the bombarded tissue results in silencing of a given gene. Although this method is relatively fast, there are several drawbacks to this delivery. First, only a few layers of tissue can be bombarded with dsRNA at a given time, negating the possibility of whole-organism analysis of gene function. This means that genes that function in development cannot be analyzed by introducing silencing constructs by this technique. Second, the technique has low efficiency because few cells actually receive the dsRNA, and many of the cells that undergo bombardment are actually killed. Third, the gene silencing is not heritable and is limited only to the tissue being bombarded. Despite these drawbacks, the technique has proven to be useful for species that are difficult to transform because there is no biological limitation on this delivery method. Bombardment is also useful for single-cell analysis where it can be used to study processes that occur at that level and are cell-autonomous, for example, biosynthetic pathways. It has also been successfully used in studies on disease resistance (Schweizer et al. 2000; Azevedo et al. 2002).


## Delivery of dsDNA by Stable Transformation

Stable transformation is the integration of a novel gene into the nuclear genome. More than 120 diverse plant species can be stably transformed (Birch 1997), including most major economic crops, vegetables, ornamentals, medicinal, fruit, tree, and pasture plants (Birch 1997). Four methods are available:

- Bombardment of DNA into various plant tissue.
- Polyethylene glycol (PEG)-mediated DNA uptake into protoplasts and cultured cells.
- Electroporation-mediated DNA delivery into protoplasts and into intact cells.
- *A. tumefaciens*-mediated transformation.

The details of these methods vary greatly for different species and are continually improved upon. The stable introduction of a dsRNA-producing construct into the genome by any of these methods will result in silencing of the gene of interest in every cell of the plant generated. The stable transformation of species is a useful tool for studying gene function; however, it is a lengthy process and therefore not the best approach for large-scale studies. In addition, this approach will not succeed if silencing the gene of interest results in embryonic lethality or death early in development.

Many species can be transformed by *A. tumefaciens* or *A. rhizogenes*, natural transformers of plants, which introduce and stably integrate their DNA into plant genomes (Tepfer 1990; Bent 2000). In nature, *A. tumefaciens* carries a Ti plasmid, which possesses *virulence* (*vir*) genes and specific signal sequences that, together with nuclear genes, mediate the excision, transfer, and integration of bacterial DNA, the T-DNA, into the plant genome (Zupan et al. 2000). The T-DNA is flanked by 25-bp direct repeat sequences, the left and right border (LB and RB). The Ti plasmid can be modified to carry a desired DNA sequence in place of bacterial sequence in the T-DNA. In fact, *Agrobacterium* will transfer any DNA situated between the LB and RB, explaining the usefulness of this system for plant transformation (Zupan et al. 2000).

The most common species transformed with *A. tumefaciens* is *A. thaliana*. Flowering *Arabidopsis* plants are simply dipped in *A. tumefaciens* suspended in an appropriate infiltration media, and seed is then collected and germinated on selection media (Clough and Bent 1998). Only the successfully transformed $T_0$ generation is able to grow on the media. After appropriate analysis of primary transformants, the phenotype of the silenced plant is observed in the $T_1$ generation.

The transformation protocols that involve protoplast isolation and callus growth are technically very challenging and time-consuming. Following successful protoplast isolation, the silencing construct can be introduced by particle bombardment, electroporation, PEG-mediated uptake, or *Agrobacterium* infection. The protoplast must grow on callus induction media with selection for the construct, and the plants are then generated from the callus. After molecular confirmation of successful transformation, the silencing phenotype can be observed in the progeny of these regenerated plants.

A procedure to introduce constructs into rice cell clusters by electroporation has been developed (Arencibia et al. 1998). Germinated embryos are grown on callus induction media for 2 months in the dark. A cell cluster fraction is then collected and resuspended in electroporation buffer that contains the plasmid of interest and is electroporated. The cell clusters are now returned to callus induction media and grown for 2 more weeks in the dark. The overall success rate for this method is 1.5% (Arencibia et al. 1998). Therefore, although most plant species can be successfully transformed, with the exception of *Arabidopsis*, these procedures are laborious, lengthy, and have a low success rate.

## Delivery of dsRNA by Agroinfiltration

The disadvantages of generating transgenics are circumvented by Agroinfiltration. Agroinfiltration, which takes advantage of the transforming capabilities of *Agrobacterium*, was first used to introduce dsRNA for silencing in plants (Schob et al. 1997). It can also be used to introduce VIGS vectors. Once the construct is made, it is simply placed in an appropriate *Agrobacterium* strain, and cultures are then grown and used for inoculation.

*Agrobacterium* cultures can be infiltrated into the plant leaves using a needless syringe. The introduced sequence is integrated into the plant genome and its transcription produces dsRNA, initiating PTGS. Agroinfiltration is a quick, reliable, and low-cost technique that facilitates the introduction of dsRNA-producing constructs into an intact plant. Thus, the effect of the knockdown of a gene of interest can be examined at the organismal level. This method is also very useful for examining multiple genes simultaneously. To do this, cultures of *Agrobacterium* carrying different silencing constructs are simply co-infiltrated.

## Delivery of dsRNA by VIGS Methodology

The following protocol performs VIGS assays in *N. benthamiana* and tomato using the TRV vector.

## Procedure

---

**MATERIALS**

**REAGENTS**

*Agrobacterium* strains GV2260 or GV3101 (see Step 2)

3´-5´ Dimethoxy 4´-hydroxy acetophenone (Acetosyringone)

Prepare 0.2 M stock solution in dimethyl formamide <!> and store at –20ºC.

Infiltration medium

10 mM 2-(*N*-morpholino) ethanesulfonic acid (MES) <!>

10 mM $MgCl_2$ <!>

150 µM Acetosyringone

Luria Broth (LB), solid and liquid media

Magnesium chloride ($MgCl_2$) (1 M stock) <!>

MES (1 M stock) <!>

pTRV1

pTRV2 with gene of interest

Relevant antibiotics

**EQUIPMENT**

Airbrush

Benchtop centrifuge

Needle-less syringe (1-ml capacity)

Razor blades

---

1. Clone a 500–700-bp region of the gene of interest into the TRV RNA 2 vector (pTRV2) (see Figure 12.2F).

   It is possible to use a smaller fragment, but the silencing effect may be reduced with less than 300 bp. If the gene of interest is a member of a gene family, then cloning the 5´-untranslated region will avoid the silencing of other members.

2. Once the gene of interest is in the silencing vector, transform *Agrobacterium* strains separately with either pTRV RNA 1 (pTRV1) or pTRV2.

   The *A. tumefaciens* strain GV2260 is used for silencing in *N. benthamiana*, and *A. tumefaciens* strain GV3101 is used for silencing in VF36 or Micro-Tom tomato.

3. Grow on selection medium and confirm the presence of the insert by PCR. Then grow 5-ml liquid cultures of *A. tumefaciens* in LB with antibiotic selection overnight at 28ºC.

4. The next day, inoculate the cultures into 50 ml of fresh medium supplemented with 10 mM MES and 20 µM Acetosyringone and containing the appropriate antibiotics. Grow the cultures overnight at 28ºC.

5. The next day, centrifuge the cultures and resuspend the pellet in infiltration medium to an OD$_{600}$ of 1.0 for *N. benthamiana* infiltrations and an OD$_{600}$ of 2.0 for tomato. Incubate for 2–3 hours at room temperature.

6. Mix the *Agrobacterium* suspensions containing pTRV1 and pTRV2 in a 1:1 ratio. Use a needle-less syringe to infiltrate the suspensions into the underside of a leaf. Make a small slit (~0.1 mm) with a razor blade and press the tip of the syringe into the leaf using a finger on the other side for support. Infiltrate the two lower leaves of tobacco plants at the four-leaf stage and infiltrate the tomato plants at the three-leaf stage after about 3 weeks of growth.

> For more efficient silencing in tomato, the *Agrobacterium* mixture can be sprayed with an artist's airbrush. Spray from ~8 inches away for about 1 second at 75 psi pressure.

> Silencing of *PDS* should become visible 5 days after infiltration in *N. benthamiana* (Liu et al. 2002b), and 10 days after infiltration in tomato (Liu et al. 2002a).

## FUTURE PERSPECTIVES

In recent years, the genomes of rice and *Arabidopsis* have been fully sequenced, and numerous EST databases are available for crop plants such as the tomato. This enormous amount of data poses a challenge because although the sequences are known, the functions of most of the proteins are unknown. Traditional genomic analysis in plants relied on the random introduction of mutations into the genome and screening for mutant phenotypes. The postgenomics era in plants will move past this single-gene focus. Already, approaches have been developed in *C. elegans* to analyze data at the whole-chromosome level (Fraser et al. 2000; Gonczy et al. 2000).

In plants, the two methods that can be used at the whole-organism level are transient VIGS and transgenic analyses. As discussed above, generation of transgenics requires a lot of time and effort. For large-scale studies involving potentially thousands of genes, transient VIGS is the only practical approach currently available. Most of the VIGS vectors were engineered for silencing in *N. benthamiana*, and although this species is particularly amenable to gene-silencing studies, limited sequence data are available. Therefore, large-scale genomic functional studies in this species would be challenging.

Currently, large-scale studies for gene silencing in plants are being conducted in tomato utilizing the Micro-Tom cultivar, which requires a smaller growing space and has a shorter generation time than other cultivars of tomato (Emmanuel and Levy 2002). With the large collection of ESTs available, it is possible to begin analyzing the entire genome of this cultivar. At present, progress is being made toward the establishment of VIGS protocols in model plants. A VIGS vector for use in *Arabidopsis* was described recently (Turnage et al. 2002) and the possibility of VIGS in monocots exists (Holzberg et al. 2002). PTGS promises to be an invaluable tool in undertaking large-scale functional genomic studies in plants.

## ACKNOWLEDGMENTS

We thank David Baulcombe, Yasunori Machida, and Peter Waterhouse for allowing use of the figures. Work on VIGS in the S.P.D.-K. lab is funded by a National Science Foundation grant (DBI-0211872).

# REFERENCES

Anadalakshmi R., Pruss G.J., Ge X., Marathe R., Smith T.H., and Vance V.B. 1998. A viral suppressor of gene silencing in plants. *Proc. Natl. Acad. Sci.* **95:** 13079–13084.

Angell S. and Baulcombe D. 1997. Consistent gene silencing in transgenic plants expressing a replicating potato virus X RNA. *EMBO J.* **16:** 3675–3684.

———. 1999. Technical advance: Potato virus X amplicon-mediated silencing of nuclear genes. *Plant J.* **20:** 357–362.

*Arabidopsis* Genome Initiative. 2000. Analysis of the genome sequence of the flowering plant *Arabidopsis thaliana*. *Nature* **408:** 796–815.

Arencibia A., Gentinetta E., Cuzzoni E., Castiglione S., Kohli A., Vain P., Leech M., Christou P., and Sala F. 1998. Molecular analysis of the genome of transgenic rice (*Oryza Sativa* L.) plants produced via particle bombardment or intact cell electroporation. *Mol. Breeding* **4:** 99–109.

Atkinson R.G., Bieleski L.R.F., Gleave A.P., Janssen B.-J., and Morris B.A.M. 1998. Post-transcriptional silencing of chalcone synthase in petunia using a geminivirus-based episomal vector. *Plant J.* **15:** 593–604.

Azevedo C., Sadanandom A., Kitagawa K., Frelaldenhoven A., Shirasu K., and Schulze-Lefert P. 2002. The Rar1 interactor SGT1, an essential component of *R* gene-triggered disease resistance. *Science* **295:** 2073–2076.

Baulcombe D.C. 1999. Fast forward genetics based on virus-induced gene silencing. *Curr. Opin. Plant Biol.* **2:** 109–113.

Beclin C., Berthome R., Palaqui J.C., Tepfer M., and Vaucheret H. 1998. Infection of tobacco or *Arabidopsis* plants by CMV counteracts systemic post-transcriptional silencing of nonviral (trans) genes. *Virology* **252:** 313–317.

Bent A.F. 2000. *Arabidopsis* in planta transformation. Uses, mechanisms, and prospects for transformation of other species. *Plant Physiol.* **124:** 1540–1547.

Birch R.G. 1997. Plant transformation: Problems and strategies for practical application. *Annu. Rev. Plant Physiol. Plant Mol. Biol.* **48:** 297–326.

Brigneti G., Voinnet O., Li W.-X., Ding S.W., and Baulcombe D.C. 1998. Viral pathogenicity determinants are suppressors of transgene silencing in *Nicotiana benthamiana*. *EMBO J.* **17:** 6739–6746.

Burton R.A., Gibeaut D.M., Bacic A., Findlay K., Roberts K., Hamilton A., Baulcombe D.C., and Fincher G.B. 2000. Virus-induced silencing of a plant cellulose synthase gene. *Plant Cell* **12:** 691–705.

Chuang C.F. and Meyerowitz E.M. 2000. Specific and heritable genetic interference by double-stranded RNA in *Arabidopsis thaliana*. *Proc. Natl. Acad. Sci.* **97:** 4985–4990.

Clough S.J. and Bent A.F. 1998. Floral dip: A simplified method for *Agrobacterium*-mediated transformation of *Arabidopsis thaliana*. *Plant J.* **16:** 735–743.

Dalmay T., Hamilton A., Mueller E., and Baulcombe D.C. 2000. Potato virus X amplicons in *Arabidopsis* mediate genetic and epigenetic gene silencing. *Plant Cell* **12:** 369–379.

Emmanuel E. and Levy A.A. 2002. Tomato mutants as tools for functional genomics. *Curr. Opin. Plant Biol.* **5:** 112–117.

Feng Q., Zhang Y., Hao P., Wang S., Fu G., Huang Y., Li Y., Zhu J., Liu Y., Hu X., Jia P., Zhang Y., Zhao Q., Ying K., Yu S., Tang Y., Weng Q., Zhang L., Lu Y., Mu J., Lu Y., Zhang L.S., Yu Z., Fan D., Liu X., Lu T., Li C., Wu Y., Sun T., Lei H., Li T., Hu H., Guan J., Wu M., Zhang R., Zhou B., Chen Z., Chen L., Jin Z., Wang R., Yin H., Cai Z., Ren S., Lv G., Gu W., Zhu G., Tu Y., Jia J., Zhang Y., Chen J., Kang H., Chen X., Shao C., Sun Y., Hu Q., Zhang X., Zhang W., Wang L., Ding C., Sheng H., Gu J., Chen S., Ni L., Zhu F., Chen W., Lan L., Lai Y., Cheng Z., Gu M., Jiang J., Li J., Hong G., Xue Y., and Han B. 2002. Sequence and analysis of rice chromosome 4. *Nature* **420:** 316–320.

Fire A., Xu S., Montgomery M.K., Kostas S.A., Driver S.E., and Mello C.C. 1998. Potent and specific genetic interference by double-stranded RNA in *Caenorhabditis elegans*. *Nature* **391:** 806–811.

Fraser A.G., Kamath R.S., Zipperlen P., Martinez-Campos M., Sohrmann M., and Ahringer J. 2000. Functional genomic analysis of *C. elegans* chromosome I by systemic RNA interference. *Nature* **408:** 325–330.

Gonczy P., Echeverri C., Oegema K., Coulson A., Jones S.J.M., Copley R.R., Duperon J., Oegema J., Brehm M., Cassin E., Hannak E., Kirkham M., Pichler S., Flohrs K., Goessen A., Leidel S., Alleaume A.-M., Martin C., Ozlu N., Bork P., and Hyman A.A. 2000. Functional genomic analysis of cell division in *C. elegans* using RNAi of genes of chromosome III. *Nature* **408:** 331–336.

Gossele V.V., Fache I.I., Meulewaeter F., Cornelissen M., and Metzlaff M. 2002. SVISS—A novel transient gene silencing system for gene function discovery and validation in tobacco. *Plant J.* **32:** 859–866.

Holzberg S., Brosio P., Gross C., and Pogue G.P. 2002. Barley stripe mosaic virus-induced gene silencing in a monocot plant. *Plant J.* **30:** 315–327.

Hull R. 2002. *Matthews' plant virology.* 4th edition. Academic Press, New York.

Jin H., Axtell M.J., Dahlbeck D., Ekwenna O., Staskawicz B., and Baker B. 2002. NPK1, an MEKK1-like mitogen-activated protein kinase kinase kinase regulates innate immunity and development in plants. *Dev. Cell* **3:** 291–297.

Kaido M., Mori M., Mise K., Okuno T., and Furusawa I. 1995. Inhibition of brome mosaic virus (BMV) amplification in protoplasts from transgenic tobacco plants expressing replicable BMV RNAs. *J. Gen. Virol.* **76:** 2827–2833.

Kasschau K.D. and Carrington J.C. 1998. A counter defensive strategy of plant viruses: Suppression of posttranscriptional gene silencing. *Cell* **95:** 461–470.

Kjemtrup S., Sampson K.S., Peele C.G., Nguyen L.V., and Conkling M.A. 1998. Gene silencing from plant DNA carried by a geminivirus. *Plant J.* **14:** 91–100.

Kumagai M.H., Donson J., Della-Cioppa G., Harvey D., Hanley K., and Grill L.K. 1995. Cytoplasmic inhibition of carotenoid biosynthesis with virus-derived RNA. *Proc. Natl. Acad. Sci.* **92:** 1679–1683.

Levin J.Z., de Frammond A.J., Tuttle A., Bauer M.W., and Heifetz P.B. 2000. Methods of double-stranded RNA-mediates gene inactivation in *Arabidopsis* and their use to define an essential gene in methionine biosynthesis. *Plant Mol. Biol.* **44:** 759–775.

Lindbo J.A., Silva-Rosales L., Proebsting W.M., and Dougherty W.G. 1993. Induction of a highly specific antiviral state in transgenic plants: Implications for regulation of gene expression and virus resistance. *Plant Cell* **5:** 1749–1759.

Liu Y., Schiff M., and Dinesh-Kumar S.P. 2002a. Virus-induced gene silencing in tomato. *Plant J.* **31:** 777–786.

Liu Y., Schiff M., Marathe R., and Dinesh-Kumar S.P. 2002b. Tobacco Rar1, EDS1 and NPR1/NIM1 like genes are required for N-mediated resistance to tobacco mosaic virus. *Plant J.* **30:** 415–429.

Liu Y., Schiff M., Serino G., Deng X.-W., and Dinesh-Kumar S.P. 2002c. Role of SCF ubiquitin-ligase and the COP9 signalosome in the *N* gene-mediated resistance response to tobacco mosaic virus. *Plant Cell* **14:** 1483–1496.

Meissner R., Chague V., Zhu Q., Emmanuel E., Elkind Y., and Levy A.A. 2000. A high throughput system for transposon tagging and promoter trapping in tomato. *Plant J.* **22:** 265–274.

Mysore K.S., Tuori R.P., and Martin G.B. 2001. *Arabidopsis* genome sequence as a tool for functional genomics in tomato. *Genome Biol.* **2:** 1003.1–1003.4.

Napoli C., Lemieux C., and Jorgensen R. 1990. Introduction of a chimeric chalcone synthase gene into petunia results in reversible co-suppression of homologous genes in *trans. Plant Cell* **2:** 279–289.

Nishihama R., Soyano T., Ishikawa M., Araki S., Tanaka H., Asada T., Irie K., Ito M., Terada M., Banno H., Yamazaki Y., and Machida Y. 2002. Expansion of the cell plate in plant cytokinesis requires a kinesin-like protein/MAPKKK complex. *Cell* **109:** 87–99.

Parinov S., Sevugan M., Ye D., Yang W.-C., Kumaran M., and Sundaresan V. 1999. Analysis of flanking sequences from dissociation insertion lines: A database for reverse genetics in *Arabidopsis. Plant Cell* **11:** 2263–2270.

Peart J.R., Cook G., Feys B.J., Parker J.E., and Baulcombe D.C. 2002a. An *EDS1* orthologue is required for *N*-mediated resistance against tobacco mosaic virus. *Plant J.* **29:** 569–579.

Peart J.R., Lu R., Sadanandom A., Malcuit I., Moffet P., Brice D.C., Schauser L., Jaggard D.A., Xiao S., Coleman M.J., Dow M., Jones J.D., Shirasu K., and Baulcombe D.C. 2002b. Ubiquitin ligase-associated protein SGT1 is required for host and non-host disease resistance in plants. *Proc. Natl. Acad. Sci.* **99:** 10865–10869.

Peele C., Jordan C.V., Muangsan N., Turnage M., Egelkrout E., Eagle P., Hanley-Bowdoin L., and Robertson D. 2001. Silencing of a meristematic gene using geminivirus-derived vectors. *Plant*

*J.* **27:** 357–366.

Ratcliff F., Martin-Hernandez A.M., and Baulcombe D.C. 2001. Tobacco rattle virus as a vector for analysis of gene function by silencing. *Plant J.* **25:** 237–245.

Romeis T., Ludwig A.A., Martin R., and Jones J.D.G. 2001. Calcium dependent protein kinases play an essential role in a plant defence response. *EMBO J.* **20:** 5556–5567.

Ruiz M.T., Voinnet O., and Baulcombe D.C. 1998. Initiation and maintenance of virus-induced gene silencing. *Plant Cell* **10:** 937–946.

Saitoh H. and Terauchi R. 2002. Virus-induced silencing of *FtsH* gene in *Nicotiana benthamiana* causes a striking bleached leaf phenotype. *Genes Genet. Syst.* **77:** 335–340.

Schob H., Kunz C., and Meins F.J. 1997. Silencing of transgenes introduced into leaves by agroinfiltration: A simple, rapid method for investigating sequence requirements for gene silencing. *Mol. Gen. Genet.* **256:** 581–585.

Schweizer P., Pokorny J., Schulze-Lefert P., and Dudler R. 2000. Technical advance. Double-stranded RNA interferes with gene function at the single-cell level in cereals. *Plant J.* **24:** 895–903.

Sharp P.A. 2001. RNA interference—2001. *Genes Dev.* **15:** 485–490.

Smith N.A., Singh S.P., Wang M.-B., Stoutjesdijk P.A., Green A.G., and Waterhouse P.M. 2000. Total silencing by intron-spliced hairpin RNAs. *Nature* **407:** 319–320.

Tepfer D. 1990. Genetic transformation using *Agrobacterium rhizogenes. Physiol. Plant.* **79:** 140–146.

Thomas C.L., Jones L., Baulcombe D.C., and Maule A.J. 2001. Size constraints for targeting post-transcriptional gene silencing and for using RNA-directed methylation in *Nicotiana benthamiana* using a potato virus X vector. *Plant J.* **25:** 417–425.

Turnage M.A., Muangsan N., Peele C.G., and Robertson D. 2002. Geminivirus-based vectors for gene silencing in *Arabidopsis. Plant J.* **30:** 107–117.

van der Krol A.R., Mur L.A., Beld M., Mol J.N., and Stuitje A.R. 1990. Flavonoid genes in petunia: Addition of a limited number of gene copies may lead to a suppression of gene expression. *Plant Cell* **2:** 291–299.

Vaucheret H., Beclin C., and Fagard M. 2001. Post-transcriptional gene silencing in plants. *J. Cell Sci.* **114:** 3083–3091.

Waterhouse P.M., Graham M.W., and Wang M.-B. 1998. Virus resistance and gene silencing in plants is induced by double stranded RNA. *Proc. Natl. Acad. Sci.* **95:** 13959–13964.

Waterhouse P.M., Wang M.-B., and Lough T. 2001. Gene silencing as an adaptive defense against viruses. *Nature* **411:** 834–842.

Wesley S.V., Helliwell C.A., Smith N.A., Wang M.B., Rouse D.T., Liu Q., Gooding P.S., Singh S.P., Abbott D., Stoutjesdijk P.A., Robinson S.P., Gleave A.P., Green A.G., and Waterhouse P.M. 2001. Construct design for efficient, effective and high-throughput gene silencing in plants. *Plant J.* **27:** 581–590.

Zupan J., Muth T.R., Draper O., and Zambryski P. 2000. The transfer of DNA from *Agrobacterium tumefaciens* into plants: A feast of fundamental insights. *Plant J.* **23:** 11–28.

# CHAPTER 13

# Mammalian RNA Interference

Thomas Tuschl

*Laboratory for RNA Molecular Biology, The Rockefeller University, New York 10021*

WHEN VIRUSES INFECT EUKARYOTIC CELLS or when transposons and transgenes randomly integrate into host genomes, double-stranded RNA (dsRNA) is frequently produced from the invading genes, either during viral replication or by aberrant transcription from promoters located near the transgene insertion site. Eukaryotes such as plants, protists, and filamentous fungi and invertebrate and vertebrate animals have evolved a cellular defense system that responds to dsRNA and protects their genomes against these invading foreign elements. The dsRNA is rapidly processed by a cellular enzyme to small dsRNA fragments of distinct size and structure (Bernstein et al. 2001), which then direct the sequence-specific degradation of the single-stranded mRNAs of the invading genes (Elbashir et al. 2001a). These short RNA duplexes were therefore named short interfering RNAs (siRNAs). The entire process of posttranscriptional dsRNA-dependent gene silencing is commonly referred to as RNA interference or RNAi (for recent reviews, see Hammond et al. 2001a; Matzke et al. 2001a; Sharp 2001; Tuschl 2001; Waterhouse et al. 2001; Hutvágner and Zamore 2002). In some instances, posttranscriptional gene silencing is also linked to transcriptional silencing (for reviews, see Wassenegger 2000; Bender 2001; Matzke et al. 2001b; Pal-Bhadra et al. 2002).

Experimental introduction of dsRNA into cells has been used to disrupt the activity of cellular genes homologous in sequence to the introduced dsRNA (Fire et al. 1998). RNAi-based reverse genetic analysis now provides a rapid link between sequence data and biological function. RNAi is particularly useful for the analysis of gene function in *Caenorhabditis elegans* (for reviews, see Hope 2001; Kim 2001), but it is also widely used in other invertebrate animals (Kennerdell and Carthew 1998; Ngo et al. 1998; Brown et al. 1999). dsRNA of several hundred base pairs in length is typically required for effective gene silencing (Parrish et al. 2000; Elbashir et al. 2001b). Its application in vertebrate animals, including mammals, has proven to be more difficult because of the presence of additional dsRNA-triggered pathways that mediate nonspecific suppression of gene expression (Caplen et al. 2000; Nakano et al. 2000; Oates et al. 2000; Zhao et al. 2001). Fortunately, these nonspecific responses to dsRNA in vertebrates are not triggered by the siRNAs (Bitko and Barik 2001; Caplen et al. 2001; Elbashir et al. 2001c; Zhou et al. 2002). siRNAs can target genes as effectively as long dsRNAs (Elbashir et al. 2001b) and are widely used today for assessing gene function in cultured mammalian cells or early developing vertebrate embryos (Harborth et al. 2001; Elbashir et al. 2002; Zhou et al. 2002). siRNAs are also promising reagents for developing gene-specific therapeutics (Tuschl and Borkhardt 2002). This chapter concentrates on RNAi as it relates to mammalian systems and on the application of siRNAs for targeting genes expressed in somatic mammalian cell lines.

## CONCEPTS AND STRATEGIES

### General Mechanism of RNA Interference

Biochemical studies are beginning to unravel the mechanistic details of RNAi. The first cell-free systems were developed using *Drosophila melanogaster* cell or embryo extracts (Tuschl et al. 1999; Hammond et al. 2000; Zamore et al. 2000) and were followed by the development of in vitro systems from *C. elegans* embryos (Ketting et al. 2001) and mouse embryonal carcinoma (EC) cell lines F9 and P19 (Billy et al. 2001). However, the latter two systems do not recapitulate all aspects of RNAi when compared to the *D. melanogaster* systems. Figure 13.1 summarizes the conserved features of the mechanism of RNAi.

### Long dsRNA is first processed by Dicer RNase III to siRNAs

Long dsRNAs are first processed to siRNAs by the ribonuclease III (RNase III)-like enzyme Dicer (Hammond et al. 2000; Billy et al. 2001; Ketting et al. 2001). Dicer has an amino-terminal DExH/DEAH RNA helicase domain, a PAZ (Piwi-Argo-Zwille/Pinhead) domain (Cerutti et al. 2000), a tandem repeat of RNase III catalytic domain sequences, and a carboxy-terminal dsRNA-binding motif. In *D. melanogaster* embryo extracts and in Dicer immunoprecipitates of *D. melanogaster* cells, the rate of siRNA formation is ATP-dependent, and siRNAs produced in the embryo lysate in the absence of ATP are one nucleotide longer than in the presence of ATP (Zamore et al. 2000; Bernstein et al. 2001).

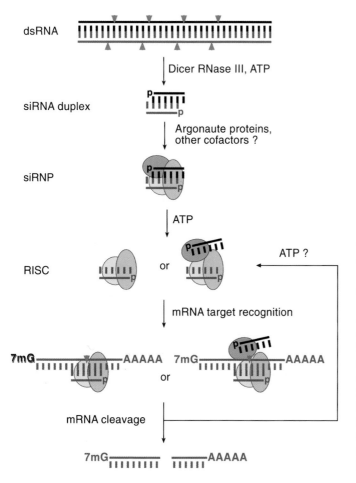

FIGURE 13.1. Model for RNA interference. dsRNA is processed to 21–23-nucleotide siRNA duplexes by Dicer RNase III and possibly other dsRNA-binding factors in an ATP-dependent manner. The siRNA duplexes are incorporated into a siRNA-ribonucleoprotein complex (siRNP) which rearranges, presumably by assistance of a member of the Argonaute protein family and other cofactors such as the catalytic subunit, to the RISC (RNA-induced silencing complex). Subsequently, the mRNA-targeting RISC is formed in an ATP-dependent fashion, which presumably reflects siRNA duplex unwinding. This step could be envisioned to occur in two forms, either by removing one of the strands of the duplex from RISC or by keeping the two siRNA strands spatially separated. After target RNA cleavage, the mRNA cleavage products are released and RISC may be reactivated for another round of catalytic target RNA cleavage.

**TABLE 13.1.** Size distribution of siRNAs in various eukaryotes

| Organism | Predominant length of siRNA (nucleotides) | References |
|---|---|---|
| Plants | 21–23 | Hamilton and Baulcombe (1999); Dalmay et al. (2000a); Hutvágner et al. (2000) |
| *Neurospora crassa* | 25[a] | Catalanotto et al. (2002) |
| *Drosophila melanogaster* | 21–22 | Elbashir et al. (2001a) |
| *Caenorhabditis elegans* | 23 | Parrish et al. (2000); Ketting et al. (2001) |
| *Trypanosoma brucei* | 24–26 | Djikeng et al. (2001) |
| *Mus musculus* | 21–22 | Yang et al. (2000); Billy (2001); Paddison et al. (2002a) |

[a]It is likely that the siRNA length was overestimated because DNA size markers that migrate faster than RNA size markers were used for analysis.

Cytoplasmic extracts from mouse EC cells also process dsRNA to siRNAs (Billy et al. 2001), but the addition of ATP only modestly stimulates the dsRNA processing reaction. The function of ATP during dsRNA processing and the role of the Dicer ATP-dependent RNA helicase domain remain to be elucidated.

Naturally produced siRNA duplexes have two- to three-nucleotide 3′ overhanging ends and contain 5′ phosphate and free 3′ hydroxyl termini (Zamore et al. 2000; Elbashir et al. 2001a,b). The presence of 5′ phosphate and 3′ hydroxyl termini after dsRNA cleavage is a characteristic of all RNase-III-processing reactions (Conrad and Rauhut 2002). In mouse, Dicer is expressed in all stages of development and in a wide variety of adult mouse organs (Nicholson and Nicholson 2002), consistent with RNAi being an innate cellular defense mechanism. Moreover, Dicer has an important role in the processing of microRNAs (miRNAs) (Grishok et al. 2001; Hutvágner et al. 2001; Knight and Bass 2001), which define a new regulatory RNA gene family (for review, see Ambros 2001; Eddy 2001; Grosshans and Slack 2002; Moss 2002; Pasquinelli 2002). Dicer is localized in the cytoplasm of mouse EC cells, indicating that long dsRNA processing as well as miRNA processing reactions occur in the cytoplasm (Billy et al. 2001). Furthermore, invertebrates and vertebrates possess an additional RNase III enzyme, Drosha, which is involved in ribosomal RNA precursor processing (Filippov et al. 2000; Wu et al. 2000). Drosha does not contain a helicase or PAZ domain, but has instead an SR (serine-arginine-rich) and a proline-rich domain.

The production of siRNAs from in-vivo-expressed dsRNAs of transgenes or from synthetic dsRNA delivered into cells is the hallmark of RNAi. Formation of siRNAs has been documented for plants (Hamilton and Baulcombe 1999; Dalmay et al. 2000a; Hutvágner et al. 2000), filamentous fungi (Catalanotto et al. 2002), *C. elegans* (Parrish et al. 2000; Ketting et al. 2001), the trypanosome *Trypanosoma brucei* (Djikeng et al. 2001), and mouse embryonic stem (ES) cells (Yang et al. 2001) and EC cells (Billy et al. 2001; Yang et al. 2001; Paddison et al. 2002a). The length of siRNAs produced varies between 21 and 28 nucleotides (Table 13.1), presumably reflecting structural differences of the various Dicer orthologs. The distinct size and structure of siRNAs presumably reflect the geometric spacing between the active sites of dsRNA-bound dimers of Dicer during dsRNA processing (Blaszczyk et al. 2001; Zamore 2001b).

Cloning and sequencing of small RNAs isolated from *D. melanogaster* and *T. brucei* indicated that siRNAs are indeed produced from dsRNA of retrotransposal origin (Djikeng et al. 2001; Elbashir et al. 2001a), providing additional evidence that RNAi is important for controlling transposable elements (Jensen et al. 1999a,b; Ketting et al. 1999; Tabara et al. 1999; Wu-Scharf et al. 2000).

## siRNAs direct sequence-specific target mRNA cleavage after assembly into an endonuclease RNP complex

Analysis of RNAi in *D. melanogaster* extracts has provided compelling evidence that siRNA duplexes, after being generated by Dicer cleavage of dsRNA, are assembled into a multi-component nuclease, which guides the sequence-specific recognition of the target mRNA (Hammond et al. 2000; Yang et al. 2000; Zamore et al. 2000; Elbashir et al. 2001a). This complex is referred to as the RNA-induced silencing complex (RISC). siRNAs in *D. melanogaster* are predominantly 21 and 22 nucleotides in size (Elbashir et al. 2001a), and when paired to contain the two-nucleotide 3′ overhanging structure are most effective for formation of RISC (Elbashir et al. 2001b). Mammalian systems produce siRNAs of similar size (Yang et al. 2000; Billy et al. 2001; Paddison et al. 2002a), and siRNAs of 21- and 22-nucleotide size represent the most effective sizes for silencing genes expressed in mammalian cells (Caplen et al. 2001; Elbashir et al. 2001c, 2002).

RISC activity formed after incubation of siRNA duplexes in *D. melanogaster* embryo lysate targets homologous sense as well as antisense single-stranded RNAs (ssRNAs) for degradation (Elbashir et al. 2001a,b). The cleavage sites for both sense and antisense single-stranded target RNAs are located in the middle of the region spanned by the siRNA duplexes. The targets are cleaved precisely ten nucleotides upstream of the target position complementary to the 5′ most nucleotide of the sequence-complementary guide siRNA. Importantly, the 5′ end, and not the 3′ end, of the guide siRNA sets the ruler for target RNA cleavage (Elbashir et al. 2001a,b). Furthermore, the presence of a 5′ phosphate at the target-complementary strand of an siRNA duplex is required for siRNA function, and ATP is used to maintain the 5′ phosphates of the siRNAs (Nykänen et al. 2001). Synthetic siRNA duplexes with free 5′ hydroxyls and two-nucleotide 3′ overhangs are so readily phosphorylated in *D. melanogaster* embryo lysates that the RNAi efficiencies of 5′-phosphorylated and nonphosphorylated siRNAs are not significantly different (Elbashir et al. 2001b). However, under certain circumstances, e.g., using 22-nucleotide siRNA duplexes in *D. melanogaster* injection experiments, 5′-phosphorylated siRNAs may show slightly enhanced properties relative to 5′ hydroxyl siRNAs (Boutla et al. 2001). In gene targeting experiments in human HeLa cells, no differences in gene targeting efficiency were observed when comparing 5′ hydroxyl or 5′-phosphorylated siRNAs (Elbashir et al. 2002). Furthermore, in-vitro-transcribed siRNAs that carry 5′ triphosphates are active in human cell gene-silencing experiments (Donzé and Picard 2002; Paddison et al. 2002b). In a recently developed HeLa cell in vitro system (Martinez et al. 2002), siRNA duplexes are also rapidly 5′-phosphorylated, and the siRNAs target RNA cleavage to exactly the same position as in *D. melanogaster* lysates. Taken together, these results indicate that the mechanism of siRNA-mediated target RNA cleavage is conserved between *D. melanogaster* and mammals.

Unwinding of the siRNA duplex must occur prior to target RNA recognition. The initially formed siRNA duplex-containing ribonucleoprotein complex is referred to as siRNP (Nykänen et al. 2001). Analysis of ATP requirements revealed that the formation of RISC on siRNA duplexes requires ATP in lysates of *D. melanogaster*, but once formed, RISC can mediate robust, sequence-specific cleavage of its target in the absence of ATP (Nykänen et al. 2001). This need for ATP probably reflects the unwinding step and probably other conformational requirements. In addition, extensively purified RISC is active in the absence of exogenous nucleotide cofactors (Hammond et al. 2000). However, it is currently unknown whether both of the unwound strands of the siRNA duplex remain associated with RISC or whether RISC only contains a single-stranded siRNA. On the basis of the observations that in *C. elegans* (1) only antisense siRNAs accumulate over time after exposure to dsRNA directed against endogenous genes (Timmons and Fire 1998) and (2) only one of the two strands constituting an miRNA precursor hairpin accumulates in a stable miRNP complex (Mourelatos et al. 2002), it may be speculated that the latter is true. The simultaneous detection of sense and antisense siRNAs during RNAi and the symmetric cleavage of sense and antisense single-stranded RNA targets may be due to the symmetry of the siRNA duplexes, which may give rise to approximately equal populations of sense and antisense strand-containing RISCs (Elbashir et al. 2001a,b). Alternatively, this observation may also indicate that most siRNAs within a cell are present in the form of duplexes or siRNPs and only a small fraction in the activated form of the RISC.

The identification of the protein components of RISC, especially the catalytic subunit, is important for understanding the function of RISC. Dicer is probably not part of RISC because RISC and Dicer activity can be separated and RISC is unable to process dsRNA to siRNAs (Hammond et al. 2000, 2001b). Furthermore, when siRNAs are used to knock down Dicer in human cells, it does not affect the ability of unrelated siRNAs to target unrelated genes, but as expected compromises the ability to process longer dsRNA and miRNA-like precursors (Hutvágner et al. 2001; Paddison et al. 2002b).

One component associated with RISC from *D. melanogaster* Schneider 2 (S2) cells was identified as Argonaute2 (Hammond et al. 2001b), a member of a large family of proteins (the Argonaute or PPD family) that are characterized by the presence of a PAZ domain and a carboxy-terminal Piwi domain, both of unknown function (Cerutti et al. 2000; Schwarz and Zamore 2002). The PAZ domain is also present in Dicer, and because Dicer and Argonaute2 interact in S2 cells, PAZ may function as a protein-protein interaction motif (Hammond et al. 2001b). Possibly, the interaction between Dicer and Argonaute2 facilitates siRNA incorporation into RISC. The catalytic subunit of RISC still remains to be identified.

Members of the Argonaute gene family have been genetically identified in various organisms and some have important roles during RNAi, whereas others are important in developmental regulation. *C. elegans* contains 24 representatives of this gene family, one of which has been shown to be required for RNAi only, and *rde-1* mutant worms, although defective for RNAi, show no developmental abnormalities. *rde-1* mutants show normal dsRNA processing to siRNAs in vitro (Ketting et al. 2001) as well as in vivo when assayed 12 hours after injection of dsRNA into the syncytial germ line of adult worms (Parrish and Fire 2001). However, in *rde-1* mutant worms exposed to dsRNA by feeding them dsRNA-expressing bacteria, siRNA accumulation was not observed (Tijsterman et al. 2002). However, *N. crassa* RNAi-defective *qde-2* mutants still accumulate siRNAs (Catalanotto et al. 2002). These observations suggest a role for these proteins downstream from dsRNA processing, possibly in stabilization of siRNAs, RISC formation, and/or mRNA targeting. In *Arabidopsis*, Argonaute1 is also involved in posttranscriptional gene silencing (PTGS) and development (Bohmert et al. 1998; Fagard et al. 2000).

In *D. melanogaster*, the Argonaute family has five members and the mRNAs coding for all the Argonaute proteins are maternally deposited (Williams and Rubin 2002). During embryonic development, Argonaute1 and Argonaute2 expression is strong and fairly ubiquitous, whereas Argonaute3, Piwi, and Aubergine zygotic transcription becomes restricted to the presumptive gonad (Williams and Rubin 2002). Argonaute1 mutant flies show defects in early embryo development (Kataoka et al. 2001) and are reduced in their ability to degrade mRNAs in response to dsRNA, although formation of siRNAs was unaffected (Williams and Rubin 2002). Thus, the function of Argonaute1 may be similar to that of Argonaute2, which is associated with RISC (Hammond et al. 2001b). Piwi is required for siRNA formation during silencing of multiple transgenic copies of the *Adh* gene and has a role in some form of transcriptional silencing (Pal-Bhadra et al. 2002). Piwi is furthermore required during *D. melanogaster* development for regulating germ-line stem cell division (Cox et al. 2000). Aubergine is required for the silencing of testis-expressed

Stellate genes by paralogous Su(Ste) tandem repeats involving an RNAi-like mechanism (Aravin et al. 2001) and translational suppression during oogenesis and embryogenesis (Wilson et al. 1996; Harris and Macdonald 2001). Argonaute3 was identified through genome sequencing and remains to be characterized (Williams and Rubin 2002). Two members of the rich Argonaute family in *C. elegans*, *alg-1* and *alg-2*, are required for maturation and stability of miRNAs, which are important regulator molecules that control development (Grishok et al. 2001). The function of most of the other members of this gene family in *C. elegans* (Grishok et al. 2001) remains to be characterized.

The mammalian members of the Argonaute family are also poorly characterized. A rabbit protein from this gene family, eIF2C (Zou et al. 1998), has been implicated in translation initiation. eIF2C was isolated as a major component of a cytoplasmic protein fraction that stimulates the formation of a ternary complex between Met-tRNA, GTP, and the eukaryotic peptide chain initiation factor 2 (eIF2) (Roy et al. 1988; Zou et al. 1998). The human ortholog, eIF2C2, was recently shown to be complexed with Gemin3 (a DEAD-box putative RNA helicase), Gemin4, and mature miRNAs (Mourelatos et al. 2002). The function of this 15S ribonucleoprotein complex (miRNP) is unknown. On the basis of the role of *alg-1/alg-2* in miRNA maturation and stability in *C. elegans* (Grishok et al. 2001), and the presence of a putative RNA helicase in the 15S complex, these miRNPs are involved either in processing miRNAs from longer precursor RNAs and/or in downstream events such as target RNA recognition (Mourelatos et al. 2002). Another member of this family, the human paralog eIF2C1, has been cloned and genetically characterized (Koesters et al. 1999). eIF2C1 is ubiquitously expressed but its function is unknown. Two other members of the mammalian Argonaute family were defined as Miwi (mouse homolog of Piwi), and its human ortholog Hiwi, as well as mouse Mili (Kuramochi-Miyagawa et al. 2001; Sharma et al. 2001). Miwi and Mili were both found in germ cells of adult testis, suggesting that these proteins may function in spermatogenesis (Kuramochi-Miyagawa et al. 2001). Hiwi, which is also expressed in adult testis, was also found expressed in human CD34(+) hematopoietic progenitor cells but not in more differentiated cell populations, again suggesting a role in development of progenitor cells (Sharma et al. 2001). The molecular function and interacting partners of these proteins are currently unknown.

### Differences between mammalian RNAi and *C. elegans* or plant RNAi

Plants and worms show systemic silencing, indicating the spread of an amplifiable sequence-specific signal throughout the organisms. The molecular nature of this signal remains to be identified. The signal is most likely RNA in the form of dsRNA or antisense RNA directing new sequence-specific dsRNA synthesis. In *C. elegans*, a putative transmembrane protein, SID-1, was shown to be important for systemic RNAi (Winston et al. 2002). The *sid-1* gene is required to spread gene-silencing information between tissues but not to initiate or maintain an RNAi response. It is possible that SID-1 is involved in endocytosis of the systemic RNAi signal, perhaps functioning as a receptor or as a channel. Consistent with the apparent lack of systemic RNAi in *D. melanogaster* (Kennerdell and Carthew 2000), *sid-1* homologs are absent from the fly genome. The strong similarity to predicted human and mouse proteins, however, suggests the possibility that RNAi could have a systemic component in mammals (Winston et al. 2002).

Screens for genes required for gene silencing in plants, fungi, and worms have identified a family of proteins whose sequences suggest they are RNA-dependent RNA polymerases (RdRPs) (Cogoni and Macino 1999; Dalmay et al. 2000b; Mourrain et al. 2000; Sijen et al. 2001) (see Chapter 9). The discovery of RdRPs in RNAi and PTGS provides a possible explanation for the remarkable efficacy of dsRNA in gene silencing in these organisms. New dsRNA could be synthesized by RdRPs and thus amplify the silencing process. In *D. melanogaster* and mammals, RdRP genes have not been identified by database analysis.

In *C. elegans*, systemic silencing and signal amplification may also cause transitive RNAi, which is a spreading of silencing outside of the locus targeted by an initiator dsRNA or dsRNA-expression construct (Sijen et al. 2001) (see also Chapter 9). Transitive RNAi is accompanied by the formation of secondary siRNAs, which derive from newly synthesized dsRNA presumably due to RdRP activity. Although this appears to have important implications for RNAi-based analysis of gene function, because silencing may spread between genes that share homologous sequences, phenotypic analysis of a large set of silenced genes in *C. elegans* suggests that transitive RNAi between naturally occurring homologous gene sequences is probably of no major concern (Fraser et al. 2000; Gönczy et al. 2000). It was also suggested that siRNAs might prime novel dsRNA synthesis (Lipardi et al. 2001; Sijen et al. 2001). However, it should be pointed out that siRNAs, in comparison to longer dsRNAs, are extremely poor initiators of gene silencing in *C. elegans* (Parrish et al. 2000; Tijsterman et al. 2002).

Biochemical evidence for RdRP acivity in *D. melanogaster* was recently reported (Lipardi et al. 2001), although classical RdRP genes presumably encoding such activity appear to be lacking from the *D. melanogaster* genome. Despite the postulated target-RNA-dependent dsRNA synthesis, which could potentially lead to amplification of the silencing signal (Lipardi et al. 2001), biochemical evidence for spreading of silencing outside of regions targeted by dsRNAs has not been observed in similar biochemical systems (Zamore et al. 2000; Elbashir et al. 2001a,b; Zamore 2001a). Our attempts to detect polymerization products upon incubation of internally radiolabeled siRNAs with target RNA and nucleoside triphosphates in *D. melanogaster* embryo lysate or HeLa cell lysate, under conditions where dsRNA or siRNAs mediated target RNA degradation, were never successful (Martinez et al. 2002). Additional evidence against propagation of gene silencing in mammalian cells is the ability of siRNAs to specifically silence various isoforms expressed at the same time in the same cell (Kisielow et al. 2002; J. Harborth, unpubl.). This suggests that gene silencing in *D. melanogaster* and mammals is due to siRNA-mediated degradation of target mRNA by RISC, which itself may well catalyze multiple turnovers.

In some instances, PTGS is also linked to transcriptional gene silencing (TGS) (for reviews, see Wassenegger 2000; Bender 2001; Matzke et al. 2001b; Pal-Bhadra et al. 2002). In plants, TGS causes chromatin modifications of the silenced locus, e.g., increased DNA methylation, and it requires promoter sequences to be targeted by PTGS. RNA-directed DNA methylation is most beautifully demonstrated in an experiment where transgene copies of a viral gene present in the nucleus only become methylated upon infection of the plant by the homologous virus, which has a dsRNA genome and which does not enter the nucleus (Jones et al. 1999; Pelissier and Wassenegger 2000). Mutations in *Arabidopsis* selected for reduced DNA methylation, *ddm1*, an SWI2/SNF chromatin component, and *met1*, the major DNA methyltransferase, relieve TGS (Jeddeloh et al. 1998; Mittelsten-Scheid et al. 1998), and in some cases, also include a stochastic reversal of PTGS (Morel et al. 2000).

Additional links between PTGS and TGS were observed. Transgene arrays in the *C. elegans* germ line are desilenced (Tabara et al. 1999) and appear less condensed in mutant backgrounds for some genes required for RNAi (Dernburg et al. 2000). A similar situation was encountered in *D. melanogaster*, where mutations of the *piwi* gene affected both posttranscriptional as well as transcriptional modes of gene silencing (Pal-Bhadra et al. 2002). Such cases of transgene-induced gene silencing appear to require Polycomb-Group proteins, which are complexes known to be involved in the maintenance of repressive chromatin structure (Pal-Bhadra et al. 1997; Kelly and Fire 1998). Whether mammalian RNA silencing systems also trigger methylation and chromatin changes remains to be resolved.

## Analysis of Gene Function in Mammalian Cells Using RNAi

Mammalian gene function has been determined traditionally by methods such as disruption of murine genes, the introduction of transgenes, the molecular characterization of human hereditary diseases, and targeting of genes by antisense or ribozyme techniques. In addition, microinjection of specific antibodies into cultured cells or binding of antibodies to cell-surface-exposed receptors may provide information on the function of the targeted protein.

It has been difficult to detect potent and specific RNAi in commonly used mammalian cell culture systems applying long dsRNA varying in size between 38 and 1662 bp (Caplen et al. 2000; Ui-Tei et al. 2000; Yang et al. 2001; Paddison et al. 2002a). On the one hand, the apparent lack of RNAi in mammalian cell culture was unexpected, because RNAi exists in mouse oocytes and early embryos (Svoboda et al. 2000; Wianny and Zernicka-Goetz 2000), and RNAi-related transgene-mediated cosuppression was also observed in cultured Rat-1 fibroblasts (Bahramian and Zarbl 1999). But, on the other hand, it is known that dsRNA in the cytoplasm of mammalian cells can trigger profound physiological reactions that lead to the induction of interferon synthesis (Lengyel 1987; Stark et al. 1998; Barber 2001). In the interferon response, dsRNA greater than 30 bp binds and activates the protein kinase PKR and 2′,5′-oligoadenylate synthetase (2′,5′-AS) (Minks et al. 1979; Manche et al. 1992). Activated PKR stalls translation by phosphorylation of the translation initiation factors eIF2$\alpha$, and activated 2′,5′-AS causes mRNA degradation by 2′,5′-oligoadenylate-activated RNase L. These responses are intrinsically sequence-nonspecific with respect to the inducing dsRNA.

In an attempt to bypass these sequence-nonspecific effects, three major strategies were employed. In the first case, cell lines were identified that preserved the characteristics of early embryonic stages and have not yet established their interferon system. In the second case, siRNAs were used that are short enough or have a specific structure to escape detection of the interferon system and do not activate PKR or 2′,5′-AS. In the third case, short miRNA-like stem-loop structures were used, which require processing by Dicer, but were short enough to go undetected by the interferon system.

### RNAi in embryonic stem cells and embryonic carcinoma cells

Several hundred base-pair-long dsRNAs, transfected or electroporated into undifferentiated mouse ES cells or mouse EC cell lines F9 and P19, induce specific gene silencing without any apparent sequence-nonspecific side-effects (Billy et al. 2001; Yang et al. 2001; Paddison et al. 2002a). Silencing by introduced dsRNA is generally transient, and mouse ES cells recover from the specific knockdown about 5 days after transfection of the dsRNA, presumably due to dilution of the dsRNA during cycles of cell division (Yang et al. 2001). Although most of the reported experiments were focused on suppression of green fluorescent protein (GFP) reporter genes, one of these studies also demonstrated specific silencing of two of the endogenous subunits of cell surface receptor proteins, integrins $\alpha$3 and $\beta$1 in F9 cells (Billy et al. 2001). These proteins turn over rapidly, and their absence at the cell surface was monitored by simple adhesion assays; the reduction of integrin mRNA or protein levels varied between 60% and 90%.

Long dsRNA can also be expressed from transfected plasmid DNA encoding an inverted repeat of a segment of the targeted mRNA. Two different dsRNA expression strategies were applied. In one study, the inverted repeat was under the control of a T7 promoter. The linearized dsRNA-encoding plasmid was cotransfected into ES cells together with a plasmid encoding T7 RNA polymerase (Yang et al. 2001). In the other case, hairpin dsRNA synthesis was driven by the strong cytomegalovirus (CMV) polymerase II promoter (Paddison et al. 2002a). Stably transformed cells carrying the G418-selectable dsRNA expression construct were expanded into clonal cell lines, some of which were able to specifically silence the transfected homologous reporter gene.

Together, these examples illustrate the ability to transiently or stably silence genes expressed in ES or EC cells, which may be useful to study aspects of cell biology or cell differentiation in undifferentiated cells.

## Analysis of gene function in somatic mammalian cells using siRNAs

As an alternative to reverse genetic approaches with long dsRNAs, siRNAs can be used that are also extremely potent elicitors of gene silencing (Caplen et al. 2001; Elbashir et al. 2001c). In contrast to long dsRNAs, siRNAs do not activate the cellular enzymes PKR and 2′,5′-AS of the interferon system established in most transformed somatic mammalian laboratory cell lines. Standard tissue culture cell lines provide starting points for mammalian functional screens because siRNAs can be effectively delivered by classical gene transfer methodologies such as electroporation or cationic liposome-mediated transfection. Transfection efficiencies greater than 90% are commonly achieved in standard laboratory cell lines provided transfection reagents are used that were specially designed for siRNA or antisense oligonucleotide applications (Elbashir et al. 2002). For small-scale applications, microinjection of siRNAs may represent an alternative. Technical problems due to low transfection efficiencies may also be partially overcome by including cell-sorting protocols after cotransfecting siRNAs together with sorting markers such as GFP expression plasmids. Alternatively, siRNAs targeting cell surface marker proteins may be cotransfected, and loss of the cotargeted cell surface marker may be used to gate knockdown cell populations by cell sorting.

The design of siRNA duplexes that interfere with the expression of a specific gene requires accurate knowledge of at least a 20-nucleotide segment of its encoded mRNA (Figure 13.2) (Elbashir et al. 2001b). Intronic sequences contained in pre-mRNAs are best neglected for targeting, because incompletely spliced mRNAs are normally retained in the nucleus and RNAi is believed to occur predominantly, if not exclusively, in the cytoplasm (Montgomery et al. 1998). Also in mammalian cells, mRNA isoforms can be individually silenced, providing further evidence that siRNA-mediated mRNA degradation is a cytoplasmic event (Kisielow et al. 2002). Sequence information about mature mRNAs may be extracted from expressed sequence tag (EST) databases or can be predicted from genomic sequences using gene prediction programs. However, sequencing errors in single-pass EST sequence data or gene predictions should be kept in mind.

siRNA duplexes composed of 21-nucleotide sense and 21-nucleotide antisense strands, paired in a manner to have a two-nucleotide 3′ overhang, are the most efficient triggers of sequence-specific mRNA degradation in tissue culture systems (Elbashir et al. 2002). The target RNA cleavage reaction guided by siRNAs is highly sequence-specific (Elbashir et al. 2001b). However, not all positions of an siRNA contribute equally to target recognition. Mismatches in the center of the siRNA duplex are most critical and essentially abolish target RNA cleavage (Elbashir et al. 2001c; Brummelkamp et al. 2002; Holen et al. 2002). It should be noted that the effect of the mismatches on the specificity of target RNA cleavage is dependent not only on the position of the mismatch relative to the target RNA cleavage site, but probably also on steric or thermodynamic effects that are dependent on the nature of the mismatch. In contrast to mismatches in the paired

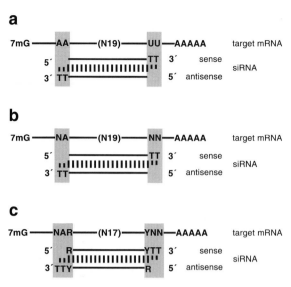

**FIGURE 13.2.** Selection of siRNA duplexes for mRNA targeting. (*a*) Design of siRNA duplexes for target mRNAs that contain the sequence AA(N19)UU. (*b*) Design of siRNA duplexes in the absence of AA(N19)UU target sequences. As long as one adenosine is present in the targeted region, siRNA duplexes with 3′-TT overhangs can be used without effect on the specificity of target recognition or RNAi efficiency. (*c*) Design of siRNA duplexes that could later be expressed by cloning the sequence into H1 or U6 polymerase III expression cassettes. R and Y indicate purine and pyrimidine nucleotides, respectively; N represents any of the four ribonucleotides.

region of siRNAs, the 3′ nucleotide of the siRNA strand (position 21) that is complementary to the single-stranded target RNA does not contribute to the specificity of target recognition (Elbashir et al. 2001b; Holen et al. 2002). As may be expected, the sequence of the unpaired two-nucleotide 3′ overhang of the siRNA strand with the same polarity as the target RNA is not critical for target RNA cleavage because only the antisense siRNA strand guides target recognition (Elbashir et al. 2001b; Holen et al. 2002). Thus, only the penultimate position of the antisense siRNA (position 20) needs to match the targeted sense mRNA.

Selection of the targeted region is currently a trial-and-error process, but with a likelihood of 80–90% success given a large enough random selection of target genes (Harborth et al. 2001). In every single case, however, the half-life of the targeted gene product, its abundance, or the regulation of its expression must be considered. For example, in an attempt to knock down the strongly expressed and stable intermediate filament protein vimentin, only two out of four randomly selected siRNAs were effective (Harborth et al. 2001). Similar difficulties in finding amenable target sites within the human coagulation trigger tissue factor (TF) mRNA were reported (Holen et al. 2002). Interestingly, there was no apparent correlation between siRNA efficacy and computer-predicted targeted mRNA secondary structure.

Our research group selects target regions such that siRNA sequences may contain uridine residues in the two-nucleotide overhangs (Figure 13.2). Uridine residues in the two-nucleotide 3′ overhang can be replaced by 2′-deoxythymidine without loss of activity, which significantly reduces the cost of RNA synthesis and may also enhance nuclease resistance of siRNA duplexes when applied to mammalian cells (Elbashir et al. 2001c). Another rationale for designing siRNA duplexes with symmetric TT overhangs is to ensure that the sequence-specific endonuclease complex (RISC) is formed with an

**TABLE 13.2.** Human and animal cell lines in which siRNA triggers silencing

| Cell line | Tissue origin | Reference |
|---|---|---|
| A-431 | human epidermoid carcinoma | Elbashir et al. (2002) |
| A549 | human lung carcinoma | Bitko and Barik (2001) |
| BV173 | human B-precursor leukemia | Tuschl and Borkhardt (2002) |
| C-33A | human papillomavirus-negative cervical carcinoma | Sui et al. (2002) |
| CA46 | human Burkitt's lymphoma | Tuschl and Borkhardt (2002) |
| Caco2 | human colon epithelial cells | Moskalenko et al. (2002) |
| CHO | Chinese hamster ovary | Elbashir et al. (2002) |
| COS-7 | African green monkey kidney | Elbashir et al. (2001c) |
| F5 | rat fibroblast | Harborth et al. (2001) |
| H1299 | human nonsmall cell lung carcinoma | Sui et al. (2002) |
| HaCaT | human keratinocyte cell | Holen et al. (2002) |
| HEK 293 | human embryonic kidney | Elbashir et al. (2001c) |
| HeLa | human papillomavirus-positive cervical carcinoma | Elbashir et al. (2001c) |
| Hep3B | human hepatocellular carcinoma | Bakker et al. (2002) |
| HUVEC | human umbilical vein endothelial cells | Ancellin et al. (2001) |
| IMR-90 | human diploid fibroblast | Paddison et al. (2002b) |
| K562 | human chronic myelogenous leukemia, blast crisis | Tuschl and Borkhardt (2002) |
| Karpas 299 | human T-cell lymphoma | Tuschl and Borkhardt (2002) |
| MCF-7 | human breast cancer | Hirai and Wang (2002) |
| MDA-MB-468 | human breast cancer | Hirai and Wang (2002) |
| MV-411 | human acute monocytic leukemia | Tuschl and Borkhardt (2002) |
| NIH-3T3 | mouse fibroblast | Elbashir et al. (2001c) |
| P19 | mouse embryonic carcinoma | Yu et al. (2002) |
| SD1 | human acute lymphoblastic leukemia | Tuschl and Borkhardt (2002) |
| SKBR3 | human breast cancer | Elbashir et al. (2002) |
| U2OS | human osteogenic sarcoma cell | Martins et al. (2002) |

approximately equal ratio of sense to antisense target RNA-cleaving complexes (Elbashir et al. 2001a,b). This is a precaution, because we do not understand the rules that govern sense versus antisense targeting RISC formation. Other sequences within the two-nucleotide overhangs are also functional and may be preferred if a specific site is targeted, for example, within the mRNA of a fusion gene or a polymorphic or mutated allele.

Analysis of gene function in cultured somatic mammalian cells using siRNAs is now being described in a rapidly growing number of independent studies (see Table 13.2). Cells that show dramatically reduced target protein levels are referred to as knockdown cells, in contrast to knockout cells that are fully deficient for the genetic locus encoding a specific protein. The first broad application of siRNAs for the analysis of cytoskeletal proteins showed that several of these proteins were essential for cell growth (Harborth et al. 2001). But even when nonessential genes were targeted, specific secondary phenotypes were observed in cultured cells that were identical to phenotypes previously observed in mouse gene knockout cells. Furthermore, using siRNAs directed against mitotic proteins, it was possible to reproduce cellular phenotypes that recapitulate the phenotype induced by small-molecule inhibitors specific to the protein encoded by the targeted mRNA (Harborth et al. 2001). These early examples illustrated the value of siRNAs for analysis of mammalian gene function. Subsequently, knockdown of proteins with siRNAs was used for studying

- DNA damage response and cell cycle control (Cortez et al. 2001; Brummelkamp et al. 2002; Mailand et al. 2002; Porter et al. 2002; Stucke et al. 2002; Zou et al. 2002),

- general cell metabolism (Ancellin et al. 2001; Bai et al. 2001),
- signaling (Habas et al. 2001; Li et al. 2001; Martins et al. 2002),
- the cytoskeleton and its rearrangement during mitosis (Du et al. 2001; Harborth et al. 2001),
- membrane trafficking (Short et al. 2001; Moskalenko et al. 2002),
- transcription (Ostendorff et al. 2002), and
- DNA methylation (Bakker et al. 2002).

siRNAs were also used to assess the role of proteins in host-virus interactions (Bitko and Barik 2001; Garrus et al. 2001) or during other disease-causing events such as the expression of polyglutamine in neurodegenerative disorders (Caplen et al. 2002). The breadth and depth of these applications emphasize the key role that siRNAs will have in the functional characterization of gene products and for defining their roles in basic cellular events and disease-related processes in the postgenomic era.

## Analysis of gene function in somatic mammalian cells expressing siRNAs or short hairpin RNAs

Until recently, siRNAs for gene targeting experiments have only been introduced into cells via classic gene transfer methods, such as liposome-mediated transfection, electroporation, or microinjection, that require chemical or enzymatic synthesis of siRNAs. Protein knockdowns mediated by exogenous siRNAs are transient because the targeted protein levels of siRNA-treated cells recover, typically between 5 and 7 days after siRNA transfection, i.e., after 7–10 rounds of cell division (Elbashir et al. 2002; Holen et al. 2002; Kisielow et al. 2002). Alternatively, small RNA molecules may also be expressed in the cell. This is possible by cloning the siRNA templates into RNA polymerase III (pol III) transcription units, which are based on the sequences of the natural transcription units of the small nuclear RNA U6 or the human RNase P RNA H1. Two approaches are available for expressing siRNAs: (1) The sense and antisense strands constituting the siRNA duplex are transcribed from individual promoters (Figure 13.3a) (Lee et al. 2002; Miyagishi and Taira 2002; Yu et al. 2002) or (2) siRNAs are expressed as fold-back stem-loop structures that give rise to siRNAs after intracellular processing by Dicer (Figure 13.3b) (Brummelkamp et al. 2002; Paddison et al. 2002b; Paul et al. 2002; Sui et al. 2002; Yu et al. 2002). The endogenous expression of siRNAs from introduced DNA templates overcomes some limitations of exogenous siRNA delivery, in particular the transient loss of phenotype.

U6 and H1 RNA promoters are members of the type III pol III promoters (Medina and Joshi 1999; Paule and White 2000). These promoters are unusual because all promoter elements, with the exception of the first transcribed nucleotide (+1 position), are located upstream of the transcribed region so that almost any inserted sequence shorter than 400 nucleotides can be transcribed. These promoters are therefore ideally suited for expression of siRNAs or approximately 50-nucleotide siRNA stem-loop precursors. The U6 promoter and the H1 promoter are different in size, but they contain the same conserved sequence elements or protein-binding sites (Myslinski et al. 2001). The +1 nucleotide of the U6-like promoters is always guanosine, and always adenosine for H1. Interestingly, changing the +1 adenosine to U, C, or G within H1-expressed stem-loop sequences did not seem to affect gene silencing, suggesting that H1 promoters may be more flexible than U6 promoters for +1 sequence changes or may be able to initiate transcription at the

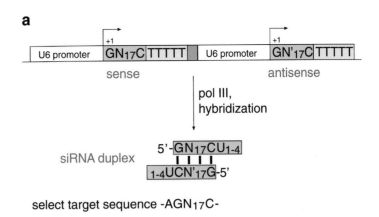

**a**

**b**

**FIGURE 13.3.** Endogenous expression of siRNAs. (*a*) Expression cassette for sense and antisense siRNAs using the U6 snRNA promoter (Lee et al. 2002; Miyagishi and Taira 2002). (*White box*) The 250-bp U6 snRNA promoter; (*blue box*) the pol III terminator signal composed of a run of thymidines; (*gray box*) the spacer between the sense and antisense expression element; (*red box*) siRNA elements. The target site preferably selected for optimal vector design is indicated at the bottom. (*b*) H1 RNA-based pol III cassette for expressing hairpin RNAs that are subsequently processed to siRNAs (Brummelkamp et al. 2002). The H1 RNA pol III promoter is only 100 bp in size, but it contains all the essential sequence motifs present in the U6 snRNA promoter (Myslinski et al. 2001). Hairpin RNAs with gene-silencing properties were also obtained by using a U6 promoter (Paddison et al. 2002b; Paul et al. 2002; Sui et al. 2002; Yu et al. 2002). In this case, transcript synthesis was initiated with a +1 guanosine, and the 3´ end of the sense strand was joined by short oligonucleotide loops with the antisense strand.

first downstream purine nucleotide encoded by the template DNA (Brummelkamp et al. 2002). RNA transcription is terminated when pol III encounters a run of four or five thymidines by incorporation of only some of the encoded uridines (Myslinski et al. 2001).

DNA constructs encoding 19-bp stem-loop sequences with 3´-overhanging uridines can silence target genes as effectively as synthetic siRNAs (Brummelkamp et al. 2002;

**TABLE 13.3.** Strategies employed for endogenous expression of siRNA or short hairpin RNAs

| Promoter | Preferred targeting structure[a] | References |
|---|---|---|
| H1 pol III | 19-bp/9-nucleotide (equivalent to 21-bp/5-nucleotide) stem-loop | Brummelkamp et al. (2002) |
| U6 pol III | 21-nucleotide sense and 21-nucleotide antisense siRNA transcribed by separate promoters | Lee et al. (2002); Miyagishi and Taira (2002) |
| U6 pol III | 4-nucleotide/19-bp stem-loop | Paul et al. (2002) |
| U6 pol III | 29-bp/8-nucleotide stem-loop | Paddison et al. (2002b) |
| U6 pol III | 21-bp/6-nucleotide (or 23-bp/2-nucleotide) stem-loop | Sui et al. (2002) |
| U6 pol III | 19-bp/2-nucleotide stem-loop | Yu et al. (2002) |

[a]The position of the loop and the length of the target-sequence-containing stem are indicated. If the length of the loop is given prior to the length of the stem, it indicates that the loop connects the 5′ end of the sense strand to the 3′ end of the antisense (targeting) strand; if the length of the stem is followed by the loop size, the 5′ end of the antisense strand is connected to the 3′ end of the sense strand.

Paul et al. 2002; Yu et al. 2002), but blunt-ended duplexes with up to 29 bp are also able to mediate RNAi in cultured cells (Paddison et al. 2002b). Intracellular processing of plasmid-encoded hairpin RNAs requires Dicer RNase III because Dicer knockdown cells do not support hairpin-mediated target gene silencing (Paddison et al. 2002b). The size, orientation, and sequence of the loop affect the efficiency of gene silencing in many cases. However, the precise processing rules for short hairpin RNAs are not fully understood, so that variations in targeting efficiency may in part be due to variation of the target cleavage site (see Table 13.3). The efficiency of target RNA cleavage, even for single-nucleotide displacements of the siRNA relative to the target, is quite variable (Elbashir et al. 2001b). Furthermore, before precise processing rules for hairpin processing have been elucidated, it should be cautioned that loop spacer elements, which are typically noncognate to the target, as well as noncognate base pairs adjacent to the central paired regions, may affect specificity or impair efficacy of the produced siRNAs.

Using siRNA expression systems, it is possible to extend the periods of persistent suppression or stable loss-of-function phenotype by producing stable cell lines propagating the siRNA expression cassettes. Miyagishi and Taira (2002) suppressed β-catenin, a protein involved in cadherin-mediated cell-cell adhesion, for more than 1 week. The β-catenin-targeting siRNA strands were expressed from a plasmid containing the Epstein-Barr virus (EBV) DNA replication origin, and the plasmid was propagated in cells stably expressing EBV nuclear antigen 1 (EBNA-1). Two other groups produced cells that stably suppressed p53 protein, an important protein involved in the cellular response to ionizing irradiation DNA damage (Brummelkamp et al. 2002; Paddison et al. 2002b). Silencing of p53 was observed for more than 2 months in antibiotic-selected, stably transfected cell clones, also indicating that long-term expression of siRNAs is nontoxic to cells (Brummelkamp et al. 2002).

In summary, stable knockdown cells of nonessential proteins are of great value for studying inducible processes such as UV irradiation damage response, host-pathogen interactions, or cell differentiation and will enable synthetic lethality screens in human cells. The establishment of clonal cell lines with inducible siRNA expression systems should add additional value to the siRNA repertoire, because it would be possible to synchronize the knockdown of entire cell populations, and because essential genes may be targeted. Strategies for the regulated expression of small RNAs have already been described (Meissner et al. 2001; Yarovoi and Pederson 2001; Miyagishi and Taira 2002) and provide a starting point for such developments.

To decide whether a transient or more long-term silencing strategy should be chosen, the following should be considered. Transfection of plasmid DNA relative to synthetic siRNAs may appear advantageous in view of the danger of RNase contamination or the current costs of chemically synthesized siRNAs or siRNA transcription kits. For practical applications, however, the additional time involved in preparing and amplifying siRNA expression vectors and the transfection efficiency of plasmids relative to siRNAs must also be considered. Furthermore, targeting of essential genes causes arrest in cell growth or cell death within 1–3 days after delivery of siRNAs, thus making long-term silencing unnecessary.

Considering all the pros and cons of expressed versus synthetic siRNAs, it is probably most effective to initiate the search for highly effective siRNAs with synthetic, ready-to-use duplex RNAs of defined sequence and length, and select the synthetic sequences such that they are already compatible with the sequence requirements for expression within U6 or H1 RNA expression cassettes. Such constraints represent (1) the +1 position of U6 snRNA for a guanosine (Paule and White 2000; Paddison et al. 2002b) and probably the +1 position for adenosine in H1 RNA and (2) the 3´-terminal uridines encoded by the oligothymidine pol III terminator signal sequence (Paule and White 2000).

In summary, the possibility for stable expression of siRNAs has paved the way to new gene therapy applications such as treatment of persistent viral infections. Incorporation of siRNA expression cassettes into (retro)viral vectors may allow targeting of primary cells previously resistant or refractory to siRNA or plasmid DNA transfection. Because of the automation developed for high-throughput sequence analysis of the various genomes, the DNA-based methodology may also provide a cost-effective alternative for automated genome-wide loss-of-function phenotypic analysis, especially when combined with miniaturized array-based phenotypic screens (Ziauddin and Sabatini 2001).

## SUMMARY

RNAi represents an evolutionarily conserved cellular defense mechanism for controlling the expression of alien genes in almost all eukaryotes including humans. RNAi is triggered by dsRNA and causes sequence-specific mRNA degradation of single-stranded target RNAs homologous in response to dsRNA. The mediators of mRNA degradation are siRNAs, which are produced from long dsRNA by enzymatic cleavage in the cell. siRNAs are approximately 21 nucleotides in length and have a base-paired structure with two-nucleotide 3' overhangs. Although they were discovered only recently, siRNAs have already revolutionized functional analysis of mammalian gene function and are rapidly moving toward genome-wide systematic analysis of gene function in cultured cells. siRNAs may soon become a valuable tool for target validation beyond classical tissue culture cell lines. Similar to humanized monoclonal antibody strategies as therapeutic platform technology, siRNAs may provide an interesting solution for gene-specific drug development, especially before the availability of highly specific small-molecule inhibitors.

A selection of protocols, modified from Elbashir et al. (2002), is presented for targeting endogenous genes in mammalian somatic cells.

# TECHNIQUES

## PROTOCOL 1: SELECTION OF siRNA SEQUENCES

To target a specific mRNA for degradation, a portion of the mRNA target sequence must be known and a segment of the target mRNA must be chosen that will be used for targeting by the cognate siRNA duplex. The siRNA selection process has recently been automated by Bingbing Yuan and Fran Lewitter at the Whitehead Institute (Cambridge, Massachusetts) and a Web Site (http://jura.wi.mit.edu/bioc/siRNA/home.php) has been made publicly available. This software allows the user to define sequence motifs and G/C content, to search siRNAs against the human and mouse genome databases to prevent mistargeting, and to exclude single-nucleotide polymorphic sites.

## Procedure

1. Select the target region from the open reading frame of a desired cDNA sequence, preferably 50–100 nucleotides downstream from the start codon.

   It is conceivable that 5´UTRs or 3´UTRs or regions close to the start codon are less effectively targeted by siRNAs, as these may be richer in regulatory protein-binding sites. UTR-binding proteins and/or translation initiation complexes could interfere with binding of RISC to the target RNA.

   If the intent, however, is to rescue a knockdown phenotype by reintroduction of a plasmid coding for a mutant or tagged form of the targeted gene, it may be preferable to target regions in the UTRs. Preparation of rescue constructs by deletion of terminal untranslated sequences is easier than the introduction of silent mutations within the targeted region of a coding segment. In a recent survey of 3´UTR-localized targeting sites of more than 40 essential genes, we found that 3´UTRs are in fact as effectively targeted as coding regions (M. Hossbach, S. Elbashir, T. Tuschl, unpubl.).

2. Search for sequences 5´-AA(N19)UU, where N is any nucleotide, in the mRNA sequence, and ideally choose those with ~50% G/C content (see Figure 13.2a). Nevertheless, 32–79% G/C content has also worked well in our hands. Highly G-rich sequences should be avoided because they tend to form G-quartet structures.

   If there are no 5´-AA(N19)TT motifs present in the target mRNA, search for 5´-AA(N21) or 5´-NA(N21) sequences (Figure 13.2b). Independent of the selection procedure described in Figure 13.2, synthesize the sense siRNA as 5´-(N19)TT, and the sequence of the antisense siRNA as 5´-(N´19)TT, where N´19 denotes the reverse complement sequence of N19. N19 and N´19 indicate ribonucleotides, and T indicates 2´-deoxythymidine.

   If the intent, however, is to also express a chemically synthesized siRNA using pol-III-based expression vectors, select the targeted sequence as 5´-NAR(N17)YNN (Figure 13.2c), where R and Y indicate purine and pyrimidine nucleotides, respectively. The sense siRNA is then

280

synthesized accordingly as 5´-R(N17)YTT and the sequence of the antisense siRNA as 5´-R´(N´17)Y´TT.

3. Perform a BLAST search (www.ncbi.nlm.nih.gov/BLAST) using the selected siRNA sequences as the input against EST libraries or mRNA sequences of the respective organism to ensure that only a single gene is targeted.

4. (*Although optional, this step is recommended*) Synthesize several siRNA duplexes to control for the specificity of the knockdown experiments.

> Those siRNA duplexes that are effective for silencing should produce exactly the same phenotype. Furthermore, a nonspecific siRNA duplex may be needed as a control. It is possible to reverse the sequence of an effective siRNA duplex or to use a siRNA duplex that targets a gene absent from the selected model organism, e.g., GFP or luciferase. We have used an siRNA duplex targeting firefly luciferase as a control for targeting endogenous genes in mammalian cells because the firefly luciferase gene was not present in the targeted cells (Elbashir et al. 2002c).

**Troubleshooting**

If the siRNA does not work, first verify that the target sequence and the cell line used are derived from the same organism. According to a recent study, there is a high probability of using the wrong cell line (Masters et al. 2001). In addition, make sure that the mRNA sequence used for selection of the siRNA duplexes is reliable; it could contain sequencing errors, mutations (e.g., in cancer cells), or polymorphisms.

## PROTOCOL 2: ANNEALING siRNAS TO PRODUCE siRNA DUPLEXES

Sense and antisense siRNA strands are annealed to form a duplex prior to transfecting them into cultured cells.

## Procedure

**▼ CAUTION**

*See Appendix for appropriate handling of materials marked with <!>.*

### MATERIALS

#### REAGENTS

2x Annealing buffer

200 mM potassium acetate
4 mM magnesium acetate <!>
60 mM HEPES-KOH (pH 7.4) <!>

Ethidium bromide solution (1% w/v) (aqueous) <!>

NuSieve GTG agarose (BMA, Rockland, Maine; www.bmaproducts.com)

Sense and antisense siRNA in $H_2O$ at a concentration >80 μM

Sucrose gel-loading buffer (Sambrook et al. 2001)

5x TBE buffer

450 mM Tris base <!>
450 mM boric acid <!>
10 mM $Na_2EDTA$

#### EQUIPMENT

Gel electrophoresis equipment

UV light source

Water bath preset to 90ºC

1. Prepare a 20 μM siRNA duplex solution by combining:

   70 μl of 2x annealing buffer

   sense siRNA to 20 μM final concentration

   antisense siRNA to 20 μM final concentration

   sterile $H_2O$ to a final volume of 140 μl.

2. Incubate the reaction for 1 minute at 90°C, followed by 1 hour at 37°C.

   > Store unused siRNA duplex solution frozen at –20°C. The siRNA duplex solution can be frozen and thawed many times and does not require any further heat shock treatments. Always keep RNA solutions on ice as much as possible to reduce the rate of RNA hydrolysis.

3. To assess the completeness of the annealing reaction:

   a. Separately load 1 μl of 20 μM sense and antisense siRNAs and 0.5 μl of 20 μM siRNA duplex onto a 4% NuSieve GTG agarose gel. When loading the samples, it is helpful to first dilute the samples with a few microliters of 0.5x TBE buffer and sucrose-loading buffer.

   b. Run the gel in 0.5x TBE buffer at 80 V for 1 hour.

      > NuSieve agarose is a low-melting-temperature agarose, which may melt if electrophoresis is performed with excessive electric current.

   c. Detect the RNA bands under UV light after ethidium bromide staining. Preferably, add the ethidium bromide to the 4% gel/0.5x TBE solution at a concentration of 0.4 mg/liter (4 μl of 1% ethidium bromide solution per 100 ml of gel solution) prior to casting the gel.

## PROTOCOL 3: CELL CULTURE AND PREPARATION OF CELLS IN 24-WELL PLATES

Transfection of cultured cells with siRNAs and downstream analysis of the knockdown cells are best performed in multiwell tissue culture plates.

## Procedure

**▼ CAUTION**
*See Appendix for appropriate handling of materials marked with <!>.*

---

### MATERIALS

### REAGENTS

Dulbecco's modified Eagle medium (DMEM) (41966-029, Life Technologies; www.lifetech.com)

Fetal bovine serum (FBS) (10500-064, Life Technologies)

Mammalian cell lines (e.g., HeLa S3, HeLa SS6, COS-7, NIH-3T3, HEK 293, CHO, A431, and SKBR3)

Penicillin and streptomycin (A2212, BioChrom; www.biochrom.com) <!>

Trypsin-EDTA solution (25300-054, Life Technologies) <!>

### EQUIPMENT

Cell culture flask (175 ml)

Cell culture plate (24 well)

Coverslips

   Optional, see Step 3.

Incubator (5% $CO_2$, humidified)

1. Grow mammalian cell lines in a 5% $CO_2$ humidified incubator at 37°C in DMEM supplemented with 10% FBS, 100 units/ml penicillin, and 100 µg/ml streptomycin. Passage cells regularly to maintain exponential growth.

   Do not exceed a passage number of 30 after unfreezing the stock culture. The number of passages may affect DNA and siRNA transfection efficiencies. Aliquots of cells with low passage number may be stored frozen and can be thawed as needed.

   For general advice on cell culture, see Spector et al. (1999).

2. At least 24 hours before plasmid/siRNA transfection, trypsinize 90% confluent cells grown in a 175-ml cell culture flask with 10 ml of trypsin-EDTA.

3. Dilute the cell suspension 1:5 with fresh DMEM without antibiotics and transfer 500-µl aliquots into each well of a 24-well plate.

   If immunofluorescence assays are planned, grow cells on coverslips placed at the bottom of the 24-well plates prior to addition of the cell suspension.

4. At least 24 hours after seeding the cells, ensure that a confluency of 50–80% is reached, which corresponds to $3 \times 10^4$ to $1 \times 10^5$ cells per well, depending on the cell line and its doubling time.

## PROTOCOL 4: COTRANSFECTION OF LUCIFERASE REPORTER PLASMIDS WITH siRNA DUPLEXES

Before siRNAs are applied to knock down an endogenous gene, it may be important to establish whether the studied cells are susceptible to RNAi. It may be possible that some cell lines have lost the ability to perform RNAi or that cells derived from certain tissues do not support RNAi.

This protocol describes a reporter assay for RNAi in mammalian cells and is based on a published procedure (Elbashir et al. 2002). The quantities of reagents given below are calculated for the transfection of one well of a 24-well plate.

## Procedure

---

**MATERIALS**

**REAGENTS**

  Dual-Luciferase Assay (E1960, Promega)

  LIPOFECTAMINE 2000 (11668-019, Invitrogen; www.invitrogen.com)

  OPTI-MEM 1 medium (31985-047, Life Technologies)

  Mammalian cells

  Plasmids

   pGL2-Control plasmid (E1611, Promega; www.promega.com)
   pGL3-Control plasmid (E1741, Promega)
   pRL-TK plasmid (E2241, Promega)

  siRNA duplexes (see Protocol 2)

  • GL2 luciferase siRNAs

    sense siRNA: 5′ CGUACGCGGAAUACUUCGAdTdT
    antisense siRNA: 5′ UCGAAGUAUUCCGCGUACGdTdT

---

- invGL2 siRNAs (inverted sequence of GL2 siRNA as nonspecific control)
  sense siRNA: 5´ AGCUUCAUAAGGCGCAUGCdTdT
  antisense siRNA: 5´ GCAUGCGCCUUAUGAAGCUdTdT
- GL3 luciferase siRNAs
  sense siRNA: 5´ CUUACGCUGAGUACUUCGAdTdT
  antisense siRNA: 5´ UCGAAGUACUCAGCGUAAGdTdT
- RL luciferase siRNAs
  sense siRNA: 5´ AAACAUGCAGAAAAUGCUGdTdT
  antisense siRNA: 5´ CAGCAUUUUCUGCAUGUUUdTdT

## EQUIPMENT

Cell culture plates (24-well)

Incubator (37°C, 5% $CO_2$, humidified)

1. On the day before transfection, culture cells in 24-well plates by completing Steps 2–4 of Protocol 3.

2. On the day of transfection, mix:

   1.0 μg of pGL2-Control plasmid or 1 μg of pGL3-Control plasmid

   0.1 μg of pRL-TK plasmid

   0.21 μg of siRNA duplex (0.75 μl of 20 μM annealed duplex; see Protocol 2)

   50 μl of OPTI-MEM 1 medium

   > Reporter plasmids may be amplified in XL-1 Blue (200249, Stratagene; www.stratagene.com) and purified using the QIAGEN EndoFree maxi plasmid kit (www.qiagen.com).

3. In a separate tube, add 2 μl of LIPOFECTAMINE 2000 to 50 μl of OPTI-MEM 1 medium. Mix the tube gently by inverting; *do not* vortex. Incubate the suspension for 5 minutes at room temperature without movement.

4. Combine the solution from Step 2 with the suspension from Step 3. Mix gently by inverting the tube, and then incubate it for 20–25 minutes at room temperature to allow for formation of liposome complexes. Do not exceed a 30-minute incubation time.

5. Add the liposome complexes to the well of cells (from Step 1) without replacing the growth medium and mix gently for 15 seconds by gently rocking the plate. Incubate the plate for 20–48 hours at 37°C in a 5% $CO_2$ humidified incubator. If cytotoxic effects are expected from the transfection reagent, change the growth medium 5 hours after transfection.

6. To monitor luciferase activity, lyse the cells and measure luciferase expression using the Dual-Luciferase Assay according to the manufacturer's instructions.

   > To estimate the transfection efficiency, it is convenient to cotransfect a GFP-coding plasmid together with 0.21 μg of a siRNA duplex noncognate to GFP (e.g., invGL2) and to count the GFP-expressing cells by fluorescence microscopy. Transfection efficiencies for most cell lines described above range from 70% to 90%.

## PROTOCOL 5: TRANSFECTION OF siRNA DUPLEXES

In the absence of reporter plasmids, siRNAs are best delivered with transfection reagents developed for delivery of antisense oligodeoxynucleotides. Such reagents are sometimes less toxic than plasmid delivery reagents and may show higher transfection efficiencies than conventional transfection reagents.

Two transfection reagents have been used predominantly in our research group: OLIGOFECTAMINE from Invitrogen and TransIT-TKO siRNA Transfection Reagent from Mirus. The quantities of reagents given below are calculated for the transfection of one well of a 24-well plate.

## Procedure

---

**MATERIALS**

**REAGENTS**

Cells (see Step 1)

OLIGOFECTAMINE (Invitrogen, www.invitrogen.com)

*or*

TransIT-TKO siRNA Transfection Reagent (Mirus, http://genetransfer.com/)

OPTI-MEM 1 medium (31985-047, Life Technologies)

siRNA duplexes (see Protocol 2)

**EQUIPMENT**

Incubator (5% $CO_2$, humidified)

---

1. The day before transfection, complete Steps 2–4 of Protocol 3, except dilute the cell suspension after trypsination of the stock culture 1:10 rather than 1:5 before transferring to the 24-well plate (see Step 3 of Protocol 3). Use a higher dilution to obtain the recommended confluency of 50% for OLIGOFECTAMINE transfection.

2. Mix 3 µl of the 20 µM siRNA duplex (0.84 µg, 60 pmoles) with 50 µl of OPTI-MEM 1.

3. In a separate tube, add 3 µl of OLIGOFECTAMINE (or 4.0 µl of TransIT-TKO) to 12 µl of OPTI-MEM 1. Mix gently and incubate it for 7–10 minutes at room temperature.

4. Slowly add the siRNA solution (Step 2) to the solution prepared in Step 3 and mix gently by inversion; *do not* vortex. Incubate the tube for 20–25 minutes at room temperature to allow for formation of lipid complexes; the solution will turn turbid. Then add 32 µl of fresh OPTI-MEM 1 medium to obtain a final volume of 100 µl and mix gently by inversion.

5. Add the 100 µl of lipid complexes from Step 4 to the well of cells (from Step 1) without replacing the growth medium and mix gently for 30 seconds by gently rocking the plate. Incubate the plate for 2–3 days at 37°C in a 5% $CO_2$ humidified incubator.

   TransIT-TKO reagent is more difficult to handle than OLIGOFECTAMINE, because the concentrations required for effective transfection also cause cytotoxic effects. Typical side effects of TransIT-TKO siRNA transfection are formation of extended lamellipodia as well as oval-shaped nuclei that appear ~2 days after transfection. These effects are observed using between 4.0 and 4.5 µl of TransIT-TKO reagent.

## PROTOCOL 6: IMMUNOFLUORESCENCE DETECTION OF PROTEIN KNOCKDOWN

The preferred way of detecting a gene knockdown is to use a specific antibody that recognizes the targeted gene product.

## Procedure

▼ **CAUTION**

*See Appendix for appropriate handling of materials marked with <!>.*

### MATERIALS

#### REAGENTS

Cells from Protocol 5

Hoechst 33342 (bisbenzimide; 15091, Serva; www.serva.com) <!>

Methanol, chilled to –10°C <!>

Moviol mounting medium (Hoechst, www.hoechst.com)

Phosphate-buffered saline (PBS) (pH 7.1)

137 mM NaCl
7 mM $Na_2HPO_4$ <!>
1.5 mM $KH_2PO_4$ <!>
2.7 mM KCl <!>

Specific primary and secondary antibodies

Dilute the antibodies with PBS buffer containing 0.5 mg/ml BSA (A 9706, Sigma) and 0.02% $NaN_3$ <!>. The secondary antibody is fluorescently labeled.

#### EQUIPMENT

Ceramic rack

Cell culture plate (24-well) carrying knockdown cells on coverslips (from Protocol 5)

Coverslips

Filter paper

Incubator (37°C)

Nail polish

See Step 10.

Petri dish (13-cm diameter)

See Step 3.

Slides

Tweezers (Dumont No. 7)

Upright light microscope

For example, a Zeiss Axiophot with an F Fluar 40x/1.30 oil objective and MetaMorph Imaging Software (Universal Imaging Corporation, West Chester, Pennsylvania).

Alternatively, a laser-scanning microscope may be used.

1. Fix and permeabilize the knockdown cells.

   a. Use tweezers to remove the coverslips carrying the knockdown cells (from Protocol 5) from the 24-well plate.

   b. Place the coverslips on a ceramic rack and then incubate them in methanol chilled to –10°C for 6 minutes.

      Methanol fixation is suitable for the detection of many cellular proteins, but the optimal fixation procedure may have to be established experimentally for each individual protein (Celis et al. 1998; Spector et al. 1999). We recommend beginning with methanol fixation, which preserves the ultrastructure of the cell and sufficiently permeabilizes the cells for penetration of the antibody.

2. Wash the methanol-fixed coverslips three times in PBS and touch filter paper to the coverslips to remove excess PBS.

3. Place the coverslips in a wet chamber with the cells side facing up. Prepare a wet chamber by soaking filter paper in $H_2O$ and placing it into a 13-cm-diameter Petri dish. Do not allow the specimens to dry out during this procedure.

4. Add 20 µl of appropriately diluted primary antibody on top of the coverslip without touching the cells. Make sure that the solution is evenly spread out over the entire surface of the coverslip. Transfer the closed wet chamber into a 37°C incubator and incubate for 45–60 minutes.

   Antibodies are diluted with PBS containing 0.5 mg/ml bovine serum albumin and 0.02% sodium azide.

5. Place the coverslips again on the ceramic rack and wash them three times with PBS, each for 5 minutes. Touch filter paper to the coverslips to remove excess PBS, and then transfer the coverslips back into the wet chamber.

6. Add 20 µl of appropriately diluted, fluorescently labeled secondary antibody to each coverslip. Incubate the cells in the closed wet chamber for 45 minutes at 37°C.

7. Repeat Step 5.

8. Detect the cell nuclei by chromatin staining. Add 20 µl of 1 µM Hoechst 33342 solution in PBS on top of the coverslip and incubate for 4 minutes at room temperature.

9. Repeat Step 5.

10. Mount two coverslips per slide by placing the coverslips with the cells side facing downward on a drop of Moviol mounting medium. Place a piece of filter paper on top of the slide and press gently on top of the paper to remove excess mounting medium. Glue coverslips to the slide with nail polish.

11. Examine the immunofluorescence staining and take pictures using an upright light microscope. Use identical exposure times for photographing both the silenced cells and the control-treated cells.

    Alternatively, a laser-scanning microscope may be used.

## PROTOCOL 7: DETECTION OF PROTEIN KNOCKDOWN BY WESTERN BLOTTING

Knockdown of proteins is frequently associated with impaired cell growth or altered cell morphology, which can be monitored by phase-contrast microscopy. If no alterations in cell growth or cell morphology are observed, immunofluorescence or western blotting can be performed to analyze the depletion of the target protein.

## Procedure

▼ CAUTION

*See Appendix for appropriate handling of materials marked with <!>.*

### MATERIALS

#### REAGENTS

Blocking solution (5% milk powder in TBST [pH 7.4])

Dulbecco's modified Eagle medium (DMEM)

ECL (enhanced chemiluminescent) detection kit (www.amersham.co.uk)

Electrotransfer buffer

25 mM Tris <!>
192 mM glycine <!>
0.01% SDS <!>
20% methanol <!>

2x Laemmli SDS sample buffer (161-073, Bio-Rad; www.bio-rad.com)

Phosphate-buffered saline (PBS) (pH 7.1)

137 mM NaCl
7 mM $Na_2HPO_4$ <!>
1.5 mM $KH_2PO_4$ <!>
2.7 mM KCl <!>

Ponceau S stain (Sigma) <!>

Primary antibody

See Step 8. If necessary, dilute the antibody in TBST.

Secondary antibody

Horseradish peroxidase (HRP)-conjugated rabbit anti-mouse *or* HRP-conjugated swine anti-rabbit antibodies (Dako Diagnostika, Hamburg, Germany; www.dako.com)

siRNA-treated cells cultivated in 24-well plates (from Protocol 5)

TBST (pH 7.4)

0.2% Tween-20
20 mM Tris-HCl <!>
150 mM NaCl

Trypsin-EDTA solution (25300-054, Life Technologies) <!>

#### EQUIPMENT

Centrifugation tube (1.5 ml)

Centrifuge

Electrotransfer equipment

Enhanced chemiluminescent detection equipment

Nitrocellulose membrane (Protran BA85 0.45 mm, 10401196, Schleicher & Schuell; www.s-und-s.de)

SDS-polyacrylamide gel electrophoresis equipment

Water bath (boiling)

1. Remove the tissue culture medium from the siRNA-treated cells cultivated in 24-well plates (from Protocol 5). Rinse the cells once with 200 μl of PBS, and add 200 μl of trypsin-EDTA. Incubate for 1 minute at 37°C; suspend the cells and add 800 μl of DMEM medium to quench the trypsin.

2. Transfer the suspended cells to a chilled 1.5-ml centrifugation tube. Collect the cells by centrifugation at 3000 rpm ($700g$) for 4 minutes at 4°C. Resuspend the cell pellet in ice-cold PBS and centrifuge again.

3. Remove the supernatant and add 25 μl of 90°C 2x concentrated Laemmli SDS sample buffer to the cell pellet obtained from one well of a 24-well plate. Incubate the sample for 3 minutes in a boiling water bath and vortex.

4. Separate the proteins by SDS-polyacrylamide gel electrophoresis using an acrylamide concentration appropriate to resolve the molecular weight of the targeted protein (Sambrook et al. 2001).

   We have separated proteins on minigels, which were run at a constant 10 mA.

5. Transfer proteins from the gel to a nitrocellulose membrane using electrotransfer buffer. Our minigels are electroblotted onto the membrane using a Bio-Rad Trans-Blot cell at 333 mA for 30 minutes in the cold room.

6. Verify the protein transfer by Ponceau S staining of the transfer membrane.

7. Incubate the membrane in blocking solution for 1 hour at 37°C.

8. Replenish the blocking solution with fresh blocking solution and add the primary antibody at the appropriate dilution. Incubate for 1–2 hours at 37°C.

9. Wash the blot four times with TBST for 10 minutes.

10. For ECL detection, incubate the blot with either HRP-conjugated rabbit anti-mouse or HRP-conjugated swine anti-rabbit antibodies at a dilution of 1:20,000 in blocking solution for 1–2 hours at 37°C.

11. Perform ECL detection according to the protocol described by the manufacturer (www.amersham.co.uk).

## ACKNOWLEDGMENTS

We acknowledge Tilmann Achsel, Alexei Aravin, Nina Dobriczikowski, Sayda Elbashir, Patrizia Fabrizio, Jens Gruber, Jens Harborth, Markus Hossbach, Klaus Weber, and Agnieszka Patkaniowska for critical comments on the manuscript.

## REFERENCES

Ambros V. 2001. microRNAs: Tiny regulators with great potential. *Cell* **107**: 823–826.

Ancellin N., Colmont C., Su J., Li Q., Mittereder N., Chae S.S., Steffansson S., Liau G., and Hla T. 2001. Extracellular export of sphingosine kinase-1 enzyme: Sphingosine 1-phosphate generation and the induction of angiogenic vascular maturation. *J. Biol. Chem.* **227**: 6667–6675.

Aravin A.A., Naumova N.M., Tulin A.V., Vagin V.V., Rozovsky Y.M., and Gvozdev V.A. 2001. Double-stranded RNA-mediated silencing of genomic tandem repeats and transposable elements in the *D. melanogaster* germline. *Curr. Biol.* **11**: 1017–1027.

Bahramian M.B. and Zarbl H. 1999. Transcriptional and posttranscriptional silencing of rodent alpha1(I) collagen by a homologous transcriptionally self-silenced transgene. *Mol. Cell. Biol.* **19**: 274–283.

Bai X., Zhou D., Brown J.R., Crawford B.E., Hennet T., and Esko J.D. 2001. Biosynthesis of the linkage region of glycosaminoglycans. Cloning and activity of galactosyltransferase II, the sixth member of the beta-1,3-galactosyltransferase family (beta 3GalT6). *J. Biol. Chem.* **276**: 48189–48195.

Bakker J., Lin X., and Nelson W.G. 2002. Methyl-CpG binding domain protein 2 represses transcription from hypermethylated p-class glutathione S-transferase gene promoters in hepatocellular carcinoma cells. *J. Biol. Chem.* **17:** 17.

Barber G.N. 2001. Host defense, viruses and apoptosis. *Cell Death Differ.* **8:** 113–126.

Bender J. 2001. A vicious cycle: RNA silencing and DNA methylation in plants. *Cell* **106:** 129–132.

Bernstein E., Caudy A.A., Hammond S.M., and Hannon G.J. 2001. Role for a bidentate ribonuclease in the initiation step of RNA interference. *Nature* **409:** 363–366.

Billy E., Brondani V., Zhang H., Muller U., and Filipowicz W. 2001. Specific interference with gene expression induced by long, double-stranded RNA in mouse embryonal teratocarcinoma cell lines. *Proc. Natl. Acad. Sci.* **98:** 14428–14433.

Bitko V. and Barik S. 2001. Phenotypic silencing of cytoplasmic genes using sequence-specific double-stranded short interfering RNA and its application in the reverse genetics of wild type negative-strand RNA viruses. *BMC Microbiol.* **1:** 34.

Blaszczyk J., Tropea J.E., Bubunenko M., Routzahn K.M., Waugh D.S., Court D.L., and Ji X. 2001. Crystallographic and modeling studies of RNase III suggest a mechanism for double-stranded RNA cleavage. *Structure* **9:** 1225–1236.

Bohmert K., Camus I., Bellini C., Bouchez D., Caboche M., and Benning C. 1998. AGO1 defines a novel locus of *Arabidopsis* controlling leaf development. *EMBO J.* **17:** 1776–1780.

Boutla A., Delidakis C., Livadaras I., Tsagris M., and Tabler M. 2001. Short 5´-phosphorylated double-stranded RNAs induce RNA interference in *Drosophila. Curr. Biol.* **11:** 1776–1780.

Brown S.J., Mahaffey J.P., Lorenzen M.D., Denell R.E., and Mahaffey J.W. 1999. Using RNAi to investigate orthologous homeotic gene function during development of distantly related insects. *Evol. Dev.* **1:** 11–15.

Brummelkamp T.R., Bernards R., and Agami R. 2002. A system for stable expression of short interfering RNAs in mammalian cells. *Science* **296:** 550–553.

Caplen N.J., Fleenor J., Fire A., and Morgan R.A. 2000. dsRNA-mediated gene silencing in cultured *Drosophila* cells: A tissue culture model for the analysis of RNA interference. *Gene* **252:** 95–105.

Caplen N.J., Parrish S., Imani F., Fire A., and Morgan R.A. 2001. Specific inhibition of gene expression by small double-stranded RNAs in invertebrate and vertebrate systems. *Proc. Natl. Acad. Sci.* **98:** 9742–9747.

Caplen N.J., Taylor J.P., Statham V.S., Tanaka F., Fire A., and Morgan R.A. 2002. Rescue of polyglutamine-mediated cytotoxicity by double-stranded RNA-mediated RNA interference. *Hum. Mol. Genet.* **11:** 175–184.

Catalanotto C., Azzalin G., Macino G., and Cogoni C. 2002. Involvement of small RNAs and role of the *qde* genes in the gene silencing pathway in *Neurospora. Genes Dev.* **16:** 790–795.

Celis J.E. 1998. *Cell biology: A laboratory handbook*, Vol. 2. Academic Press, San Diego.

Cerutti L., Mian N., and Bateman A. 2000. Domains in gene silencing and cell differentiation proteins: The novel PAZ domain and redefinition of the piwi domain. *Trends Biochem. Sci.* **25:** 481–482.

Cogoni C. and Macino G. 1999. Gene silencing in *Neurospora crassa* requires a protein homologous to RNA-dependent RNA polymerase. *Nature* **399:** 166–169.

Conrad C. and Rauhut R. 2002. Ribonuclease III: New sense from nuisance. *Int. J. Biochem. Cell Biol.* **34:** 116–129.

Cortez D., Guntuku S., Qin J., and Elledge S.J. 2001. ATR and ATRIP: Partners in checkpoint signaling. *Science* **294:** 1713–1716.

Cox D.N., Chao A., and Lin H. 2000. piwi encodes a nucleoplasmic factor whose activity modulates the number and division rate of germline stem cells. *Development* **127:** 503–514.

Dalmay T., Hamilton A., Mueller E., and Baulcombe D.C. 2000a. Potato virus X amplicons in *Arabidopsis* mediate genetic and epigenetic gene silencing. *Plant Cell* **12:** 369–380.

Dalmay T., Hamilton A., Rudd S., Angell S., and Baulcombe D.C. 2000b. An RNA-dependent RNA polymerase gene in *Arabidopsis* is required for posttranscriptional gene silencing mediated by a transgene but not by a virus. *Cell* **101:** 543–553.

Dernburg A.F., Zalevsky J., Colaiacovo M.P., and Villeneuve A.M. 2000. Transgene-mediated cosuppression in the *C. elegans* germ line. *Genes Dev.* **14:** 1578–1583.

Djikeng A., Shi H., Tschudi C., and Ullu E. 2001. RNA interference in *Trypanosoma brucei:* Cloning of small interfering RNAs provides evidence for retroposon-derived 24–26-nucleotide RNAs. *RNA* **7:** 1522–1530.

Donzé O. and Picard D. 2002. RNA interference in mammalian cells using siRNAs synthesized with

T7 RNA polymerase. *Nucleic Acids Res.* **30**: e46.

Du Q., Stukenberg P.T., and Macara I.G. 2001. A mammalian partner of inscuteable binds NuMA and regulates mitotic spindle organization. *Nat. Cell Biol.* **3**: 1069–1075.

Eddy S.R. 2001. Non-coding RNA genes and the modern RNA world. *Nat. Rev. Genet.* **2**: 919–921.

Elbashir S.M., Lendeckel W., and Tuschl T. 2001a. RNA interference is mediated by 21 and 22 nt RNAs. *Genes Dev.* **15**: 188–200.

Elbashir S.M., Harborth J., Weber K., and Tuschl T. 2002. Analysis of gene function in somatic mammalian cells using small interfering RNAs. *Methods* **26**: 199–213.

Elbashir S.M., Martinez J., Patkaniowska A., Lendeckel W., and Tuschl T. 2001b. Functional anatomy of siRNAs for mediating efficient RNAi in *Drosophila melanogaster* embryo lysate. *EMBO J.* **20**: 6877–6888.

Elbashir S.M., Harborth J., Lendeckel W., Yalcin A., Weber K., and Tuschl T. 2001c. Duplexes of 21-nucleotide RNAs mediate RNA interference in mammalian cell culture. *Nature* **411**: 494–498.

Fagard M., Boutet S., Morel J.B., Bellini C., and Vaucheret H. 2000. AGO1, QDE-2, and RDE-1 are related proteins required for post-transcriptional gene silencing in plants, quelling in fungi, and RNA interference in animals. *Proc. Natl. Acad. Sci.* **97**: 11650–11654.

Filippov V., Solovyev V., Filippova M., and Gill S.S. 2000. A novel type of RNase III family proteins in eukaryotes. *Gene* **245**: 213–221.

Fire A., Xu S., Montgomery M.K., Kostas S.A., Driver S.E., and Mello C.C. 1998. Potent and specific genetic interference by double-stranded RNA in *Caenorhabditis elegans*. *Nature* **391**: 806–811.

Fraser A.G., Kamath R.S., Zipperlen P., Martinez-Campos M., Sohrmann M., and Ahringer J. 2000. Functional genomic analysis of *C. elegans* chromosome I by systematic RNA interference. *Nature* **408**: 325–330.

Garrus J.E., von Schwedler U.K., Pornillos O.W., Morham S.G., Zavitz K.H., Wang H.E., Wettstein D.A., Stray K.M., Cote M., Rich R.L., Myszka D.G., and Sundquist W.I. 2001. Tsg101 and the vacuolar protein sorting pathway are essential for HIV-1 budding. *Cell* **107**: 55–65.

Gönczy P., Echeverri C., Oegema K., Coulson A., Jones S.J.M., Copley R.R., Duperon J., Oegema J., Brehm M., Cassin E., Hannak E., Kirkham M., Pichler S., Flohrs K., Goessen A., Leidel S., Alleaume A.-M., Martin C., Özlü N., Bork P., and Hyman A.A. 2000. Functional genomic analysis of cell division in *C. elegans* using RNAi of genes on chromosome III. *Nature* **408**: 331–336.

Grishok A., Pasquinelli A.E., Conte D., Li N., Parrish S., Ha I., Baillie D.L., Fire A., Ruvkun G., and Mello C.C. 2001. Genes and mechanisms related to RNA interference regulate expression of the small temporal RNAs that control *C. elegans* developmental timing. *Cell* **106**: 23–34.

Grosshans H. and Slack F.J. 2002. Micro-RNAs: Small is plentiful. *J. Cell Biol.* **156**: 17–21.

Habas R., Kato Y., and He X. 2001. Wnt/Frizzled activation of Rho regulates vertebrate gastrulation and requires a novel Formin homology protein Daam1. *Cell* **107**: 843–854.

Hamilton A.J. and Baulcombe D.C. 1999. A species of small antisense RNA in posttranscriptional gene silencing in plants. *Science* **286**: 950–952.

Hammond S.M., Caudy A.A., and Hannon G.J. 2001a. Post-transcriptional gene silencing by double-stranded RNA. *Nat. Rev. Genet.* **2**: 110–119.

Hammond S.M., Bernstein E., Beach D., and Hannon G.J. 2000. An RNA-directed nuclease mediates post-transcriptional gene silencing in *Drosophila* cells. *Nature* **404**: 293–296.

Hammond S.M., Boettcher S., Caudy A.A., Kobayashi R., and Hannon G.J. 2001b. Argonaute2, a link between genetic and biochemical analyses of RNAi. *Science* **293**: 1146–1150.

Harborth J., Elbashir S.M., Bechert K., Tuschl T., and Weber K. 2001. Identification of essential genes in cultured mammalian cells using small interfering RNAs. *J. Cell Sci.* **114**: 4557–4565.

Harris A.N. and Macdonald P.M. 2001. Aubergine encodes a *Drosophila* polar granule component required for pole cell formation and related to eIF2C. *Development* **128**: 2823–2832.

Hirai I. and Wang H.G. 2002. A role of the C-terminal region of hRad9 in nuclear transport of the hRad9-hRad1-hHus1 checkpoint complex. *J. Biol. Chem.* **277**: 25722–25727.

Holen T., Amarzguioui M., Wiiger M.T., Babaie E., and Prydz H. 2002. Positional effects of short interfering RNAs targeting the human coagulation trigger Tissue Factor. *Nucleic Acids Res.* **30**: 1757–1766.

Hope I.A. 2001. Broadcast interference—Functional genomics. *Trends Genet.* **17**: 297–299.

Hutvágner G. and Zamore P.D. 2002. RNAi: Nature abhors a double-strand. *Curr. Opin. Genet. Dev.* **12**: 225–232.

Hutvágner G., Mlynarova L., and Nap J.P. 2000. Detailed characterization of the posttranscription-al gene-silencing-related small RNA in a GUS gene-silenced tobacco. *RNA* **6:** 1445–1454.

Hutvágner G., McLachlan J., Bálint É., Tuschl T., and Zamore P.D. 2001. A cellular function for the RNA interference enzyme Dicer in small temporal RNA maturation. *Science* **93:** 834–838.

Jeddeloh J.A., Bender J., and Richards E.J. 1998. The DNA methylation locus DDM1 is required for maintenance of gene silencing in *Arabidopsis. Genes Dev.* **12:** 1714–1725.

Jensen S., Gassama M.P., and Heidmann T. 1999a. Cosuppression of I transposon activity in *Drosophila* by I-containing sense and antisense transgenes. *Genetics* **153:** 1767–1774.

Jensen S., Gassama M.P., and Heidmann T. 1999b. Taming of transposable elements by homology-dependent gene silencing. *Nat. Genet.* **21:** 209–212.

Jones L., Hamilton A.J., Voinnet O., Thomas C.L., Maule A.J., and Baulcombe D.C. 1999. RNA-DNA interactions and DNA methylation in post-transcriptional gene silencing. *Plant Cell* **11:** 2291–2302.

Kataoka Y., Takeichi M., and Uemura T. 2001. Developmental roles and molecular characterization of a *Drosophila* homologue of *Arabidopsis* Argonaute1, the founder of a novel gene superfami-ly. *Genes Cells* **6:** 313–325.

Kelly W.G. and Fire A. 1998. Chromatin silencing and the maintenance of a functional germline in *Caenorhabditis elegans. Development* **125:** 2451–2456.

Kennerdell J.R. and Carthew R.W. 1998. Use of dsRNA-mediated genetic interference to demon-strate that *frizzled* and *frizzled 2* act in the wingless pathway. *Cell* **95:** 1017–1026.

———. 2000. Heritable gene silencing in *Drosophila* using double-stranded RNA. *Nat. Biotechnol.* **18:** 896–898.

Ketting R.F., Haverkamp T.H., van Luenen H.G., and Plasterk R.H. 1999. Mut-7 of *C. elegans,* required for transposon silencing and RNA interference, is a homolog of Werner syndrome helicase and RNaseD. *Cell* **99:** 133–141.

Ketting R.F., Fischer S.E., Bernstein E., Sijen T., Hannon G.J., and Plasterk R.H. 2001. Dicer func-tions in RNA interference and in synthesis of small RNA involved in developmental timing in *C. elegans. Genes Dev.* **15:** 2654–2659.

Kim S.K. 2001. Functional genomics: The worm scores a knockout. *Curr. Biol.* **11:** R85–87.

Kisielow M., Kleiner S., Nagasawa M., Faisal A., and Nagamine Y. 2002. Isoform-specific knock-down and expression of adaptor protein ShcA using small interfering RNA. *Biochem. J.* **363:** 1–5.

Knight S.W. and Bass B.L. 2001. A role for the RNase III enzyme DCR-1 in RNA interference and germ line development in *C. elegans. Science* **2:** 2.

Koesters R., Adams V., Betts D., Moos R., Schmid M., Siermann A., Hassam S., Weitz S., Lichter P., Heitz P.U., von Knebel Doeberitz M., and Briner J. 1999. Human eukaryotic initiation factor EIF2C1 gene: cDNA sequence, genomic organization, localization to chromosomal bands 1p34-p35, and expression. *Genomics* **61:** 210–218.

Kuramochi-Miyagawa S., Kimura T., Yomogida K., Kuroiwa A., Tadokoro Y., Fujita Y., Sato M., Matsuda Y., and Nakano T. 2001. Two mouse piwi-related genes: miwi and mili. *Mech. Dev.* **108:** 121–133.

Lee N.S., Dohjima T., Bauer G., Li H., Li M.J., Ehsani A., Salvaterra P., and Rossi J. 2002. Expression of small interfering RNAs targeted against HIV-1 rev transcripts in human cells. *Nat. Biotechnol.* **20:** 500–505.

Lengyel P. 1987. Double-stranded RNA and interferon action. *J. Interferon Res.* **7:** 511–519.

Li L., Mao J., Sun L., Liu W., and Wu D. 2001. Second cysteine-rich domain of Dickkopf-2 activates Canonical Wnt signaling pathway via LRP-6 independently of dishevelled. *J. Biol. Chem.* **12:** 12.

Lipardi C., Wei Q., and Paterson B.M. 2001. RNAi as random degradative PCR. siRNA primers con-vert mRNA into dsRNAs that are degraded to generate new siRNAs. *Cell* **107:** 297–307.

Mailand N., Lukas C., Kaiser B.K., Jackson P.K., Bartek J., and Lukas J. 2002. Deregulated human Cdc14A phosphatase disrupts centrosome separation and chromosome segregation. *Nat. Cell Biol.* **4:** 318–322.

Manche L., Green S.R., Schmedt C., and Mathews M.B. 1992. Interactions between double-strand-ed RNA regulators and the protein kinase DAI. *Mol. Cell. Biol.* **12:** 5238–5248.

Martinez J., Patkaniowska A., Urlaub H., Lührmann R., and Tuschl T. 2002. Single-stranded anti-sense siRNAs guide target RNA cleavage in RNAi. *Cell* **110:** 563–574.

Martins L.M., Iaccarino I., Tenev T., Gschmeissner S., Totty N.F., Lemoine N.R., Savopoulos J., Gray C.W., Creasy C.L., Dingwall C., and Downward J. 2002. The serine protease Omi/HtrA2 regulates apoptosis by binding XIAP through a reaper-like motif. *J. Biol. Chem.* **277:** 439–444.

Masters J.R., Thomson J.A., Daly-Burns B., Reid Y.A., Dirks W.G., Packer P., Toji L.H., Ohno T., Tanabe H., Arlett C.F., Kelland L.R., Harrison M., Virmani A., Ward T.H., Ayres K.L., and Debenham P.G. 2001. Short tandem repeat profiling provides an international reference standard for human cell lines. *Proc. Natl. Acad. Sci.* **98:** 8012–8017.

Matzke M., Matzke A.J.M., and Kooter J.M. 2001a. RNA: Guiding gene silencing. *Science* **293:** 1080–1083.

Matzke M.A., Matzke A.J.M., Pruss G.J., and Vance V.B. 2001b. RNA-based silencing strategies in plants. *Curr. Opin. Genet. Dev.* **11:** 2221–2227.

Medina M.F. and Joshi S. 1999. RNA-polymerase III-driven expression cassettes in human gene therapy. *Curr. Opin. Mol. Ther.* **1:** 580–594.

Meissner W., Rothfels H., Schafer B., and Seifart K. 2001. Development of an inducible pol III transcription system essentially requiring a mutated form of the TATA-binding protein. *Nucleic Acids Res.* **29:** 1672–1682.

Minks M.A., West D.K., Benvin S., and Baglioni C. 1979. Structural requirements of double-stranded RNA for the activation of 2′,5′-oligo(A) polymerase and protein kinase of interferon-treated HeLa cells. *J. Biol. Chem.* **254:** 10180–10183.

Mittelsten-Scheid O., Afsar K., and Paszkowski J. 1998. Release of epigenic gene silencing by transacting mutations in *Arabidopsis. Proc. Natl. Acad. Sci.* **95:** 632–637.

Miyagishi M. and Taira K. 2002. U6 promoter driven siRNAs with four uridine 3′ overhangs efficiently suppress targeted gene expression in mammalian cells. *Nat. Biotechnol.* **20:** 497–500.

Montgomery M.K., Xu S., and Fire A. 1998. RNA as a target of double-stranded RNA-mediated genetic interference in *Caenorhabditis elegans. Proc. Natl. Acad. Sci.* **95:** 15502–15507.

Morel J.B., Mourrain P., Beclin C., and Vaucheret H. 2000. DNA methylation and chromatin structure affect transcriptional and posttranscriptional transgene silencing in *Arabidopsis. Curr. Biol.* **10:** 1591–1594.

Moskalenko S., Henry D.O., Rosse C., Mirey G., Camonis J.H., and White M.A. 2002. The exocyst is a Ral effector complex. *Nat. Cell Biol.* **4:** 66–72.

Moss E.G. 2002. MicroRNAs: Hidden in the genome. *Curr. Biol.* **12:** R138–140.

Mourelatos Z., Dostie J., Paushkin S., Sharma A., Charroux B., Abel L., Rappsilber J., Mann M., and Dreyfuss G. 2002. miRNPs: A novel class of ribonucleoproteins containing numerous microRNAs. *Genes Dev.* **16:** 720–728.

Mourrain P., Beclin C., Elmayan T., Feuerbach F., Godon C., Morel J.B., Jouette D., Lacombe A.M., Nikic S., Picault N., Remoue K., Sanial M., Vo T.A., and Vaucheret H. 2000. *Arabidopsis* SGS2 and SGS3 genes are required for posttranscriptional gene silencing and natural virus resistance. *Cell* **101:** 533–542.

Myslinski E., Amé J.-C., Krol A., and Carbon P. 2001. An unusually compact external promoter for RNA polymerase III transcription of the human H1 RNA gene. *Nucleic Acids Res.* **29:** 2502–2509.

Nakano H., Amemiya S., Shiokawa K., and Taira M. 2000. RNA interference for the organizer-specific gene *Xlim-1* in *Xenopus* embryos. *Biochem. Biophys. Res. Commun.* **274:** 434–439.

Ngo H., Tschudi C., Gull K., and Ullu E. 1998. Double-stranded RNA induces mRNA degradation in *Trypanosoma brucei. Proc. Natl. Acad. Sci.* **95:** 14687–14692.

Nicholson R.H. and Nicholson A.W. 2002. Molecular characterization of a mouse cDNA encoding Dicer, a ribonuclease III ortholog involved in RNA interference. *Mamm. Genome* **13:** 67–73.

Nykänen A., Haley B., and Zamore P.D. 2001. ATP requirements and small interfering RNA structure in the RNA interference pathway. *Cell* **107:** 309–321.

Oates A.C., Bruce A.E., and Ho R.K. 2000. Too much interference: Injection of double-stranded RNA has nonspecific effects in the zebrafish embryo. *Dev. Biol.* **224:** 20–28.

Ostendorff H.P., Peirano R.I., Peters M.A., Schluter A., Bossenz M., Scheffner M., and Bach I. 2002. Ubiquitination-dependent cofactor exchange on LIM homeodomain transcription factors. *Nature* **416:** 99–103.

Paddison P.J., Caudy A.A., and Hannon G.J. 2002a. Stable suppression of gene expression by RNAi in mammalian cells. *Proc. Natl. Acad. Sci.* **99:** 1443–1448.

Paddison P.J., Caudy A.A., Bernstein E., Hannon G.J., and Conklin D.S. 2002b. Short hairpin RNAs (shRNAs) induce sequence-specific silencing in mammalian cells. *Genes Dev.* **16:** 948–958.

Pal-Bhadra M., Bhadra U., and Birchler J.A. 1997. Cosuppression in *Drosophila:* Gene silencing of *Alcohol dehydrogenase* by *white-Adh* transgenes is *Polycomb* dependent. *Cell* **90:** 479–490.

———. 2002. RNAi related mechanism affect both transcriptional and posttranscriptional transgene silencing in *Drosophila. Mol. Cell* **9:** 315–327.

Parrish S. and Fire A. 2001. Distinct roles for RDE-1 and RDE-4 during RNA interference in *Caenorhabditis elegans. RNA* **7:** 1397–1402.

Parrish S., Fleenor J., Xu S., Mello C., and Fire A. 2000. Functional anatomy of a dsRNA trigger: Differential requirement for the two trigger strands in RNA interference. *Mol. Cell* **6:** 1077–1087.

Pasquinelli A.E. 2002. MicroRNAs: Deviants no longer. *Trends Genet.* **18:** 171–173.

Paul C.P., Good P.D., Winer I., and Engelke D.R. 2002. Effective expression of small interfering RNA in human cells. *Nat. Biotechnol.* **20:** 505–508.

Paule M.R. and White R.J. 2000. Transcription by RNA polymerase I and III. *Nucleic Acids Res.* **28:** 1283–1298.

Pelissier T. and Wassenegger M. 2000. A DNA target of 30 bp is sufficient for RNA-directed DNA methylation. *RNA* **6:** 55–65.

Porter L.A., Dellinger R.W., Tynan J.A., Barnes E.A., Kong M., Lenormand J.L., and Donoghue D.J. 2002. Human Speedy: A novel cell cycle regulator that enhances proliferation through activation of Cdk2. *J. Cell Biol.* **157:** 357–366.

Roy A.L., Chakrabarti D., Datta B., Hileman R.E., and Gupta N.K. 1988. Natural mRNA is required for directing Met-tRNA(f) binding to 40S ribosomal subunits in animal cells: Involvement of Co-eIF-2A in natural mRNA-directed initiation complex formation. *Biochemistry* **27:** 8203–8209.

Sambrook J., Fritsch E., and Maniatis T. 2001. *Molecular cloning: A laboratory manual,* 2nd edition. Cold Spring Harbor Laboratory Press, Cold spring Harbor, New York.

Schwarz D.S. and Zamore P.D. 2002. Why do miRNAs live in the miRNP? *Genes Dev.* **16:** 1025–1031.

Sharma A.K., Nelson M.C., Brandt J.E., Wessman M., Mahmud N., Weller K.P., and Hoffman R. 2001. Human CD34(+) stem cells express the *hiwi* gene, a human homologue of the *Drosophila* gene *piwi. Blood* **97:** 426–434.

Sharp P.A. 2001. RNA interference 2001. *Genes Dev.* **15:** 485–490.

Short B., Preisinger C., Korner R., Kopajtich R., Byron O., and Barr F.A. 2001. A GRASP55-rab2 effector complex linking Golgi structure to membrane traffic. *J. Cell Biol.* **155:** 877–883.

Sijen T., Fleenor J., Simmer F., Thijssen K.L., Parrish S., Timmons L., Plasterk R.H., and Fire A. 2001. On the role of RNA amplification in dsRNA-triggered gene silencing. *Cell* **107:** 465–476.

Spector D.L., Goldman R.D., and Leinwand L.A. 1999. *Cells: A laboratory manual,* Vols. I–III. Cold Spring Harbor Laboratory Press, Cold Spring Harbor, New York.

Stark G.R., Kerr I.M., Williams B.R., Silverman R.H., and Schreiber R.D. 1998. How cells respond to interferons. *Annu. Rev. Biochem.* **67:** 227–264.

Stucke V.M., Sillje H.H., Arnaud L., and Nigg E.A. 2002. Human Mps1 kinase is required for the spindle assembly checkpoint but not for centrosome duplication. *EMBO J.* **21:** 1723–1732.

Sui G., Soohoo C., Affar el B., Gay F., Shi Y., and Forrester W.C. 2002. A DNA vector-based RNAi technology to suppress gene expression in mammalian cells. *Proc. Natl. Acad. Sci.* **99:** 5515–5520.

Svoboda P., Stein P., Hayashi H., and Schultz R.M. 2000. Selective reduction of dormant maternal mRNAs in mouse oocytes by RNA interference. *Development* **127:** 4147–4156.

Tabara H., Sarkissian M., Kelly W.G., Fleenor J., Grishok A., Timmons L., Fire A., and Mello C.C. 1999. The *rde-1* gene, RNA interference, and transposon silencing in *C. elegans. Cell* **99:** 123–132.

Tijsterman M., Ketting R.F., Okihara K.L., and Plasterk R.H. 2002. RNA helicase MUT-14-dependent silencing triggered in *C. elegans* by short antisense RNAs. *Science* **295:** 694–697.

Timmons L. and Fire A. 1998. Specific interference by ingested dsRNA. *Nature* **395:** 854.

Tuschl T. 2001. RNA interference and small interfering RNAs. *ChemBioChem.* **2:** 239–245.

Tuschl T. and Borkhardt A. 2002. Small interfering RNAs—A revolutionary tool for analysis of gene function and gene therapy. *Mol. Intervent.* **2:** 42–51.

Tuschl T., Zamore P.D., Lehmann R., Bartel D.P., and Sharp P.A. 1999. Targeted mRNA degradation by double-stranded RNA in vitro. *Genes Dev.* **13:** 3191–3197.

Ui-Tei K., Zenno S., Miyata Y., and Saigo K. 2000. Sensitive assay of RNA interference in *Drosophila* and Chinese hamster cultured cells using firefly luciferase gene as target. *FEBS Lett.* **479:** 79–82.

Wassenegger M. 2000. RNA-directed DNA methylation. *Plant Mol. Biol.* **43:** 203–220.

Waterhouse P.M., Wang M.B., and Lough T. 2001. Gene silencing as an adaptive defence against viruses. *Nature* **411:** 834–842.

Wianny F. and Zernicka-Goetz M. 2000. Specific interference with gene function by double-stranded RNA in early mouse development. *Nat. Cell Biol.* **2:** 70–75.

Williams R.W. and Rubin G.M. 2002. ARGONAUTE1 is required for efficient RNA interference in *Drosophila* embryos. *Proc. Natl. Acad. Sci.* **99:** 6889–6894.

Wilson J.E., Connell J.E., and Macdonald P.M. 1996. aubergine enhances oskar translation in the *Drosophila* ovary. *Development* **122:** 1631–1639.

Winston W.M., Molodowitch C., and Hunter C.P. 2002. Systemic RNAi in *C. elegans* requires the putative transmembrane protein SID-1. *Science* **295:** 2456–2459.

Wu H., Xu H., Miraglia L.J., and Crooke S.T. 2000. Human RNase III is a 160 kDa protein involved in preribosomal RNA processing. *J. Biol. Chem.* **275:** 36957–36965.

Wu-Scharf D., Jeong B., Zhang C., and Cerutti H. 2000. Transgene and transposon silencing in *Chlamydomonas reinhardtii* by a DEAH-Box RNA helicase. *Science* **290:** 1159–1163.

Yang D., Lu H., and Erickson J.W. 2000. Evidence that processed small dsRNAs may mediate sequence-specific mRNA degradation during RNAi in *Drosophila* embryos. *Curr. Biol.* **10:** 1191–1200.

Yang S., Tutton S., Pierce E., and Yoon K. 2001. Specific double-stranded RNA interference in undifferentiated mouse embryonic stem cells. *Mol. Cell. Biol.* **21:** 7807–7816.

Yarovoi S.V. and Pederson T. 2001. Human cell lines expressing hormone regulated T7 RNA polymerase localized at distinct intranuclear sites. *Gene* **275:** 73–81.

Yu J.Y., DeRuiter S.L., and Turner D.L. 2002. RNA interference by expression of short-interfering RNAs and hairpin RNAs in mammalian cells. *Proc. Natl. Acad. Sci.* **99:** 6047–6052.

Zamore P.D. 2001a. RNA interference: Listening to the sound of silence. *Nat. Struct. Biol.* **8:** 746–750.

———. 2001b. Thirty-three years later, a glimpse at the ribonuclease III active site. *Mol. Cell* **8:** 1158–1160.

Zamore P.D., Tuschl T., Sharp P.A., and Bartel D.P. 2000. RNAi: Double-stranded RNA directs the ATP-dependent cleavage of mRNA at 21 to 23 nucleotide intervals. *Cell* **101:** 25–33.

Zhao Z., Cao Y., Li M., and Meng A. 2001. Double-stranded RNA injection produces nonspecific defects in zebrafish. *Dev. Biol.* **229:** 215–223.

Zhou Y., Ching Y.-P., Kok K.H., Kung H., and Jin D.-J. 2002. Post-transcriptional suppression of gene expression in *Xenopus* embryos by small interfering RNAs. *Nucleic Acids Res.* **30:** 1664–1669.

Ziauddin J. and Sabatini D.M. 2001. Microarrays of cells expressing defined cDNAs. *Nature* **411:** 107–110.

Zou C., Zhang Z., Wu S., and Osterman J.C. 1998. Molecular cloning and characterization of a rabbit eIF2C protein. *Gene* **211:** 187–194.

Zou L., Cortez D., and Elledge S.J. 2002. Regulation of ATR substrate selection by Rad17-dependent loading of Rad9 complexes onto chromatin. *Genes Dev.* **16:** 198–208.

# CHAPTER 14

# RNAi in Avian Embryos

Esther T. Stoeckli

*Institute of Zoology, University of Zurich, CH-8057 Zurich, Switzerland*

THE POSSIBILITY OF BEING ABLE TO TURN OFF A GENE of interest in a temporally and spatially controlled manner is the key to the analysis of gene function in developing organisms. Although both loss-of-function and gain-of-function experiments can be used to investigate the function of a particular gene, loss-of-function phenotypes are often more informative (Hudson et al. 2002). The choice of specific promoters allows both the temporal and spatial control of ectopic gene expression in many organisms. However, the induction of controlled loss-of-function phenotypes has been more difficult. With most loss-of-function approaches, targeted genes are already turned off at the single-cell stage. This can be a problem when gene function must be analyzed during later stages of development, for instance, during the development of the nervous system. If a gene has an additional role in early development, its loss of function can lead to embryonic lethality and thus prevent functional analysis in the structure of interest. Alternatively, if embryos survive and develop into mature animals, redundant genes may have masked the loss of a particular gene's function. Both of these possibilities have been a problem faced by neuroscientists interested in gene function during nervous system development. In addition, there is the limitation that development and function of the nervous system can only be studied in a living animal. Thus, good experimental model systems for functional gene analysis are rare.

Because technologies for gene manipulation are available in mice based on homologous recombination in embryonic stem (ES) cells, the mouse has become the most widely used vertebrate model system (Müller 1999; Jackson 2001). However, the generation of knockout mice is time-consuming and expensive. Often, the resulting phenotype is disappointing, and more sophisticated gene-silencing strategies are required for functional gene analysis, multiplying the expense of time and money (Porter 1998; Ihle 2000).

For decades, the chicken embryo was a classical model system for developmental studies in vertebrates due to its easy accessibility for in vivo manipulations (for a review, see Swartz et al. 2001). A major disadvantage has been the lack of genetic tools, a drawback that has hampered its use in functional gene analysis. Gene transfer with viral vectors was used successfully for lineage analysis (Petropoulos and Hughes 1991; Leber et al. 1996; Morgan and Fekete 1996), but it has been of limited usefulness for interference with gene expression. Gene transfer with viral vectors is tedious and is only successful when both the vector and cDNA of interest fulfill certain requirements of specificity and length. On the other hand, gene transfer with viral vectors is often not efficient enough to allow functional gene analysis.

Similarly, lack of efficiency has been the major problem associated with antisense oligonucleotides (Morales and de Pablo 1998; Lebedeva and Stein 2001). Morpholinos, an improved version of antisense oligonucleotides, suffer from the same problem, although their usefulness in zebrafish has been reported in several studies (for review, see Heasman 2002). Because morpholinos block translation only when they are designed to be complementary to the 5′ leader sequence or the translation start site, their usefulness for many applications is further restricted.

The problem of transfection efficiency was addressed by Muramatsu et al. (1997), who compared nonviral methods for gene transfer in ovo. These authors introduced in ovo electroporation as an effective method for gene transfer in chicken embryos. Subsequently, in ovo electroporation has been used successfully in different studies, as summarized in a review by Itasaki et al. (1999). However, the problem of obtaining loss-of-function phenotypes remained. Unless dominant-negative mutants of proteins could be expressed, loss-of-function approaches are not feasible, limiting the usefulness of this novel technique for most applications.

In the postgenomic era, the elucidation of the physiological function of individual genes has become the rate-limiting step in our quest for understanding the development and the function of living organisms. Although genes expressed under given conditions, e.g., healthy versus diseased or treated versus control conditions, can be identified with high-throughput methods, their functional analysis often requires the context of the entire organism and is therefore far more complicated and time-consuming.

As mentioned above, good vertebrate model systems for functional gene analysis are rare. This is especially true when the demands of functional genomics are added to the list of required traits. An ideal model system for functional genomics should allow efficient induction of loss-of-function phenotypes, be easy to handle, and be comparable in its development and function to the mammalian system, or better yet, humans. With the possibility of gene silencing by in ovo RNA interference (RNAi), the chicken embryo fulfills many of these requirements for investigating developmental issues and thus makes a good vertebrate model system for functional gene analysis.

## THE CHICKEN EMBRYO AS A MODEL TO STUDY GENE FUNCTION DURING THE DEVELOPMENT OF THE NERVOUS SYSTEM

Due to its easy accessibility, we have always favored the chicken embryo as a model system in studies of the development of the central nervous system and, in particular, of the molecular mechanisms of axonal pathfinding (Stoeckli 1997; Stoeckli and Landmesser 1998a,b; Sonderegger et al. 2000; Perrin et al. 2001). To study the functional role of individual candidate molecules as guidance cues, we have used function-blocking antibodies to perturb the interaction of guidance cues with their binding partners in vivo (Stoeckli and Landmesser 1995; Burstyn-Cohen et al. 1999; Fitzli et al. 2000; Perrin and Stoeckli 2000; Perrin et al. 2001). Using this approach to study pathfinding of commissural axons across the ventral midline of the spinal cord, we have shown that the cell adhesion molecules (CAM) axonin-1 and NrCAM are responsible for midline crossing (Stoeckli and Landmesser 1995; for review, see Stoeckli and Landmesser 1998a) and that F-spondin is required for the restriction of the axons' turn into the longitudinal axis (Burstyn-Cohen et al. 1999). All of these in vivo experiments were possible because function-blocking antibodies of candidate guidance cues or recombinant proteins were available.

To tackle the question of why commissural axons turn rostrally rather than caudally after midline crossing, however, we needed a different approach because no candidate guidance cues were known. Thus, we used a screen based on subtractive hybridization (V. Pekarik et al., in prep.) to identify candidate guidance cues for the longitudinal axis. We dissected floor-plate cells from embryos of different stages to isolate mRNA from the two pools. The subtractive hybridization of the derived cDNAs resulted in a pool of differentially expressed genes containing potential guidance cues needed by the commissural axons to turn rostrally rather than caudally along the contralateral floor-plate border. In a first step, we used in situ hybridization to select the cDNA fragments with an expression pattern that was compatible with a role as guidance cue. The second step was to sequence these cDNA fragments to obtain some information about their identity. As expected, the comparison of our sequence data with the NCBI database gave us no information for a number of our candidates. But, to include novel genes in our detailed analysis and to reduce the number of genes to be cloned, we needed to develop an assay that allowed us to select those genes among the candidates that are functionally involved in commissural axon guidance. RNAi as discovered in *Caenorhabditis elegans* (Fire et al. 1998) seemed to be a perfect tool for this purpose, although the mechanism of gene silencing by RNAi is still not fully understood (Carthew 2001; Caplen 2002; Hannon 2002; Hutvágner and Zamore 2002). Although results from zebrafish were controversial (Wargelius et al. 1999; Li et al. 2000; but see Oates et al. 2000; Zhao et al. 2001) and results from mouse (Svoboda et al. 2000; Wianny and Zernicka-Goetz 2000; Billy et al. 2001) were not applicable to our situation where genes had to be silenced in a comparatively much older embryo, we decided to try to use RNAi as the basis for gene silencing in the chicken embryo and to use this method to screen for genes involved in commissural axon pathfinding.

Due to its tube-like structure, the neural tube forms a perfect reservoir for nucleic acids injected into its lumen. The consistency of the tissue in early chicken embryos allows for diffusion of the injected molecules throughout the neural tube, but the basal lamina that forms a tight barrier after the third day of incubation (around stage 20 according to Hamburger and Hamilton [1951]) prevents the injected molecules from leaving the neural tube. Therefore, injected molecules are kept at relatively high concentrations around the cells that are to be transfected. Electroporation has been demonstrated to be a very efficient method for transfection in vivo (Muramatsu et al. 1997; Momose et al. 1999; Swartz et al. 2001). Using yellow fluorescent protein (YFP) as a marker, we determined that 60% of the cells in the electroporated area of the spinal cord expressed the transgene (Pekarik et al. 2003). Furthermore, the expression of the transgene could be targeted to specific areas of the spinal cord and to selected cell types by varying the position of the electrodes and the time point of electroporation (Figure 14.1). As expected, the injection of the plasmid encoding YFP without electroporation did not result in efficient transfection of cells. Only very few, if any, cells took up the plasmid and expressed YFP (Figure 14.1f).

## Injection and Electroporation of dsRNA Results in Specific Down-regulation of the Targeted Gene

In contrast to reports regarding mammalian cells, where the transfection of double-stranded RNA (dsRNA) resulted in unspecific effects on protein synthesis (for references, see Billy et al. 2001; Elbashir et al. 2001), no such effects were seen in chicken embryos after in ovo RNAi (Pekarik et al. 2003). In embryos treated with axonin-1 dsRNA, only axonin-1 was down-regulated on the electroporated side compared to the control side of

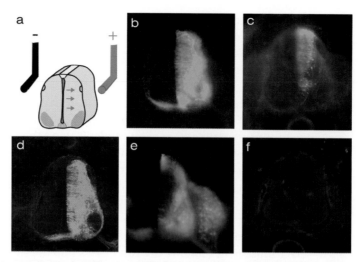

**FIGURE 14.1.** In ovo electroporation is an efficient method for selective cell transfection in vivo. The expression of yellow fluorescent protein (YFP) was used as a measurement for the efficiency of in ovo electroporation. Embryos were injected in ovo with a plasmid encoding YFP under the β-actin promoter. For electroporation, platinum electrodes were placed on either side of the embryo at the lumbosacral level of the spinal cord (*a*). Depending on the time course and the position of the electrodes, the plasmid can be targeted to different population of cells. (*b*) We determined that ~60% of the cells in the electroporated area of the spinal cord take up the plasmid and express the transgene. Placing the electrodes in a dorsal position targets YFP expression to dorsal cell populations; motorneurons are not affected (*c*). Using embryos at stage 20 (*d*) with the electrodes positioned at an intermediate dorsoventral level excludes motorneurons from taking up the plasmid (compared to *b*, where electroporation was at stage 17). Using even younger stages for injection and electroporation results in YFP expression in motorneurons and sensory neurons in the dorsal root ganglia (*e*). The injection of the YFP plasmid without electroporation results in no or very few green cells (*f*).

the spinal cord (Figure 14.2a), but the expression of the closely related molecules NrCAM and NgCAM was not affected (Figure 14.2b,c). Their expression levels were the same on both sides of the spinal cord. Similarly, when we isolated the electroporated area of the spinal cord and cut it into the two halves, we could show a decrease of axonin-1 on western blots between the electroporated side and the control side of at least 21%. The signal for NrCAM did not differ between the two sides.

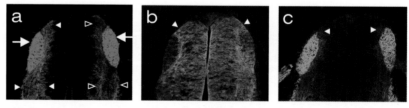

**FIGURE 14.2.** In ovo RNAi results in specific down-regulation of the targeted gene product. The injection of dsRNA derived from the cDNA encoding axonin-1 resulted in a marked decrease of axonin-1 protein on the electroporated (*a; right side, open arrowheads*) compared to the control side (*left side, arrowheads*). In the same section, staining for a related cell adhesion molecule, NrCAM, showed no difference between the electroporated and the control side (*b*). Similarly, the expression of another cell adhesion molecule, NgCAM, was not affected by axonin-1 dsRNA injection and electroporation (*c*). The section shown in *c* is an adjacent section to the one shown in *a* and *b*. Note that the expression of axonin-1 in the dorsal root entry zone (arrows in *a*) is the same on both sides of the spinal cord because the sensory neurons of the dorsal root ganglia were not affected under the experimental conditions used here. (Adapted, with permission, from Pekarik et al. 2003.)

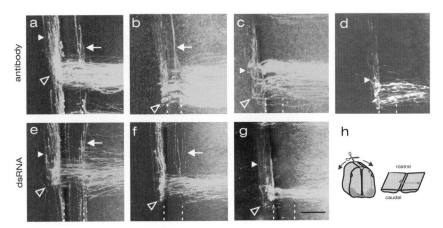

**FIGURE 14.3.** In ovo RNAi results in specific loss-of-function phenotypes. Due to our previous in vivo studies (Stoeckli and Landmesser 1995), we knew the loss-of-function phenotypes resulting from perturbation of axonin-1, NrCAM, and NgCAM interactions during commissural axon pathfinding. To demonstrate the specificity of in ovo RNAi, we compared the phenotypes resulting from pertur- bation at the protein level (*a–c*) with those resulting from gene silencing with dsRNA (*e–g*). Embryos were injected repeatedly with function-blocking antibodies against axonin-1 (*a*), NrCAM (*b*), or NgCAM (*c*) during the time of commissural axon pathfinding. The trajectories of commissural axons were analyzed in so-called open-book preparations (*h*). This is a whole-mount preparation of the spinal cord that is cut at the roof plate, as indicated by the schematic drawing in *h*. Trajectories of commissural axons are visualized by application of the lipophilic dye Fast-DiI to the cell bodies as detailed in the Techniques section. The perturbation of axonin-1 interactions by either the injection of function-blocking antibodies (*a*) or in ovo RNAi (*e*) resulted in pathfinding errors (*arrows*). Commissural axons turned erroneously along the ipsilateral floor-plate border and failed to cross the floor plate (indicated by dotted lines in *a* through *g*). In addition, commissural axons grew in a defas- ciculated manner along the contralateral floor-plate border (*arrowhead*) and turned within a wider area (*open arrowheads*). The perturbation of NrCAM interactions by function-blocking antibodies (*b*) and in ovo RNAi (*f*) resulted in pathfinding errors, but did not affect the fasciculation of commissur- al axons compared to control embryos (*d*). The perturbation of NgCAM interaction by function- blocking antibodies (*c*) and by in ovo RNAi (*g*) resulted in a defasciculated growth pattern but was without effect on pathfinding. Bar, 50 μm. (Adapted, with permission, from Pekarik et al. 2003.)

## Gene Silencing by In Ovo RNAi Results in Specific Loss-of-function Phenotypes

The specificity of in ovo RNAi was best demonstrated by the functional analysis of known guidance cues for commissural axons (Figure 14.3) (Pekarik et al. 2003). We confirmed the roles of axonin-1, NrCAM, and NgCAM in commissural axon guidance in the embry- onic chicken spinal cord that were known based on our in vivo loss-of-function assays at the protein level (Stoeckli and Landmesser 1995; Stoeckli et al. 1997; Fitzli et al. 2000). The loss-of-function phenotypes induced by function-blocking antibodies and by in ovo RNAi were identical (Figure 14.3). Axonin-1 loss-of-function resulted in the failure of commissural axons to cross the midline and in their premature turn along the ipsilateral floor-plate border (Figure 14.3a,e). In addition, axons grew in a defasciculated manner both toward the floor plate (not shown) and along the contralateral floor-plate border (arrowheads in Figure 14.3a,e). The perturbation of NrCAM function resulted in pathfinding errors but had no effect on fasciculation (Figure 14.3b,f). NgCAM loss-of- function resulted in a strong defasciculation of commissural axons but did not affect their pathfinding (Figure 14.3c,g). The fact that axonin-1, NrCAM, and NgCAM are related cell adhesion molecules, together with the finding that in ovo RNAi was capable of reflecting relatively subtle differences in growth behavior of commissural axons when related genes were targeted, best demonstrates the specificity of gene silencing.

The localization of the derived dsRNA with respect to the full-length cDNA should not matter in order to make in ovo RNAi a valuable tool for functional genomics. In particular, the translation start site should not have to be part of the targeted sequence, because the 5′ end of cDNAs is often not part of the cDNAs retrieved in screens. Screens based either on subtractive hybridization or differential display yield cDNA fragments of variable length and origin. Therefore, it was important to show that the same loss-of-function phenotypes were obtained with different fragments produced from the full-length cDNAs of our target genes. For this purpose, we used at least three different fragments of each full-length cDNA encoding either axonin-1, NrCAM, or NgCAM (see Pekarik et al. 2003). No difference in efficiency or nature of the induced phenotypes was detectable.

## What Does In Ovo RNAi Do for Me?

The possibility of gene silencing in chicken embryos by in ovo RNAi is rejuvenating the chicken embryo as a classical model system for experimental biology. The drawback of not having any tools for genetic manipulations in chickens has been overcome by in ovo RNAi and electroporation, two methods that allow for the temporal and spatial control of gene expression. Therefore, the chicken embryo itself is a good choice as a model system for developmental studies. However, the usefulness of the chicken embryo is not limited to projects focusing on its own development. Rather, the chicken embryo may be used more generally as a model organism for functional genomics.

Investigators may ultimately want to work with mice, but it may be advantageous to start out with chicken embryos. One possibility is to use chicken tissue as starting material for screens and only later clone the orthologs of the identified candidate genes in rodents after their number has been reduced by functional analysis. Another possibility is using rodent tissue for the screen and getting homologous chicken expressed sequence tags (ESTs) for functional analysis. Switching species will become more and more easy with the genome projects under way or near completion and with the compilation of EST databases for the chick as well (Boardman et al. 2002; http://www.chick.umist.ac.uk). No matter what strategy is used, the results of the functional in vivo screen in the chicken embryo can be used to decide on the most promising gene knockout strategy in mice (Porter 1998; Mueller 1999).

In addition to being a powerful tool for functional genomics, in ovo RNAi can be the method of choice for the evaluation of an individual gene's function during development, for example, for the analysis of genes that have been identified on the basis of their function in the adult nervous system. Because chicken embryos have been used for a variety of studies addressing patterning, cell migration, axon guidance, and apoptosis, the number of assays available for functional gene analysis is too large to be included in this chapter. For a source for detailed protocols, see the specialized literature (Bronner-Fraser 1996). We provide the experimental procedures used to study commissural axon pathfinding in the embryonic chicken spinal cord (see Pekarik et al. 2003), along with related protocols for staining of sections/slices and whole embryos (Perrin and Stoeckli 2000).

# TECHNIQUES

## PROTOCOL 1: PRODUCTION OF DsRNA

dsRNA is produced from different fragments of cDNAs encoding axonin-1, NrCAM, NgCAM, or candidate genes cloned in plasmids with an SP6 and a T7 promoter flanking the insert (e.g., pSP72 or pCRII).

### Procedure

**▼ CAUTION**

*See Appendix for appropriate handling of materials marked with <!>.*

---

**MATERIALS**

**REAGENTS**

Acidic phenol:chloroform (phenol:chloroform:isoamyl alcohol 25:24:1) <!>

Ammonium acetate (7.5 M)

cDNA fragment of the gene of interest in plasmid with SP6 and T7 promoters

Chloroform:isoamyl alcohol (24:1) <!>

DEPC-treated $H_2O$

DNase I (10 units/µl) ( Roche, Basel, Switzerland)

EDTA (0.5 M, pH 8.0)

Ethanol (absolute and 70%) <!>

dNTPs (100 mM) (Roche)

Phosphate-buffered saline (PBS)

     137 mM NaCl

     2.7 mM KCl

     8 mM $Na_2HPO_4$ <!>

     1.5 mM $NaH_2PO_4$ (pH 7.4) <!>

RNase I

RNAsin (30 units/µl) (Promega)

SP6 polymerase (10–20 units/µl) (Roche)

T7 polymerase (15 units/µl) (Promega)

**EQUIPMENT**

Centrifuge

Gel electrophoresis equipment

Vortex

Water bath

---

1. Linearize the cDNA with proper restriction endonucleases, and then carry out RNA in vitro synthesis from T7 and SP6 promoters. For each direction, mix 2 µg of the lin-

earized plasmid DNA with a final concentration of 4 mM dNTPs, 2 μl of SP6 polymerase (10–20 units/μl) or T7 polymerase (15 units/μl), and 0.5 μl of RNAsin (30 units/μl) in the appropriate transcription buffer (total volume for transcription is 20 μl). Incubate for 2 hours at 37°C. Collect 1 μl of the samples and keep for analysis.

2. Stop the transcription and add 2 μl of DNase I to remove template DNA. Incubate for 30 minutes at 37°C. Collect 1 μl of the samples and keep for analysis.

3. Add 20 μl of DEPC-treated H$_2$O and mix well. Add a mixture of 2 μl of 0.5 M EDTA and 22 μl of ammonium acetate. Extract the RNA with 1 volume of acidic phenol:chloroform (phenol:chloroform:isoamyl alcohol 25:24:1) followed by extraction with 1 volume of chloroform:isoamyl alcohol (24:1).

4. Precipitate with ethanol (by adding 2.5 volumes of absolute ethanol). Wash the pellet with 70% ethanol. Air dry the pellet and then dissolve the RNA in 20 μl of PBS. For quality control, analyze 1 μl of each sample by gel electrophoresis.

5. Mix equal amounts of sense and antisense RNA, heat to 95°C for 5 minutes, and reanneal by slowly cooling over several hours by simply turning off the power to the water bath. Monitor the annealing and quality of the dsRNA by electrophoresis under nondenaturing conditions and after digestion with 0.5 μg/ml RNase I.

6. Store samples at –70°C for several weeks.

## PROTOCOL 2: INJECTION OF DsRNA AND ELECTROPORATION IN OVO

### Procedure

**▼ CAUTION**
*See Appendix for appropriate handling of materials marked with <!>.*

**MATERIALS**

**REAGENTS**

Ethanol (70%) <!>
Fertilized chicken eggs
Paraffin
Phosphate-buffered saline (PBS, see Protocol 1)
Trypan Blue (0.4%) (Invitrogen)

**EQUIPMENT**

Brush
Coverslip
Electrode puller
Egg incubator
Forceps, sterile
Injection needles pulled from borosilicate glass on a Narashige PC-10 electrode puller
Hot plate to melt paraffin
Light source for egg candling
Platinum electrodes (BTX, Genetronics, San Diego, California)
Pointed scissors
Polyethylene tubing
Scalpel
Scotch Tape
Spring scissors, sterile
Square pulse electroporator (BTX, Genetronics, San Diego, California)
Syringe with 18-gauge needle

1. Place fertilized eggs in an incubator at 38.5–39.5°C with at least 45% humidity.

   It is usually sufficient to place a water-filled tray with maximal surface area at the bottom of the incubator.

2. On the third day of incubation (or earlier, if required by the experiment), cut a window into the eggshell. To do this, place the egg in a stable position on its side to allow the embryo to float on top of the yolk.

   The trays that are used for the delivery of eggs by most suppliers are well suited for this purpose.

3. Use a strong light source to candle the egg after ~20 minutes, when the embryo has acquired its stable position. Mark the position of the embryo on the shell with a pencil (Figure 14.4). Wipe the egg with 70% ethanol. Use a scalpel to make a small hole at the blunt end of the egg. Outline the position of the window with four holes marking the corners.

4. Remove 2–3 ml of albumin through the hole at the blunt end of the egg so that the embryo falls down onto the yolk and does not stick to the shell. Use pointed scissors to cut a window into the eggshell above the embryo.

   Place a piece of Scotch Tape onto the shell before cutting to prevent small pieces of eggshell from falling into the egg (Figure 14.4b).

5. Seal the egg by applying melted paraffin to the edge of the window with a brush and by covering the hole with a coverslip (Figure 14.4c). Alternatively, use Scotch Tape to close the window.

   The method using paraffin and a coverslip is more time-consuming than sealing eggs with tape, but reopening and closing the window are easier and faster with this method. Proper sealing is important to prevent dehydration of the embryo.

6. Stage the embryos according to Hamburger and Hamilton (1951) at the time of injection and analysis.

7. Before injections, carefully remove the extraembryonic membranes above the embryo using spring scissors and forceps.

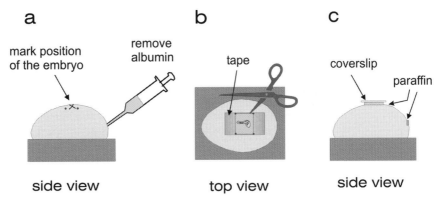

FIGURE 14.4. Chicken embryos can easily be accessed through a window in the eggshell. Fertilized eggs are incubated in a stable position on their side before opening (*a*). The position of the embryo is marked with a pencil. Through a hole in the shell at the blunt end of the egg, 2–3 ml of albumin is removed (*a*). To prevent pieces of the shell from falling inside the egg while cutting the window with pointed scissors, a piece of tape is placed onto the eggshell (*b*). The egg is closed with melted paraffin and a coverslip (*c*).

## a

## b

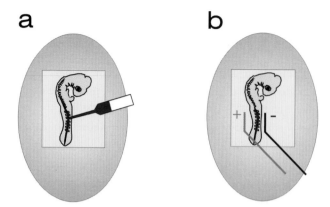

**FIGURE 14.5.** In ovo RNAi. Using a hot iron, melt the paraffin used to seal the egg with a coverslip, remove the coverslip, and carefully remove the extraembryonic membranes covering the embryo. A glass needle with a tip diameter of ~5 μm is used to inject the dsRNA. Adding Trypan Blue to the solution helps to control the site of injection and allows an estimate of the injected volume (*a*). For electroporation, electrodes are positioned on either side of the embryo at the desired level of the spinal cord (*b*). The position of the electrodes and the time point of electroporation can be varied to affect different types of cells (see Figure 14.1).

8. Break off the tip of the injection needle to obtain a tip diameter of ~5 μm.

   The tip diameter of the injection needle should be kept as small as possible to prevent leakage of the injected solution and/or damage to the embryo. To maximize the effect, avoid leakage of the injected dsRNA (see the panel on Troubleshooting at the end of this chapter).

9. For injections, attach the injection needle to a piece of polyethylene tubing and fill it with the solution containing the dsRNA and 0.04% Trypan Blue.

   The Typan Blue is used to visualize the distribution of the injected solution and to estimate the injected volume (Figure 14.5). In addition, the blue color helps to spot potential leakage of the injected solution from the embryo.

10. Inject the dsRNA into the central canal of the spinal cord (Figure 14.5a).

    Depending on the age of the embryos, the injected volumes should range from 0.1 μl to 0.5 μl. Add a few drops of cold PBS to the egg.

11. Place the platinum electrodes on either side of the embryo, parallel to its longitudinal axis for electroporation (see Figure 14.5b). The number of pulses and the voltage delivered by the electroporator is determined according to the age of the embryos.

    For three 1-day-old embryos, we routinely use 5 pulses of 50 msec duration and 26 V with a distance of 4 mm between the two electrodes.

12. Add a few more drops of cold PBS. Reseal the egg and place it in the incubator.

13. Rinse the electrodes immediately after electroporation with a generous amount of water to remove denatured proteins from their surface.

## PROTOCOL 3: ANALYSIS OF AXONAL PATHFINDING

### Method 1: Dissection of Spinal Cords for Open-book Preparations and DiI Labeling

This method was used to analyze commissural axon guidance (see Figure 14.3).

## Procedure

**MATERIALS**

**REAGENTS**

Fast-DiI

Dissolve at 5 mg/ml in methanol (Molecular Probes) or use a comparable lipophilic dye.

Methanol <!>

Paraformaldehyde (4%) in PBS <!>

Phosphate-buffered saline (PBS, see Protocol 1)

**EQUIPMENT**

Dissection tools (spring scissors, forceps)

Injection needles (see Protocol 2)

Insect pins

Petri dish

Coat the dish with Sylgard (World Precision Instrument, Berlin) according to manufacturer's instructions.

1. Sacrifice embryos 2 days after electroporation (see Protocol 2).

2. Remove the spinal cord from the embryos. Cut the roof plate and flip the cord open. Pin it down with insect pins in a Petri dish coated with Sylgard.

3. Fix the spinal cord in 4% paraformaldehyde for 2.5 hours at room temperature.

4. Rinse the spinal cords with PBS.

5. Apply Fast-DiI dissolved in methanol (5 mg/ml) to the cell bodies of commissural axons (for further details, see Perrin and Stoeckli 2000).

   Use glass needles with a tip diameter of less than 5 μm to inject very small volumes of the dye.

### Method 2: Analysis of Whole-mount Preparations

For the analysis of the innervation pattern of the hindlimb (Figure 14.6), we have used embryos at different stages up to about stage 33 (7 days of incubation). Using an antibody against the 160-kD subunit of neurofilaments to stain nerve fibers allows for the comparison of the branching pattern of motor and sensory neurons between control and experimental embryos.

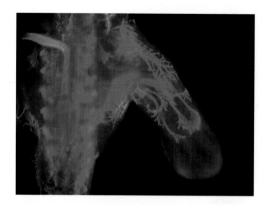

**FIGURE 14.6.** Whole-mount staining of chicken embryos is an efficient method to check for axon guidance phenotypes induced by RNAi. To assess whether a candidate gene has an effect on axon growth or guidance, the entire embryo can be stained with an antibody against the 160-kD isoform of neurofilaments (for details, see Protocol 3, Method 2). Potential phenotypes can easily be detected by comparing the two sides of the body after in ovo RNAi. Whole-mount neurofilament staining followed by tissue clearing can be used to visualize motor and sensory nerves in a developing limb bud. The left limb bud of an embryo at stage 26 is shown in a ventral view.

## Procedure

### MATERIALS

**REAGENTS**

Antibodies

- Monoclonal antibody against 160-kD neurofilament subunit (RMO 270; Zymed, South San Francisco, California)
- Appropriate secondary antibody, e.g., goat anti-mouse IgG Cy3 (Jackson ImmunoResearch, West Grove, Pennsylvania)

Benzyl alcohol:benzyl benzoate (BABB) (1:2) <!>

Blocking buffer (10% fetal calf serum in PBS)

Lysine (20 mM) in 0.1 M sodium phosphate (pH 7.4)

Methanol, graded series (25%, 50% 75% in $H_2O$; see Step 11) <!>

Paraformaldehyde (4%) in PBS <!>

Phosphate-buffered saline (PBS) (see Protocol 1)

Triton X-100 (1%) in PBS

**EQUIPMENT**

Forceps

Horizontal shaker

Petri dish

   Coat with Sylgard (World Precision Instruments) according to manufacturer's instructions.

Plate (24-well)

1. Sacrifice the embryos by decapitation and remove the internal organs with forceps.

2. Rinse the embryos with PBS, pin them down with insect pins in a Petri dish with Sylgard coating (see Method 1), and fix them in 4% paraformaldehyde for 2.5 to 3 hours at room temperature.

3. Rinse the embryos with PBS.

4. Incubate the embryos in 1% Triton-X-100 in PBS for 1 hour to permeabilize the tissue.

   For this and all subsequent steps, keep the embryos in a 24-well plate or small glass vials on a horizontal shaker at low speed to prevent tissue damage.

5. Incubate the embryos in 20 mM lysine in 0.1 M sodium phosphate (pH 7.4) for 1 hour to quench background fluorescence.

6. Rinse the tissue in PBS three times for 10 minutes each, followed by incubation in blocking buffer for 2 hours.

   The recipe for the blocking buffer may have to be adjusted to the requirements of the specific antibodies used.

7. Incubate the embryos on a horizontal shaker for at least 48 hours at 4°C with the first antibody (monoclonal anti-neurofilament antibody) diluted 1:1500 in blocking buffer.

8. Rinse the tissue with five changes of PBS followed by an incubation in PBS for 12 hours at 4°C.

9. Incubate the embryos in blocking buffer for at least 2 hours before adding the secondary antibody, goat anti-mouse IgG Cy3, diluted 1:250 in blocking buffer. Incubate for 5 hours at room temperature in the dark.

10. Rinse the embryos with at least five changes of PBS, followed by incubation in PBS for at least 1 day at 4°C.

    To avoid background, it is extremely important to rinse the unbound secondary antibody before clearing the tissue.

11. To clear the tissue, dehydrate the embryos in a graded methanol series (25%, 50%, 75%, 2x 100%, 30 minutes each step). Then incubate in BABB.

    Clearing takes only a few minutes, but embryos can be stored in BABB for weeks at 4°C in the dark.

## Method 3: Staining of Tissue Slices

If the branching pattern must be analyzed in older embryos or in regions that are not accessible by whole-mount analysis, the tissue can be cut in slices using a vibratome or tissue slicer. Slices can also be used for tracing of axon tracts with lipophilic dyes rather than immunohistochemical detection of all neurites with anti-neurofilament antibodies (for details, see Perrin and Stoeckli 2000).

## Procedure

▼ **CAUTION**
*See Appendix for appropriate handling of materials marked with <!>.*

---

**MATERIALS**

**REAGENTS**

Antibodies and solutions for staining as detailed above (Method 2)

Embryos dissected and fixed as detailed in Method 2

Gelatin (15%) (high strength, Sigma) in PBS

Ultra-low gelling agarose (6.6%) (Sigma) in PBS

**EQUIPMENT**

Plate (24-well)

Tissue slicer (Electron Microscopic Sciences) or vibratome

---

**Low Survival Rate**

- *Technical skills.* Handling chicken embryos in ovo requires some manual skills that must be acquired by training. Obviously, it is crucial to avoid physical damage to the embryos.

- *Incubation conditions.* If the survival rate is low, even for those embryos that are not injected, the incubation conditions are most likely the cause. The temperature of the incubator should be between 38.5°C and 39.5°C. Lower temperatures can be tolerated much better than high temperatures. Make sure that the incubator is capable of regulating the temperature properly. Humidity is important! Usually water is added in a tray with a maximal surface area at the bottom of the egg incubator. To ensure that the incubation conditions for the chicken embryos are correct, sacrifice embryos at different time points and compare their development with the stages as described by Hamburger and Hamilton (1951). Note that the development of chicken embryos is strongly dependent on the temperature of the incubator.

- *Developmental stages.* If the development of the embryos is too fast or too slow, change the settings of the egg incubator. Eggs set at the same time should yield embryos of comparable stages. Large differences between embryos indicate poor incubation conditions or poor quality of the fertilized eggs. Fertilized eggs should be stored at 15°C for less than 8 days before incubation. It is advisable to keep track of developmental stages throughout the experiment as deviations from the normal time course may provide important information for troubleshooting. For proper analysis of the changes resulting from gene manipulations, stage the embryos according to the method of Hamburger and Hamilton (1951). Reporting the embryonic age as days of incubation is not acceptable as small differences in incubation parameters may lead to considerable differences in development.

## Problems with the Analysis

- *High background.* Insufficient washing of the tissue (number of changes of washing buffer, or total time of washing) or overfixation of the tissue are the most common reasons for high background. Furthermore, especially for whole-mount staining, the quality of the antibodies is crucial.

- *No staining.* As a positive control, use the same antibodies to stain tissue sections where permeabilization of the tissue is not a problem.

- *Low penetrance or strong variability of the effect.* A variability of the phenotype of choice may be due to suboptimal RNAi conditions. Make sure that the injected solution does not leak out of the neural tube when the glass needle is pulled back and that the electroporation is effective, e.g., by co-injecting a plasmid encoding GFP or another marker. If necessary, increase the voltage, the position of the electrodes, or reduce the distance between them. Make sure to compare the control-injected embryos with untouched control embryos.

- *Controls.* The comparison between control-injected and untouched embryos allows control for the absence of artifacts introduced by embryo handling and/or the buffer used for RNAi.

## Additional Issues

- *Embryonic stage.* The stage of the embryo at the time of injection and electroporation may have to be changed to improve the effectiveness of in ovo RNAi. The sensitivity of cells for electroporation changes with differentiation.

- *Expression pattern of target gene.* A detailed analysis of the expression pattern of the target gene is important. In particular, the half-life of the protein product of the target gene may affect the time point of injection and electroporation. Remember, proteins already synthesized and inserted into the membrane are not affected by RNAi. Therefore, proteins with slow turnover may require RNAi at earlier stages, preferably before the onset of expression.

- *Quality of the dsRNA.* Check the quality of the dsRNA used for injections. Make sure that the buffer used for injection does not have an effect on development and survival of the embryo. In particular, Tris buffers and buffers containing glycerol are not compatible with in vivo injections. Salt concentrations and pH values should be in the physiological range. Always use embryos injected and electroporated with the same buffer as that of the controls in addition to untouched control embryos.

1. Use a tissue slicer to cut tissue slices of 250-μm thickness.

    For chicken embryos up to stage 26, embedding in 6.6% ultra-low gelling agarose works very well. For older embryos, embedding in 15% gelatin in PBS works better.

2. Remove residual agarose or gelatin from the tissue slices before staining in 24-well plate.

3. Staining is the same as for whole-embryo staining (Method 2), but the incubation times can be shortened considerably. Sufficient permeabilization is achieved within 30 minutes. Incubate the first antibody overnight, rinse in PBS for a total of 1 hour, and then block for 1 hour. Incubate the secondary antibody for 5 hours at room temperature.

4. Rinse off the secondary antibody with at least five changes of PBS with a total incubation time of 2 hours.

## ACKNOWLEDGMENTS

I thank the members of my lab who have contributed to the development of in ovo RNAi, in particular Dimitris Bourikas and Vlad Pekarik, as well as Monika Mielich for her excellent technical assistance. We were supported by grants from the Swiss National Science Foundation, the Ott Foundation, and the Human Frontier Science Program Organization.

## REFERENCES

Billy E., Brondani V., Zhang H., Mueller U., and Filipowicz W. 2001. Specific interference with gene expression induced by long, double-stranded RNA in mouse embryonal teratocarcinoma cell lines. *Proc. Natl. Acad. Sci.* **98:** 14428–14433.

Boardman P.E., Sanz-Ezquerro J., Overton I.M., Burt D.W., Bosch E., Fong W.T., Tickle C., Brown W.R.A., Wilson S.A., and Hubbard S.J. 2002. A comprehensive collection of chicken cDNAs. *Curr. Biol.* **12:** 1965–1969.

Bronner-Fraser M., ed. 1996. Methods in avian embryology. In *Methods in cell biology*, vol. 51. Academic Press, San Diego.

Burstyn-Cohen T., Tzarfaty V., Frumkin A., Feinstein Y., Stoeckli E.T., and Klar A. 1999. F-spondin is required for accurate pathfinding of commissural axons at the floor plate. *Neuron* **23:** 233–246.

Caplen N.J. 2002. A new approach to the inhibition of gene expression. *Trends Biotechnol.* **20:** 49–51.

Carthew R.W. 2001. Gene silencing by double-stranded RNA. *Curr. Opin. Cell Biol.* **13:** 244–248.

Elbashir S.M., Harborth J., Lendeckel W., Yalcin A., Weber K., and Tuschl T. 2001. Duplexes of 21-nucleotide RNAs mediate RNA interference in cultured mammalian cells. *Nature* **411:** 494–498.

Fire A., Xu S., Montgomery M.K., Kostas S.A., Driver S.E., and Mello C.C. 1998. Potent and specific genetic interference by double-stranded RNA in *Caenorhabditis elegans*. *Nature* **391:** 806–811.

Fitzli D., Stoeckli E.T., Kunz S., Siribour K., Rader C., Kunz B., Kozlov S.V., Buchstaller A., Lane R.P., Suter D.M., Dreyer W.J., and Sonderegger P. 2000. A direct interaction of axonin-1 with NgCAM-related cell adhesion molecule (NrCAM) results in guidance, but not growth of commissural axons. *J. Cell Biol.* **149:** 951–968.

Hamburger V. and Hamilton H.L. 1951. A series of normal stages in the development of the chick embryo. *J. Morphol.* **88:** 49–92.

Hannon G.J. 2002. RNA interference. *Nature* **418:** 244–251.

Heasman J. 2002. Morpholino oligos: Making sense of antisense? *Dev. Biol.* **243:** 209–214.

Hudson D.F., Morrison C., Ruchaud S., and Earnshaw W.C. 2002. Reverse genetics of essential genes in tissue-culture cells: "Dead cells talking." *Trends Cell Biol.* **12:** 281–287.

Hutvágner G. and Zamore P.D. 2002. RNAi: Nature abhors a double-strand. *Curr. Opin. Genet. Dev.* **12:** 225–232.

Ihle J.N. 2000. The challenges of translating knockout phenotypes into gene function. *Cell* **102:** 131–134.

Itasaki N., Bel-Vialar S., and Krumlauf R. 1999. "Shocking" developments in chick embryology: Electroporation and in ovo gene expression. *Nat. Cell Biol.* **1:** E203–E207.

Jackson I.J. 2001. Mouse mutagenesis on target. *Nat. Genet.* **28:** 198–200.

Lebedeva I. and Stein C.A. 2001. Antisense oligonucleotides: Promise and reality. *Annu. Rev. Pharmacol. Toxicol.* **41:** 403–419.

Leber S.M., Yamagata M., and Sanes J.R. 1996. Gene transfer using replication-defective retroviral and adenoviral vectors. *Methods Cell Biol.* **51:** 161–183.

Li Y.-X., Farrell M.J., Liu R., Mohanty N., and Kirby M.L. 2000. Double-stranded RNA injection produces null phenotypes in zebrafish. *Dev. Biol.* **217:** 394–405.

Momose T., Tonegawa A., Takeuchi J., Ogawa H., Umesono K., and Yasuda K. 1999. Efficient targeting of gene expression in chick embryos by microelectroporation. *Dev. Growth Differ.* **41:** 335–344.

Morales A.V. and de Pablo F. 1998. Inhibition of gene expression by antisense oligonucleotides in chick embryos *in vitro* and *in vivo. Curr. Top. Dev. Biol.* **36:** 37–49.

Morgan B.A. and Fekete D.M. 1996. Manipulating gene expression with replication-competent retroviruses. *Methods Cell Biol.* **51:** 185–218.

Müller U. 1999. Ten years of gene targeting: Targeted mouse mutants, from vector design to phenotype analysis. *Mech. Dev.* **82:** 3–21.

Muramatsu T., Mizutani Y., Ohmori Y., and Okumura J. 1997. Comparison of three nonviral transfection methods for foreign gene expression in early chicken embryos in ovo. *Biochem. Biophys. Res. Commun.* **230:** 376–380.

Oates A.C., Bruce A.E.E., and Ho R.K. 2000. Too much interference: Injection of double-stranded RNA has nonspecific effects in the zebrafish embryo. *Dev. Biol.* **224:** 20–28.

Pekarik V., Bourikas D., Miglino N., Joset P., Preiswerk S., and Stoeckli E.T. 2003. Gene silencing by RNAi as an in vivo screen for gene function. *Nat. Biotechnol.* **21:** 93–96.

Perrin F.E. and Stoeckli E.T. 2000. Use of lipophilic dyes in studies of axonal pathfinding in vivo. *Microsc. Res. Technol.* **48:** 25–31.

Perrin F.E., Rathjen F.G., and Stoeckli E.T. 2001. Distinct subpopulations of sensory afferents require F11 or axonin-1 for growth to their target layers within the spinal cord of the chick. *Neuron* **30:** 707–723.

Petropoulos C. and Hughes S. 1991. Replication-competent retrovirus vectors for the transfer and expression of gene cassettes in avian cells. *J. Virol.* **65:** 3728–3737.

Porter A. 1998. Controlling your losses: Conditional gene silencing in mammals. *Trends Genet.* **14:** 73–79.

Sonderegger P., Welte W., and Stoeckli E.T. 2000. Sensing cues for axon guidance—From extracellular protein conformation to intracellular signalling. In *The ELSO Gazette: e-magazine of the European Life Scientist Organization,* Issue 1 (at http://www.the-elso-gazette.org/magazines/reviews/review1.asp).

Stoeckli E.T. 1997. Molecular mechanisms of growth cone guidance: Stop and go? *Cell Tissue Res.* **290:** 441–449.

Stoeckli E.T. and Landmesser L.T. 1995. Axonin-1, NrCAM, and NgCAM play different roles in the *in vivo* guidance of chick commissural neurons. *Neuron* **14:** 1165–1179.

———. 1998a. Axon guidance at choice points. *Curr. Opin. Neurobiol.* **8:** 73–79.

———. 1998b. Molecular mechanisms of growth cone guidance in the vertebrate nervous system. *Cell Commun. Adhesion* **6:** 161–181.

Stoeckli E.T., Sonderegger P., Pollerberg G.E., and Landmesser L.T. 1997. Interference with axonin-1 and NrCAM interactions unmasks a floor plate activity inhibitory for commissural axons. *Neuron* **18:** 209–221.

Svoboda P., Stein P., Hayashi H., and Schultz R.M. 2000. Selective reduction of dormant maternal mRNAs in mouse oocytes by RNA interference. *Development* **127:** 4147–4156.

Swartz M., Eberhart J., Mastick G.S., and Krull C.E. 2001. Sparking new frontiers: Using in vivo electroporation for genetic manipulations. *Dev. Biol.* **233:** 13–21.

Wargelius A., Ellingsen S., and Fjose A. 1999. Double-stranded RNA induces specific developmental effects in zebrafish embryos. *Biochem. Biophys. Res. Commun.* **263:** 156–161.

Wianny F. and Zernicka-Goetz M. 2000. Specific interference with gene function by double-stranded RNA in early mouse development. *Nat. Cell Biol.* **2:** 70–75.

Zhao Z., Cao Y., Li M., and Meng A. 2001. Double-stranded RNA injection produces nonspecific defects in zebrafish. *Dev. Biol.* **229:** 215–223.

CHAPTER **15**

# Guide to RNAi in Mouse Oocytes and Preimplantation Embryos

## Paula Stein and Petr Svoboda

*Department of Biology, University of Pennsylvania, Philadelphia, Pennsylvania 19104-6018*

B Y THE MECHANISM OF RNA INTERFERENCE (RNAi), double-stranded RNA (dsRNA) efficiently ablates gene function by promoting selective mRNA degradation (Fire et al. 1998). RNAi has been observed in unicellular organisms, plants, and in many distant animal taxa including bilaterals, acoelomates, pseudocoelomates, protostomes, and deuterostomes (for review, see Bosher and Labouesse 2000; Sijen and Kooter 2000). RNAi is closely related to other silencing mechanisms such as posttranscriptional gene silencing (PTGS) in plants and quelling in fungi. The RNAi field is rapidly developing in two major areas: (1) the analysis of this intriguing phenomenon and its role in living systems and (2) the use of RNAi as an experimental tool to silence genes of interest. This chapter reviews existing data on RNAi in mammals. In particular, the biology of RNAi in mouse oocytes and preimplantation embryos is analyzed and its potential as a method for studying development is evaluated.

## CONCEPTS AND STRATEGIES

### Response to dsRNA in Mammalian Cells

It has been known for more than a quarter of century that the exposure of mammalian cells to dsRNA, regardless of its sequence, triggers a global repression of protein synthesis (Hunter et al. 1975). In most mammalian somatic cells, dsRNA activates protein kinase PKR (Figure 15.1), which catalyzes phosphorylation of target molecules, such as the translation initiation factor eIF2$\alpha$, which in turn inhibits translation. PKR is also involved in the regulation of NF-$\kappa$B, which has a key role in interferon induction. Interferon and dsRNA also activate 2′,5′-oligoadenylate synthetase (2′,5′-OAS), leading to the production of 2′,5′-oligoadenylates with 5′-terminal triphosphate residues. These subsequently induce activation of RNase L, which is responsible for general RNA degradation (for review, see Barber 2001). PKR and 2′,5′-OAS are essential for the apoptotic response to dsRNA that has been demonstrated in null mutant mice (Der et al. 1997; Zhou et al. 1997). This response to dsRNA caused many investigators to doubt that RNAi functions in mammalian systems following the first reports on RNAi in *Caenorhabditis elegans* and *Drosophila* (Fire et al. 1998; Kennerdell and Carthew 1998). However, the initial reports about RNAi in the mouse showed that it can be efficiently used as a tool in oocytes and early embryos and that long dsRNAs do not cause nonspecific effects at these stages (Svoboda et al. 2000; Wianny and Zernicka-Goetz 2000). A deeper understanding of the

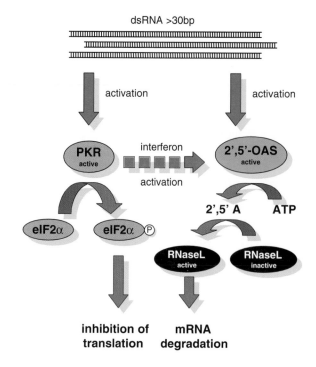

dsRNA >30bp

activation                    activation

PKR                interferon        2',5'-OAS
active                                active

activation

eIF2α        eIF2α ℗              2',5' A        ATP

RNaseL        RNaseL
active        inactive

inhibition of        mRNA
translation        degradation

## APOPTOSIS

**FIGURE 15.1.** Nonspecific response to dsRNA in mammalian somatic cells.

RNAi mechanism led to the development of an RNAi approach in somatic cells that bypasses the apoptotic response to dsRNA. During RNAi, dsRNA is processed to short 21–23-mers (termed short interfering RNA or siRNA), which induce RNAi but are too short to trigger the PKR/2′,5′-OAS pathway in somatic cells (Zamore et al. 2000; Elbashir et al. 2001a,b). Although the RNAi response in mammalian systems appears not to be as robust as in *C. elegans* or *Drosophila* (Svoboda et al. 2000; Ui-Tei et al. 2000), it can be efficiently used as a tool to analyze gene function and, moreover, has the potential to become a very useful gene-targeting method.

## Mammalian RNAi Pathway

The mechanism of dsRNA interference has been analyzed using genetic and biochemical approaches in nonmammalian species. The generation and subsequent analysis of RNAi-resistant *C. elegans* mutants (Ketting et al. 1999; Tabara et al. 1999) and the development of an in vitro RNAi system (Tuschl et al. 1999; Zamore et al. 2000) were crucial initial steps toward deciphering the RNAi mechanism (see also Chapters 4, 5, and 13). Numerous genes have been associated with RNAi and related silencing mechanisms in various organisms thus far. Currently, the RNAi mechanism (for review, see Bass 2000; Sharp 2001; Zamore 2001) is viewed as a multistep process initiated by the processing of dsRNA into siRNA by the ATP-dependent nuclease Dicer (Zamore et al. 2000; Bernstein et al. 2001). Dicer is also implicated in generating small temporal RNAs (stRNAs), which are regulatory RNA molecules that control the timing of development in *C. elegans* by inhibiting translation (Grishok et al. 2001; Hutvagner et al. 2001; Knight and Bass 2001). Homologs of stRNAs were found among animals with bilateral symmetry, suggesting that regulation of development by stRNA is a conserved mechanism (for review, see Rougvie 2001).

In the RNAi pathway, the siRNAs produced by Dicer serve as guide sequences for the RNA-induced silencing complex (RISC), a multicomponent nuclease that cleaves the target mRNA. Purified RISC is a large ribonucleoprotein complex biochemically separable from Dicer (Hammond et al. 2001). Protein microsequencing revealed that one constituent of this complex is AGO2, a member of the Argonaute family of proteins essential for gene silencing in other species (Fagard et al. 2000; Hammond et al. 2001). Members of the Argonaute family are highly similar to the translation initiation factor eIF4C, and they are involved in other processes such as posttranscriptional silencing (Fagard et al. 2000) and in stRNA-mediated translational repression (Grishok et al. 2001). In the final step, the RISC complex recognizes and cleaves the targeted mRNA one or more times at sites corresponding to the middle of the guiding nucleotide sequences, and this cleaved mRNA is ultimately degraded (Elbashir et al. 2001a). The RISC complex can efficiently target sense as well as antisense RNA strands but not dsRNA homologous to the "triggering" sequence (Elbashir et al. 2001a). The exact mechanisms of recognition of the targeted mRNA and its cleavage are still unknown.

Another silencing element well-conserved among *Arabidopsis, Neurospora*, and *C. elegans* is an RNA-dependent RNA polymerase (RdRP) (Mourrain et al. 2000). RdRP may serve as an amplifier of the RNAi response, extending both the duration and intensity of interference (Sijen et al. 2001). Alternatively, it may actually be a part of the mRNA degradation mechanism (Lipardi et al. 2001).

Although the RNAi mechanism in mammals has not yet been analyzed in great detail, there is some experimental evidence that the RNAi pathway in mammals is conserved and in principle very similar to that of *Drosophila* and *C. elegans* (Figure 15.2). Gene silencing by dsRNA in mammals is also achieved via sequence-specific mRNA degradation. The human Dicer family member is capable of generating siRNA from dsRNA substrates (Bernstein et al. 2001), and Dicer activity has been detected in several cell lines including embryonic carcinoma (EC) cells (Billy et al. 2001), embryonic stem (ES) cells, CHO-K1, mouse embryonic fibroblasts (S. Yang et al. 2001), as well as mouse oocytes and preimplantation embryos (P. Svoboda et al., unpubl.). In fact, siRNA can induce an RNAi effect in mammalian cell lines as well as in oocytes and early embryos (Elbashir et al. 2001b; Harborth et al. 2001; P. Svoboda et al., unpubl.). Highly similar mammalian protein sequences can be found for most of the genes known to be involved in RNAi (including Dicer and Argonaute family). Whether these similarities also reflect functional conservation remains to be verified. Our lab recently tested several RecQ family proteins in search of mammalian Mut-7 and Qde-3 orthologs (Mut-7 is essential for RNAi in *C. elegans* and Qde-3 is essential for quelling in *Neurospora*). Analysis of knockout mice revealed that none of the candidate *recQ* genes are involved in sequence-specific RNA degradation in mouse oocytes (Stein et al. 2003a), suggesting that possible differences exist among RNAi pathways in evolutionarily more distant species.

RNAi in mammals also exhibits slower kinetics and lower efficiency compared to RNAi in *Drosophila* and *C. elegans* (Svoboda et al. 2000; Ui-Tei et al. 2000). Interestingly, mammals most likely lack the RdRP ortholog. In general, the lack of an ortholog would not necessarily mean that an RdRP activity is not involved in RNAi in mammals, because an RdRP could be acquired via horizontal transfer from a viral genome and could replace the RdRP protein involved in RNAi. In fact, a putative hepatitis-virus-related mammalian RdRP has been described recently (Sam et al. 1998). However, our analysis of the RdRP activity in

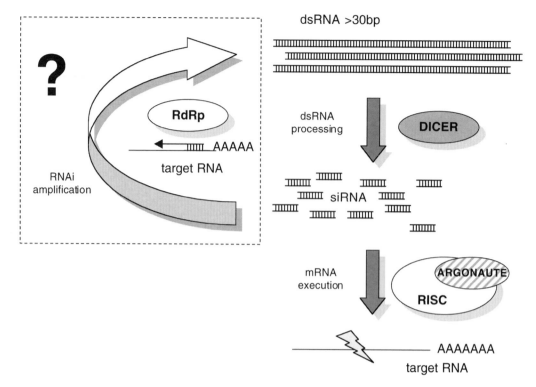

dsRNA >30bp

**FIGURE 15.2.** RNAi mechanism in mammals.

mouse oocytes indicates that RdRP activity is not involved in RNAi in this model system (Stein et al. 2003b). Whether the lack of RdRP in mammals results in the lower efficiency of RNAi remains unknown.

## Role of RNAi in Early Mammalian Embryos

RNAi and related mechanisms seem to be involved in silencing "parasitic" sequences such as those from viruses or transposable elements. In plants, there is a growing body of evidence that RNAi and other closely related silencing mechanisms are an essential part of a complex defense mechanism against transposable elements and RNA viruses (for review, see Waterhouse et al. 2001). There are currently more than 200 classified genera with more than 4000 species of viruses. RNA viruses (excluding reverse-transcribing viruses, which do not reproduce through a dsRNA stage during their cycle) constitute 7 of 12, 60 of 79, 6 of 22, and 50 of 89 described viral genera in fungi, plants, invertebrates and vertebrates, respectively (http://www.ncbi.nlm.nih.gov/ICTV/). The number of classified genera is not a very accurate estimation of the frequency of RNA viruses in the environment for obvious reasons; however, it can be inferred that RNA viruses which infect eukaryotic organisms are very abundant. One of the possible roles of RNAi could thus be an ancient and conserved form of antiviral defense. However, mammals evolved a specialized and highly complex immune system capable of dealing with a wide range of pathogens including viruses (in fact, the previously mentioned PKR/2′,5′-OAS pathways belong to this antiviral system), and our current knowledge suggests that the role of RNAi in the antiviral response in mammals is most likely minor, if any. RNAi in mammals is likely used to silence transposable elements and perhaps to control development

(Grishok et al. 2001), rather than serve as an antiviral defense. However, RNAi could still have a protective antiviral role in preimplantation embryos that are, in principle, exposed to viruses present in the female reproductive tract.

The first evidence that RNAi suppresses transposable elements was found during analysis of RNAi-resistant strains in *C. elegans* that showed mobilization of transposable elements (Ketting et al. 1999). Another independent line of evidence came from studies in *Drosophila* on transposon taming (transposable elements can invade virgin genomes within a few generations, after which the elements are "tamed" and retain only limited transpositional activity). It has been demonstrated that this taming is homology-dependent and mediated by RNA (Jensen et al. 1999). Finally, in *Trypanosoma*, isolated siRNAs were derived from a retrotransposon (Djikeng et al. 2001).

Mobile elements contribute greatly to genome plasticity and thus have a crucial role in the evolution of genomes. However, at the level of the organism, transposable elements are unwelcome guests because they can cause insertional mutations and chromosomal aberrations. Transposable elements comprise a large portion of mammalian genomes; for example, in humans, it is estimated that approximately 50% of the genome is occupied by four major classes of interspersed repetitive elements related to transposable elements (for more information, see Lander et al. 2001). Among the most abundant (and most active) mammalian classes of transposable elements are long interspersed nuclear elements (LINEs) and retrovirus-like elements, such as intracisternal A particles (IAPs). Both are autonomous elements transposing through RNA intermediates by a "copy and paste" mechanism, and hence are putative targets for an RNA-based repressive mechanism. Despite the large number of copies of many different types of transposable elements, the number of active transposable elements in a given mammalian genome is rather small due to mutations (Kazazian and Moran 1998; Lander et al. 2001).

It should be noted that organisms evolved mechanisms to repress transposable elements other than RNAi and that these mechanisms likely act in concert. Parasitic or invading sequences can be silenced posttranscriptionally by RNAi and related mechanisms, as well as at the transcriptional level by DNA methylation and/or histone deacetylation. Recent data from plants suggest that transcriptional and posttranscriptional silencing are mechanistically related (Mette et al. 2000; Morel et al. 2000; Sijen et al. 2001). It has been demonstrated that expression of IAPs, one of the most aggressive mobile sequences known in the mouse genome (Kuff and Lueders 1988; Kazazian 1998), is repressed by methylation (Walsh et al. 1998). Chromatin structure most likely also has a role in this repression because DNA methyl-binding proteins such as MeCP2 recruit complexes with histone deacetylase activity that alter chromatin structure (Jones et al. 1998; Nan et al. 1998). Whether methylation of transposable elements in mammalian genomes is somehow connected to RNAi remains unknown.

Interestingly, during preimplantation development, DNA methylation, including methylation of some transposable elements (Howlett and Reik 1991), decreases and reaches its minimum at the blastocyst stage (Reik et al. 2001; Santos et al. 2002) while several different transposable elements are expressed during this time period. Moreover, dsRNA derived from IAP and L1 retrotransposons was isolated from mouse cells (Kramerov et al. 1985). There is a high probability that transposition during preimplantation development would be transmitted through the germ line into future generations. Furthermore, reprogramming of the zygotic genome during zygotic gene activation provides a window of opportunity where parasitic sequences can temporarily escape transcriptional repression (Ma et al. 2001). Although transposable elements cannot be removed from the genome, they can be pacified by losing the ability to transpose due to mutations. Therefore, repression of their expression is an essential factor in increasing the probability that a given active transposable element mutates before it transposes. From this perspective, the requirement for protection of the germ line and preimplantation embryos against transposition correlates well with the presence of a specific RNAi response in oocytes, preimplantation embryos, ES cells, and EC cells. However, whether RNAi is also present in the male germ line is still unknown, and the experimental evidence that RNAi in mammals suppresses transposable elements has not yet been provided.

## RNAi as an Experimental Tool in Mammals

RNAi has become a widely used method to study gene function, especially in *C. elegans* and *Drosophila melanogaster*. In these two model systems, RNAi is currently a standard experimental tool that can be used even in a very large-scale analysis (for review, see Bargmann 2001; Barstead 2001) (see also Chapter 11). In contrast, there are not many reports on RNAi in vertebrate systems, particularly mammals. The potential for using RNAi to study gene function in nonmammalian vertebrates is unclear, as the data are contradictory (Wargelius et al. 1999; Li et al. 2000; Oates et al. 2000; Zhao et al. 2001). However, published results from mammalian systems as well as our unpublished data demonstrate that many genes in mammals can be inhibited efficiently by RNAi.

The first RNAi studies in mammals showed that microinjection of dsRNA greater than 500 bp is capable of inhibiting gene function in mouse oocytes and early embryos (Svoboda et al. 2000; Wianny and Zernicka-Goetz 2000). It was demonstrated that dsRNA directed toward *Mos* and tissue plasminogen activator (tPA) mRNAs in oocytes effectively results in the specific destruction of the targeted mRNA in both a time-dependent and concentration-dependent manner. Nonspecific effects of dsRNA in mouse embryos and oocytes were not observed, and it was shown that dsRNA is a more potent silencing agent than antisense RNA (Svoboda et al. 2000), which has been previously used in mammalian systems with varying success. More recently, it was shown that transfection of siRNA efficiently blocks gene function in several somatic cell lines without inducing the PKR/2′,5′-OAS pathway (Elbashir et al. 2001a; Harborth et al. 2001). Recently, it has also been shown that dsRNA expressed from a plasmid can induce the RNAi effect in mouse oocytes (Svoboda et al. 2001), providing an important step toward the development of a transgenic RNAi approach in mammals (Stein et al. 2003a), which is a very attractive tool for studying gene function in mouse oocytes and early embryos.

### Important factors influencing RNAi phenotypes

*RNAi trigger.* RNAi techniques useful in mammalian systems fall into three categories based on the form of dsRNA used.

- **Long dsRNA molecules.** Long dsRNA molecules can be made by annealing separated sense and antisense strands in vitro ("traditional dsRNA," see Protocol 1) or by in vitro transcription of an inverted repeat that results in an RNA hairpin. The inhibitory efficiencies of traditional dsRNA and hairpin dsRNA are very similar (Svoboda et al. 2001). The generation of RNA by in vitro transcription is relatively inexpensive compared to the cost of synthesizing siRNAs. Long dsRNA is probably most useful for studies of mammalian preimplantation embryos that can be microinjected and do not exhibit the nonspecific response to dsRNA. Long dsRNA should be used for studies with tissue culture cells with great caution, because it might cause nonspecific effects (Yang et al. 2001). Long dsRNAs have been transfected into several cell lines with varying success. EC cells and undifferentiated ES cells exhibited a specific RNAi response (Billy et al. 2001; Yang et al. 2001), whereas other cell lines (including differentiated ES cells) showed either nonspecific effects or no effect at all (Yang et al. 2001).

- **Short interfering RNA: siRNA.** The observation that dsRNA is processed into short dsRNA molecules was followed by the discovery that these short RNA molecules (siRNA) can themselves induce RNAi (Elbashir et al. 2001b). This led to the idea that siRNA could mediate efficient RNAi in mammalian somatic cells and bypass the PKR/2´,5´-OAS pathway that is activated by longer dsRNA molecules, which has allowed siRNAs to be used to target numerous genes in cultured mammalian cells (Elbashir et al. 2001a; Harborth et al. 2001). Efficient interference with siRNA has also been found in mouse oocytes and preimplantation embryos (P. Svoboda et al., unpubl.). Details about the appropriate design of siRNA can be found elsewhere (Elbashir et al. 2001c). The only disadvantage of siRNA-mediated RNAi is cost, because synthesis of RNA is expensive (for details, see http://www.dharmacon.com/sirna.html). Ideally, a gene should be targeted with several dsRNA oligonucleotides to assure maximum efficiency, because different siRNAs may have different efficiencies, as has been shown during analysis of vimentin (Elbashir et al. 2001a; Harborth et al. 2001). However, methods are being developed that would significantly cut the cost of siRNAs.

  One of these methods is the synthesis of siRNAs using short DNA templates and T7 polymerase (Donze and Picard 2002). Similarly, it is possible to transcribe in vitro short hairpin RNA (shRNA) instead of individual siRNA strands (Paddison et al. 2002b; Yu et al. 2002). Finally, long dsRNA can be processed in vitro into small fragments using RNase III from *E. coli* (Yang et al. 2002). The last approach generates a pool of siRNAs, which assures better targeting than an individual siRNA with unpredictable efficiency.

- **RNA expressed from a plasmid or transgene.** Expression of dsRNA is a promising strategy that could allow for the screening of large numbers of genes at a reasonable price in oocytes, embryos, and tissue culture cells, as well as for controlled timing of interference. dsRNA can be generated from an inverted repeat (IR) transcribed from a single promoter; one gene fragment can be transcribed in both directions using a dual promoter system; or a bidirectional promoter can be used (Figure 15.3).

There are reports describing induction of RNAi in mammalian cells by the expression of long dsRNA (Svoboda et al. 2001; Yang et al. 2001; Paddison et al. 2002a) and several reports describing various systems expressing short hairpins or siRNAs in mammalian cells (for review, see McManus and Sharp 2002). Both the long and short hairpin-expressing constructs were also successfully used to obtain an RNAi effect in transgenic mice (Hasuwa et al. 2002; Stein et al. 2003a).

shRNA-expressing systems were designed to avoid possible problems with the non-specific PKR-mediated response to long dsRNA in somatic cells. Despite their successful use in tissue culture, these systems have two disadvantages. First, there is an unpredictable variability in efficiency of different siRNAs or shRNAs. Second, current siRNA and shRNA expression systems can hardly be used in a tissue-specific manner because no tissue-specific promoters are available for them. This is a significant disadvantage because, for example, genes with a lethal phenotype could not be studied in oocytes or early embryos using this approach.

For expression of long dsRNA hairpins, reports from other model systems describe a variety of targeting constructs, which should be taken into account when considering expression of dsRNA in cells (see Figure 15.3). More systematic comparisons of different types of plasmids have been done in plants (Smith et al. 2000; Wesley et al. 2001).

In general, inverted-repeat-bearing plasmids are more difficult to generate, but they better assure formation of dsRNA. Piccin et al. (2001) included a green fluorescent protein (GFP) spacer between repeats to facilitate cloning (Piccin et al. 2001). It should be noted that transgenic RNAi in metazoan species seems to produce weaker phenotypes than regular RNAi mediated by dsRNA (Tavernarakis et al. 2000; Martinek and Young 2000; Piccin et al. 2001). Recent reports from plants and *Drosophila* suggest that the tar-

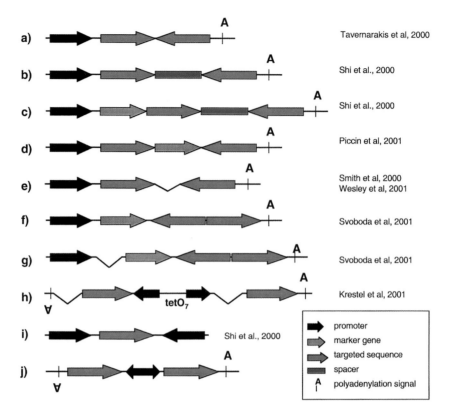

**FIGURE 15.3.** Design of constructs expressing dsRNA.

geting efficiency of expressed dsRNA can be increased by employing constructs in which an inverted repeat is interrupted by an intron. The intron acts as a spacer to increase cloning efficiency, but because it is spliced following transcription, the product is a loop-less hairpin dsRNA (Smith et al. 2000; Wesley et al. 2001; Kalidas and Smith 2002).

With dual-promoter plasmids, cloning of the target sequence is simple, but these promoters express sense and antisense strands separately, and their efficiency therefore depends on the annealing of single-stranded RNAs in vivo. In addition, competition of transcription of sense and antisense strands from the same DNA should lead to decreased levels of expression compared to inverted-repeat-based plasmids. The main advantages of the plasmid-driven RNAi are that it allows for the following:

- Use of various selective markers. In fact, EGFP expression can be driven together with dsRNA hairpin from a single promoter (Svoboda et al. 2001).

- Induction of dsRNA expression at a given time via an inducible expression system.

- Screening of a large number of genes by RNAi at relatively low cost. Whether plasmid-driven RNAi can overcome the problem of a nonspecific response to dsRNA in somatic cells has not yet been addressed.

***Characteristics of the targeted gene/protein.*** In general, the RNAi pathway in mammalian cells seems to be less efficient than that in *C. elegans* or *Drosophila* (Svoboda et al. 2000; Ui-Tei et al. 2000). However, more than 30 mammalian genes have been targeted successfully with RNAi, most of them in tissue culture (Table 15.1). Mammalian genes efficiently targeted by RNAi code for proteins with many different functions, including signaling proteins (Mos, CDK1), nuclear and cytoplasmic structural proteins (β and γ actin, A/C, B1, and B2 lamin, vinculin), secreted enzymes (Plat), and ion channels (inositol 1,

**TABLE 15.1.** Mammalian genes targeted by RNAi

| Gene | Function | Method | Tissue | Silencing[a] mRNA | Silencing[a] protein | Phenotype | References |
|---|---|---|---|---|---|---|---|
| E-cadherin | adhesion | dsRNA | embryo | n.d. | reduced | yes | Wianny and Zernicka-Goetz (2000) |
| Egfp (transgene) | fluorescent marker | dsRNA | embryo | n.d. | reduced | yes | Wianny and Zernicka-Goetz (2000) |
| Mos | kinase | dsRNA, plasmid | oocyte | reduced | reduced | yes | Wianny and Zernicka-Goetz (2000); Svoboda et al. (2000, 2001) |
| Plat | protease | dsRNA | oocyte | reduced | reduced | yes | Svoboda et al. (2000) |
| Msy2 | RNA binding | dsRNA | oocyte | reduced | n.d. | n.d. | J.Y. Yu et al. (unpubl.) |
| CaMKII | Ca dep. kinase | dsRNA | oocyte | reduced | reduced | n.d. | Michaut et al. (unpubl.) |
| Iptr1 | Ca channel | dsRNA | oocyte | reduced | reduced | yes | Z. Xu et al. (2003) |
| Basonuclin | transcription factor | dsRNA | oocyte | reduced | n.d. | n.d. | J. Ma et al. (2003) |
| Acrogranin | cell adhesion | dsRNA | embryo | reduced | n.d. | no | J. Jun et al. (unpubl.) |
| MDicer | dsRNase | dsRNA, siRNA | embryo | reduced | reduced | yes | P. Svoboda et al. (unpubl.) |
| NuMA | nuclear protein | siRNA | HeLa | n.d. | reduced | yes | Elbashir et al. (2001b); Haborth et al. (2002) |
| GAS41 | nuclear protein | siRNA | HeLa | n.d. | reduced | yes | Haborth et al. (2002) |
| SV40 T antigen | nuclear protein | siRNA | rat fibroblast | n.d. | reduced | n.d. | Haborth et al. (2002) |
| Lamin A/C | nuclear envelope | siRNA | HeLa | n.d. | reduced | yes | Elbashir et al. (2001b); Haborth et al. (2002) |
| Lamin B1 | nuclear envelope | siRNA | HeLa | n.d. | reduced | yes | Elbashir et al. (2001b); Haborth et al. (2002) |
| Lamin B2 | nuclear envelope | siRNA | HeLa | n.d. | reduced | n.d. | Haborth et al. (2002) |
| LAP2 | nuclear envelope | siRNA | HeLa | n.d. | reduced | n.d. | Haborth et al. (2002) |
| Emerin | nuclear envelope | siRNA | HeLa | n.d. | reduced | n.d. | Haborth et al. (2002) |
| Nup153 | nuclear envelope | siRNA | HeLa | n.d. | reduced | n.d. | Haborth et al. (2002) |
| β-actin | cytopl. cytoskeleton | siRNA | HeLa | n.d. | reduced | yes | Haborth et al. (2002) |
| γ-actin | cytopl. cytoskeleton | siRNA | HeLa | n.d. | n.d. | yes | Haborth et al. (2002) |
| ARC21 | cytopl. cytoskeleton | siRNA | HeLa | n.d. | reduced | n.d. | Haborth et al. (2002) |
| VASP | cytopl. cytoskeleton | siRNA | HeLa | n.d. | reduced | n.d. | Haborth et al. (2002) |
| Vinculin | cytopl. cytoskeleton | siRNA | mouse 3T3 | n.d. | reduced | n.d. | Haborth et al. (2002) |
| Zyxin | cytopl. cytoskeleton | siRNA | mouse 3T3 | n.d. | reduced | yes | Haborth et al. (2002) |
| Vimentin | cytopl. cytoskeleton | siRNA | HeLa | n.d. | reduced | n.d. | Haborth et al. (2002) |
| Keratin18 | cytopl. cytoskeleton | siRNA | HeLa | n.d. | reduced | n.d. | Haborth et al. (2002) |
| Eg5 | cytopl. cytoskeleton | siRNA | HeLa | n.d. | n.d. | yes | Haborth et al. (2002) |
| CENP-E | centromere | siRNA | HeLa | n.d. | n.d. | yes | Haborth et al. (2002) |
| cytopl.dynein | spindle | siRNA | HeLa | n.d. | n.d. | yes | Haborth et al. (2002) |
| CdK1 | kinase | siRNA | HeLa | n.d. | n.d. | yes | Haborth et al. (2002) |
| Mad2 | spindle checkpoint | siRNA | HeLa | n.d. | reduced | yes | Luo (2002) |
| Mst1 | signaling | siRNA | HeLa | n.d. | no | no | Luo (2002) |

[a] n.d. indicates not determined.

4,5-trisphosphate receptor). However, not every gene can be targeted using RNAi. Typically, the targeted mRNA is specifically degraded, and the corresponding protein product is depleted over a period of time that depends on the stability of the protein. In such cases, the RNAi phenotype resembles a loss-of-function phenotype. However, mRNA and protein elimination simply may not fit into the experimental time frame and thus a phenotype might not be obtained. This can be a problem, especially for studies in preimplantation embryos that develop in a rather limited time period, because the RNAi treatment cannot be extended as needed. In *Drosophila*, it has been observed that genes coding for more stable proteins (such as cytoplasmic actin) are less amenable to the application of RNAi than proteins with relatively short lives (~10 hours), which are efficiently silenced (Wei et al. 2000). Interestingly, Harborth et al. (2001) showed in cultured mammalian cells that targeting of cytoplasmic β- and γ-actin can lead to a dramatic but incomplete elimination of protein and cause an observable phenotype.

Because RNAi often does not reduce protein levels to zero, the residual amount of targeted protein may be enough to perform the gene's function without resulting in a phenotypic change. For example, one unpublished but well-documented case in preimplantation embryos involved RNAi targeting of acrogranin. RNAi failed to phenocopy the phenotype obtained by blocking antibodies, although the mRNA level decreased by approximately 90% (J. Jun et al., unpubl.). There might be other cases where RNAi did not yield the expected phenotype, but it is important to distinguish when RNAi did not work because the chosen gene has been "RNAi-resistant" or because of some technical problem, for example, inefficiency of a given siRNA.

## RNAi as a tool to study early mammalian development

In mammals, RNAi can, in principle, be used as an alternative or complement to genetic studies. For studies in preimplantation embryos, long dsRNAs, siRNAs, or RNAi driven from a plasmid can be used. However, the targeting RNA must be microinjected, which limits the timing of dsRNA introduction and the number of oocytes and embryos that can be analyzed for a phenotype in a single experiment. Transfection of dsRNA (as was used in tissue culture studies) has not been successful in oocytes and preimplantation embryos (P. Stein et al., unpubl.). Similarly, plasmid transfection into preimplantation embryos has been inefficient (P. Stein et al., unpubl.). A recently published transgenic RNAi approach, in which the *mos* gene was targeted in oocytes using an inverted repeat transcribed from an oocyte-specific promoter, provides an attractive alternative to dsRNA microinjection (Stein et al. 2003a).

RNAi is an excellent tool for studying the role of maternal transcripts recruited during either oocyte maturation or embryo development. Recruited transcripts accumulate in the oocyte but are not translated; therefore, the stability of the coded protein does not affect the efficiency of RNAi. Inhibition of oocyte maturation with compounds that elevate cAMP levels, such as 3-isobutyl-l-methylxanthine (IBMX), extends the exposure time to dsRNA, so transcripts recruited during oocyte maturation can be efficiently degraded. It should be noted that our previously unpublished experiments in the oocyte showed decreased RNAi efficiency when two genes were targeted simultaneously (Figure 15.4), suggesting that the oocyte RNAi pathway can be readily saturated so that simultaneous targeting of several genes might be difficult at this stage.

Genes activated during early embryo development can also be targeted, but the efficiency of targeting depends on the mRNA level, the role of the protein, and its stability, as discussed above. Another problem is that the development of early embryos cannot be stopped or slowed down significantly to extend the exposure time to dsRNA. The most suitable stage for microinjection is the one-cell embryo; therefore, it might be more difficult to target genes activated at later stages because the trigger dsRNA might be diluted, degraded, or outcompeted by endogenous dsRNA molecules during development.

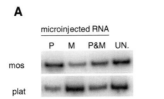

| | | B | relative mRNA level (%) | |
| --- | --- | --- | --- | --- |
| | | | mos | tPA |
| | | uninjected | 100 | 100 |
| | | Mos dsRNA | **22** | 78 |
| | | tPA dsRNA | 102 | **35** |
| | | Mos&tPA dsRNA | **72** | **70** |

**FIGURE 15.4.** Simultaneous targeting of two genes by RNAi in mouse oocytes. (*A*) Radiolabeled semiquantitative RT-PCR analysis of microinjected oocytes. (M) Microinjection of $10^6$ molecules of Mos dsRNA; (P) microinjection of $10^6$ molecules of Plat dsRNA; (P&M) microinjection of $10^6$ molecules of each Mos and Plat; (UN) uninjected control. (*B*) Average mRNA levels from two independent experiments. Oocytes were microinjected with $10^6$ molecules of each dsRNA. Shown in bold are mRNA levels in oocytes microinjected with Mos or Plat dsRNA or with their combination.

Perhaps these problems are the reasons why the literature contains only one report of a bona fide endogenous mammalian gene inhibited successfully by RNAi in early embryos (Wianny and Zernicka-Goetz 2000).

## FUTURE DIRECTIONS

Analysis of more genes by RNAi in mammals will help us to better understand what genes are suitable for RNAi targeting. Simultaneously, various RNAi approaches need to be compared and standard protocols must be developed. Besides the "mainstream" siRNA- and shRNA-based methods used typically in tissue culture, another promising approach is the transgenic expression of long dsRNA. Currently, this approach has been developed only for oocytes. However, it should be possible to use it in other tissues even if the expression of long dsRNA would cause nonspecific effects due to PKR-mediated responses, because this problem could be overcome by using PKR-knockout animals that are viable and fertile, but do not exhibit the nonspecific response to dsRNA (Der et al. 1997). In addition, it is expected that many different inducible expression systems will be combined with transgenic RNAi to allow for precise timing of target gene inhibition.

The RNAi transgenic approach is very attractive for studies of early mammalian development because it solves the problem of the elimination of the maternal pool of targeted protein as well as the problem with the need for extensive microinjection. Transgenic RNAi has several advantages over traditional gene knockout technology for the following reasons:

- It is simpler, because there is no need for complicated ES cell targeting and manipulation and no need to screen large numbers of mutant lines and animals.

- It is faster, because there is no need to screen mosaic animals and subsequent extensive crossing to obtain hetero- and homozygotes.

- It is cheaper, mainly for reasons stated above.

Because RNAi usually does not totally eliminate gene function, this would provide a range of phenotypes depending on the level of interference, which could be, in fact, an advantage, because it yields information about threshold effects that cannot be obtained by the classical gene knockout experiments (Stein et al. 2003a).

Better understanding of the RNAi mechanism in mammals will also lead to attempts to modify and enhance the RNAi response. One can imagine that if some step in the RNAi mechanism is identified to limit the efficiency of RNAi in mammals, this would lead to the production of transgenic animals expressing higher levels of that factor, hence enhancing the efficiency of RNAi. Finally, as RNAi emerges as a useful silencing tool for studies in mammals, its therapeutic use can be assessed, most likely in fields using tissue cultures extensively, such as hematology or in the treatment of parasitic infections.

# TECHNIQUES

## PROTOCOL 1: PREPARATION OF LONG DsRNA MOLECULES

This protocol is based on the methods described by Fire et al. (1998) and Kennerdell and Carthew (1998). In summary, dsRNA is obtained after annealing sense and antisense RNA strands, or spontaneously during in vitro transcription of an inverted repeat. We usually include RNase T1 treatment of the annealed RNA prior to the purification step in order to remove unannealed single-stranded RNA (ssRNA), which can interfere with the quantification of dsRNA in a nondenaturing agarose gel (see Figure 15.5c). Unannealed ssRNA should not, however, cause any nonspecific effects during RNAi experiments.

The preparation of dsRNA should take less than 1 week after the targeted gene has been chosen. The exact time depends on the availability of suitable sequences, primers, and clones. The preparation of dsRNA itself from in vitro transcription can be done in 1 day if necessary.

## Procedures

### Part A: Preparation of templates for in vitro transcription

Sense and antisense DNA templates from the gene of interest can be prepared by polymerase chain reaction (PCR) or from an appropriate plasmid carrying a piece of the gene and appropriate promoter for in vitro transcription using standard procedures (see Sambrook and Russell 2001).

#### Preparation of template by PCR

- Combine PCR primers carrying an SP6 or T7 promoter sequence at their 5′ ends with promoterless primers to generate sense and antisense templates (Figure 15.6). Add three to four additional nucleotides at the 5′ end of the promoter.

- As a template, use the highly diluted (500–1000-fold) gel-purified reverse transcription (RT)-PCR product generated by identical primers lacking promoters.

- In general, perform gel extraction of the PCR product whenever there is any sign of nonspecific band(s) after PCR.

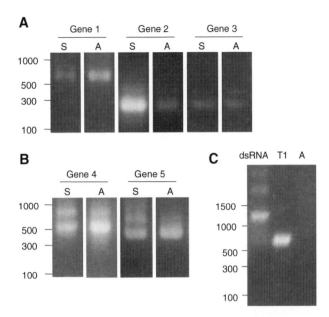

**FIGURE 15.5.** Analysis of in vitro transcription products using ethidium-bromide-stained 1.5% non-denaturing agarose gels in 1× TAE buffer. (*A*) Different yields of in vitro transcription reactions using SP6 polymerase and equal amounts of different purified PCR-generated DNA templates. (*B*) Single-stranded RNA (shown here) and dsRNA with single-stranded overhangs often form multiple bands in nondenaturing agarose, presumably by forming duplexes. (*C*) Effect of RNase T1 on mobility of dsRNA hairpin with a long (800 nucleotides) single-stranded overhang. Note the mobility shift of the main band and elimination of secondary bands. (S) Sense RNA; (A) antisense RNA; (dsRNA) dsRNA hairpin with a long single-stranded RNA overhang after in vitro transcription; (T1) same sample as "dsRNA" but treated with RNase T1; (A) in part C, same sample as dsRNA but treated with RNase A.

- After purification, estimate the amount and quality of the template by spectrophotometry and gel electrophoresis. Be aware that the quality of the templates depends on the quality of the primer synthesis. In our experience, the use of different primers with identical promoters may result in quite a variable yield. This may sometimes be a complication in the generation of equimolar amounts of sense and antisense RNAs, because there might be a severalfold difference in the yield of each strand (see Figure 15.5a).

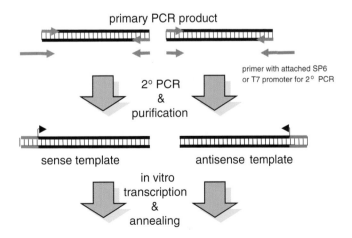

**FIGURE 15.6.** Templates for in vitro transcription.

• The advantages of this method are speed and the absence of foreign sequence in the dsRNA. The latter is true because the template does not carry any sequence other than the promoter and target sequences. When used for a larger number of genes, two disadvantages of the PCR method are that it is more expensive than using plasmid templates and it fails to provide templates of consistent quality.

## Preparation of template from a plasmid

• Prepare the templates from suitable plasmids carrying the chosen target sequence. Either (1) generate two plasmids with inserts in both directions (sense and antisense) when the plasmid carries only one promoter or only one RNA polymerase is available or (2) use one plasmid that has appropriate promoters flanking the multiple cloning site from both sites, such as pCRII (Invitrogen). Briefly, amplify the desired sequence by PCR and insert it into this plasmid using TA or TOPO-TA Cloning (Invitrogen) on day 1. Positive colonies are transferred into liquid culture on day 2. On day 3, use standard procedures (Sambrook and Russell 2001) for the isolation of DNA, plasmid linearization, and purification.

• A very convenient method of dsRNA generation involves transcribing RNA from a plasmid carrying an inverted repeat of the target sequence. Although cloning of an inverted repeat is not simple, the advantage of such a template is that it produces dsRNA during in vitro transcription without any need for estimating equimolar ratios of sense and antisense strands and annealing.

  On day 1, generate inverted repeats by ligating two digested PCR products of different sizes. These products should contain the same restriction site in the 5′ forward primer sequence and one of the PCR products should be somewhat longer (20–50 bp) at the 5′ end. The excess sequence will form a loop in the RNA molecule. This sequence also serves as a short spacer in the middle of the inverted repeat and enhances efficiency of inverted repeat cloning (Svoboda et al. 2001). After ligation, purify the appropriate size band from the gel. Add up to several hundred nanograms of this DNA to a PCR mixture with *Taq* polymerase and incubate for 10 minutes at 72°C (in this step, *Taq* polymerase adds A overhangs). After this step, process the product as a regular PCR product (see above).

• The plasmid carrying the template must be completely digested downstream from the insert, preferentially with an enzyme that does not leave 3′ overhangs (commonly used enzymes leaving 3′ overhangs are, for example, *Apa*I, *Bst*X1, *Kpn*I, *Pst*I, *Pvu*I, *Sac*I, or *Sac*II). If there is no alternative restriction site, the 3′ overhang should be blunt-ended using the Klenow fragment prior to in vitro transcription. For digestion and subsequent purification, it should be considered that the amount of plasmid used for in vitro transcription will be 5–10 μg. Therefore, a standard plasmid miniprep (high-copy plasmid from a 1.5-ml bacterial culture) will provide enough template for only a few transcription reactions.

  On the other hand, one in vitro transcription reaction provides enough material to produce dsRNA for many rounds of microinjection. After restriction digestion, purify the digested plasmid by standard phenol extraction, precipitate (for a more detailed protocol, see Sambrook and Russell 2001), and resuspend in the appropriate amount of TE. Keep the final DNA concentration >500 ng/μl.

**Part B:** In vitro transcription of sense and antisense strands

The following protocol describes in vitro transcription with the SP6 polymerase. For other polymerases (T3, T7), simply use the appropriate buffer and desired RNA polymerase instead of the SP6 polymerase.

---

## MATERIALS

### REAGENTS

5x Buffer (usually provided with SP6 polymerase)

Dithiothreitol (0.1 M) (DTT) (provided with SP6 polymerase) <!>

DNA template (5–10 µg)

NTPs (25 mM)

RNase T1

　　See Step 2.

RNAsin (40 units/µl) (Promega)

RQ1 DNase

SP6 RNA polymerase (20 units/µl)

$H_2O$ (sterile, DEPC-treated)

### EQUIPMENT

Agarose gel (1.5%), nondenaturing in 1x TAE buffer, containing ethidium bromide <!>

　　Use a 1.5% agarose gel in 1x TAE buffer to check the quality of ssRNA and dsRNA.

　　**IMPORTANT:** To avoid problems with RNA degradation, keep all material for these gels separate from the reagents and equipment used for other gels, and never touch the gels without gloves. DEPC-treated $H_2O$ is unnecessary; Milli Q $H_2O$ is fine for washing.

Microfuge tubes (1.5 ml)

---

1. Set up the reaction mixture:

   | | |
   |---|---|
   | 5x buffer | 20 µl |
   | 0.1 M DTT | 10 µl |
   | 40 units/µl RNAsin | 5 µl |
   | 25 mM NTPs | 20 µl |
   | SP6 RNA polymerase (20 units/µl) | 5 µl |
   | DNA template (5–10 µg) | |
   | $H_2O$ (sterile, DEPC-treated) | up to 100 µl |

   Incubate the reaction for 2–4 hours at 37°C.

2. Add 5 units of RQ1 DNase to the reaction, and incubate the tube for 15 minutes at 37°C (for more than 5 µg of template, add ~6 minutes of incubation time for each additional microgram). If an inverted repeat was transcribed, also add 5 µl of RNase T1 and incubate the mixture with RQ1 DNase and RNase T1 for 30 minutes.

3. Estimate the amount and quality of transcribed RNA by electrophoresis. Load 1, 2, and 4 µl of the reaction (from Step 2) into a 1.5% nondenaturing agarose gel (in 1x TAE buffer, containing ethidium bromide).

   Usually more than one band appears, with the fastest-migrating band corresponding to half the size of the RNA compared to the DNA marker. The more slowly migrating bands are dimers, trimers, etc., that form under nondenaturing conditions. The pattern is dependent on the RNA sequence. Some sequences produce only the lowest band, whereas some others give two or more bands (see Figure 15.5B). Loading different amounts of reaction product helps in estimating the ratio of sense to antisense reaction mixture volumes to be used for annealing; loading various amounts is not necessary when transcribing an inverted repeat in vitro. In this case, load 4 μl to estimate the amount and quality of RNA and proceed to the purification.

**Resolution of In Vitro Transcription Products**

When resolving in vitro transcription products, usually two or more bands appear on the nondenaturing agarose gel. These bands are imperfect duplexes of ssRNA, rarely products of cryptic SP6 promoters (see Figure 15.5).

**Part C:** Annealing

**MATERIALS**

**REAGENTS**

H₂O (sterile, DEPC-treated)

Nondenaturing agarose gel (see Part B)

RNase T1

Optional, see Step 6.

Sense and antisense RNAs (from Part B)

**EQUIPMENT**

Hybridization oven preset to 65°C

Optional, see Step 5.

Microfuge tubes (1.5 ml)

Microfuge tube cap locks

Water bath (boiling) in a 1-liter beaker

1. On the basis of the result of the gel from Part B, Step 3, in a 1.5-ml microfuge tube, mix roughly equimolar amounts of sense and antisense RNAs (the ratio does not have to be exact; a slight excess of one ssRNA will be removed during RNase T1 treatment after annealing). Keep the concentration of RNA in the annealing mixture at ~200–500 ng/ml in 100–200 μl of the annealing mixture. Adjust the volume with DEPC-treated sterile H₂O.

It is unnecessary to use perfectly equimolar ratios for annealing, because the unannealed ssRNA does not affect RNAi.

2. Secure the lid of the microfuge tube with a lock

Locks are available for a single tube or use whole lockable racks. Unsecured tubes often open and become easily contaminated with H₂O from the bath.

3. Incubate the annealing mixture for 3 minutes in a 1-liter beaker of boiling water. If the lid of the microfuge tube is unsecured, boil the tube for only 1 minute.

4. Remove the beaker from the heat, and place the tube in the beaker of H₂O for 3 hours at room temperature to allow it to cool down.

5. Check an aliquot (1–4 μl) of the reaction in a 1.5% nondenaturing agarose gel (in 1x TAE buffer, stained with ethidium bromide). If the annealing results are unsatisfacto-

ry (i.e., a smear or a large amount of ssRNA can be seen), incubate the reaction tube—still in the beaker—in a preheated 65°C hybridization oven for 1 hour, followed by a gradual cooling down period for 2 hours at room temperature.

6. (Optional) After annealing, add 5 μl of RNase T1 and incubate the mixture for 30 minutes at 37°C.

## Part D: dsRNA purification

**MATERIALS**

**REAGENTS**

Chloroform <!>

dsRNA (prepared in Part C)

Ethanol (95%) <!>

$H_2O$ (sterile, but not DEPC-treated)

Phenol (pH 4.5) <!>

Phenol:chloroform (1:1 v/v) <!>

Potassium acetate (3 M)

**EQUIPMENT**

Agarose gel (1.5%), nondenaturing in 1x TAE buffer

Use a 1.5% agarose gel in 1x TAE buffer to check the quality of ssRNA and dsRNA.

**IMPORTANT:** To avoid problems with RNA degradation, keep all material for these gels separate from the reagents and equipment used for other gels, and never touch the gels without gloves. DEPC $H_2O$ is unnecessary; Milli Q water is fine for washing.

Microcentrifuge (e.g., Eppendorf 5145C)

Microfuge tubes (0.5 and 1.5 ml)

1. Add 1 volume of phenol (pH 4.5) to the dsRNA (from Part C, Step 4 or 6), vortex, and centrifuge the tube in a microfuge at maximum speed for 10 minutes at 4°C.

    All centrifugations in this procedure are performed at 4°C.

2. Transfer the aqueous phase into a new microfuge tube. Add 1 volume of phenol:chloroform (1:1 v/v) to the aqueous phase and extract as in Step 1.

3. Repeat Step 2, but substitute chloroform for the phenol:chloroform.

4. Precipitate the dsRNA with 0.1 volume of 3 M potassium acetate and 3 volumes of 95% ethanol for >1 hour at –20°C.

    The RNA may be precipitated overnight if it is convenient.

5. Centrifuge the tube in a microfuge at maximum speed for 20 minutes at 4°C to pellet RNA.

6. Remove the supernatant and wash the pellet twice with 400 μl each of 75% ethanol. Air-dry the pellet. Do not dry the pellet under vacuum.

7. Resuspend the pellet in 20 μl of $H_2O$, and estimate the quality and the amount of dsRNA.

8. Dilute the dsRNA to the desired concentration with sterile $H_2O$ (100 ng/μl of 0.5 kb dsRNA is ~200,000 molecules/pl), aliquot it into 0.5-ml microfuge tubes, and freeze the samples at –80°C.

    dsRNA is relatively stable. In our hands, there was no degradation of the dsRNA after a 3-hour run in a vacuum centrifuge (e.g., SpeedVac), during storage for 1 month at –20°C, or after a day or two at room temperature.

**No Transcription**

**Possible cause:** Inappropriate reaction components.

**Solution:** Verify that the RNA polymerase used recognizes the template promoter. SP6 polymerase will not start synthesis from a T7 promoter and vice versa. Confirm that NTPs, not dNTPs, were used.

**Possible cause:** Enzyme lost its activity or NTPs are hydrolyzed.

**Solution:** Try a different batch of enzyme and prepare fresh NTPs.

## After In Vitro Transcription, Electrophoresis Yields Smears but No Discrete Bands

**Possible cause:** Overloaded electrophoresis. Sometimes the reaction mixture may be really dense and, in combination with small wells, causes "plugging" of the wells, resulting in poor separation of the RNA.

**Solution:** Dilute the reaction mixture five- to tenfold depending on the intensity of the smear, and repeat the electrophoresis.

**Possible cause:** DNase has lost its activity. The smear comes from a partially degraded DNA template that masks the RNA. Add RNase A to a small aliquot of the reaction mixture, incubate for 30 minutes at 37°C, and analyze by electrophoresis. Do not use the same electrophoresis equipment that was used for RNA analysis because of RNase A contamination.

**Solution:** Dilute the reaction mixture five- to tenfold and repeat the electrophoresis.

**Possible cause:** RNA is degraded due to RNase contamination.

**Solution:** Use gloves at all times and change them frequently. Use RNase-free material and reagents (tips, RQ DNase, $H_2O$, etc.). Keep separate solutions just for RNA gel electrophoresis in the cold and dark. Repeat in vitro transcription using a smaller volume; wash the gel box with detergent and rinse thoroughly with distilled and Milli Q $H_2O$. Usually, strong RNA degradation is caused during gel electrophoresis, not during in vitro transcription. If this problem does not disappear, some reaction components are likely contaminated with RNases. Depending on time considerations, either test the reaction components for the presence of RNases or exchange any components for new ones and repeat the small-scale test. If the problem disappears, repeat the large-scale in vitro transcription.

## An Additional Discrete Band of the Same Size as the Template

**Possible cause:** RQ1 DNase has lost its activity or was not added at all.

**Solution:** Test the RQ1 DNase. If the problem persists, purify the in vitro transcription mixture by acidic phenol extraction (pH 4.0–4.5), precipitate the RNA, dissolve it in 5% RNAsin, and repeat the RQ1 DNase treatment. Alternatively, after purification, proceed to the annealing and carry out the RQ1 DNase treatment after annealing.

## Multiple Bands After Annealing and RNase T1 Treatment

**The lowest band is the same size as dsRNA and other bands are at least twice that size**

**Possible cause:** RNase T1 has lost its activity. When transcribing RNA from plasmids such as pCRII, there are often overhanging sequences of the plasmid origin that have partial homology and can hybridize leading to the formation of dimers.

**Solution:** This does not affect the RNAi procedure. However, RNase T1 treatment and purification can be repeated.

**The lowest band is the size of ssRNA and other bands are at least twice that size**

**Possible cause:** RNase T1 has lost its activity and did not remove unannealed ssRNA.

**Solution:** Repeat RNase T1 treatment and purification.

## PROTOCOL 2: MICROINJECTION INTO OOCYTES AND EARLY EMBRYOS

Introduce the dsRNA of choice (prepared in Protocol 1) into mouse oocytes or fertilized one-cell embryos by microinjection (for specific protocols and information for preparation of experiments, see Nagy et al. 2003). The microinjected cells are cultured for the desired period of time until testing for the RNAi effect is performed. The first assay to conduct with microinjected oocytes or embryos should always be RT-PCR to check that the cognate mRNA has been degraded, i.e., to prove that the dsRNA is producing an RNAi effect. Specific biological assays can range from watching the development or morphology of the cells under the microscope to techniques such as western blot, immunofluorescence, and enzyme activity determination. A typical experimental outline of an RNAi experiment in mouse oocytes is shown in Figure 15.7.

The duration of microinjection experiments depends on both the type of cells used (oocytes or early embryos) and the phenotypic characteristic studied. Cell collection and microinjection of dsRNA are completed in 1 day, but in most experiments, the oocytes/embryos are cultured for 1–4 days until assayed for RNAi effect. In addition, female mice must be primed with gonadotropins (and mated when using embryos) prior to cell collection.

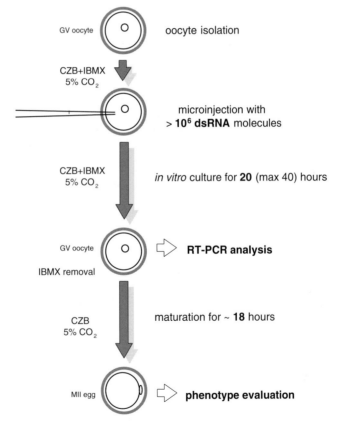

FIGURE 15.7. An outline of the typical experimental design used in RNAi experiments in mouse oocytes. Culturing conditions are shown on the left. (CZB) Culture medium; (IBMX) 0.2 mM 3-isobutyl-1-methylxanthine (inhibitor of meiotic resumption). Microinjected oocytes are cultured in an IBMX-containing medium. Using IBMX allows for extension of the exposure of oocytes to dsRNA and, especially in the case of dormant maternal mRNAs, allows for mRNA degradation prior to its translation after resumption of meiosis. Oocytes are typically cultured in IBMX-containing medium for 20 hours. If needed, this culture can be extended up to 40 hours. However, it should be kept in mind that longer culture times can affect the developing potential of cultured oocytes.

**Part A:** Collection of oocytes or early embryos

## MATERIALS

### REAGENTS

*For oocyte collection*

CZB medium:

| | |
|---|---|
| 81.62 mM NaCl | 0.27 mM sodium pyruvate |
| 4.83 mM KCl <!> | 1 mM L-glutamine |
| 1.18 mM $KH_2PO_4$ <!> | 0.11 mM EDTA |
| 1.18 mM $MgSO_4$ <!> | 3 mg/ml bovine serum albumin (BSA) |
| 1.7 mM $CaCl_2$ <!> | 10 µg/ml gentamicin |
| 25.12 mM $NaHCO_3$ | 10 µg/ml phenol red <!> |
| 31.3 mM sodium lactate | |

MEM/PVP medium:

Bicarbonate-free minimal essential medium (Earle's salts) supplemented with sodium pyruvate (100 µg/ml), gentamicin (10 µg/ml), polyvinylpyrrolidone (PVP; 3 mg/ml) <!>, and 25 mM HEPES (pH 7.2).

*For embryo collection*

Hyaluronidase

IBMX (3-isobutyl-1-methylxanthine) (0.2 M stock solution in DMSO)

KSOM medium:

| | |
|---|---|
| 95 mM NaCl | 25 mM $NaHCO_3$ |
| 2.5 mM KCl | 1.71 mM $CaCl_2$ |
| 0.35 mM $KH_2PO_4$ | 1 mM L-glutamine |
| 0.2 mM $MgSO_4$ | 0.01 mM EDTA |
| 10 mM sodium lactate | 1 mg/ml BSA |
| 0.2 mM glucose | 10 µg/ml gentamicin |
| 0.2 mM sodium pyruvate | |

### EQUIPMENT

Forceps

Glass pipettes (mouth-operated)

Incubator (humidifed 5% $CO_2$ in air at 37°C)

Needle (27-gauge)

Paraffin oil (light)

Syringe (1 ml)

Tissue culture dishes (sterile plastic, 35 mm and 60 mm)

Watch glasses

### ANIMALS

CF1 female mice, 6 weeks old (Harlan, Indianapolis, Indiana)

B6D2F1/J male mice (Jackson Laboratories, Bar Harbor, Maine)

Fully grown germinal vesicle (GV)-stage mouse oocytes are obtained from 6-week-old CF1 female mice. To improve the yield of antral follicles containing fully grown oocytes, mice are given a single injection of 5 IU of pregnant mare's serum gonadotropin (PMSG) intraperitoneally 48 hours prior to sacrifice.

One-cell embryos are obtained from 6-week-old CF1 female mice mated to B6D2F1/J males. Females are superovulated by an intraperitoneal (i.p.) injection of PMSG (5 IU) followed by an i.p. injection of human chorionic gonadotropin (hCG; 5 IU) 48 hours later. Females are then mated and one-cell embryos are collected 18–20 hours after hCG.

## Mouse oocyte collection

Set up the culture dishes in advance (Steps 1 and 2) to allow for temperature and $CO_2$ equilibration.

1. Prepare the culture medium. Add 2 μl of 0.2 M IBMX to 2 ml of CZB medium (Chatot et al. 1989).

   **IMPORTANT:** IBMX, a phosphodiesterase inhibitor, or cAMP analogs (typically dibutyryl cAMP) must be added to all collection and culture media used for oocytes to prevent spontaneous meiotic maturation and keep the oocytes at the GV stage.

2. Set up microdrop culture dishes. Place several 75-μl drops of CZB+IBMX on the bottom of a 60-mm sterile plastic tissue culture dish and cover the dish with light paraffin oil. Place the dish in a humidified incubator containing 5% $CO_2$ in air at 37°C.

3. Sacrifice the females by cervical dislocation and remove the ovaries.

4. Place the ovaries in a watch glass with MEM/PVP containing 0.2 mM IBMX.

5. Release the antral follicles from the ovaries by puncturing them several times with a 27-gauge needle attached to a 1-ml syringe. Use a mouth-operated glass pipette to collect the oocyte–cumulus cell complexes and transfer them to a clean watch glass containing 0.2–2 ml of MEM/PVP+IBMX. Select large antral follicles containing cumulus-enclosed oocytes and avoid picking the smaller pre-antral follicles or denuded oocytes.

6. Use a pipette whose tip diameter is about the size of the oocytes to pipette the complexes up and down to detach the cumulus cells. Transfer the cumulus-free oocytes to the culture dish and place in the incubator.

## Mouse one-cell embryo collection

Set up the culture dishes in advance (Step 1) to allow for temperature and $CO_2$ equilibration.

1. Set up microdrop culture dishes. Place several 75-μl drops of KSOM on the bottom of a 60-mm sterile plastic tissue culture dish and cover the dish with light paraffin oil. Place the dish in a humidified incubator containing 5% $CO_2$ in air at 37°C.

2. Sacrifice the females by cervical dislocation and remove the oviducts. Place them in a 35-mm plastic culture dish containing MEM/PVP.

3. Transfer the individual oviducts to individual drops of MEM/PVP containing 3 mg/ml hyaluronidase.

4. With the help of a pair of forceps, hold the upper part of the oviduct (ampulla) and pierce it with a needle to release the zygotes. Place them in the hyaluronidase solution for a couple of minutes until the cumulus cells detach.

5. Wash the cumulus-free zygotes through several drops of MEM/PVP and then transfer them to the culture dish and place in the incubator.

**Part B:** Microinjection

---

**MATERIALS**

**REAGENTS**

dsRNA solution

Injection media

*For oocytes:* MEM/PVP containing 0.2 mM IBMX

*For embryos:* MEM/PVP

See Part A for MEM/PVP recipe.

**EQUIPMENT**

Gas supply

Incubator

Injection and holding pipettes

Injection pipettes are made by pulling borosilicate-glass capillary tubing on a mechanical pipette puller. They can be prepared in advance or as microinjection proceeds. Holding pipettes are pulled the same way, but they must be cut to a diameter of 80–120 μm and the tip melted using a microforge. They are prepared in advance. (For detailed instructions, see Nagy et al. 2003.)

Micromanipulator

Paraffin oil (light)

Tissue culture dishes (sterile plastic 100 mm)

**BIOLOGICAL SAMPLE**

Oocytes or embryos

After collection, oocytes and embryos should be kept in culture for at least 30 minutes to allow recovery before microinjection.

---

1. Set up the micromanipulator for injection.

   - Place a 5-μl drop of injection medium (microinjection drop) on the top of a 100-mm plastic tissue culture dish.

   - Place a 1-μl drop of dsRNA solution as close as possible to the other drop and then flood the dish with light paraffin oil.

   - Place the dish in the stage of the micromanipulator, position the injection and holding pipettes, and connect the gas supply.

   The injection medium is MEM/PVP, supplemented with 0.2 mM IBMX in the case of oocytes.

   When more than one dsRNA is injected or vehicle (usually $H_2O$) injection is used as a control, set up a new set of drops (a 5 μl-drop of MEM/PVP [+IBMX] and a 1 μl-drop of dsRNA or $H_2O$) for each substance to be microinjected.

2. Transfer a group of oocytes/embryos from the incubator to the microinjection drop and inject 5–10 pl of dsRNA into their cytoplasm. Place them back in the incubator. Repeat the procedure with another group until all oocytes/embryos are microinjected. Check all microinjected cells under the microscope and remove those that did not survive.

**Part C:** Oocyte or embryo culture

---

## MATERIALS

### REAGENTS

CZB medium

> See Part A for CZB recipe.

IBMX (3-isobutyl-1-methylxanthine) (0.2 M stock solution in DMSO)

KSOM medium

> See Part A for KSOM recipe.

### BIOLOGICAL SAMPLE

Oocytes or embryos, microinjected in Part B

---

### Oocyte culture

1. Culture the oocytes in CZB+IBMX for 20–24 hours. They can then be lysed and assayed by RT-PCR (proceed directly to Part D).

2. If meiotic maturation is required, wash the oocytes with several drops of IBMX-free CZB and culture them in CZB for 16–18 hours. The in-vitro-matured metaphase II eggs can be processed for immunofluorescence, western blot, or enzyme activity measurements, or they can be fertilized in vitro.

### Embryo culture

1. Culture one-cell embryos in KSOM up to the desired stage.

   > If one-cell embryos will be cultured only up to the two-cell stage, the culture can be done in CZB. For longer cultures, however, KSOM is preferred, as it enhances embryo development to later stages. In addition, the development is improved when the culture is performed in at atmosphere of 5% $CO_2$/5% $O_2$/90% $N_2$.

**Part D:** RNA isolation and RT-PCR

---

**▼ CAUTION**

*See Appendix for appropriate handling of materials marked with <!>.*

## MATERIALS

**IMPORTANT:** All of the solutions used during RNA isolation and reverse transcription must be RNase-free. Lysis buffer should be stored, foiled wrapped, at room temperature.

### REAGENTS

Acetic acid 1 M <!>

DEPC-treated $H_2O$

*E. coli* rRNA

Ethanol (75% and absolute) <!>

---

Lysis buffer

    4 M guanidine thiocyanate <!>
    1 M 2-mercaptoethanol <!>
    0.1 M Tris-HCl (pH 7.4) <!>

Phenol, saturated (pH 4.3) (Amresco or Fisher) <!>

Potassium acetate (2 M and 3 M, pH 5.2) <!>

Rabbit β-globin mRNA

Reverse transcriptase (e.g., SUPERSCRIPT II, Life Technologies)

RNA resuspension solution

    40 mM Tris-HCl (pH 7.9) <!>
    10 mM NaCl
    6 mM $MgCl_2$ <!>

RQ1 DNase

*For oligo(dT)-primed reverse transcription*

    0.1 M dithiothreitol <!>
    10 mM dNTPs
    5x first-strand buffer
    oligo(dT) (12–18) (500 ng/μl)
    RNAsin (40 units/μl)

*For random hexamer-primed reverse transcription*

    0.1 M dithiothreitol <!>
    10 mM dNTPs
    5x first-strand buffer
    random hexamers (1 μg/μl)
    RNAsin (40 units/μl)

*For PCR amplification*

    $[\alpha\text{-}^{32}P]$dCTP (1 mCi/ml) (Amersham) <!>
    DNA polymerase (5 units/μl) (e.g., AmpliTaq, Roche)
    $MgCl_2$ (25 mM)
    10x PCR buffer
    3′ and 5′ primers (2 μM each)

*For gel electrophoresis*

    acrylamide (30%)/bis 37.5:1 solution <!>
    ammonium persulfate (10%) <!>
    6x electrophoresis sample-loading buffer
    10x TBE <!>
    TEMED <!>

## EQUIPMENT

Dry ice

Microcentrifuge (e.g., Eppendorf 5145C)

Microfuge tubes (0.5 ml and 1.5 ml)

Phosphorimaging equipment

Water baths preset to 42ºC, 65ºC, and 70ºC

## BIOLOGICAL SAMPLE

Oocytes or embryos, microinjected

## RNA isolation

1. Transfer oocytes or embryos (controls and dsRNA-injected) in a minimal volume directly from the culture dish into a 0.5-ml microfuge tube containing 100 μl of lysis buffer. Add 0.125 pg of rabbit β-globin mRNA per embryo/oocyte and 20 μg of *E. coli* rRNA, vortex well, and store at –80°C.

   > β-globin mRNA serves as an internal control for the efficiency of RNA isolation and RT-PCR.

   > To achieve reproducible results, use at least 20 oocytes/embryos for RNA isolation. Depending on the gene product under study, 15–20 embryos might work as well, but this is not always the case.

2. Thaw the tubes on ice.

3. Add to each tube:

   8 μl of 1 M acetic acid

   5 μl of 2 M potassium acetate

   60 μl of absolute ethanol

   Vortex the tubes *thoroughly*. Precipitate the RNA overnight at –80°C or for 2 hours on dry ice.

4. Pellet the RNA by centrifugation in a microfuge at maximum speed for 20 minutes at 4°C.

5. Wash the pellet twice at room temperature with 200 μl each of ice-cold 75% ethanol. Remove as much of the supernatant as possible. Briefly air dry the pellet at room temperature and then place the open tube on ice. This will prevent overdrying the pellet, which makes it impossible to resuspend.

6. Resuspend the RNA pellet in 20 μl of RNA resuspension solution. Make sure that the pellets have dissolved completely.

7. Add 1 unit of RQ1 DNase and incubate for 30 minutes at 37°C.

8. Place the tube on ice, add an additional 20 μl of RNA resuspension solution and 1 final volume (40 μl) of saturated phenol (pH 4.3). Vortex well to mix the organic and aqueous phases.

9. Centrifuge the tubes for 10 minutes at 4°C. Transfer the aqueous phase to a 1.5-ml microfuge tube.

10. Add 5 μl of 3 M potassium acetate (pH 5.2) and 3 final volumes (135 μl) of absolute ethanol. Precipitate overnight at –80°C or for 2 hours on dry ice.

11. Pellet the RNA by centrifugation for 20 minutes at 4°C.

12. Discard the supernatant and wash the pellet with 200 μl of 75% ethanol. Incubate for 10 minutes at room temperature, spin for 5 minutes, discard the supernatant, and air dry the pellet.

13. Resuspend the pellet in DEPC-treated $H_2O$ (or in 5% RNAsin). Resuspension volume depends on the number of lysed embryos and number of negative controls). Store the RNA at –80°C or keep it on ice if the reverse transcription will be performed immediately.

## Reverse transcription reaction

14. Use either oligo(dT)-primed RT or random-hexamer-primed RT reactions described below.

Oligo(dT)-primed RT

Prepare the RT master mix (on ice):

| | |
|---|---|
| 5x first-strand buffer | 4 µl |
| 0.1 M dithiothreitol | 2 µl |
| RNAsin (40 units/µl) | 0.5 µl |
| 10 mM dNTPs | 1 µl |
| oligo(dT) (12–18, 500 ng/µl) | 0.4 µl |

RNA (containing at least 15 embryo-equivalents)

Add DEPC-treated $H_2O$ to a final volume of 19 µl

*Or*

Random-hexamer-primed RT

Mix the RNA with 0.4 µl of random hexamers (1 µg/µl) and add $H_2O$ to a final volume of 11.5 µl. Incubate for 5 minutes at 65°C and quickly chill on ice. Add:

| | |
|---|---|
| 5x first-strand buffer | 4 µl |
| 0.1 M dithiothreitol | 2 µl |
| RNAsin (40 units/µl) | 0.5 µl |
| 10 mM dNTPs | 1 µl |

15. Incubate the RT reaction for 2 minutes at 42°C. Add 1 µl of reverse transcriptase and further incubate for 1 hour at 42°C.

16. Inactivate the reaction by heating for 15 minutes at 70°C and chill on ice. This is the cDNA template to be amplified by PCR.

## PCR amplification

17. Set up the PCR in 0.2-ml tubes:

| | |
|---|---|
| 10x buffer | 5 µl |
| 25 mM $MgCl_2$ | 5 µl |
| 2 µM 3′ primer | 5 µl |
| 2 µM 5′ primer | 5 µl |
| 10 mM dNTPs | 0.75 µl |
| [α-$^{32}$P]dCTP (1 mCi/ml) | 0.25 µl |
| cDNA template | 2–4 oocyte/<br>embryo equivalents |
| AmpliTaq DNA polymerase<br>  (5 units/µl) | 0.25 µl |
| $H_2O$ | up to 50 µl |

18. Run the PCR using conditions that are optimal for the DNA template. The PCR should be semiquantitative. For each set of gene-specific primers used in the PCR, determine the linear region of semilog plots of the amount of PCR product as a function of cycle number. Select a number of cycles for each primer pair that is in this linear range.

The amount of PCR product under these conditions is proportional to the number of cells used (Manejwala et al. 1991). This method permits comparison of relative changes in the abundance of a particular transcript (Latham et al. 1994; Temeles et al. 1994; Ho et al. 1995).

19. After the PCR is complete, store the tubes at –20°C in an appropriate container (the sample is radioactive).

20. Cast a 1.5-mm-thick medium size (10–15 cm) 6% polyacrylamide gel:

| | |
|---|---|
| 30% acrylamide/bis 37.5:1 solution | 10.7 ml |
| 10x TBE | 4 ml |
| $H_2O$ | 25 ml |
| 10% ammonium persulfate | 0.25 ml |
| TEMED | 0.05 ml |

21. Add 10 µl of 6x electrophoresis sample-loading buffer to the sample, mix, and load 20 µl of it in the gel. Carry out electrophoresis in 1x TBE for 3 hours at 25 mA (constant current).

22. Dry the gel under vacuum for 1 hour at 80°C, and expose it overnight in a phosphorimager cassette. Scan the cassette screen in a phosphorimager to quantify the signal (ImageQuant software, Molecular Dynamics). Compare the intensity of the band corresponding to the DNA fragment in dsRNA-injected oocytes or embryos to the controls.

**Troubleshooting**

### Low Survival after Microinjection

**Possible cause:** The volume injected is too large.

**Solution:** Prepare injection pipettes with a smaller diameter opening.

**Possible cause:** Something in the dsRNA solution is toxic to the cells.

**Solution:** Precipitate the dsRNA again, wash it with 75% ethanol, air-dry, and resuspend it in $H_2O$.

**Possible cause:** The microinjection took too long.

**Solution:** Transfer fewer oocytes/embryos to the micromanipulator. The less time the cells spend outside the incubator, the better their survival and development.

### Poor In Vitro Development of the Embryos

**Possible cause:** Embryos were outside the incubator for too long.

**Solution:** Transfer fewer embryos to the micromanipulator, inject them quickly, and expose them to as little light as possible.

### No RNAi Effect

**Possible cause:** The dsRNA is not effective.

**Solution:** Try a more concentrated solution of dsRNA or make a longer dsRNA molecule. Alternatively, try longer culture times.

## ACKNOWLEDGMENTS

The authors thank Dr. Richard M. Schultz, in whose lab all of the research and development of protocols presented in this work were conducted and supported by a grant from the National Institutes of Health (HD 22681).

## REFERENCES

Barber G.N. 2001. Host defense, viruses and apoptosis. *Cell Death Differ.* **8:** 113–126.

Bargmann C.I. 2001. High-throughput reverse genetics: RNAi screens in *Caenorhabditis elegans. Genome Biol.* **2:** REVIEWS1005.

Barstead R. 2001. Genome-wide RNAi. *Curr. Opin. Chem. Biol.* **5:** 63–66.

Bass B.L. 2000. Double-stranded RNA as a template for gene silencing. *Cell* **101:** 235–238.

Bernstein E., Caudy A.A., Hammond S.M., and Hannon G.J. 2001. Role for a bidentate ribonuclease in the initiation step of RNA interference. *Nature* **409:** 363–366.

Billy E., Brondani V., Zhang H., Muller U., and Filipowicz W. 2001. Specific interference with gene expression induced by long, double-stranded RNA in mouse embryonal teratocarcinoma cell lines. *Proc. Natl. Acad. Sci.* **98:** 14428–14433.

Bosher J.M. and Labouesse M. 2000. RNA interference: Genetic wand and genetic watchdog. *Nat. Cell Biol.* **2:** E31–E36.

Chatot C.L., Ziomek C.A., Bavister B.D., Lewis J.L., and Torres I. 1989. An improved culture medium supports development of random-bred 1-cell mouse embryos in vitro. *J. Reprod. Fertil.* **86:** 679–688.

Der S.D., Yang Y.L., Weissmann C., and Williams B.R. 1997. A double-stranded RNA-activated protein kinase-dependent pathway mediating stress-induced apoptosis. *Proc. Natl. Acad. Sci.* **94:** 3279–3283.

Djikeng A., Shi H., Tschudi C., and Ullu E. 2001. RNA interference in *Trypanosoma brucei:* Cloning of small interfering RNAs provides evidence for retroposon-derived 24-26-nucleotide RNAs. *RNA* **7:** 1522–1530.

Donze O. and Picard D. 2002. RNA interference in mammalian cells using siRNAs synthesized with T7 RNA polymerase. *Nucleic Acids Res.* **30:** e46.

Elbashir S.M., Lendeckel W., and Tuschl T. 2001a. RNA interference is mediated by 21- and 22-nucleotide RNAs. *Genes Dev.* **15:** 188–200.

Elbashir S.M., Martinez J., Patkaniowska A., Lendeckel W., and Tuschl T. 2001c. Functional anatomy of siRNAs for mediating efficient RNAi in *Drosophila melanogaster* embryo lysate. *EMBO J.* **20:** 6877–6888.

Elbashir S.M., Harborth J., Lendeckel W., Yalcin A., Weber K., and Tuschl T. 2001b. Duplexes of 21-nucleotide RNAs mediate RNA interference in cultured mammalian cells. *Nature* **411:** 494–498.

Fagard M., Boutet S., Morel J.B., Bellini C., and Vaucheret H. 2000. AGO1, QDE-2, and RDE-1 are related proteins required for post-transcriptional gene silencing in plants, quelling in fungi, and RNA interference in animals. *Proc. Natl. Acad. Sci.* **97:** 1650–1654.

Fire A., Xu S., Montgomery M.K., Kostas S.A., Driver S.E., and Mello C.C. 1998. Potent and specific genetic interference by double-stranded RNA in *Caenorhabditis elegans. Nature* **391:** 806–811.

Grishok A., Pasquinelli A.E., Conte D., Li N., Parrish S., Ha I., Baillie D.L., Fire A., Ruvkun G., and Mello C.C. 2001. Genes and mechanisms related to RNA interference regulate expression of the small temporal RNAs that control *C. elegans* developmental timing. *Cell* **106:** 23–34.

Hammond S.M., Boettcher S., Caudy A.A., Kobayashi R., and Hannon G.J. 2001. Argonaute2, a link between genetic and biochemical analyses of RNAi. *Science* **293:** 1146–1150.

Harborth J., Elbashir S.M., Bechert K., Tuschl T., and Weber K. 2001. Identification of essential genes in cultured mammalian cells using small interfering RNAs. *J. Cell Sci.* **114:** 4557–4565.

Hasuwa H., Kaseda K., Einarsdottir T., and Okabe M. 2002. Small interfering RNA and gene silencing in transgenic mice and rats. *FEBS Lett.* **532:** 227–230.

Ho Y., Wigglesworth K., Eppig J.J., and Schultz R.M. 1995. Preimplantation development of mouse embryos in KSOM: Augmentation by amino acids and analysis of gene expression. *Mol. Reprod. Dev.* **41:** 232–238.

Howlett S.K. and Reik W. 1991. Methylation levels of maternal and paternal genomes during preimplantation development. *Development* **113:** 119–127.

Hunter T., Hunt T., Jackson R.J., and Robertson H.D. 1975. The characteristics of inhibition of protein synthesis by double-stranded ribonucleic acid in reticulocyte lysates. *J. Biol. Chem.* **250:** 409–417.

Hutvagner G., McLachlan J., Pasquinelli A.E., Balint E., Tuschl T., and Zamore P.D. 2001. A cellular function for the RNA-interference enzyme Dicer in the maturation of the let-7 small temporal RNA. *Science* **293:** 834–838.

Jensen S., Gassama M.P., and Heidmann T. 1999. Taming of transposable elements by homology-dependent gene silencing. *Nat. Genet.* **21:** 209–212.

Jones P.L., Veenstra G.J., Wade P.A., Vermaak D., Kass S.U., Landsberger N., Strouboulis J., and Wolffe A.P. 1998. Methylated DNA and MeCP2 recruit histone deacetylase to repress transcription. *Nat. Genet.* **19:** 187–191.

Kalidas S. and Smith D.P. 2002. Novel genomic cDNA hybrids produce effective RNA interference in adult *Drosophila*. *Neuron* **33:** 177–184.

Kazazian H.H., Jr. 1998. Mobile elements and disease. *Curr. Opin. Genet. Dev.* **8:** 343–350.

Kazazian H.H., Jr. and Moran J.V. 1998. The impact of L1 retrotransposons on the human genome. *Nat. Genet.* **19:** 19–24.

Kennerdell J.R. and Carthew R.W. 1998. Use of dsRNA-mediated genetic interference to demonstrate that frizzled and frizzled 2 act in the wingless pathway. *Cell* **95:** 1017–1026.

Ketting R.F., Haverkamp T.H., van Luenen H.G., and Plasterk R.H. 1999. Mut-7 of *C. elegans*, required for transposon silencing and RNA interference, is a homolog of Werner syndrome helicase and RNaseD. *Cell* **99:** 133–141.

Knight S.W. and Bass B.L. 2001. A role for the RNase III enzyme DCR-1 in RNA interference and germ line development in *Caenorhabditis elegans*. *Science* **293:** 2269–2271.

Kramerov D.A., Bukrinsky M.I., and Ryskov A.P. 1985. DNA sequences homologous to long double-stranded RNA. Transcription of intracisternal A-particle genes and major long repeat of the mouse genome. *Biochim. Biophys. Acta* **826:** 20–29.

Krestel H.E., Mayford M., Seeburg P.H., and Sprengel R. 2001. A GFP-equipped bidirectional expression module well suited for monitoring tetracycline-regulated gene expression in mouse. *Nucleic Acids Res.* **29:** E39.

Kuff E.L. and Lueders K.K. 1988. The intracisternal A-particle gene family: Structure and functional aspects. *Adv. Cancer Res.* **51:** 183–276.

Lander E.S., Linton L.M., Birren B., Nusbaum C., Zody M.C., Baldwin J., et al. 2001. Initial sequencing and analysis of the human genome. *Nature* **409:** 860–921.

Latham K.E., Doherty A.S., Scott C.D., and Schultz R.M. 1994. Igf2r and Igf2 gene expression in androgenetic, gynogenetic, and parthenogenetic preimplantation mouse embryos: Absence of regulation by genomic imprinting. *Genes Dev.* **8:** 290–209.

Li Y.X., Farrell M.J., Liu R., Mohanty N., and Kirby M.L. 2000. Double-stranded RNA injection produces null phenotypes in zebrafish. *Dev. Biol.* **217:** 394–405. [Published erratum appears in *Dev. Biol.* 2000. **220:** 432.]

Lipardi C., Wei Q., and Paterson B.M. 2001. RNAi as random degradative PCR: siRNA primers convert mRNA into dsRNAs that are degraded to generate new siRNAs. *Cell* **107:** 297–307.

Luo X., Tang Z., Rizo J., and Yu H. 2002. The Mad2 spindle checkpoint protein undergoes similar major conformational changes upon binding to either Mad1 or Cdc20. *Mol. Cell* **9:** 59–71.

Ma J., Svoboda P., Schultz R.M., and Stein P. 2001. Regulation of zygotic gene activation in the preimplantation mouse embryo: Global activation and repression of gene expression. *Biol. Reprod.* **64:** 1713–1721.

Ma J., Zhou H.L., Su L., and Ji W.Z. 2002. Effects of exogenous double-stranded RNA on the basonuclin gene expression in mouse oocytes. *Sci. China Ser. C* **45:** 593.

Manejwala F.M., Logan C.Y., and Schultz R.M. 1991. Regulation of hsp70 mRNA levels during oocyte maturation and zygotic gene activation in the mouse. *Dev. Biol.* **144:** 301–308.

Martinek S. and Young M.W. 2000. Specific genetic interference with behavioral rhythms in *Drosophila* by expression of inverted repeats. *Genetics* **156:** 1717–1725.

McManus M.T. and Sharp P.A. 2002. Gene silencing in mammals by small interfering RNAs. *Nat. Rev. Genet.* **3:** 737–747.

Mette M.F., Aufsatz W., van der Winden J., Matzke M.A., and Matzke A.J. 2000. Transcriptional silencing and promoter methylation triggered by double-stranded RNA. *EMBO J.* **19:** 5194–5201.

Morel J.B., Mourrain P., Beclin C., and Vaucheret H. 2000. DNA methylation and chromatin structure affect transcriptional and post-transcriptional transgene silencing in *Arabidopsis*. *Curr. Biol.* **10:** 1591–1594.

Mourrain P., Beclin C., Elmayan T., Feuerbach F., Godon C., Morel J.B., Jouette D., Lacombe A.M., Nikic S., Picault N., Remoue K., Sanial M., Vo T.A., and Vaucheret H. 2000. *Arabidopsis* SGS2 and SGS3 genes are required for posttranscriptional gene silencing and natural virus resistance. *Cell* **101**: 533–542.

Nagy A., Gertsenstein M., Vintersten K., and Behringer R. 2003. *Manipulating the mouse embryo: A laboratory manual,* 3rd. Edition. Cold Spring Harbor Laboratory Press, Cold Spring Harbor, New York.

Nan X., Ng H.H., Johnson C.A., Laherty C.D., Turner B.M., Eisenman R.N., and Bird A. 1998. Transcriptional repression by the methyl-CpG-binding protein MeCP2 involves a histone deacetylase complex [see comments]. *Nature* **393**: 386–389.

Oates A.C., Bruce A.E., and Ho R.K. 2000. Too much interference: Injection of double-stranded RNA has nonspecific effects in the zebrafish embryo. *Dev. Biol.* **224**: 20–28.

Paddison P.J., Caudy A.A., and Hannon G.J. 2002a. Stable suppression of gene expression by RNAi in mammalian cells. *Proc. Natl. Acad. Sci.* **99**: 1443–1448.

Paddison P.J., Caudy A.A., Bernstein E., Hannon G.J., and Conklin D.S. 2002b. Short hairpin RNAs (shRNAs) induce sequence-specific silencing in mammalian cells. *Genes Dev.* **16**: 948–958.

Piccin A., Salameh A., Benna C., Sandrelli F., Mazzotta G., Zordan M., Rosato E., Kyriacou C.P., and Costa R. 2001. Efficient and heritable functional knock-out of an adult phenotype in *Drosophila* using a GAL4-driven hairpin RNA incorporating a heterologous spacer. *Nucleic Acids Res.* **29**: E55.

Reik W., Dean W., and Walter J. 2001. Epigenetic reprogramming in mammalian development. *Science* **293**: 1089–1093.

Rougvie A.E. 2001. Control of developmental timing in animals. *Nat. Rev. Genet.* **2**: 690–701.

Sam M., Wurst W., Kluppel M., Jin O., Heng H., and Bernstein A. 1998. Aquarius, a novel gene isolated by gene trapping with an RNA-dependent RNA polymerase motif. *Dev. Dyn.* **212**: 304–317.

Sambrook J. and Russell D.W. 2001. *Molecular cloning: A laboratory manual,* 3rd edition. Cold Spring Harbor Laboratory Press, Cold Spring Harbor, New York.

Santos F., Hendrich B., Reik W., and Dean W. 2002. Dynamic reprogramming of DNA methylation in the early mouse embryo. *Dev. Biol.* **241**: 172–182.

Sharp P.A. 2001. RNA interference—2001. *Genes Dev.* **15**: 485–490.

Sijen T., Fleenor J., Simmer F., Thijssen K.L., Parrish S., Timmons L., Plasterk R.H., and Fire A. 2001. On the role of RNA amplification in dsRNA-triggered gene silencing. *Cell* **107**: 465–476.

Shi H., Djikeng A., Mark T., Wirtz E., Tschudi C., and Ullu E. 2000. Genetic interference in *Trypanosoma brucei* by heritable and inducible double-stranded RNA. *RNA* **6**: 1069–1076.

Sijen T. and Kooter J.M. 2000. Post-transcriptional gene-silencing: RNAs on the attack or on the defense? *BioEssays* **22**: 520–531.

Sijen T., Vijn I., Rebocho A., van Blokland R., Roelofs D., Mol J.N., and Kooter J.M. 2001. Transcriptional and posttranscriptional gene silencing are mechanistically related. *Curr. Biol.* **11**: 436–440.

Smardon A., Spoerke J.M., Stacey S.C., Klein M.E., Mackin N., and Maine E.M. 2000. EGO-1 is related to RNA-directed RNA polymerase and functions in germ-line development and RNA interference in *C. elegans. Curr. Biol.* **10**: 169–178.

Smith N.A., Singh S.P., Wang M.B., Stoutjesdijk P.A., Green A.G., and Waterhouse P.M. 2000. Total silencing by intron-spliced hairpin RNAs. *Nature* **407**: 319–320.

Stein P., Svoboda P., and Schultz R.M. 2003a. Transgenic RNAi in mouse oocytes: A simple and fast approach to study gene function. *Dev. Biol.* **256**: 187–193.

Stein P., Svoboda P., Anger M., and Schultz R.M. 2003b. RNAi: Mammalian oocytes do it without RNA-dependent RNA polymerase. *RNA* **9**: 187–192.

Svoboda P., Stein P., and Schultz R.M. 2001. RNAi in mouse oocytes and preimplantation embryos: Effectiveness of hairpin dsRNA. *Biochem. Biophys. Res. Commun.* **287**: 1099–1104.

Svoboda P., Stein P., Hayashi H., and Schultz R.M. 2000. Selective reduction of dormant maternal mRNAs in mouse oocytes by RNA interference. *Development* **127**: 4147–4156.

Tabara H., Sarkissian M., Kelly W.G., Fleenor J., Grishok A., Timmons L., Fire A., and Mello C.C. 1999. The *rde-1* gene, RNA interference, and transposon silencing in *C. elegans. Cell* **99**: 123–132.

Tavernarakis N., Wang S.L., Dorovkov M., Ryazanov A., and Driscoll M. 2000. Heritable and inducible genetic interference by double-stranded RNA encoded by transgenes. *Nat. Genet.* **24**: 180–183.

Temeles G.L., Ram P.T., Rothstein J.L., and Schultz R.M. 1994. Expression patterns of novel genes during mouse preimplantation embryogenesis. *Mol. Reprod. Dev.* **37:** 121–129.

Tuschl T., Zamore P.D., Lehmann R., Bartel D.P., and Sharp P.A. 1999. Targeted mRNA degradation by double-stranded RNA in vitro. *Genes Dev.* **13:** 3191–3197.

Ui-Tei K., Zenno S., Miyata Y., and Saigo K. 2000. Sensitive assay of RNA interference in *Drosophila* and Chinese hamster cultured cells using firefly luciferase gene as target. *FEBS Lett.* **479:** 79–82.

Walsh C.P., Chaillet J.R., and Bestor T.H. 1998. Transcription of IAP endogenous retroviruses is constrained by cytosine methylation. *Nat. Genet.* **20:** 116–117.

Wargelius A., Ellingsen S., and Fjose A. 1999. Double-stranded RNA induces specific developmental defects in zebrafish embryos. *Biochem. Biophys. Res. Commun.* **263:** 156–1561.

Waterhouse P.M., Wang M.B., and Lough, T. 2001. Gene silencing as an adaptive defence against viruses. *Nature* **411:** 834–842.

Wei Q., Marchler G., Edington K., Karsch-Mizrachi I., and Paterson B.M. 2000. RNA interference demonstrates a role for nautilus in the myogenic conversion of Schneider cells by daughterless. *Dev. Biol.* **228:** 239–255.

Wesley S.V., Helliwell C.A., Smith N.A., Wang M.B., Rouse D.T., Liu Q., Gooding P.S., Singh S.P., Abbott D., Stoutjesdijk P.A., Robinson S.P., Gleave A.P., Green A.G., and Waterhouse, P.M. 2001. Construct design for efficient, effective and high-throughput gene silencing in plants. *Plant J.* **27:** 581–590.

Wianny F. and Zernicka-Goetz M. 2000. Specific interference with gene function by double-stranded RNA in early mouse development. *Nat. Cell Biol.* **2:** 70–75.

Xu Z., Williams C.J., Kopf G.S., and Schultz R.M. 2003. Maturation-associated increase in IP(3) receptor type 1: Role in conferring increased IP(3) sensitivity and CA(2+) oscillatory behavior in mouse eggs. *Dev. Biol.* **254:** 163–171.

Yang D., Buchholz F., Huang Z., Goga A., Chen C.Y., Brodsky F.M., and Bishop J.M. 2002. Short RNA duplexes produced by hydrolysis with *Escherichia coli* RNase III mediate effective RNA interference in mammalian cells. *Proc. Natl. Acad. Sci.* **99:** 9942–9947.

Yang S., Tutton S., Pierce E., and Yoon K. 2001. Specific double-stranded RNA interference in undifferentiated mouse embryonic stem cells. *Mol. Cell. Biol.* **21:** 7807–7816.

Yu J.Y., DeRuiter S.L., and Turner D.L. 2002. RNA interference by expression of short-interfering RNAs and hairpin RNAs in mammalian cells. *Proc. Natl. Acad. Sci.* **99:** 6047–6052.

Zamore P.D. 2001. RNA interference: Listening to the sound of silence. *Nat. Struct. Biol.* **8:** 746–750.

Zamore P.D., Tuschl T., Sharp P.A., and Bartel D.P. 2000. RNAi: Double-stranded RNA directs the ATP-dependent cleavage of mRNA at 21 to 23 nucleotide intervals. *Cell* **101:** 25–33.

Zhao Z., Cao Y., Li M., and Meng A. 2001. Double-stranded RNA injection produces nonspecific defects in zebrafish. *Dev. Biol.* **229:** 215–223.

Zhou A., Paranjape J., Brown T.L., Nie H., Naik S., Dong B., Chang A., Trapp B., Fairchild R., Colmenares C., and Silverman R.H. 1997. Interferon action and apoptosis are defective in mice devoid of 2′,5′-oligoadenylate-dependent RNase L. *EMBO J.* **16:** 6355–6363.

# CHAPTER 16

# RNAi Technologies in *Drosophila* Cell Culture

## Scott M. Hammond

*Department of Cell and Developmental Biology, University of North Carolina,
Chapel Hill, North Carolina 27599*

THE USE OF THE MODEL SYSTEM *DROSOPHILA MELANOGASTER* is well-established, and the ability to do forward genetics, transgenics, and RNAi (RNA interference)-mediated gene knockdown using this organism has made it a powerful biological tool. However, the use of *Drosophila* cell lines has been traditionally underutilized.

As annotated genomic sequence becomes available, the need to assign function to genes is apparent. One of the most convincing approaches to determine gene function is to block the expression of the gene in question and assay the phenotype, an approach recently been made possible by the discovery of RNAi. Targeted gene silencing in cultured cells is especially powerful because of the relative low cost and simplicity of many assays. Additionally, basic cytological processes, for example, signal transduction, cell cycle, and cytoskeletal regulation, can be most easily assayed in cultured cells.

The recent development of RNAi in *Drosophila* cultured cells is making this system attractive for these types of functional studies. A large fraction of human genes have homologs in *Drosophila*. For example, a cursory study of these genomes reveals that of 67 genes mutated in human cancers, 46 have homologs in *Drosophila* (Rubin et al. 2000), including key tumor suppressors such as ATM, p53, and pRB. An increasing number of papers using *Drosophila* cells to understand basic cell biology, including the use of RNAi in *Drosophila* cells, are appearing in the literature. RNAi in the more popular mammalian cell lines is becoming well-established. However, *Drosophila* cells are a powerful system in their own right, and their value in this respect is demonstrated in Figure 16.1. *Drosophila*

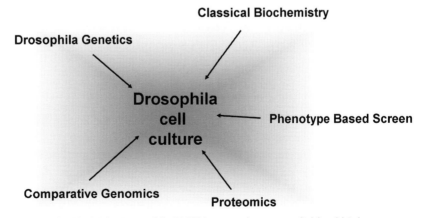

**FIGURE 16.1.** *Drosophila* RNAi is central to many fields of biology.

cultured cells reside at the nexus of a varied collection of experimental approaches, and they can bridge classical genetic and biochemical systems with increasingly important genomic approaches.

RNAi is extremely effective in *Drosophila* cells, with the degree of silencing achieved being much stronger than typically seen in mammalian cells; total knockout is often achieved. The protocol of soaking cells with RNA-polymerase-prepared double-stranded RNA (dsRNA) makes *Drosophila* RNAi simple to perform and inexpensive. This chapter discusses the characteristics of RNAi in *Drosophila* cells as well as the variety of applications that are amenable to this system.

## CONCEPTS AND STRATEGIES

## Characteristics of the *Drosophila* S2 Cell System

A number of continuous cell lines and primary cell cultures have been developed from *D. melanogaster*, but by far the most commonly used system is the S2 cell line. This chapter focuses on the S2 system; however, other *Drosophila* lines and cultures will be briefly mentioned.

### Culture

S2 cells were derived from late-stage *Drosophila* embryos (Schneider 1972). They were originally described as epithelium-like; however, recent evidence suggests that they may have hematopoietic origins (Ramet et al. 2002). Whatever their original cell type, extended time in culture, under probably nonideal conditions, has promoted the outgrowth of a poorly differentiated, highly proliferative cell type. These cells grow loosely attached to a plastic substrate, but they do not spread out on the substrate, probably due to low levels of integrin expression (Bunch and Brower 1992). The cells are released from the culture dish by shaking; trypsin is not needed.

In addition to monolayer growth, S2 cells can be easily adapted to suspension growth; this is the preferred way to maintain stock cultures, in which cells are seeded into plates for transfection. Additionally, suspension growth allows large-scale cultures to be grown, and with the addition of aeration, high cell densities are possible. We have grown 10-liter cultures with ease, using aeration, for preparative biochemical work. Protein purified from S2 cells lacks high-abundance contaminating proteins found in embryo extracts, albeit at a higher cost (i.e., media).

S2 cells grow quite well at room temperature, in the absence of $CO_2$, permitting benchtop growth, although refrigerated incubators should be used in laboratories where ambient temperature fluctuates. Various insect media can be used, supplemented with fetal bovine serum. Alternatively, serum-free media is available from a number of vendors.

### Transient transfection

The ability of S2 cells to grow as a monolayer allows standard transient transfection methods to be used to a high degree of efficiency. We have seen efficiencies as high as 50% using the calcium phosphate method (Hammond et al. 2000). Lipid-based reagents also work well. A variety of expression vectors are available that have constitutive or inducible promoters. Transgene expression levels are typically high. We have had success with Invitrogen's pMT vector, which contains the metallothionein promoter that is induced with copper sulfate. It is also available as a "topo" vector for enzyme-free cloning. As discussed below, introduction of dsRNA into S2 cells is especially efficient and can be done by transfection or by simply soaking cells in dsRNA.

## Stable transfection

It is possible to generate stable S2 cell lines, although not with the simplicity of mammalian cells. The most common selectable marker is the hygromycin resistance gene, which is available from Invitrogen. Alternatively, green fluorescent protein (GFP) can be used as a marker and cells selected by fluorescence-activated cell sorting (FACS). Obtaining clonal lines is difficult because S2 cells, and *Drosophila* cultured cells in general, do not grow at low density (Echalier 1997). Presumably, bovine serum is not a totally adequate medium supplement; thus, the cells require secreted autocrine growth factors. Growth of cells below 500,000 per milliliter is not possible, but this limitation can be circumvented by growth in conditioned media. The poor attachment of S2 cells requires that clones be selected via growth in 96-well plates, or in soft agar. Using these methods, stable clonal lines can be obtained. However, we have seen an unusual effect in which colonies of cells expressing a GFP reporter developed "sectoring," or regions of negative expression (S.M. Hammond and A.A. Caudy, unpubl.). This effect was possibly due to specific loss of the transgene. An alternative explanation can be forwarded, however, based on the properties of RNAi: Stable S2 cell lines result from integration of large arrays of head-to-tail concatamers of linearized plasmid (Echalier 1997). Transcription through these arrays yields hairpin RNAs that could enter the RNAi pathway, leading to silencing of the transgene. Indeed, deliberate expression of hairpin RNAs is an established method for stable gene silencing in a number of organisms, including *Drosophila*. This effect raises concerns about the stability of transgenic S2 cell lines.

## Biological assays in S2 cells

Historically, *Drosophila* cells have not been widely used to study biological pathways. This was due to several reasons, most notably because *Drosophila* as an organism was principally a genetic tool. Additionally, continuous cell lines, such as the S2 line, are somewhat poorly characterized; indeed, the primary use of S2 cells was as an "inert" expression tool (i.e., mammalian biological pathways that did not exist in S2 cells were studied there because there was no contaminating influence of endogenous activities). S2 cells were used to study the properties of the Sp1 transcription factor (Suske 1999) and various ion channels (Towers and Sattelle 2002).

More recently, it has become apparent that S2 cells are usable as a general biochemical and cytological tool for the following reasons:

- Many fundamental processes, such as growth, viability, and death, are readily measured.
- Growth curves are straightforward.
- Apoptosis can be measured using standard methodology.
- S2 cells display classic apoptotic features, including membrane blebbing, nuclear condensation, and DNA fragmentation (Jones et al. 2000; Li et al. 2002; Muro et al. 2002).
- Cell cycle function can be assayed by propidium iodide FACS. This method has been used to probe the function of cyclins and anaphase-promoting complex (APC) proteins on S-phase induction, $G_2$ arrest, and endoreduplication (Cornwell et al. 2002; Mihaylov Iet al. 2002; Pile et al. 2002).

Standard suspension cell methods can be used for immunolocalization of internal proteins. Cytocentrifugation can be used to attach cells to slides, an approach that has been used to visualize the mitotic (Warren et al. 2000) and cytokinetic (Somma et al. 2002) machinery.

The chief disadvantage with *Drosophila* cell culture in general is the lack of "traditional" cell lines, i.e., epithelial or fibroblast cell lines. The easily grown immortalized cell lines, including S2 and Kc, do not have typical adherent cell morphology, thus limiting their use for many cytoskeletal studies. Intriguingly, S2 cells grown on plates coated with concanavalin A spread on the substrate, displaying a microtubular network resembling adherent mammalian cell lines (Rogers et al. 2002). Additionally, several groups have expressed *Drosophila* integrins in S2 cells, causing normal attachment and spreading on matrix proteins or tissue culture plastic (Bunch and Brower 1992; Gotwals et al. 1994).

### Other *Drosophila* cell lines

*Drosophila* primary cultures can be generated from embryos ranging from 6 to 24 hours. These cultures are highly unstable, with different cell types proliferating as the cultures develop. Primary cultures from the larval central nervous system and imaginal discs have also been prepared. However, the small amount of source material, combined with the short proliferative lifespan, limits their usefulness (for general information on *Drosophila* cell cultures and continuous lines, see Echalier 1997).

A more viable approach is the use of immortalized lines. In addition to the popular S2 and Kc cell lines, a variety of other *Drosophila* cell lines have been developed. Several lines have been derived from tumorous blood cells from *lethal malignant blood neoplasm* (*l(2)mbn* and *l(3)mbn*) mutants. Cell lines have also been developed from the "normal" larval central nervous system and wing, leg, and antenna imaginal discs. These cells are typically grown in a medium supplemented with adult fly extract, insulin, and fetal bovine serum.

With the exception of Kc and the neuronal line BG2-C6, RNAi has not been reported to work in cell lines other than S2, although there have been no reports of failure (Clemens et al. 2000). Given the universality of RNAi in the fly, it is expected that these nonembryonal lines would be competent for RNAi.

## Development of RNAi in Cultured S2 Cells

The use of RNAi in S2 cells was reported concurrently by three groups (Caplen et al. 2000; Clemens et al. 2000; Hammond et al. 2000). Our interest in the system was largely as a tool to study the mechanism of RNAi. We were interested in developing a biochemical system that was derived from cultured cells. At the time, *Caenorhabditis elegans* and *Drosophila* were the only organisms in which RNAi had been reported to work. Human cells were not expected to be amenable to RNAi because of the nonspecific dsRNA responses, an opinion that predated the groundbreaking work of Tuschl using short interfering RNAs (siRNAs) (Elbashir et al. 2001c).

We chose the *Drosophila* S2 cell line as a model system for the study of RNAi mechanistic pathways even though introduction of dsRNA into S2 cells had not been reported previously. We reasoned that dsRNA, being similar chemically to dsDNA, would transfect to a similar degree of efficiency. This appeared to be the case, because transfection of various dsRNAs could induce silencing in a large fraction of cells. Indeed, dsRNA seemed to transfect better than dsDNA plasmid constructs. In retrospect, this could be explained by the soaking phenomenon reported by Dixon (see below).

We have successfully used calcium-phosphate- and liposome-based transfection methods for RNAi in S2 cells. We can easily achieve more than 95% silencing with either method (for a representative experiment, see Figure 16.2). RNAi in S2 cells displayed the

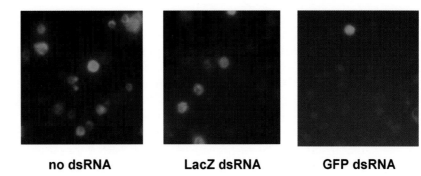

no dsRNA          LacZ dsRNA          GFP dsRNA

**FIGURE 16.2.** RNAi in *Drosophila* S2 cells is highly effective. S2 cells were transfected with a GFP reporter plasmid and with dsRNA to GFP or with a control dsRNA. GFP expression was detected by epifluoresence 48 hours posttransfection.

hallmarks of RNAi seen in *C. elegans*, including loss of targeted mRNA, high sequence specificity, and dsRNA trigger-length dependence (Hammond et al. 2000).

Concurrent with our work, two other groups were developing S2 RNAi technologies. Morgan and colleagues reported silencing of a GFP marker after transfection with GFP dsRNA, using lipid transfection reagents (Caplen et al. 2000). These authors also reported spreading of silencing from one population of cells to another, analogous to in vivo spreading seen in *C. elegans*. Cells were transfected with dsRNA, washed, and used to condition medium with a "silencing entity." This medium was capable of inducing moderate silencing in a different population of cells. Whether this silencing entity was leftover dsRNA from the transfection or a substance released from cells was not determined. The discovery of the soaking phenomenon in S2 cells (see below) strengthens the possibility that some leftover dsRNA was simply absorbed by the target cell population.

Dixon's group (Clemens et al. 2000), who co-discovered RNAi in S2 cells, developed a novel approach. They simply soaked cells with dsRNA, reasoning that cells would take it up in a manner similar to what occurs with in vivo spreading in *C. elegans* (Fire et al. 1998). This method proved to be highly efficient, eliciting silencing in essentially all cells; it is also simple to perform. For these reasons, soaking has become the preferred method. Dixon's laboratory has reported the successful knockdown of 18 genes (Clemens et al. 2000; Worby et al. 2001b, 2002; Muda et al. 2002).

## TECHNIQUES

### Basic S2 Cell RNAi Protocol

As mentioned above, two basic protocols for inducing silencing have been reported: transfection and soaking. Both methods yield essentially 100% efficiency in the delivery of dsRNA. The choice of method largely depends on whether plasmid DNA is also being delivered. If a plasmid is being used, for instance, as a reporter, transfection is the preferred method. If no plasmid is needed, soaking is technically easier.

### Transfection

We originally adapted calcium-phosphate-based transfection for introduction of dsRNA into S2 cells, based on the protocol reported by Di Nocera and Dawid (1983), and readers are directed to this reference for recipes. We generally prepare a set of transfection reagents spanning a pH range of 6.8–7.8 and empirically determine the optimum pH. Simple substitution of dsRNA for plasmid DNA yielded highly efficient nucleic acid transfer. We found that dsRNA could be mixed with plasmid at a 1:10 ratio without significant loss in silencing efficiency, although we usually use a 1:3 ratio. The duration of silencing continues at least 10 days, measured by reporter gene expression (S.M. Hammond, unpubl.).

We have also used lipid-based transfection reagents, including Fugene 6 (Roche), Lipofectamine (Invitrogen), and Superfect (QIAGEN), with success. We followed the manufacturers' instructions, substituting dsRNA for plasmid DNA. This method is somewhat easier to use and not as reagent-sensitive. However, it is more expensive, prohibitively so for preparative-scale transfections.

### Soaking

In *C. elegans*, it is possible to initiate RNAi in worms by merely soaking the animals in dsRNA (Tabara et al. 1998). By an unknown transport mechanism, a silencing entity spreads throughout the worm, effecting gene knockdown in essentially all cell types (neurons are somewhat resistant). It is not clear whether this transport mechanism exists in *Drosophila*; however, this did not prevent Dixon's group from attempting it with S2 cells. It works surprisingly well, achieving gene-silencing levels similar to those using transfection. For details, see Dixon's protocol (Worby et al. 2001a). Cells are washed in serum-free medium, and resuspended in medium containing dsRNA. The use of serum-free medium is essential, presumably to prevent RNase-mediated degradation of the dsRNA. Possibly, the serum-free condition enhances active uptake of dsRNA. After 10 minutes of incubation in serum-free medium plus dsRNA, serum is reintroduced, and the cells are cultured for 1–3 days, to allow turnover of target protein. The milder conditions that prevail in the soaking protocol may permit earlier time point analysis; however, one is ultimately limited by target protein turnover rates (see below).

Interestingly, there has been a report of gene knockdown in *Drosophila* embryos by a soaking procedure (see Eaton et al. 2002), whereby embryos are dechorionated, soaked in dsRNA, and allowed to hatch on standard medium. Gene expression in adult flies was suppressed. This further implies the existence of an active uptake mechanism in *Drosophila*.

## dsRNA Production

With the development of high-yield transcription kits, it is easy to make large quantities of dsRNA. We have used Megascript (Ambion) and Ribomax (Promega) kits with equal success. Transcription templates are polymerase chain reaction (PCR) products containing T7 promoters on each end. Transcription with T7 polymerase yields dsRNA. The duplex nature can be confirmed by native agarose gel electrophoresis or, more reliably, with RNase III treatment; however, we find that this is not necessary. After simple phenol extraction and ethanol precipitation, RNA is essentially fully duplex. We have also explored the use of crude dsRNA for the purpose of high-throughput RNAi. Calcium phosphate transfection with dsRNA taken directly from the transcription reaction was highly effective at initiating RNAi (J.M Silva and D. Siolas, pers. comm.).

Transcription templates can be derived from any region in the coding sequence, which is in contrast to the positional sensitivity seen with siRNA-mediated RNAi in mammalian cells (Holen et al. 2002). Untranslated region (UTR) sequences have been reported to be effective in *C. elegans* (Fire et al. 1998). They have not been directly tested in *Drosophila* cell culture; however, it is expected that they will be effective. dsRNAs in the range of 300–1000 nucleotides are most effective. Longer dsRNAs are prone to developing secondary structure, and thus do not form complete duplexes. dsRNAs shorter than 200 nucleotides become less effective silencing agents, as seen in *C. elegans* (Hammond et al. 2000).

## Further Considerations

The onset of gene silencing is based on target protein turnover. Because the RNAi machinery targets mRNA only, existing protein levels must decay before knockdown is observed. We have targeted D and E cyclins with the expectation that knockdown will be rapid due to rapid protein turnover. Indeed, cell cycle arrest is observed within 12 hours of transfection. As time is required to synchronize this population in late $G_1$, the destruction of mRNA is extremely rapid (E. Bernstein and S.M. Hammond, unpubl.). This is expected based on in vitro measurements of mRNA degradation (Hammond et al. 2000). Therefore, the time to a knockdown phenotype is almost entirely due to protein decay. In some cases, stable proteins may be more effectively targeted through repeated soaking or transfection (A.A Caudy, pers. comm.). We have monitored silencing of a reporter as long as 10 days after transfection and seen little recovery of expression. Morgan and colleagues reported approximately 50% recovery after 10 days (Caplen et al. 2000).

Silencing can be assayed by northern blotting or reverse transcriptase-PCR (RT-PCR), or by western blotting. If antibodies are available, then western blotting is preferable because it more directly measures the desired effect of RNAi silencing, i.e., protein level reduction. This method also guards against the remote possibility that silencing will act via translational blockade, instead of the expected mRNA turnover mechanism.

### Stable hairpin transgenes

The aforementioned RNAi protocols deliver a transient silencing affect. Silencing can be introduced stably using a hairpin expression cassette, a plasmid construct that has an inverted repeat of the targeted sequence. Transcription yields a hairpin RNA molecule that enters the RNAi pathway. This approach was first reported in *Drosophila* by Carthew's group (Kennerdell and Carthew 2000). P-element-mediated introduction of a *lacZ* hairpin into flies resulted in suppression of β-galactosidase expression in flies carrying the gene. Silencing was inducible via the heat shock promoter that was driving expression of the hairpin RNA. Subsequently, several groups reported technological improvements, including splicing of the hairpin RNA transcript (Reichhart et al. 2002) and improvements in hairpin plasmid construction (Piccin et al. 2001; Paddison et al. 2002).

This system is adaptable to tissue-specific silencing using the GAL4/UAS system, which is particularly valuable for essential genes. *sec10*, a secretory pathway gene essential for embryonal development, was silenced in adult flies by tissue-specific hairpin RNAi (Andrews et al. 2002), and it was shown to be nonessential for general secretion in somatic cells. Lethality arose from a hormone secretion defect in development.

Although this approach is popular with *Drosophila* geneticists, it has not been reported in *Drosophila* cell culture. In principle, it could be adapted to this system by stable integration of a hairpin plasmid construct.

### Use of siRNAs

The use of long dsRNAs in mammalian cells is fraught with problems due to nonspecific responses. Tuschl and colleagues overcame this limitation by using synthetic 21-nucleotide dsRNAs (siRNAs; see Elbashir et al. 2001c). These RNAs, essentially synthetic mimics of the natural products of the Dicer RNase III protein, are too short to activate nonspecific response pathways, but they are still able to enter the RNAi pathway. This technology has revolutionized mammalian cell culture systems. Although siRNAs are highly effective in *Drosophila* cells (Elbashir et al. 2001c) and embryos (Boutla et al. 2001), their use has not had a significant impact on the field. Long dsRNAs are preferred. They can be made enzymatically with standard transcription kits, which is less expensive than synthetic siRNAs. Additionally, long dsRNAs typically do not suffer from positional effects; i.e., siRNAs tend to have variable effectiveness, depending on the sequence of the mRNA that is being targeted (Holen et al. 2002), which is thought to be due to currently unknown structural features of the mRNA target. Long dsRNAs cover a large enough region of the mRNA to avoid this problem.

The major contribution that *Drosophila* systems have made to the siRNA field is as a discovery tool. Tuschl and colleagues used an in vitro *Drosophila* system to define the most effective siRNA structure for silencing. They screened approximately 100 siRNA structures of varying lengths, overhang composition, and target mismatches (Elbashir et al. 2001a,b), which led to the prototype 21-nucleotide siRNA with the 3′ overhangs that is used in mammalian cells.

## More Comments on the Targeting Sequence

The basic RNAi protocol is based on targeting large regions of the coding sequence to generate total knockdown of expression. By tailoring the dsRNA trigger sequence, more subtle effects can be achieved.

### Targeting of mutant alleles

One potential use of siRNAs is their ability to target specific alleles. By nature of their short length, they can be designed to differentiate between two transcripts that vary by a single point mutation. Using their *Drosophila* cell-free system, Tuschl's group demonstrated that a single mismatch in the center of the siRNA significantly reduced the effectiveness of silencing (Elbashir et al. 2001a). Some mutations, however, were found to have a less deleterious effect (Boutla et al. 2001; Amarzguioui et al. 2003). Validation is especially important for these experiments. Several different siRNAs, with the mutation at different locations, may need to be tested to identify allele-specific silencing triggers.

This approach was tested functionally in a human system by blocking expression of the dominant V12 allele of Ki-*ras* while not affecting expression of the wild-type allele. Such treatment reversed the tumorigenicity of human tumor cells (Brummelkamp et al. 2002).

### Targeting of splice variants

One complexity of interpreting genomic information is the presence of multiple mRNA isoforms for many genes. The function of each mRNA isoform can be ascertained by exon-specific RNAi. A panel of dsRNAs corresponding to individual exons can be used to deplete specific mRNA isoforms, and the function of the remaining variants can be determined. An example of this approach was used to study the *Drosophila Dscam* gene, which has 95 alternative exons capable of generating 38,016 mRNA isoforms (Schmucker et al. 2000). RT-PCR demonstrated that individual isoforms could be depleted by RNAi without affecting the total pool of *Dscam* mRNA (Celotto and Graveley 2002). The function of mRNA isoforms of STAT92 was similarly studied, although these isoforms arise from an alternative promoter (Henriksen et al. 2002). These methods have also found success in *C. elegans* (Cho et al. 2000) and human cells (Kisielow et al. 2002).

## Allele replacement

In a manner similar to targeting a specific allele, RNAi can be used to deplete the endogenous allele for the purpose of replacing it with a transgenic one. This approach is based on the ability to suppress gene expression by directing dsRNAs against untranslated regions of the mRNA (Parrish et al. 2000). This allows one to eliminate expression of the endogenous gene without targeting the coding sequence. Mutant alleles that lack the targeted untranslated region can then be introduced for expression. In this way, the properties of the mutant allele can be ascertained without contaminating the influence of the endogenous wild-type alleles.

In a related method, epitope-tagged alleles of a gene can be expressed in cells that have been depleted of endogenous protein via targeting of its untranslated region. This approach has a high potential value as a proteomic tool. For example, it has been used to deplete the endogenous *Drosophila* exosome component Rrp4 while introducing a tagged allele of the human ortholog (Forler et al. 2003), thus greatly improving the efficiency in incorporation of the tagged protein into complexes with endogenous exosomal proteins. Affinity purification directed at the tag yielded extremely clean preparations of ten endogenous exosome components. In this case, the targeted gene was replaced with an orthologous gene; however, by targeting the UTR, the same gene could be reintroduced.

## Combinatorial RNAi

In *C. elegans*, only a few dsRNA molecules are required per cell to silence gene expression effectively. The high potency of RNAi begs the question of how many genes can be targeted simultaneously. This approach would be particularly valuable for defining the function of genes that reside in families with several apparently redundant isoforms. Although the maximum number has not been rigorously defined, it has been reported that minimally four genes can be targeted at one time in *Drosophila* cells. Virshup and colleagues targeted combinations of four isoforms of PP2A B subunits and examined the effect on PP2A complex formation and resultant effects on apoptotic pathways (Li et al. 2002). Even when targeting four genes, protein levels for each were reduced as efficiently as when targeting each gene alone. Similarly, in *Drosophila* embryos, four genes could be targeted with only a modest loss of efficiency (Schmid et al. 2002). In a *Drosophila* embryo lysate, titrating competitor dsRNA into the RNAi reaction resulted in approximately 50% recovery of reporter gene expression when using 20-fold excess of competitor (Tuschl et al. 1999). This demonstrates that there is a finite limit on the amount of dsRNA the system can handle. Therefore, the combinatorial limit would be expected in the range of five to ten genes.

# Applications of S2 Cell RNAi

Although most investigators will be interested in targeting one specific gene for the purpose of identifying its function, a range of possible experimental approaches is available. Some of the approaches are more generally applicable than others, as will be apparent. It should be noted that these techniques are not necessarily comprehensive. RNAi in *Drosophila* cells is a basic experimental platform on which a host of end uses can be built.

## Somatic cell genetics: Epistatic analysis

Originally the domain of geneticists, gene epistasis experiments are being incorporated into cultured cell experiments. Previous work has relied on dominant-negative and dominant-activated alleles to demonstrate genetic pathways. RNAi knockdown is a significant improvement over dominant-negative alleles, if only because it can be reliably performed for any gene, not only those whose protein products have well-characterized activities. The combination of RNAi knockdown and overexpression has the ability to recapitulate classical genetics in cultured cells.

These cultured cell genetic methods have been used successfully to probe several biological pathways in S2 cells. For example, to identify a role for human geminin in $G_2$ checkpoint control, the *Drosophila* ortholog was suppressed by RNAi (Mihaylov et al. 2002). The resultant polyploid cells were treated with dsRNA to a variety of cell cycle regulators to define their genetic relationship. Epistasis (recovery of the wild-type phenotype as a result of a second gene knockdown) was observed with the *Drosophila* ortholog of Ctd1, whereas the Chk1 ortholog partially rescued the geminin phenotype.

Similar experiments have been used to define the genetic relationships among apoptosis regulators. Suppression of the *Drosophila* anti-apoptotic proteins called inhibitors of apoptosis (IAP) resulted in extensive apoptosis. This phenotype was rescued by knockdown of either the caspase homolog DRONC or the APAF homolog DARK (Igaki et al. 2002). In another study, knockdown of DARK did not rescue cells from activation of the *reaper*/caspase-8 pathway (Quinn et al. 2000).

## High-throughput/Forward genetics

The logical extension to genetic studies on individual genes is to expand to larger sets of genes. Thus, cultured cells would approach the status of a "true" genetic system. The ultimate experiment would be a high-throughput phenotype-based screen against the entire genome using arrayed dsRNAs. This system is depicted in Figure 16.3A. dsRNA transcription templates are generated by PCR using a nonredundant cDNA set. The *Drosophila* set is about 70% complete and is commercially available from Research Genetics (Stapleton et al. 2002). Transcription of the templates in vitro yields dsRNA, which is used directly to transfect 96-well plates of S2 cells. A high-throughput assay is used to screen the desired phenotype. Although this system remains untested as a whole, each step has been validated individually (J.M. Silva, D. Siolas, and G.H. Hannon, pers. comm.).

An alternative protocol is based on cDNAs derived from a library (Figure 16.3B). Clones are picked at random, templates are made, and dsRNAs are transcribed. These dsRNAs can then be pooled, or arrayed individually, and used for high-throughput phenotype assays. This method was successfully used to identify genes required for phagocytosis in S2 cells (Ramet et al. 2002). A total of 1000 random dsRNAs were screened, of which 34 had an effect on phagocytosis of *Escherichia coli*. The major drawback of this approach is the likelihood of redundancy for abundantly transcribed genes. With the availability and relatively low cost of nonredundant gene sets, library-based dsRNAs are generally not recommended.

In addition to these large-scale approaches, a number of reports have been published using RNAi in smaller-scale genetic studies, which generally target a small set of genes known to be involved in a specific biological pathway. For instance, ten core proteosomal components were depleted, and their effects on proteosome function, apoptosis, and the cell cycle were determined (Wojcik and DeMartino 2002). In another study, 17 genes known to be involved in the apoptotic machinery were silenced and their genetic relationship was defined (Li et al. 2002). Genes involved in cytokinesis (Somma et al. 2002) and protein phosphorylation/dephosphorylation (Muda et al. 2002; Worby et al. 2002) were similarly identified.

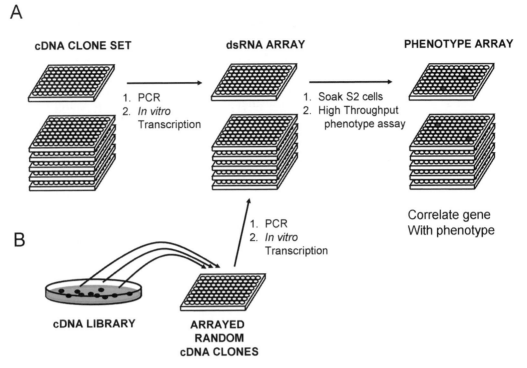

**FIGURE 16.3.** High-throughput phenotype screens in S2 cells. (*A*) Model for generation of dsRNA arrays from a cDNA set. The dsRNAs are soaked into arrayed S2 cells, which are screened for the desired phenotype. The cells displaying the phenotype are correlated back to the dsRNA that caused it. (*B*) dsRNAs are derived from random clones instead of from cDNA sets. The phenotype is screened similarly.

## In conjunction with classical forward genetics

Although RNAi has the ability to transform cultured cells into a genetic system, it is also a valuable validation tool for classical genetic screens in flies. This is particularly the case with overexpression screens, where loss-of-function phenotypes are unknown. This idea was pursued with an overexpression screen for genes involved in cell cycle progression in the developing eye (Tseng and Hariharan 2002). Of the 32 loci identified in the screen, 3 were further characterized as putative negative regulators of the cell cycle. RNAi knockdown in S2 cells established a role for INCENP in maintaining cell ploidy, whereas knockdown of the other identified genes, *elbowB* and *CG11518*, had no effect on the cell cycle.

## RNAi of RNAi

RNAi has been used as a tool to dissect a host of cellular pathways. The one pathway, in principle, that might be resistant to RNAi knockdown is the RNAi pathway itself. This was unfortunate for our group, and others in the field, who study RNAi. It was out of necessity that we attempted RNAi of RNAi, despite the logic against it, only to find success in the approach. We have had success in targeting five genes in the RNAi pathway (Bernstein et al. 2001; Hammond et al. 2001; Caudy et al. 2002), and another group successfully targeted a sixth (Ishizuka et al. 2002). RNAi of RNAi has also found success in *C. elegans* (Grishok et al. 2001; Dudley et al. 2002) and human cells (Hutvagner et al. 2001). We have used RNAi of RNAi to confirm the requirement of specific genes in the

RNAi pathway, whereas other investigators have used the approach as a discovery tool to identify previously unknown players in the pathway (Dudley et al. 2002). It remains to be seen if a whole-genome RNAi of RNAi screen will be attempted.

## Confirmation of antibody specificity

One experimental detail that often goes unaddressed is the characterization of a new antibody. A band of the expected size on a western blot is often the only criterion for the specificity of an antibody, even if several cross-reacting proteins are evident. Even peptide competition cannot guarantee that the observed protein is correct, because other proteins in the source material might have similar epitopes. This has been resolved for an EB1 antibody by RNAi depletion of the protein, eliminating the band in question on a western blot (Rogers et al. 2002). This approach provided a unique way to correlate protein antigen with DNA sequence.

## Perturbation of protein complexes by the targeting of individual subunits

One characteristic of biological systems is the presence of multiprotein complexes. Methods to identify complex components by replacement of a wild-type allele with a tagged allele were described in an earlier section. Another approach is to disrupt complex assembly by RNAi depletion of an essential component. This has been used in at least one instance to characterize the assembly of laminin heterotrimers (Goto et al. 2001). Each subunit was individually depleted, and the properties of the remaining components were analyzed.

## Large-scale biochemical approaches

We originally developed RNAi in S2 cells as a factory for making extracts for in vitro studies. The ability to grow liter quantities of cells that were undergoing efficient RNAi was essential to our research program. Cells were soaked in large batches with dsRNA and transferred to an aerated suspension flask for 6 days. This provided approximately $10^{11}$ cells at a time. Lysates were routed into a purification scheme, resulting in sufficient purified material for protein sequencing. Using this system, we identified four novel components of the RNAi machinery (Hammond et al. 2001; Caudy et al. 2002).

Our interest in this case was restricted to studies on the mechanism of RNAi; however, this approach could be adapted to other systems. As described earlier, RNAi has great potential as an adjunct to proteomic studies by allele replacement with a tagged gene. The scalability of S2 cells would be essential for these projects, because mass spectrometric protein sequencing is the end goal. Disruption, or partial disruption of multiprotein complexes (above), in combination with preparative-scale protein purification is another method.

## CONCLUDING REMARKS

RNAi in *Drosophila* cultured cells is a valuable experimental tool. This chapter has described some of these approaches. It should be noted that not all possible adaptations of the technology are described. Rather, an overview of current approaches has been summarized. The field of RNAi is in its infancy, and as it develops, more creative experiments will be devised.

RNAi has the ability to bridge genetics, cell biology, and biochemistry. As it and other methods are developed, the three areas of biology will flow more seamlessly into each other. The completion of the genome sequence, along with functional genomic, proteomic, and bioinformatic approaches, will further integrate all areas of biology. This promises to change biology fundamentally, perhaps to the extent that the discovery of the double helix did 50 years ago.

## NOTE ADDED IN PROOF

A genome-wide screen in the *Drosophila* imaginal disc cell line c1-8 has recently been reported. dsRNAs were generated to 43% of all predicted genes in *Drosophila* and used to screen for regulators of the hedgehog signaling pathway (Lum et al. 2003). Four genes previously shown to be in the hedgehog pathway were positively identified, in addition to four novel components.

## ACKNOWLEDGMENTS

The author thanks Greg Hannon and Amy Caudy for comments on the manuscript. The author also thanks Greg Hannon, Amy Caudy, and Jose Maria Silva for sharing unpublished results.

## REFERENCES

Amarzguioui M., Holen T., Babaie E., and Prydz H. 2003. Tolerance for mutations and chemical modifications in a siRNA. *Nucleic Acids Res.* **31:** 589–595.

Andrews H.K., Zhang Y.Q., Trotta N., and Broadie K. 2002. *Drosophila* Sec10 is required for hormone secretion but not general exocytosis or neurotransmission. *Traffic* **3:** 906–921.

Bernstein E., Caudy A.A., Hammond S.M., and Hannon G.J. 2001. Role for a bidentate ribonuclease in the initiation step of RNA interference. *Nature* **409:** 363–366.

Boutla A., Delidakis C., Livadaras I., Tsagris M., and Tabler M. 2001. Short 5′ phosphorylated double stranded RNAs induce RNA interference in *Drosophila. Curr. Biol.* **11:** 1776–1780.

Brummelkamp T.R., Bernards R., and Agami R. 2002. Stable suppression of tumorigenicity by virus-mediated RNA interference. *Cancer Cell* **2:** 243–247.

Bunch T.A. and Brower D.L. 1992. *Drosophila* PS2 integrin mediates RGD-dependent cell-matrix interactions. *Development* **116:** 239–247.

Caplen N.J., Fleenor J., Fire A., and Morgan R.A. 2000. dsRNA-mediated gene silencing in cultured *Drosophila* cells: A tissue culture model for the analysis of RNA interference. *Gene* **252:** 95–105.

Caudy A.A., Myers M., Hannon G.J., and Hammond S.M. 2002. Fragile X-related protein and VIG associate with the RNA interference machinery. *Genes Dev.* **16:** 2491–2496.

Celotto A.M. and Graveley B.R. 2002. Exon-specific RNAi: A tool for dissecting the functional relevance of alternative splicing. *RNA* **8:** 718–724.

Cho J.H., Bandyopadhyay J., Lee J., Park C.S., and Ahnn J. 2000. Two isoforms of sarco/endoplasmic reticulum calcium ATPase (SERCA) are essential in *Caenorhabditis elegans. Gene* **261:** 211–219.

Clemens J.C., Worby C.A., Simonson-Leff N., Muda M., Maehama T., Hemmings B.A., and Dixon J.E. 2000. Use of double-stranded RNA interference in *Drosophila* cell lines to dissect signal transduction pathways. *Proc. Natl. Acad. Sci.* **97:** 6499–6503.

Cornwell W.D., Kaminski P.J., and Jackson J.R. 2002. Identification of *Drosophila* Myt1 kinase and its role in Golgi during mitosis. *Cell Signal* **14:** 467–476.

Di Nocera P.P. and Dawid I.B. 1983. Transient expression of genes introduced into cultured cells of *Drosophila. Proc. Natl. Acad. Sci.* **80:** 7095–7098.

Dudley N.R., Labbe J.C., and Goldstein B. 2002. Using RNA interference to identify genes required for RNA interference. *Proc. Natl. Acad. Sci.* **99:** 4191–4196.

Eaton B.A., Fetter R.D., and Davis G.W. 2002. Dynactin is necessary for synapse stabilization. *Neuron* **34:** 729–741.

Echalier G. 1997. Drosophila *cells in culture,* Chapters 2 and 3. Academic Press, New York.

Elbashir S.M., Lendeckel W., and Tuschl T. 2001a. RNA interference is mediated by 21- and 22-nucleotide RNAs. *Genes Dev.* **15:** 188–200.

Elbashir S.M., Martinez J., Patkaniowska A., Lendeckel W., and Tuschl T. 2001b. Functional anatomy of siRNAs for mediating efficient RNAi in *Drosophila melanogaster* embryo lysate. *EMBO J.* **20:** 6877–6888.

Elbashir S.M., Harborth J., Lendeckel W., Yalcin A., Weber K., and Tuschl T. 2001c. Duplexes of 21-nucleotide RNAs mediate RNA interference in cultured mammalian cells. *Nature* **411:** 494–498.

Fire A., Xu S., Montgomery M.K., Kostas S.A., Driver S.E., and Mello C.C. 1998. Potent and specific genetic interference by double-stranded RNA in *Caenorhabditis elegans. Nature* **391:** 806–811.

Forler D., Kocher T., Rode M., Gentzel M., Izaurralde E., and Wilm M. 2003. An efficient protein complex purification method for functional proteomics in higher eukaryotes. *Nat. Biotechnol.* **21:** 89–92.

Goto A., Aoki M., Ichihara S., and Kitagawa Y. 2001. alpha-, beta- or gamma-chain-specific RNA interference of laminin assembly in *Drosophila* Kc167 cells. *Biochem. J.* **360:** 167–172.

Gotwals P.J., Fessler L.I., Wehrli M., and Hynes R.O. 1994. *Drosophila* PS1 integrin is a laminin receptor and differs in ligand specificity from PS2. *Proc. Natl. Acad. Sci.* **91:** 11447–11451.

Grishok A., Pasquinelli A.E., Conte D., Li N., Parrish S., Ha I., Baillie D.L., Fire A., Ruvkun G., and Mello C.C. 2001. Genes and mechanisms related to RNA interference regulate expression of the small temporal RNAs that control *C. elegans* developmental timing. *Cell* **106:** 23–34.

Hammond S.M., Bernstein E., Beach D., and Hannon G.J. 2000. An RNA-directed nuclease mediates post-transcriptional gene silencing in *Drosophila* cells. *Nature* **404:** 293–296.

Hammond S.M., Boettcher S., Caudy A.A., Kobayashi R., and Hannon G.J. 2001. Argonaute2, a link between genetic and biochemical analyses of RNAi. *Science* **293:** 1146–1150.

Henriksen M.A., Betz A., Fuccillo M.V., and Darnell J.E., Jr. 2002. Negative regulation of STAT92E by an N-terminally truncated STAT protein derived from an alternative promoter site. *Genes Dev.* **16:** 2379–2389.

Holen T., Amarzguioui M., Wigler M.T., Babaie E., and Prydz H. 2002. Positional effects of short interfering RNAs targeting the human coagulation tissue factor. *Nucleic Acids Res.* **30:** 1757–1766.

Hutvagner G., McLachlan J., Pasquinelli A.E., Balint E., Tuschl T., and Zamore P.D. 2001. A cellular function for the RNA-interference enzyme Dicer in the maturation of the *let-7* small temporal RNA. *Science* **293:** 834–838.

Igaki T., Yamamoto-Goto Y., Tokushige N., Kanda H., and Miura M. 2002. Down-regulation of DIAP1 triggers a novel *Drosophila* cell death pathway mediated by Dark and DRONC. *J. Biol. Chem.* **277:** 23103–23106.

Ishizuka A., Siomi M.C., and Siomi H. 2002. A *Drosophila* fragile X protein interacts with components of RNAi and ribosomal proteins. *Genes Dev.* **16:** 2497–2508.

Jones G., Jones D., Zhou L., Steller H., and Chu Y. 2000. Deterin, a new inhibitor of apoptosis from *Drosophila melanogaster. J. Biol. Chem.* **275:** 22157–22165.

Kennerdell J.R. and Carthew R.W. 2000. Heritable gene silencing in *Drosophila* using double-stranded RNA. *Nat. Biotechnol.* **18:** 896–898.

Kisielow M., Kleiner S., Nagasawa M., Faisal A., and Nagamine Y. 2002. Isoform-specific knockdown and expression of adaptor protein ShcA using small interfering RNA. *Biochem. J.* **363:** 1–5.

Li X., Scuderi A., Letsou A., and Virshup D.M. 2002. B56-associated protein phosphatase 2A is required for survival and protects from apoptosis in *Drosophila melanogaster. Mol. Cell Biol.* **22:** 3674–3684.

Lum L., Yao S., Mozer B., Rovescalli A., Von Kessler D., Nirenberg M., and Beachy P.A. 2003. Identification of Hedgehog pathway components by RNAi in *Drosophila* cultured cells. *Science* **299:** 2039–2045.

Mihaylov I.S., Kondo T., Jones L., Ryzhikov S., Tanaka J., Zheng J., Higa L.A., Minamino N., Cooley L., and Zhang H. 2002. Control of DNA replication and chromosome ploidy by geminin and cyclin A. *Mol. Cell Biol.* **22:** 1868–1880.

Muda M., Worby C.A., Simonson-Leff N., Clemens J.C., and Dixon J.E. 2002. Use of double-stranded RNA-mediated interference to determine the substrates of protein tyrosine kinases and phosphatases. *Biochem. J.* **366:** 73–77.

Muro I., Hay B.A., and Clem R.J. 2002. The *Drosophila* DIAP1 protein is required to prevent accumulation of a continuously generated, processed form of the apical caspase DRONC. *J. Biol. Chem.* **277:** 49644–49650.

Paddison P.J., Caudy A.A., and Hannon G.J. 2002. Stable suppression of gene expression by RNAi in mammalian cells. *Proc. Natl. Acad. Sci.* **99:** 1443–1448.

Parrish S., Fleenor J., Xu S., Mello C., and Fire A. 2000. Functional anatomy of a dsRNA trigger: Differential requirement for the two trigger strands in RNA interference. *Mol. Cell.* **6:** 1077–1087.

Piccin A., Salameh A., Benna C., Sandrelli F., Mazzotta G., Zordan M., Rosato E., Kyriacou C.P., and Costa R. 2001. Efficient and heritable functional knock-out of an adult phenotype in *Drosophila* using a GAL4-driven hairpin RNA incorporating a heterologous spacer. *Nucleic Acids Res.* **29:** E55–E55.

Pile L.A., Schlag E.M., and Wassarman D.A. 2002. The SIN3/RPD3 deacetylase complex is essential for G(2) phase cell cycle progression and regulation of SMRTER corepressor levels. *Mol. Cell Biol.* **22:** 4965–4976.

Quinn L.M., Dorstyn L., Mills K., Colussi P.A., Chen P., Coombe M., Abrams J., Kumar S., and Richardson H. 2000. An essential role for the caspase Dronc in developmentally programmed cell death in *Drosophila*. *J. Biol. Chem.* **275:** 40416–40424.

Ramet M., Manfruelli P., Pearson A., Mathey-Prevot B., and Ezekowitz R.A. 2002. Functional genomic analysis of phagocytosis and identification of a *Drosophila* receptor for *E. coli*. *Nature* **416:** 644–648.

Reichhart J.M., Ligoxygakis P., Naitza S., Woerfel G., Imler J.L., and Gubb D. 2002. Splice-activated UAS hairpin vector gives complete RNAi knockout of single or double target transcripts in *Drosophila melanogaster*. *Genesis* **34:** 160–164.

Rogers S.L., Rogers G.C., Sharp D.J., and Vale R.D. 2002. *Drosophila* EB1 is important for proper assembly, dynamics, and positioning of the mitotic spindle. *J. Cell Biol.* **158:** 873–884.

Rubin G.M., Yandell M.D., Wortman J.R., Gabor Miklos G.L., Nelson C.R., Hariharan I.K., Fortini M.E., Li P.W., Apweiler R., Fleischmann W., Cherry J.M., Henikoff S., Skupski M.P., Misra S., Ashburner M., Birney E., Boguski M.S., Brody T., Brokstein P., Celniker S.E., Chervitz S.A., Coates D., Cravchik A., Gabrielian A., Galle R.F., Gelbart W.M., George R.A., Goldstein L.S., Gong F., Guan P., Harris N.L., Hay B.A., Hoskins R.A., Li J., Li Z., Hynes R.O., Jones S.J., Kuehl P.M., Lemaitre B., Littleton J.T., Morrison D.K., Mungall C., O'Farrell P.H., Pickeral O.K., Shue C., Vosshall L.B., Zhang J., Zhao Q., Zheng X.H., and Lewis S. 2000. Comparative genomics of the eukaryotes. *Science* **287:** 2204–2215.

Schmid A., Schindelholz B., and Zinn K. 2002. Combinatorial RNAi: A method for evaluating the functions of gene families in *Drosophila*. *Trends Neurosci.* **25:** 71–74.

Schmucker D., Clemens J.C., Shu H., Worby C.A., Xiao J., Muda M., Dixon J.E., and Zipursky S.L. 2000. *Drosophila* Dscam is an axon guidance receptor exhibiting extraordinary molecular diversity. *Cell* **101:** 671–684.

Schneider I. 1972. Cell lines derived from late embryonic stages of *Drosophila melanogaster*. *J. Embryol. Exp. Morphol.* **27:** 353–365.

Somma M.P., Fasulo B., Cenci G., Cundari E., and Gatti M. 2002. Molecular dissection of cytokinesis by RNA interference in *Drosophila* cultured cells. *Mol. Biol. Cell.* **13:** 2448–2460.

Stapleton M., Liao G., Brokstein P., Hong L., Carninci P., Shiraki T., Hayashizaki Y., Champe M., Pacleb J., Wan K., Yu C., Carlson J., George R., Celniker S., and Rubin G.M. 2002. The *Drosophila* gene collection: Identification of putative full-length cDNAs for 70% of *D. melanogaster* genes. *Genome Res.* **12:** 1294–1300.

Suske G. 1999. The Sp-family of transcription factors. *Gene* **238:** 291–300.

Tabara H., Grishok A., and Mello C.C. 1998. RNAi in *C. elegans:* Soaking in the genome sequence. *Science* **282:** 430–431.

Towers P.R. and Sattelle D.B. 2002. A *Drosophila melanogaster* cell line (S2) facilitates post-genome functional analysis of receptors and ion channels. *Bioessays* **24:** 1066–1073.

Tseng A.S. and Hariharan I.K. 2002. An overexpression screen in *Drosophila* for genes that restrict growth or cell-cycle progression in the developing eye. *Genetics* **162:** 229–243.

Tuschl T., Zamore P.D., Lehmann R., Bartel D.P., and Sharp P.A. 1999. Targeted mRNA degradation by double-stranded RNA in vitro. *Genes Dev.* **13:** 3191–3197.

Warren W.D., Steffensen S., Lin E., Coelho P., Loupart M., Cobbe N., Lee J.Y., McKay M.J., Orr-

Weaver T., Heck M.M., and Sunkel C.E. 2000. The *Drosophila* RAD21 cohesin persists at the centromere region in mitosis. *Curr. Biol.* **10:** 1463–1466.

Wojcik C. and DeMartino G.N. 2002. Analysis of *Drosophila* 26S proteasome using RNA interference. *J. Biol. Chem.* **277:** 6188–6197.

Worby C.A., Simonson-Leff N., Clemens J.C., Huddler D., Jr., Muda M., and Dixon J.E. 2002. *Drosophila* Ack targets its substrate, the sorting nexin DSH3PX1, to a protein complex involved in axonal guidance. *J. Biol. Chem.* **277:** 9422–9428.

Worby C.A., Simonson-Leff N., and Dixon J.E. 2001a. RNA interference of gene expression (RNAi) in cultured *Drosophila* cells. *Sci. STKE* **2001:** L1.

Worby C.A., Simonson-Leff N., Clemens J.C., Kruger R.P., Muda M., and Dixon J.E. 2001b. The sorting nexin, DSH3PX1, connects the axonal guidance receptor, Dscam, to the actin cytoskeleton. *J. Biol. Chem.* **276:** 41782–41789.

# CHAPTER 17

# RNAi Applications in *Drosophila melanogaster*

Richard W. Carthew

*Department of Biochemistry, Molecular Biology, and Cell Biology, Northwestern University, Evanston, Illinois 60208*

For many experiments in modern biomedicine, knowledge of a gene's function in vivo is a prerequisite for its further manipulation. The genome sequencing project for the fruit fly *Drosophila melanogaster* has yielded the bulk of genes within the fly genome, approximately 13,744 annotated sequences in Release 2.0 (Adams et al. 2000). Computer-assisted identification of protein-coding regions within the DNA sequence, followed by computer-assisted similarity searches of protein databases, can lead to important insights about the structure and function of any sequenced gene and its product. However, this approach has its limitations. Approximately 7576 genes from Release 1.0 of the sequenced *Drosophila* genome cannot be classified into a functional group on the basis of this approach (Adams et al. 2000). Moreover, the functional properties assigned to the protein products of genes are centered on what might be called molecular functions. These can be specified whenever a protein is found to be similar to one for which the function has been determined by biochemical methods. Quite often, there is little connection between these molecular functions and the classical assignment of function by phenotype, which is at the level of the cell or organism. The middle ground—the participation of genes in the physiology of cells and how cells contribute to the function of the organism—is a gap that remains to be closed. The problem faced by post-genomic biology involves bridging that gap between genotype and phenotype.

RNA interference (RNAi) is a natural phenomenon common across Eukaryota that has the potential to be harnessed for functional understanding of genes at the phenotype level. RNAi is based on specific silencing of gene expression using double-stranded RNA (dsRNA) as both guide and trigger for the silencing reaction (Fire et al. 1998). Its theoretical and mechanistic properties are the subjects of other chapters within this book. This chapter concerns the practical aspects of using RNAi as a genetic tool in *Drosophila*. The basic presumption of using RNAi in such a manner extends from the classical notion that existence of a wild-type gene cannot be inferred until a mutant version with an altered function has been isolated. For T.H. Morgan to say that there was a *Drosophila* gene for eye pigmentation, he had to find heritable eye-color variants that had an eye color different from red. This genetics provides, if not knowledge, then at least a classification of the functions of genes. RNAi specifically alters gene function by silencing gene expression, thereby creating conditions in which a gene functions rather poorly or not at all. The resulting change in phenotype can be construed as a consequence of one poorly functioning gene. A strong link is thus forged between a gene, as defined as any piece of expressed sequence, and function, as defined by a phenotype. For the library of thousands of expressed *Drosophila* sequences, this kind of approach may eventually tell us what it all means.

## CONCEPT AND STRATEGIES

The two basic methods that are widely used to perform RNAi experiments on *Drosophila* differ primarily in the technique used to generate dsRNA. One method uses dsRNA that is synthesized in vitro and then introduced into *Drosophila* cells. The other method uses a DNA expression vector that is introduced into *Drosophila* cells and, when expressed, synthesizes dsRNA in vivo. These two methods are further described below.

## Exogenous RNAi

The exogenous method of generating RNAi using in vitro dsRNA was originally developed by Fire, Mello, and co-workers (Fire et al. 1998). The dsRNA corresponding to a particular gene was synthesized in vitro and injected into adult nematodes. Over the years, this basic experimental paradigm has been adapted in many different ways for *Drosophila*. There are variations in how the dsRNA is made, how it is introduced into *Drosophila*, and what cells actually receive the dsRNA. The various methodologies for each are outlined below.

### Synthesis of dsRNA

dsRNAs can be synthesized either enzymatically or chemically. The enzymatic method can make dsRNAs of virtually any useable length. It employs a DNA-directed RNA polymerase to synthesize sense and antisense RNA copies from a DNA template. These RNA polymerases cannot initiate RNA chains de novo. Rather, they initiate RNA synthesis at a promoter sequence within the template DNA. Chain extension from 5′ to 3′ is controlled by base-pair matching to the template DNA until the 3′ end of the DNA template molecule is reached. A combined or separate annealing reaction then allows the complementary RNA strands to base-pair and form dsRNA. Two protocols for enzymatic dsRNA synthesis are provided. The original method was developed for use with plasmid DNA templates that contain particular promoters for prokaryotic RNA polymerases. The second method was developed for use with polymerase chain reaction (PCR) DNA templates and uses the bacteriophage T7 RNA polymerase for RNA synthesis. This second method is faster to use and more adaptable for a variety of templates and, not surprisingly, has become the method of choice to synthesize dsRNA enzymatically. The chemical synthesis method makes dsRNAs of precisely 19 bp in length with each strand containing a two-nucleotide single-strand 3′ overhang. These are the "classic" short interfering RNAs (siRNAs) that were pioneered by Tuschl and co-workers (Elbashir et al. 2001a,b).

### Sources of *Drosophila*

A variety of *Drosophila* cells can be presented with dsRNA. Perhaps the simplest cell sources are from tissue culture. Many *Drosophila* immortalized cell lines exist that are derived from a number of different tissue sources (Ashburner 1989). Methods to induce RNAi in cell lines have been established and are highly successful in studying gene function within immortalized cells (Caplen et al. 2000; Clemens et al. 2000; Hammond et al. 2000). Generating RNAi in primary *Drosophila* cells in culture is also possible. Andres and colleagues (Biyasheva et al. 2001) successfully silenced gene expression in cells within dissected salivary glands that were incubated with dsRNA in culture. Although other examples of RNAi of primary cells are forthcoming, this approach promises to be a powerful technique for primary cell and organ studies.

The approaches for introducing dsRNA into whole organisms are more problematic. Unlike the nematode *Caenorhabditis elegans*, *D. melanogaster* does not appear to allow spreading of gene silencing from cell to cell in any systemic manner. Three lines of evidence suggest this to be the case (J. Kennerdell and R. Carthew, unpubl.): (1) Injection of dsRNA into one cell does not lead to gene silencing within nearby cells. (2) Injection of dsRNA into the hemolymph or intestine does not lead to gene silencing within an individual. (3) *Drosophila* larvae raised by feeding on yeast (*Saccharomyces cerevisiae*) that express *Drosophila* dsRNA exhibit no evidence of gene silencing. Therefore, systemic RNAi effects cannot be generated at any life cycle stage by introduction of dsRNA in one region and having it spread. Instead, this type of approach is only feasible at the earliest stages of embryonic development. After fertilization, the *Drosophila* egg undergoes a series of rapid mitotic divisions, one about every 9 minutes, but, unlike most animal embryos, there is no cleavage of the cytoplasm. The result is a syncytium in which many nuclei are present in a common cytoplasm (Ashburner 1989). The embryo essentially remains a single cell during its early development. After nine divisions, the nuclei migrate to the periphery and membranes grow in to enclose each nucleus and form cells. Because of the syncytium, even large molecules such as dsRNA can diffuse throughout the entire embryo in the first hour after egg laying. This feature is of great benefit to researchers who wish to generate RNAi in *Drosophila* with exogenous dsRNA.

If dsRNA is introduced into syncytial embryos, what are the kinetic properties of gene silencing? First, the dsRNA itself rapidly diffuses throughout the syncytium within minutes after injection. If the target mRNA is already present in the embryo at the time of injection, the kinetics of mRNA destruction are rapid. Levels of mRNA transcripts are reduced 80–90% within 40 minutes after injection for two genes that our lab has studied, *bicoid* and *hunchback* (Kennerdell et al. 2002). If the target mRNA is not present in the embryo at the time of injection, RNAi will still occur when the target mRNA begins to be synthesized, possibly many hours later in embryogenesis (Misquitta and Paterson 1999; Kennerdell and Carthew 2000). Thus, the dsRNA perdures in cells of the developing fly and a silencing potential is maintained. This feature is important if one is interested in silencing genes that are expressed later in development.

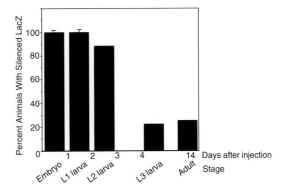

**FIGURE 17.1.** Time course of RNAi perdurance over the *Drosophila* life cycle. A 745-bp dsRNA corresponding to a portion of the *lacZ* gene (Kennerdell and Carthew 2000) was injected into syncytial *Drosophila* embryos carrying the *engrailed-lacZ* gene. This transgene is expressed in an engrailed pattern throughout the life cycle (Hama et al. 1990). At particular stages after injection, individuals were collected, fixed, and stained for *lacZ* expression. Individuals were scored as exhibiting RNAi silencing if more than one body segment was missing *lacZ* expression. Shown is the percentage of individuals at each stage that were scored as silenced. Bars represent standard deviations.

How long through development can the potential for RNAi silencing last? At room temperature, a fly typically takes 12 days to develop from fertilized egg to mature adult. Embryogenesis itself lasts for 24 hours. The changes in silencing activity for a target gene during development are shown in Figure 17.1. Silencing activity is very robust up to the end of embryogenesis and even through the first larval instar stage. However, silencing activity decreases as development proceeds through subsequent larval instar stages, such that by pupariation, only one quarter of the treated animals exhibit an RNAi phenotype. This modest activity is maintained up to the adult stage and disappears altogether in the next generation of flies produced from silenced adults. Thus, a problem in the practice of *Drosophila* RNAi is obtaining a significant silenced phenotype at post-embryonic stages of the life cycle. As discussed below, this drawback has been overcome by using endogenous RNAi.

### Methods of dsRNA delivery

Two techniques are currently available to introduce dsRNA into syncytial embryos. One is to microinject the dsRNA into the syncytial cytoplasm (see Protocol 2). The second is to bombard embryos with dsRNA coated on gold microparticles using a gene gun (see Protocol 3). Either technique works well to generate RNAi silencing throughout the organism providing the dsRNA is introduced prior to cellularization. To introduce dsRNA into immortalized *Drosophila* cells in culture, either the dsRNA is added directly to the culture medium for cells to take up or it is transfected into cells (see Protocol 5). The efficiency and potency of gene silencing are very strong in many cell lines, and these methods are being used extensively to study gene function (Worby et al. 2001).

One method to introduce dsRNA into primary cells in organ culture is to use a cell-loading technique based on the osmotic lysis of pinocytic vesicles, a technique introduced by Okada and Rechsteiner (1982). Water-soluble polar compounds can be introduced into many cells simultaneously by induced pinocytosis without significantly altering normal cell function. It is a more gentle cell-loading method than the typical cell-loading techniques of microinjection, transfection, or electroporation. Moreover, it has been shown to rapidly and effectively silence gene expression in cells within dissected larval salivary glands (Biyasheva et al. 2001).

### Endogenous RNAi

The underlying assumption of endogenous RNAi is that cells which themselves synthesize dsRNA will induce self-silencing of target gene expression. This necessarily requires that the gene or genes that synthesize the dsRNA are in *trans* to the target gene. Fortunately, there is a well-established method to introduce foreign genes into the germ-line genome of *Drosophila*: P-element-mediated transformation (Spradling and Rubin 1982). Hence, endogenous RNAi has the potential to work in the whole organism and to be stably inherited within a purebred stock for many generations. For this reason and others, endogenous RNAi is more powerful and flexible than exogenous RNAi as a means to silence genes in individual flies.

The inspiration for endogenous *Drosophila* RNAi originated in studies of posttranscriptional gene silencing (PTGS) in plants (for review, see Sijen and Kooter 2000). Many features of plant PTGS resemble RNAi, indicating a common mechanism operating in each situation. Expression of sense or antisense RNA in transgenic plants conferred a form of PTGS on the plant, typified by degradation of both the transgenic mRNA and target mRNA, which contained the same or complementary sequences. Transgenic plants that produced RNAs capable of duplex formation into dsRNA conferred much more potent PTGS on the plants (Waterhouse et al. 1998). This was accomplished by three independent methods: (1) using transcripts from a sense gene and an antisense gene colocated within a transformation vector, (2) bringing together sense and antisense genes in an $F_1$ plant from parents that contained one or the other transgene, and (3) using a transgene with an inverted repeat such that the self-complementary transcript would form an snapback structure. These procedures provided a possible menu of methods for inducing endogenous *Drosophila* RNAi.

Most of the methods for *Drosophila* RNAi employ transgenes having an inverted repeat configuration that are able to produce dsRNA as snapback RNA (Fortier and Belote 2000; Kennerdell and Carthew 2000; Lam and Thummel 2000; Martinek and Young 2000). An alternative method uses a transgene that is symmetrically transcribed from opposing promoters (Giordano et al. 2002). Most of the transformation vectors used for endogenous RNAi are derived from the *Drosophila* transformation plasmid vector pUAST (Brand and Bourbon 1993) (see Protocol 7). On the 5′ side of the multicloning site, pUAST contains a *Drosophila* promoter linked to GAL4-responsive UAS enhancer repeats, and on the 3′ side of the multicloning site, pUAST contains a polyadenylation signal sequence. Cloning into a pUAST vector allows one to use the modular design of the GAL4/UAS system in *Drosophila* for expressing transgenes. Many useful lines of *Drosophila* express the yeast GAL4 protein in a variety of cells/tissues at various stages of the fly life cycle (Perrimon 1998). GAL4 acts as a sequence-specific transcription activator in *Drosophila*. The GAL4 line is crossed to a target UAS line carrying a single target P element inserted at a unique and random position in the genome. The target element carries a GAL4-responsive UAS enhancer, and progeny that contain both GAL4 and UAS elements express the snapback RNA in cells expressing GAL4. Phenotypes due to the presence of dsRNA in these cells can then be scored directly in flies. The approach takes advantage of two very useful techniques in *Drosophila:* P-element transformation and the modular GAL4/UAS system. The modular design makes analysis by RNAi flexible because dsRNA can be produced in any spatial or temporal pattern. Moreover, RNAi is conditional, dependent on the presence of both UAS and GAL4 elements in the same individual. Thus, RNAi that might induce lethal or sterile phenotypes is conditionally generated in selected flies, and *Drosophila* lines stably carrying the UAS element alone can be propagated without deleterious RNAi effect.

To date, most RNAi transgenes have been constructed by insertion of a particular target gene into pUAST, often in an inverted repeat configuration. However, with the increasing usage of RNAi techniques in *Drosophila*, transformation vectors explicitly devoted to RNAi have been developed. These vectors are based on the GAL4/UAS system. One plasmid vector, SympUAST, contains a multicloning site flanked on both sides by two identical but oppositely oriented UAS promoters coupled with an inversely oriented polyadenylation site (Giordano et al. 2002). It is able to drive transcription of sense and antisense strands of a given DNA inserted in the multicloning site. Another vector, pWIZ, contains a functional *Drosophila* intron flanked on both sides by two multicloning sites, all located downstream from a UAS promoter (Lee and Carthew 2003). A given DNA is inserted into each multicloning site in opposing orientation, and the resulting RNA transcripts are spliced in *Drosophila* to yield loopless snapback RNA. A similar type of vector has been developed by Reichhart, Gubb, and colleagues (Reichhart et al. 2002). The reasoning behind making intron-spliced snapback RNAs is based on earlier work done studying plant PTGS. It was observed that constructs encoding intron-spliced RNA with a hairpin structure induce PTGS in tobacco with almost 100% efficiency (Smith et al. 2000). In *Drosophila*, hybrid transgenes have been made that are composed of cDNA/genomic inverted repeats such that intron splicing of the hairpin RNAs occurs (Kalidas and Smith 2002). These transgenes also exhibit superior silencing activity, indicating that intron-spliced hairpin RNA is also highly effective in *Drosophila*.

## Choosing between Exogenous and Endogenous RNAi Methods

### Exogenous RNAi

As described above, the exogenous RNAi method is based on the ability of dsRNA made in vitro to induce gene silencing when introduced into *Drosophila* cells.

- *Major advantages.* This method is rapid. The synthesis of dsRNA can be completed within a few hours. In addition, a large number of dsRNA samples can be prepared simultaneously. Moreover, the techniques to introduce dsRNA into cells are rapid, taking at most 1 or 2 days, and they can be scaled up for large numbers of dsRNA samples. Currently, a realistic year-project scale for one research worker to analyze embryonic RNAi phenotypes by this method is approximately 200 genes. To analyze RNAi phenotypes in cell lines, the target gene number is considerably higher. Automation of dsRNA synthesis and cell delivery by the use of robotics is under development (Kiger et al. 2002). As this technology is further developed, more rapid and less tedious acquisition of functional data by RNAi for large-scale projects will be possible. Thus, exogenous RNAi has the potential to be used on a genome-wide scale.

- *Major disadvantages.* The major disadvantage of exogenous RNAi is its limitation for silencing gene expression within individual flies. It is limited to gene silencing during the early stages of the life cycle—embryos and L1 larvae—and rather impotent at later developmental stages. Thus, RNAi applications to study metamorphosis, behavior, and adult physiology are not possible with this method. Moreover, it will systemically silence early gene expression and induce early phenotypes, possibly at the expense of phenotypes that may be manifested at later stages. This lack of conditionality restricts the study of gene function to early stages exclusively.

### Endogenous RNAi

Endogenous RNAi is based on the ability of *Drosophila* to synthesize dsRNA and induce its own gene silencing.

- *Major advantages.* The major advantage of the endogenous method is that problems associated with introducing dsRNA into individual flies are eliminated, permitting RNAi in cell-specific and tissue-specific patterns at any time during the life cycle. It is the only method currently available for inducing RNAi phenotypes at late stages of the life cycle. It is important to note that this approach appears to work in most tissue types (Fortier and Belote 2000; Kennerdell and Carthew 2000; Lam and Thummel 2000; Martinek and Young 2000; Kalidas and Smith 2002; Reichhart et al. 2002; Lee and Carthew 2003), including neurons, which are resistant to RNAi in *C. elegans*. The conditional targeting of RNAi to specific cells and tissues provides a means to analyze the phenotype in targeted tissues without obfuscating phenotypes in other tissues. It is important to note that unlike plants and nematodes, *Drosophila* does not appear to allow spreading of gene silencing between cells (Kennerdell and Carthew 2000; Kalidas and Smith 2002). Additionally, organismal phenotypes produced by endogenous RNAi are more uniform than phenotypes produced by exogenous RNAi, which makes it easier to interpret the phenotype and its root causes.

- *Major disadvantages.* The primary disadvantage of the endogenous method is the time and effort required. To construct a transgene, make stable transformants, and test those transformants require 2–3 months. The labor invested is considerable enough that a realistic year-project scale for one research worker is approximately 20 genes or an order of magnitude fewer than by the exogenous method.

## Establishing the Specificity of RNAi

Classical genetics has at its disposal a substantial bag of experimental tricks to establish rigorously a causal link between a mutation at a particular locus and a phenotype. Clearly, RNAi is a different phenomenon altogether from DNA mutagenesis, and new ways are needed to prove causality between RNAi treatment and a phenotype. Several issues arise, each with its necessary experimental controls.

One issue is whether the dsRNA actually knocks down expression of the target gene. This issue is easily addressed in *Drosophila* because of the wealth of techniques for assaying gene expression. If antibody reagents are available, immunohistochemistry or western blotting experiments will establish whether protein levels are reduced. Otherwise, in situ hybridization, northern blotting, or quantitative reverse transcriptase (RT)-PCR will determine whether transcript levels are reduced. This experiment can be expanded to include analysis of heterologous gene expression as one control for specificity.

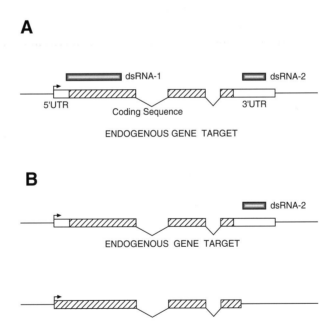

**FIGURE 17.2.** Control experiments to determine RNAi specificity. (*A*) Perform two or more independent RNAi experiments using nonoverlapping dsRNAs toward the same gene. If all dsRNAs generate the same phenotype, it is strong evidence that the observed phenotype is caused by specific loss of target gene expression. (*B*) Perform transgenic rescue of RNAi. Generate a transgene that contains DNA segments corresponding to a gene, but with one important modification. Specific sequence changes in the transgene should be made that will not perturb its proper function, but will render the transgene resistant to dsRNA that nevertheless silences the endogenous gene. Such a transgenic line can be tested for its ability to rescue a phenotype generated by the dsRNA.

A drawback of comparing expression of the target gene to a heterologous gene is that one cannot look at expression of every heterologous gene in the genome, short of performing microarray analysis. The issue then remains whether RNAi treatment is directly silencing one or more genes other than the intended target. Moreover, if a heterologous gene is affected by RNAi, one cannot rule out that the gene is downstream from the target gene in a regulatory pathway.

Two solutions to this problem are available. One solution is an RNAi version of using independent alleles to link genotype and phenotype: Perform RNAi using two or more dsRNAs that are directed against the same gene but do not overlap and have little or no sequence similarity (Figure 17.2A). If these dsRNAs independently generate an identical phenotype, then it can be argued that each dsRNA disrupts the same process by silencing the same gene. Because the dsRNAs correspond to a common gene locus but are otherwise unrelated, it is highly unlikely that each RNA would also silence the same set of heterologous genes. Of course, this interpretation is built upon the premise that a given dsRNA trigger does not induce synthesis of siRNAs directed against other regions of a gene, akin to the so-called transitive RNA seen in *C. elegans* (Alder et al. 2003). Although it has been reported that transitive RNA can be produced from *Drosophila* extracts (Lipardi et al. 2001), other data have indicated that transitive RNA is not generated in *Drosophila* (Schwarz et al. 2002). Moreover, *D. melanogaster* lacks the RNA-dependent RNA polymerase gene responsible for transitive RNA (Adams et al. 2000).

The second solution for the specificity problem is an RNAi version of using transformation rescue to link genotype and phenotype. Suppose that a dsRNA homologous to a given gene produces a given phenotype. One places the target gene into a P-element transformation vector, creates transgenic lines, and challenges these lines with the same dsRNA (Figure 17.2B). If the transgene carries sequences complementary to the dsRNA, it should be silenced along with its endogenous counterpart, and the phenotype should be the same. However, if the transgene is missing the sequences complementary to the dsRNA, it should be resistant to silencing and the phenotype should revert to wild type. This then is the basis for the specificity experiment. A transgene is constructed in such a way that the RNAi target sequence is missing or scrambled. The simplest method to do this with the least impact on transgene functionality is to use the target sequence in one of the transcript's untranslated regions (UTR). This avoids changing the protein-coding sequence. To minimize change even further, using an siRNA for the experiment allows one to simply delete or scramble 20 nucleotides of sequence, rather than 200 nucleotides or more if using a long dsRNA. If the modified transgene completely rescues the RNAi phenotype, then the RNAi phenotype must be specific to the endogenous target gene.

# TECHNIQUES

Protocols 1 through 6 provide exogenous methods for RNAi, whereas Protocols 7 and 8 provide endogenous methods for RNAi. In the basic procedure, dsRNA is synthesized in vitro corresponding to a particular *Drosophila* sequence. The RNA is then introduced into cells by one of a variety of techniques.

## PROTOCOL 1: SYNTHESIS OF DsRNA

The following are three variations for the preparation of dsRNA for interference: (1) Synthesize RNA strands separately from linearized plasmid templates and then anneal the strands together, (2) synthesize both RNA strands simultaneously from a PCR fragment which contains a T7 promoter on each end, and (3) synthesize 21-nucleotide RNAs and anneal them to form siRNA. The most prudent approach is to make a dsRNA corresponding to a UTR sequence (either UTR can mediate interference) or a unique coding sequence, since a dsRNA to one gene can potentially silence another gene if the dsRNA is sufficiently similar in sequence to the other gene. Therefore, compare the sequence of the target gene of interest to other genes within the *Drosophila* genome and locate regions in which there are no stretches of contiguous sequence identity of 21 bp or greater between the target gene and any other gene. This avoids potential cross-reaction between a particular siRNA and transcripts from other genes. Keep in mind that minor mismatches between an siRNA and sequence within the 3′UTR of transcripts can lead to gene silencing by translational repression rather than transcript cleavage.

There are no theoretical guidelines for choosing the most potent dsRNA against a particular gene. A prudent course is to pick two or more regions from which to make dsRNA. These regions should not be overlapping. If they are independently tested, the chances are better that at least one will produce a potent silencing effect. Moreover, if two or more nonoverlapping dsRNAs generate a similar phenotype, there is greater assurance that the target gene is specifically silenced.

The first two methods essentially produce long dsRNA chains. We have not systematically determined the optimum length of a dsRNA for maximum interference activity. However, circumstantial evidence suggests that lengths of 500–800 bp are most active. We have found that dsRNAs as short as 200 bp and as long as 1000 bp have potent interfering activities. The dsRNA can be made from cDNA or genomic DNA templates, as long as most of the dsRNA corresponds to presumptive exon sequence. dsRNAs with two or more exon regions interrupted by introns will also work.

## Procedures

**Method 1:** Plasmid Template Method

**▼ CAUTION**
*See Appendix for appropriate handling of materials marked with <!>.*

**MATERIALS**

**REAGENTS**

Annealing buffer
   1 mM Tris (pH 7.5) <!>
   1 mM EDTA
Ethanol (absolute) <!>
Gene of interest cloned in plasmid polylinker
NH₄OAc (5 M)
Phage RNA polymerase
Polymerase promoters flanking the plasmid polylinker (e.g., pBlueScript)
RNase-free DNase
TE buffer

**EQUIPMENT**

Beaker (150 ml)
Spectrophotometer

1. Subclone a fragment of the gene of interest into a plasmid vector with T7, T3, and/or SP6 RNA polymerase promoters flanking the polylinker (e.g. pBlueScript). Linearize the plasmid on either side of the insertion site to create two preparations of linear

DNA template, one to synthesize the sense strand and the other to synthesize the antisense strand. Remove enzymes and restriction buffer, and dissolve DNA in TE buffer.

2. Perform RNA synthesis reactions in a 50-µl volume with 1 µg of DNA template using appropriate phage RNA polymerase. Follow standard protocols as described in general lab manuals.

3. Remove DNA template with RNase-free DNase. Extract and precipitate the RNA with NH$_4$OAc and ethanol.

4. Dissolve RNA in 5 µl of annealing buffer and measure the yield spectrophotometrically.

Typical yields of RNA from 1 µg of DNA template are in the 40–50-µg range.

5. To anneal, mix equimolar quantities of sense and antisense RNAs in annealing buffer to a final concentration of 0.45 µM each.

6. Heat small aliquots (11 µl) of the mixture in a 150-ml beaker of boiling water for 1 minute. Then remove the beaker from the heat source and allow it to cool for 18 hours to room temperature.

The 18 hours is critical to produce high yields of dsRNA. Shorter time periods are not effective for some reason.

7. Store RNA aliquots as an NaOAc/ethanol precipitate at –80°C until immediately before use.

The RNA is stable for years when stored in this condition.

## Method 2: PCR Template Method

▼ CAUTION
*See Appendix for appropriate handling of materials marked with <!>.*

**MATERIALS**

**REAGENTS**

Ethanol (absolute) <!>

NH$_4$OAc (5 M)

PCR purification kit (QIAGEN)

Phenol:chloroform <!>

Polymerase buffer

Ribonucleotides (2 mM)

RNase-free DNase (RQ Promega)

RNAsin

T7 RNA polymerase (25 units)

T7-linked primers and template

TE buffer

**EQUIPMENT**

PCR machine

Primer3 software

Web site: http://www-genome.wi.mit.edu/cgi-bin/primer/primer3_www.cgi.

Spectrophotometer

1. Choose primer sequences which will amplify that region of the DNA used to encode the dsRNA.

   We use MIT's Primer3 to choose optimal primer sequences for a given region. Complementary sequences should be 20–24 nucleotides in length with a 22-nucleotide optimum and 60°C optimum melting temperature.

   The 5′ end of each primer should correspond to a 23-nucleotide T7 RNA polymerase promoter sequence (GCTTCTAATACGACTCACTATAG). Thus, each primer will be ~43–47 nucleotides long (23 + [20 to 24]).

2. Perform a 50-μl PCR amplification with T7-linked primers and suitable template.

   *Drosophila* DNA cloned in plasmids, BACs, or phage are optimal, but the method will also work on total cDNA or genomic DNA. The first 10 cycles should have a 40°C annealing step, followed by 35 cycles with a 55°C annealing step.

3. Use a QIAGEN PCR purification kit to purify the PCR product. Dissolve in TE buffer and measure the concentration spectrophotometrically.

4. Perform an RNA synthesis reaction in a 50-μl volume with 1 μg of PCR DNA template and 25 units of T7 RNA polymerase, 2 mM ribonucleotides, and RNAsin in standard T7 RNA polymerase buffer for 90 minutes at 37°C.

   The RNA becomes double-stranded during the synthesis reaction.

5. Remove the DNA template with a RNase-free DNase incubation for 20 minutes at 37°C. Extract with phenol:chloroform and precipitate the RNA with $NH_4OAc$ and ethanol.

6. Dissolve the dsRNA in TE buffer and measure yield by spectrophotometer.

   Typical yields of RNA from 1 μg of DNA template are in the 30–100-μg range.

7. Aliquot and store dsRNA as an NaOAc/ethanol precipitate at –80°C until use.

   The RNA is stable for years when stored in this condition.

**Testing the dsRNA**  No matter how the dsRNA is prepared, test its condition by native agarose gel electrophoresis in TBE. Electrophorese 3–5 μg of RNA and stain with ethidium bromide. Alternatively, trace-label RNA with [$^{32}$P]ATP during synthesis, and use autoradiography to detect electrophoresed products. Use only preparations in which the electrophoretic mobility of most of the RNA corresponds to that expected for dsRNA (very close to duplex DNA mobility) of the appropriate length. Sometimes we observe a smear of higher-order RNAs migrating slower than the dsRNA species. Although this may constitute over half of the RNA present, we find that the interfering activities of these preparations are similar to that of the more homogeneous preparations.

## Method 3: siRNA Synthesis

siRNA can be highly useful when attempting specifically to silence a gene whose sequence is not very different from other gene sequences. Another advantage of siRNAs is that much higher doses of siRNA can be delivered into *Drosophila* without nonspecific toxic side effects. This allows application of stronger silencing to target genes, and silencing continues for a longer period of time.

Selection of siRNA sequence depends on several criteria. In general, we target the coding sequence of a gene whose sequence is not similar to other gene sequences. The sequence should contain ~50% GC content. The target mRNA sequence should begin at the 5′ end with the triplet AAG. The 18 bases immediately following the AAG are an excellent potential target sequence, in particular if the last base is a C. Scan all AAGs in coding mRNA sequence. Examine all candidate 21-mer sequences to see how well they fit the above criteria and then choose the one that fits the best. Below is the design:

```
                         Cleavage site on mRNA


                  1 2 3  4 5  6 7  8 9 10 11 12 13 14 15 16 17 18 19
     mRNA  5´ A A G R R  R R  R R R  R  R  R  R  R  R  R  R  C X X 3´

     siRNA 5´    G R R  R R  R R R  R  R  R  R  R  R  R  R  C U U 3´
            3´ U U C R´R´R´R´R´R´R´R´R´R´ R´ R´ R´ R´ R´ R´ R´ G´    5´
```

Each oligonucleotide has 21 ribonucleotides. The sequence of the sense strand is R and the complementary sequence of the antisense strand is R′. A number of companies now synthesize siRNA oligonucleotides on demand. Full-length oligonucleotides are purified using one of several standard procedures. Oligonucleotides are shipped with protecting groups, which must be removed following the manufacturer's instructions prior to use. Deprotection is followed by oligonucleotide annealing as follows.

**▼ CAUTION**
*See Appendix for appropriate handling of materials marked with <!>.*

---

**MATERIALS**

**REAGENTS**

    Annealing buffer

        30 mM HEPES-KOH (pH 7.4) <!>

        100 mM potassium acetate

        2 mM magnesium acetate <!>

    Ethanol (asbsolute) <!>

    RNA oligonucleotides (see above)

    Sodium acetate (3 M) at neutral pH <!>

---

1. Dissolve each RNA oligonucleotide separately in annealing buffer to a final concentration of 20 μM.

2. Mix 0.25 ml of each complementary RNA oligonucleotide together (0.5 ml total). Incubate for 1 minute at 90°C and then for 1 hour 37°C.

3. Add sodium acetate (neutral pH) to 0.3 M final concentration and 2.5 volumes of ethanol. Store ethanol precipitate at −70°C until ready to use.

    The siRNAs are stable for years under these conditions.

## PROTOCOL 2: EMBRYO MICROINJECTION AND ANALYSIS

Introduction of siRNA or dsRNA into *Drosophila* embryos is basically an adaptation of the standard protocol used for P-element transformation. The first section describes a basic protocol, although each laboratory should use procedures with which they are familiar pertaining to transformation. The second section outlines guidelines and troubleshooting advice for RNAi by microinjection.

Several days of preparation are required before injections into *Drosophila* embryos begin. Flies must be in abundant supply for egg collection.

## Procedure

**▼ CAUTION**
*See Appendix for appropriate handling of materials marked with <!>.*

### MATERIALS

### REAGENTS

Bleach (50%) with $H_2O$ <!>

*Drosophila* eggs (see below)

dsRNA or siRNA (from Protocol 1)

Drierite

Egg-laying plates

1110 ml of $H_2O$
44 g of Bacto-agar
180 ml of molasses

Autoclave for 30 minutes with a stir bar. After autoclaving, cool to 50ºC. While stirring, add 0.25 ml of 10% (w/v) *p*-hydroxy–benzoic acid methyl ester in ethanol, 10 ml of ethyl acetate <!>, and 6.35 ml of proprionic acid <!>. Pour into 60 x 15-mm tissue culture dishes.

Ethanol (70%) <!>

Halocarbon oil

Injection buffer

0.1 mM sodium phosphate (pH 7.8) <!>
5 mM KCl <!>

Halocarbon oil

Purchase halocarbon oil Series 700, CAS# 9002-83-9 from Halocarbon Products Corporation, New Jersey.

Phosphate-buffered saline (PBS)

Rubber glue

2 ml of rubber cement (Sanford Corporation, Bellwood, Illinois)
2 ml of acetone <!>
Mix on a rotator in a 5-ml screw-capped glass vial.

Tape glue

3-cm length Scotch Brand double-stick tape
1 ml of heptane <!>

To extract, place in 5-ml screw-capped glass vial and mix on a rotator overnight. An alternative double-stick tape is 3M (Type 415), which has supposedly been tested to be nontoxic (Flybase/news/DIN/vol11.txt).

**EQUIPMENT**

Agar strips

Borosilicate glass capillaries with an internal filament (World Precision Instruments TWF100-4)

Coverslips (22-mm square)

Double-stick tape

Egg-laying cage

> Set up cages for egg collection by taping an egg-laying plate smeared with yeast paste to the bottom of a plastic beaker that has had air holes punched in it with a 20-gauge or smaller needle. Add 100–200 flies to a cage. To improve the number of eggs being laid, move the cages to a day-for-night schedule at least 2 days before collections begin.

Egg-strainer

> Slice the bottom from a 50-ml flat-topped polypropylene centrifuge tube (Corning) such that the screw cap and a one-inch open tube remains. Cut out the flat top of the screw cap, leaving a screw-cap ring. Place a square of Nitex mesh (size 3-85/44 from Tetko Inc.) over the top of the open tube, and screw tightly into place with the screw-cap ring.

Humidifier/mister

Microfuge

Microinjection needles

Microinjector attached to inverted microscope

Microslides

Needle puller

Paint brush (fine bristle)

Petri dish

Pneumatic pump (PicoPump850, World Precision Instruments)

Push-pin

Stereomicroscope with fiber optic light source

Syringe (10 ml)

1. Centrifuge siRNA or dsRNA to pellet from ethanol, and wash the pellet with ice-cold 70% ethanol. Carefully remove all wash solution and let the pellet air-dry for 8 minutes at room temperature.

2. Dissolve the dsRNA or siRNA into injection buffer solution. For dsRNA, dissolve to a final concentration of no greater than 5 μM, which is 1.65 mg/ml for a 500-bp dsRNA. For siRNA, dissolve to a final concentration of between 0.1 and 5 mg/ml.

   > If the dsRNA solution is too viscous or makes many bubbles in the needle, try a lower concentration of dsRNA. This is not a problem with siRNAs. RNAs can be frozen and thawed for re-use a couple of times, and RNAs dissolved in injection buffer are stable months at –70°C.

3. Pull needles from 1-mm borosilicate glass capillaries with an internal filament. Microfuge the RNA for 10 minutes at 23°C. Back-fill the needle with 0.5 μl of the RNA solution.

   **CAUTION:** Make sure to wear gloves to avoid contamination.

4. Place the needle on the microinjector and move the needle over the inverted microscope stage so that the tip of the needle is in the center of the field of view.

   a. Move needle up in the *z* dimension.

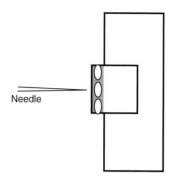

**FIGURE 17.3.** Placement of dechorionated embryos on a 22-mm-square coverslip for microinjection. Shown is the orientation of embryos on the coverslip relative to the injection needle and a microslide on which the coverslip rests. In this diagram, the embryos are oriented so that the needle penetrates midway along the anteroposterior axis. Embryos adhere to the coverslip by a thin coat of tape glue or rubber glue.

Needle

b. Place a microslide at a 45° angle on the microscope stage and bring it into the field of view such that the microscope focuses on the edge of slide.

c. Move the needle down in the $z$ dimension until in focus (at 20x) and slowly move the microslide toward the needle tip until the tip just breaks slightly to make a tip diameter of 0.5–2.5 μm.

d. Immediately raise and lower the needle tip into a puddle of halocarbon oil on a slide. Blow out some solution to ensure that the tip is open.

e. Calibrate the volume output.

5. The eggs are typically collected every 30–60 minutes at 23°C and injected within the next 45 minutes so that the RNA is introduced before cellularization takes place.

a. Before the first round of injection, place a fresh egg-laying plate with yeast paste into the egg-laying cage for 1 hour to induce the flies to lay any overdeveloped eggs.

b. Collect eggs over 30–60 minutes by exchanging a fresh plate for the older plate in the cage.

c. Once collected on a plate, release the eggs from the agar by brushing lightly with a soft, wet paint brush. Brush eggs into the egg strainer. Wash the eggs with $H_2O$ or PBS, and blot the excess wash onto a paper towel.

d. Remove the chorion, or eggshell, by dechorionation. Place the egg strainer in a Petri dish of 50% bleach for 3 minutes, with occasional swirling, followed by thorough rinsing and washing of the eggs with $H_2O$. Use a fine paint brush to transfer the embryos to a thin strip of agar cut from an egg-laying plate.

An alternative method of dechorionation is to brush the embryos onto a piece of double-stick tape. Use a relatively sharp-tipped needle instrument (a push-pin stuck on a pencil will do) to rub the side of each embryo gently until it rolls over and pops out of its eggshell.

6. Pick up the embryos with a needle instrument. Place them very near the edge of a 22-mm-square coverslip that has been previously brushed with a solution of either tape glue or rubber glue (Figure 17.3), which serves to stick the embryo to the coverslip.

Embryo transfer is usually done using a stereomicroscope and should be done under a fine vapor mist to prevent uneven desiccation. We use small room humidifier/misters purchased from a department store.

7. When 40–50 embryos have been lined up, place the coverslip in an air-tight container (~0.5–1-liter volume) with a 1-cm layer of Drierite on the bottom for 5–15 minutes.

> The optimum time depends on many factors. Judge when the embryos are optimally desiccated so that the majority do not leak after injection, but make sure that they are not so desiccated as to appear "wrinkled" or bag-like.

8. After desiccation, adhere the coverslip to a microslide such that the side of the coverslip containing the embryos hangs 2–3 mm over the edge of the slide. Use a 10-ml syringe to cover the embryos with a *thin* layer of halocarbon oil (2-mm wide).

9. Use a pneumatic pump to inject the dsRNA solution into the embryos. Insert the needle gently with the tip precisely orthogonal to the membrane at insertion site. Inject the solution into the middle of embryo. Pull out the needle gradually to avoid plasm leakage.

**Location of Injection** — Injection location is typically on the ventral side extending from 20% to 80% egg length. However, any location seems to work since varied locations give similar RNAi effects on the same target gene. Anterior injections are less likely to cause cytoplasmic leakage, whereas posterior injections tend to leak more frequently. However, this matter is very much dependent on the investigator, and the best orientation should be determined empirically. The dsRNA or siRNA diffuses uniformly throughout syncytial embryos within minutes after injection. The average injection volume that we use is 85 pl, but it ranges from 65 to 110 pl, as determined by measuring the diameter of droplets injected into halocarbon oil. Typically, we try to deliver ~0.2 fmoles of dsRNA per embryo.

Perform a dose-response titration experiment with each RNA. Low doses of dsRNA produce weak or no RNAi phenotype, whereas high doses produce a nonspecific lethal phenotype. This phenotype is typically one in which embryos develop without head structures and with entirely naked cuticle. Other artifactual phenotypes are also seen. A major advantage of using siRNAs is that these nonspecific phenotypes are rarely observed even at very high doses of siRNA (10 mg/ml).

10. Place the slide/coverslip sandwiches of injected embryos in a humidified chamber at 18–25°C. To prevent the oil from spreading over the entire coverslip and exposing the embryos to the air, slightly tilt the slide/coverslip sandwiches (2°) so that the edge of the coverslip with oil and embryos is facing slightly downward. Check the embryos every 12 hours to verify that they are properly humidified and covered with oil. Add as much oil as is necessary to keep the embryos covered, but not too much.

Typically, the penetrance of a successful RNAi experiment is when an RNAi phenotype is seen in 30–100% of injected embryos. The expressivity of the phenotype also varies. Some animals exhibit a totally systemic phenotype, whereas others exhibit a partial phenotype. This partiality can be seen as an overall weak phenotype or a strong but localized phenotype (mosaicism). Typically, a successful RNAi experiment produces a partial RNAi phenotype in 20–50% embryos and a strong phenotype in 20–50% embryos. Note that different specimens along this phenotypic series should be qualitatively similar to each other if the phenotype is caused by RNAi. Thus, RNAi by injection must be quantitated. A statistically significant number of animals must be injected with RNA and their phenotypes classified. An equivalent number of animals must be injected with buffer alone or a heterologous RNA and classified. To ensure that a phenotype is caused by the specific RNA, there must be a statistically significant difference between the two treated populations.

There is variability in the interference activities of different dsRNAs toward the same gene. Some dsRNAs systemically generate null phenotypes with high efficiency, but others generate localized or weaker phenotypes at the same dose. Differences in sequence compositions may affect silencing activity, stability, or transport of dsRNA. There may be unequal sensitivity to interference based on relative accessibility of the dsRNA to find its target sequence. Variability also exists between genes in their sensitivity to RNAi by this method. Some genes produce strong RNAi phenotypes, whereas others appear to be insensitive to RNAi. Several factors may have a role in this variability. Perhaps the most important factor is the relationship between a gene's phenotype and its transcript abundance. Some genes exhibit a wild-type phenotype even with only 10% of their normal transcript, whereas the phenotypes of other genes are very sensitive to transcript abundance. Because RNAi by embryo injection usually reduces but does not abolish all gene transcripts in *Drosophila* (Kennerdell et al. 2002), this relationship between phenotype and dosage is important.

Another reason for failing to detect an RNAi phenotype by injection is when the investigator is unskilled at embryo injection. Embryo injection is a difficult technique to acquire, and when it is not mastered, embryo death rates are very high. Beginners frequently kill 90–100% of the embryos. Thus, buffer control and RNA treatments are indistinguishable because most embryos will have a "nonspecific" lethal phenotype. An important criterion to consider *before* starting an RNAi experiment is the baseline skill of the investigator. If greater than 50% survival to larval hatching can be achieved, then the investigator is ready and able to start an RNAi experiment. If this "gold-standard" has not been met, then more instruction and practice is needed. There are several problems that will tend to arise, each with its own solution:

- Simply try dechorionating and transferring embryos to coverslips. Omit desiccation and injection. Cover with oil. How many embryos survive? Beginners must achieve a light touch when handling embryos and this only comes with experience.

- Avoid too much oil as it causes hypoxia, but too little oil causes the embryos to dry out when exposed to air.

- When embryos leak immediately upon being punctured, they have been under-desiccated. Increase the desiccation time by a few minutes next time.

- If many of the embryos are flaccid, reduce the desiccation time.

- If the embryos are not easily punctured by the needle or leak often when the needle is pulled out, the needle is not sharp enough. Try rebreaking the tip on the edge of the slide, or make a new needle.

- If the embryo leaks while performing injections, too much RNA solution is being injected.

- If the needle jams frequently, either there is an excess of particulate matter in the injection solution or the needle's opening is too small. Remember to spin down the solution before back-filling the needle.

## PROTOCOL 3: DELIVERY INTO EMBRYOS BY GENE GUN

An alternative method to deliver dsRNA into embryos is by particle bombardment. In this method, subcellular-sized particles are accelerated to high velocity to carry dsRNA into embryos. The technique was first described as a method of gene transfer into plants (Klein et al. 1992). The major advantage of this procedure over microinjection is that particle bombardment is easier and faster to perform. In addition, the mechanical trauma received is far less than by microinjection, allowing better survival of embryos and fewer phenotypic artifacts. The principle of particle bombardment is to employ a high-velocity stream of helium gas to accelerate gold particles coated with dsRNA to sufficient velocities to penetrate embryos. Compressed helium held at high pressure is momentarily discharged by a solenoid valve through a narrow bore containing the gold particles. When the helium enters the bore, it picks up the gold particles, which become entrained in the helium stream. Past the acceleration channel, the gold particles disperse into a cone-like projection through the air until they reach the target.

## Procedures

### Step 1: Coating Particles with dsRNA

▼ **CAUTION**
*See Appendix for appropriate handling of materials marked with <!>.*

---

**MATERIALS**

**REAGENTS**

Calcium chloride ($CaCl_2$) (1 M) <!>

dsRNA or siRNA (from Protocol 1)

Ethanol (absolute) <!>

Gold particles (1.0-µm diameter) (Bio-Rad)

Polyvinylpyrrolidone (360 kD) (Sigma) <!>

Spermidine (0.05 M)

**EQUIPMENT**

Microfuge

Microfuge tube

Ultrasonic cleaner

---

1. Weigh the gold particles into a microfuge tube.

   The quantity depends on particle delivery size (usually 0.5–1.0 mg per shot) and the number of shots planned in the experiment. For simplicity, assume that 10 mg is to be prepared.

2. Add 100 µl of 0.05 M spermidine. Vortex the gold and spermidine mixture briefly. Sonicate for 3–5 seconds using an ultrasonic cleaner.

3. Add dsRNA dissolved in distilled $H_2O$ to the spermidine-gold mixture. The ratio of dsRNA to gold should be 20 µg of dsRNA per 10 mg of gold. The volume of the dsRNA solution added should be less than 100 µl. Vortex the dsRNA-spermidine-gold mixture at a moderate rate briefly.

4. While vortexing the mixture, add 100 μl of 1 M CaCl$_2$ dropwise to the mixture. Make sure that the volume added is equal to that of the spermidine in Step 2.

5. Allow the mixture to precipitate for 10 minutes at room temperature.

6. Microfuge for 15 seconds and then remove the supernatant.

7. Wash the pellet three times with 1 ml of ethanol each time, vortex to disperse the pellet, and microfuge for 5 seconds to collect the pellet.

8. To 10 mg of the dsRNA-coated gold, add 100–200 μl of either ethanol or 0.05 mg/ml polyvinylpyrrolidone in ethanol (prepared fresh before use).

> The final concentration of the coated gold suspension is 50–100 mg/ml. The choice of which concentration to use depends on whether 0.5 or 1 mg of gold per shot is delivered because each shot uses 10 μl of the suspension.

> Polyvinylpyrrolidone serves as an adhesive and can increase the number of particles delivered at higher discharge pressures.

9. Use the suspension immediately in a gene gun or store up to 2 months at –20°C.

## Step 2: Particle Bombardment with Gene Gun

Particle delivery is carried out with pressurized helium. Commercial delivery systems include the Helios Gene Gun and Biolistic PDS-1000 from Bio-Rad. The PDS-1000 delivers particles to samples within a vacuum chamber. The Helios is more portable and requires no vacuum. Either system will work on *Drosophila* embryos. Method 1 (below) describes the protocol for RNA bombardment with the Helios system. In addition to commercial gene guns, several lab-made gene guns have been designed for particle bombardment. Method 2 (below) describes construction and operation of one design that works for *Drosophila* embryos. All bombardment systems use the following procedure for particle delivery into syncytial *Drosophila* embryos.

**▼ CAUTION**
*See Appendix for appropriate handling of materials marked with <!>.*

**MATERIALS**

**REAGENTS**

Bleach (50%) with H$_2$O <!>

*Drosophila* embryos

dsRNA-gold-ethanol slurry <!>

Rubber glue (see Protocol 2)

Tape glue (see Protocol 2)

**EQUIPMENT**

Aluminum rod frame

Helios Gene Gun or Biolistic PDS-100 (Bio-Rad)

Helium tank <!>

> Use compressed helium of grade 4.5 or higher and pressurized to 2600 psi.

Microslide

Paint brush (fine-bristle)

Petri dish

Tefzel tubing (Bio-Rad)

1. Collect embryos 0–60 minutes after egg laying. Dechorionate the embryos with 50% bleach, wash with $H_2O$, and blot dry.

2. Use a fine-bristle paint brush to spread the embryos in a 20-mm-diameter lawn on a microslide or Petri dish that has been coated with either rubber glue or tape glue (see Protocol 2).

3. Desiccate the embryos for 5 minutes in the open air at room temperature. Bombard them with a single shot of gold-RNA.

4. Immediately cover the embryos with a thin layer of halocarbon oil. Incubate in a humidified chamber at 18–25°C until analysis.

## Method 1: Helios Gene Gun System

Although the Helios system is a hand-held gun-like instrument, more reproducible particle delivery is provided by clamping the gun to a stationary aluminum rod frame.

1. Clamp the gun to a stationary aluminum rod frame. Orient the gun vertically such that the distance from nozzle to embryo plate below it is 2.5 cm.

   CAUTION: Use hearing and eye protection for this procedure.

2. Attach the gun to the flexible hose leading from the helium tank.

   The helium pressure regulator allows adjustment of pressure from 0 to 700 psi.

3. Unlock the cartridge holder from the gun. Place on a flat surface with numbered edge facing up. Load an empty cartridge into position 1.

   Cartridges are 0.5-inch pieces of Tefzel tubing (Bio-Rad) cut square with a razor blade.

4. Vigorously mix the dsRNA-gold-ethanol slurry and pipette 10 µl of slurry into the tubing cartridge. This will contain 0.5–1.0 mg of gold particles and 1–2 µg of dsRNA.

5. Insert the loaded cartridge holder into the gene gun such that position 1 is in the firing position.

6. Open the helium regulator and set the helium pressure to between 300 and 450 psi. Activate the safety interlock switch and press the trigger button to discharge the gold-dsRNA.

## Method 2: Lab-built Gene Gun

The gun is essentially a solenoid valve triggered by a relay switch to fire for 50 msec. The gun was designed by Drs. Michael Nonet and Paul Bridgman (Washington University, St Louis) as a modification from two previous gene gun designs: one design made for transformation of *Volvox* (Schiedlmeier et al. 1994) and another design made for transformation of *C. elegans* (Wilm et al. 1999). The system, including the regulator and helium tank, can be purchased and constructed for about $500. Details of the gun's design and construction can be seen at Dr. Nonet's Web Site at http://thalamus.wustl.edu/nonetlab/NMimages/genegun/Genegun.htm. Other lab-designed gene guns are available and can be found by accessing local Internet newsgroups.

| PARTS LIST | | | | |
|---|---|---|---|---|
| Catalog number | Quantity | Company* | Description | Cost ($) |
| **GUN** | | | | |
| 3UL42 | 1 | Grainger | ASCO solenoid valve (750 psi) | 37.95 |
| 60417 | 2 | US Plastics | Parker O-ring tube fitting (1/4"OD/1/8 NPT) | 0.93 |
| 5X0001300 | 1 box (10) | Fisher | Swinnex 13-mm filter holder | 33.84 |
| XX30 02561 | 1 | Millipore | 1/8" NPTM-to-M Luer locking adaptor | 40.58 |
| 2P237 | 1 bag (10) | Grainger | Male connectors (1/4 x 1/8 NPT) | 6.16 |
| **TRIGGER** | | | | |
| 6A855 | 1 | Grainger | Multi-time range/multi-function relay | 45.84 |
| 6X156 | 1 | Grainger | Socket for relay | 6.54 |
| 275-709 | 1 | RadioShack | Momentary switch | 3.29 |
| | 1 | | Wire (12 gauge) | |
| | 1 | | Electric cord with grounded plug | |
| | 1 | | Electric box | |
| **GAS DELIVERY** | | | | |
| 56073 | 5–10 ft | US Plastics Corp. | TFE Teflon tubing (559 psi) | 1.67/ft |
| KEKP200+ | 1 | PuritanBennett | H-size tank of high-purity helium | 75.00 |
| VIC0784-2319 | 1 | Airgas | 580 CGA 1/4 female fitting outlet (500 psi delivery pressure helium regulator) | 258.00 |
| 2P238 | 1 bag | Grainger | Male connectors (1/4 x 1/4 NPT) | 10.70 |

*Airgas (www.airgas.com); Fisher (www.fishersci.com); Grainger (www.grainger.com); Millipore (www.millipore.com); RadioShack (www.radioshack.com); US Plastic Corp. (www.usplastic.com).

### The gun

The gun consists of a solenoid valve with 1/8-inch NPT female fittings. On the input side, the solenoid is attached via tubing to connect to the helium regulator, whereas on the output side, the solenoid is attached to a Swinnex filter holder, which acts as the support to load the dsRNA-coated gold particles. One way to connect the Swinnex filter to the solenoid is to shave off the threading on the Swinnex filter. It then fits tightly into the Parker O-ring fitting. A more elegant and expensive approach is to purchase a 1/8-inch NPTM-to-M Luer locking adaptor from Millipore (see Parts List). Secure the solenoid by using a mounting clamp attached to an aluminum rod frame so that the height can be adjusted.

### The trigger mechanism

The solenoid is wired to a trigger that consists of a momentary switch, which fires a one-shot relay switch. The wiring is outlined at http://thalamus.wustl.edu/nonetlab/NMimages/wiring.jpg. The 120-V AC current powers both the relay and the solenoid. Set the relay to fire at 10-msec intervals from 0.01 to 10.00 seconds. To ensure that the solenoid fires reliably, use at least 50 msec (005 on the relay). Set the relay to one-shot mode. The trigger mechanism can be mounted into any type of plastic electric box (Starsted blue tip boxes work fine). The only purpose of the relay is to have consistency in the helium burst from one shot to the next.

### The helium source

The regulator should provide up to 500 psi delivery pressure. The purity of the helium should be greater than grade 4.5 to ensure that there are no mold spores in the gas. Use high-pressure nylon tubing to connect the helium tank to the solenoid valve. Connect the tubing to the solenoid via one of the Parker O-ring connectors. A different type of O-ring connector may be used to attach tubing to the helium regulator, depending on the fitting present on the regulator.

### Operation of gun

1. Vigorously mix the dsRNA-gold-ethanol slurry and pipette 10 μl of slurry onto one of the Swinnex filters (male end). This will contain 0.5–1.0 mg of gold particles and 1–2 μg of dsRNA.

   The gold will spread relatively evenly onto the plastic support using this volume. The ethanol does not harm the embryos, and most of it probably evaporates before hitting the embryos during firing. The Swinnex filter holder male end is pre-cut with a saw below the threading so that the bottom is wide open. Consequently, there is a diffuse gold spray that projects widely when the device is fired.

   CAUTION: Use hearing and eye protection for this procedure.

2. Fit the male end of the Swinnex filter into the female end. Place the embryos on a benchtop immediately below the solenoid/filter unit and position the filter so that it is ~7 cm from the embryos.

3. Set the relay switch to fire for 50 msec (0.005 or 9.995 on the relay setting). Set the helium regulator pressure to between 300 and 450 psi.

4. Wearing hearing and eye protection, hit the momentary switch (firing trigger) to eject the gold.

## PROTOCOL 4: PHENOTYPE ANALYSIS OF EMBRYOS

Some standard experimental protocols used by the *Drosophila* community do not work successfully with RNAi-treated embryos. Thus, modifications have been made to these procedures so that they will work. Some of these procedures might be used in analysis of an RNAi phenotype, and, consequently, they have been included in this section.

## Procedures

### Method 1: Semiquantitative RT-PCR of *Drosophila* Embryo mRNA

It is often desirable to monitor gene expression by mRNA abundance, and this is easily accomplished in *Drosophila* by RT-PCR (Kennerdell and Carthew 2000).

**Step 1: RNA Purification from Embryos**

**MATERIALS**

**REAGENTS**

Chloroform <!>

Embryos treated with dsRNA or siRNA (from Protocol 2 or 3)

Ethanol (absolute) <!>

Heptane <!>

Isopropanol <!>

Trizol (GIBCO BRL)

**EQUIPMENT**

Dish (60 mm)

Microfuge tube

Micropipettor

Pasteur pipette

Razor blade

Tape glue (see Protocol 2)

1. Attach the embryos to coverslips with tape glue. Incubate the embryos under oil in a humidified chamber at 18–25°C.

2. Collect the embryos for staining at the appropriate stage. Use a razor blade to scrape excess oil from coverslip, taking care not remove embryos. Remove as much oil as possible.

3. Remove the coverslip from the slide and hold the coverslip over a 60-mm dish of heptane. Use a Pasteur pipette to wash the embryos on the slide with heptane until the oil is washed away (~6 sprays). Continue washing until all of the embryos are washed into the dish.

4. Transfer the embryos to a microfuge tube. Withdraw as much heptane as possible without exposing the embryos. Store the embryos under heptane for up to 3 days at –70°C.

5. Scrape the inside of the tube with a pipette tip to break open embryos. Add 5 μl of Trizol per embryo (e.g., 100 μl to 20 embryos) and triturate.

6. Vortex the Trizol-embryo mixture for 15 seconds. Incubate for 5 minutes at room temperature. Add 1 μl of chloroform per 10 μl of Trizol. Vortex for 15 seconds. Incubate on ice for 15 minutes. Microfuge the mixture for 15 minutes at 4°C. Transfer the clear upper phase to a new tube. Add 6.4 μl of isopropanol for every 10 μl of Trizol that was initially used. Vortex for 2 seconds. Incubate for 15 minutes at –70°C.

7. Microfuge for 15 minutes at 4°C. Wash the alcohol pellet with 0.4 ml of 70% ice-cold ethanol. Microfuge for 5 minutes and remove the supernatant. Centrifuge in a microfuge for 2 seconds to force all of the liquid to the bottom of the tube. Carefully pipette off the liquid with a micropipettor.

8. Air-dry the pellet for 8 minutes at room temperature and then dissolve the pellet in nuclease-free $H_2O$. Add 0.5 μl of $H_2O$ per embryo-equivalent of RNA.

## Step 2: Reverse Transcriptase (RT) Reaction

1. Prepare 2x RT Mix on ice as follows:

| | |
|---|---|
| $H_2O$ | 3.3 μl |
| 2.5 mM random hexamer oligonucleotides (Pharmacia) | 0.2 μl |
| 5 mM dNTPs | 1 μl |
| 5x Reverse transcriptase first-strand synthesis buffer (GIBCO-BRL) | 4 μl |
| RNAsin (Promega) | 0.5 μl |
| MoMLV reverse transcriptase (200 units/μl) (GIBCO BRL) | 1 μl |
| Total volume | 10 μl |

2. Add equal volume of 2x RT Mix to RNA solution from Step 1. Incubate for 2 hours at 37°C.

3. Store products at –20°C (they remain stable for weeks).

## Step 3: PCR Amplification

Primers should have ~50% GC content, be 16–18 nucleotides long, and have a melting temperature of ~60°C. They should be spaced apart to generate a PCR product 100–200 bp in length. Use the Primer3 program (MIT Genome Center Web Site) for choosing primers.

▼ **CAUTION**
*See Appendix for appropriate handling of materials marked with <!>.*

**MATERIALS**

**REAGENTS**

Ethanol (absolute) <!>

Formamide loading buffer

   20 mM EDTA
   0.05% (w/v) bromophenol blue<!>
   0.5% (w/v) xylene cyanol <!>
   in 90% deionized formamide <!>

Polyacrylamide (6%), urea (7 M), 0.5x TBE denaturing gel (0.4-mm thick)

RT cDNA sample (from Step 2)

Stop buffer

   16.7 μg/ml denatured DNA
   2.8 M $NH_4OAc$

**EQUIPMENT**

Electrophoresis equipment

Microfuge

Microfuge tubes

Micropipettor

1. Prepare PCR Master Mix on ice as follows:

| | |
|---|---|
| $H_2O$ | 5.84 μl |
| dNTPs (2 mM) | 1 μl |
| $MgCl_2$ (25 mM) | 0.6 μl |
| 10x *Taq* polymerase buffer | 1 μl |
| [$^{32}$P]dCTP or dATP (3000 Ci/mmole; 10 μCi/μl) | 0.1 μl |
| *Taq* DNA polymerase | 0.06 μl |
| Total volume per reaction | 8.6 μl |

2. Prepare PCR Mix on ice as follows:

| | |
|---|---|
| PCR Master Mix | 8.6 μl |
| 10 μM forward primer | 0.2 μl |
| 10 μM reverse primer | 0.2 μl |
| Total volume per reaction | 9.0 μl |

3. Add 1 μl of RT cDNA sample from Step 2 to 9 μl of PCR Mix.

4. PCR program:

    Initiate: 2 minutes at 94°C

    Cycle: 30 seconds at 94°C

    Cycle: 30 seconds at 55°C

    Cycle: 30 seconds at 72°C

**Number of Cycles** End the reaction in the geometric phase of the PCR. To determine the geometric phase for a particular mRNA-tissue-primer pair, perform a cycle-course experiment with given RT cDNA and primers. Choose PCR samples from over a range of cycle points. Usually, pick a sample from every other cycle point ranging from 11 to 30 cycles. Run PCR products on a denaturing gel and quantitate radioactive product formation. Plot the product abundance versus cycle number on semi-log graph. The geometric phase of the PCR is the cycle range where the plotted curve is linear. Ideally, product should increase fourfold with every other cycle during the geometric phase, an exponent of 2. In reality, the exponent is usually less than 2. Choose a cycle point in the middle of the geometric phase for future experiments using that particular mRNA-tissue-primer pair. This point depends on abundance of each transcript, and so is empirically independent for each type of transcript assayed.

5. Add 90 μl of stop buffer.

6. Add 250 μl of cold ethanol, and incubate on ice for 15 minutes. Microfuge for 15 minutes at 4°C. Wash the pellet with 0.4 ml of ice-cold ethanol. Microfuge for 10 minutes at 4°C, remove the supernatant, and centrifuge the pellet in a microfuge for 2 seconds to force all of the liquid to the bottom of the tube. Carefully pipette off the last liquid with a micropipettor.

7. Air-dry for 8 minutes at room temperature and dissolve the pellet in 3 μl of $H_2O$. Add 3 μl of 2x formamide loading buffer. Boil the mixture for 3 minutes and place immediately on ice. Load 2 μl onto a 6% polyacrylamide/7 M urea/0.5x TBE denaturing gel. Electrophorese, fix, and dry the gel for exposure.

**Method 2:** Transcript In Situ Hybridization of Whole-mount Embryos

This protocol fixes and prepares embryos for in situ hybridization to visualize transcript expression patterns. It is a modification of the method developed by Tautz and Pfeifle (1989) for whole-mount in situ analysis of embryos. Use of the standard hybridization protocol on RNAi-treated embryos results in high background staining, which makes visualization of transcript expression patterns practically impossible. The following modifications eliminate this problem and allow visualization of transcript expression after RNAi injections.

▼ **CAUTION**
*See Appendix for appropriate handling of materials marked with <!>.*

## MATERIALS

### REAGENTS

Embryos (from Protocol 2 or 3)

Fixative A (10:1:9 v/v)

> *n*-heptane <!>
> 37% formaldehyde <!>
> PEM buffer

> Vortex for 30 seconds to saturate both phases.

Fixative B

> 5% (w/v) paraformaldehyde <!>
> 25 mM EGTA
> 0.05% (v/v) Tween-20

> Dissolve in phosphate-buffered saline. <!>

Formamide/TE (1:1 v/v) <!>

Glycine block

> Dissolve 2 mg/ml glycine in PTW buffer. <!>

Hybe buffer

> 50 ml of deionized 100% formamide <!>
> 25 ml of 20x SSC
> 2 ml of denatured/sonicated salmon testis DNA (10 mg/ml)
> 0.5 ml of tRNA (20 mg/ml)
> 50 µl of heparin (100 mg/ml)

> Adjust distilled $H_2O$ to 100 ml. Adjust pH to 5.0 with HCl. Store at –70°C.

Hybe/PTW (1:1 v/v)

> Mix equal volumes of Hybe and PTW buffers.

Methanol (100%) <!>

Methanol/PTW buffer (1:1 v/v) <!>

PEM buffer

> 0.1 M PIPES (pH 6.95)
> 2 mM EGTA
> 1 mM $MgSO_4$ <!>

Phosphate-buffered saline (PBS)

Proteinase K <!>

> Dissolve 10 µg/ml proteinase K in PTW buffer.

PTW buffer (0.1% [v/v] Tween-20 in PBS)

---

**EQUIPMENT**

Double-stick tape

Egg strainer (Nitex mesh)

Eppendorf tubes

Forceps

Kimwipes

Needle (21 gauge)

P1000 pipettor

Plastic dish (60 mm)

Spot dish (glass or porcelain)

Stereomicroscope

---

1. Harvest the embryos following Step 1 of Method 1 (Protocol 4). Transfer embryos to a well in the spot dish. At this point, the embryos are normally shrunken due to dehydration. Remove the heptane and incubate the embryos in 0.5 ml of $H_2O$ (embryos float at the water surface). Allow them to rehydrate for 30 seconds to several minutes, monitoring under a stereomicroscope.

2. Replace the $H_2O$ with Fixative A. Ensure that the well has both phases present. Cover with a slide and incubate for 20 minutes at room temperature.

3. Use a tip-cut P1000 pipettor to remove the embryos from the fix interface. Pipette the embryos onto the bottom of an upside-down egg strainer stuffed inside with Kimwipes. Let the solvent blot through the mesh leaving the embryos on top of the mesh.

4. Immediately pick up the embryos from the mesh with a strip of double-stick tape.

5. Flip the tape over and apply it to the bottom of a 60-mm plastic dish (embryo side up). Submerge tape and embryos under PBS.

6. Manually devitellinize the embryos with the tip of a 21-gauge needle and forceps. Popping the embryos with a nudge from the posterior is often sufficient to release them.

7. Remove the PBS from the dish and replace with PTW buffer. Pipette the embryos with PTW into an Eppendorf tube.

8. Remove PTW and incubate in 1 ml of PTW for 2 minutes.

9. Remove PTW and incubate in 1 ml of methanol/PTW (1:1) for 2 minutes.

10. Remove methanol/PTW and incubate in 1 ml of methanol for 2 minutes.

11. Remove methanol and incubate in 1 ml of methanol/PTW for 2 minutes.

12. Remove Methanol/PTW and incubate in 1 ml of PTW for 2 minutes.

13. Wash with PTW three more times.

14. Remove PTW and incubate in 1 ml of Fixative B for 20 minutes.

15. Remove Fixative B and wash embryos three times with PTW.

16. Remove PTW wash and digest embryos with proteinase K for 2 minutes.

17. Replace proteinase K solution with 1 ml of glycine block. Incubate for 10 minutes.

18. Remove glycine block and wash embryos twice with PTW.

19. Fix embryos a second time with Fixative B for 20 minutes.

20. Remove Fixative B and wash embryos five times with PTW.

21. Incubate embryos for 10 minutes in 1 ml of Hybe/PTW (1:1).

22. Replace with 100 μl of formamide/TE (1:1) and incubate for 10 minutes at 75ºC.

23. Plunge tube into ice-water slush to quick-chill the embryos. Allow to stand for 5 minutes.

24. Remove formamide/TE and incubate in 1 ml of Hybe buffer for 10 minutes.

25. Replace with a fresh aliquot of Hybe buffer and incubate for 60 minutes at 48ºC.

26. Hybridize embryos with an in situ probe at 48ºC as described by Tautz and Pfeifle (1989). Perform washes, antibody reactions, and color reactions as described in Tautz and Pfeifle (1989).

## Method 3: Immunohistochemistry of RNAi-treated Embryos

This protocol prepares and fixes embryos for standard antibody treatments. Because of the microinjection procedure and pre-fix heptane treatment, many embryos become morphologically deformed. This protocol will help to regain normal morphology, although the procedure is not 100% effective.

▼ **CAUTION**
*See Appendix for appropriate handling of materials marked with <!>.*

**MATERIALS**

**REAGENTS**

Embryos (from Protocol 2 or 3)

Fixative A (10:1:9 v/v/v)

    *n*-heptane <!>
    37% formaldehyde <!>
    PEM buffer
    Vortex for 30 seconds to saturate both phases.

PBT buffer (0.1% [v/v] Triton X-100 in PBS)

PEM buffer

    0.1 M PIPES (pH 6.95)
    2 mM EGTA
    1 mM $MgSO_4$ <!>

Phosphate-buffered saline (PBS)

**EQUIPMENT**

Double-stick tape

Fine-mesh basket

Forceps

Kimwipes

Needle (21 gauge)

P1000 pipettor

Plastic dish (60 mm)

Slides

Spot dish (glass or porcelain)

Stereomicroscope

1. Follow Steps 1–6 in Protocol 4 (Method 2).

2. Remove the PBS and add PBT to the dish. Incubate for 10 minutes.

3. Use a tip-cut P1000 pipettor to transfer the embryos with PBT to an Eppendorf tube.

4. Carry out standard antibody incubations (Sullivan et al. 2000).

### Method 4: Larval/Embryo Cuticle Analysis

## MATERIALS

### REAGENTS

CMCP-10/lactic acid (3:1 v/v) <!>

> Make fresh before each use. CMCP-10 is purchased from Masters Chemical Company, Bensenville, Illinois.

Embryos (from Protocol 2 or 3)

Glycerol/acetic acid (1:4 v/v) <!>

Phosphate-buffered saline (PBS)

Rinse buffer (0.5% [v/v] Triton X-100 in PBS)

### EQUIPMENT

Microscope slides

Slide coverslips

Spot dish (glass or porcelain)

1. Isolate hatched larvae. Dissect unhatched embryos that have secreted a cuticle from their vitelline membranes. Transfer embryos and larvae to a well in the spot dish.

2. Wash embryos thoroughly in rinse buffer.

3. Remove rinse buffer and incubate the embryos in 1 ml of glycerol/acetic acid overnight in a humidified chamber at 60°C.

4. Mount embryos on a microscope slide in CMCP-10/lactic acid and cover with coverslip. Bake for 1–2 days at 70°C.

## PROTOCOL 5: CELL CULTURE RNAi

The *Drosophila* research community tends to use a limited number of immortalized cell lines for most purposes, including Schneider line 2 (S2), Kc, and Clone-8 cells (Ashburner 1989). The former two lines are derived from embryonic sources, whereas Clone-8 cells are derived from larval imaginal disc tissue. Cell lines grow well at 25°C with air as the gaseous phase. S2 and Kc lines grow loosely, adhering to the surface or in suspension, making possible mechanical removal (rather than enzymatic removal) of cells from a surface. Cell-doubling times in serum-supplemented medium are on the order of 20–24 hours. Overall, these cells are easy to care for and maintain. Two methods of performing RNAi have been developed and are generally used. Their relative effectiveness is fairly equivalent. The second method—dsRNA Soaking—is simpler and less toxic to cells, and therefore is the method of choice.

# Procedures

## Method 1: Calcium Phosphate Transfection

This method, developed by Greg Hannon's laboratory, is very effective for S2 cells (Hammond et al. 2000). Approximately 90% of cells are affected. Its effectiveness for other cell lines is less characterized, although DNA transformation by calcium phosphate is ineffective in some cell lines (Ashburner 1989).

### MATERIALS

### REAGENTS

$CaCl_2$ (0.25 M)

  3.68 g of $CaCl_2\cdot 2H_2O$, double-distilled $H_2O$ to 100 ml. Sterile-filter and aliquot into 15-ml polypropylene tubes. Store at –20°C.

Carrier plasmid DNA

dsRNA or siRNA from (Protocol 1)

2x HEBS

  16 g of NaCl
  0.7 g of KCl <!>
  0.4 g of $Na_2HPO_4$ <!>
  2 g of dextrose
  10 g of HEPES Free Acid

  Adjust pH to 7.1 with NaOH. <!> Add double-distilled $H_2O$ to 1000 ml. Sterile-filter, aliquot in 50 ml, and store at –20°C. When thawing for a transfection, again adjust pH to 7.1 and re-filter before use.

S2 cell line

Schneider's medium (Sigma) supplemented with 10% (v/v) heat-inactivated
  fetal bovine serum (FBS)

### EQUIPMENT

Polycarbonate tubes (17 x 100 mm)

Polypropylene tubes (15 ml)

Tissue culture dishes (60 mm)

1. Seed 5 ml of Schneider's medium + 10% (v/v) FBS in a 60-mm tissue culture dish with S2 cells to a final density of $2 \times 10^5$ to $4 \times 10^5$ cells/ml.

2. Incubate the cells for 6–18 hours at 25°C.

3. Mix 5 μg of dsRNA and 5 μg of carrier plasmid DNA with 0.4 ml of sterile 0.25 M $CaCl_2$. Place 0.4 ml of 2x HEBS in a 17 x 100-mm polycarbonate tube (sterile). Add the $RNA/CaCl_2$ solution dropwise to HEBS while constantly swirling.

4. Incubate the mixture for 20 minutes at room temperature.

5. Add 0.8 ml of the mixture dropwise to each 60-mm dish. Swirl and incubate for 24 hours at 25°C.

6. Split the cells 1:4 in fresh medium and plate onto four new dishes.

7. Incubate a further 48 hours at 25°C.

8. An alternative transfection protocol uses siRNAs in liposomes. Formulate 0.03–1.0 μg of siRNA into liposomes with Oligofectamine (GIBCO), and apply in a 0.6-ml volume to each well of a 24-well plate. Remove after 15–20 hours and replace with Schneider's medium supplemented with serum. Transfection can be repeated one or two more times (15 hours each), although this may be unnecessary.

## Method 2: dsRNA Soaking

This method was developed by the Dixon laboratory (Clemens et al. 2000). It is 90–99% effective for a number of cell lines including S2, Kc, BG2-C6, and Shi (Worby et al. 2001).

---

**MATERIALS**

**REAGENTS**

Cells

dsRNA or siRNA (from Protocol 1)

Schneider's medium (Sigma) supplemented with 10% (v/v) heat-inactivated fetal bovine serum (FBS)

Ultimate Insect Serum-free Medium (Invitrogen)

**EQUIPMENT**

Six-well cell culture dishes

---

1. Within one well of a six-well cell culture dish, seed 1 ml of serum-free medium (Invitrogen) with cells to a final density of $1 \times 10^6$ cells/ml.

2. Add dsRNA directly to the medium to a final concentration of 40 nM. This corresponds to 16 μg of RNA for a 700-bp dsRNA. Mix thoroughly.

3. Incubate for 30 minutes at 25°C.

4. Add 2 ml of Schneider's medium supplemented with 10% FBS.

5. Incubate for 3 days at 25°C.

**Alternative Method for S2 Cells**

1. Seed Ultimate Insect Serum-free Medium (Invitrogen) with S2 cells to $3 \times 10^5$ cells/ml. Add dsRNA to the medium to a final concentration of 3 μg/ml.

2. Incubate for 7 days at 25°C. If necessary, split cells in fresh medium supplemented with 3 μg/ml dsRNA to maintain cells in logarithmic growth phase.

## PROTOCOL 6: PRIMARY CELL RNAi

This method has been demonstrated to silence gene expression in cells of dissected larval salivary glands (Biyasheva et al. 2001). It is not yet clear whether it can be applied to other primary cell types, either dissociated or within organ culture. The method is based on a commercially available kit for introducing large water-soluble molecules into dissected tissues by pinocytosis. The technique, pioneered by Okada and Rechsteiner (1982), is as follows. Compounds to be loaded are mixed with a hypertonic medium, allowing the material to be carried into the cells via pinocytic vesicles. The cells are then transferred to a hypotonic medium, which results in the release of trapped material from the pinocytic vesicles within the cells, filling the cytoplasm with the compound. Endosomal compartments containing the hypertonic loading medium do not fuse with lysosomes. Therefore, materials introduced into cells are not exposed to lysosomal enzymes. Furthermore, lysosomal components are not released into the cytoplasm as a consequence of the procedure.

## Procedure

---

**MATERIALS**

**REAGENTS**

Dissected larval salivary glands

dsRNA or siRNA (from Protocol 1)

Hypertonic Loading Medium (Molecular Probes)

Hypotonic Medium (60% v/v) (Schneider's serum-free medium diluted with $H_2O$)

Influx Pinocytic Cell-loading Reagent (Molecular Probes)

---

1. Use Influx Pinocytic Cell-loading Reagent and prepare Hypertonic Loading Medium according to manufacturer's instructions.

2. Add 2 μg of dsRNA or siRNA to 10 μl of Hypertonic Loading Medium and mix thoroughly.

3. Immediately place dissected tissue into Hypertonic Medium and incubate for 10 minutes at 25°C.

4. Transfer the tissue (blot or wipe to remove as much adhering medium as possible) to 1 ml of Hypotonic Medium.

5. Incubate for 2 minutes and no more. Longer treatment causes cell death.

6. Transfer tissue to 1 ml of fresh Schneider's medium supplemented with 10% FBS.

7. Incubate for at least 6 hours at 25°C.

## PROTOCOL 7: FIRST-GENERATION TRANSGENIC RNAi

Several groups have developed methods to express dsRNA as hairpin-loop snapback RNA from transgenes (Fortier and Belote 2000; Kennerdell and Carthew 2000; Lam and Thummel 2000; Martinek and Young 2000). The methods are modeled on the successful application of snapback RNAs in generating PTGS in plants by expression of transgenes with inverted repeat sequences. The length of the repeat generally varies between 250 and 1500 bp, with 500–700 bp being optimum. The repeats can be oriented head-to-head or tail-to-tail with respect to the sense orientation of the target gene. When using a head-to-head approach, caution must be taken to ensure that spurious transcription and translation of the RNA product could not theoretically synthesize a partial or complete polypeptide.

Most of the methods insert a nonpalindromic spacer between the repeats to increase the stability of the repeat DNA during plasmid replication. Inverted repeats are often deleted in *E. coli* because cruciform intermediates form during replication of plasmid DNA and are excised by the *sbcBC* gene products. Insertion of nonrepetitive sequence greater than 4 bp in length between repeats can inhibit cruciform excision during replication (Davison and Leach 1994). Plasmids with inverted repeats will also replicate with more fidelity in recombination-deficient strains. The SURE strain (Stratagene) is deficient in *recBC sbcBC* and eliminates all known restriction systems. Other strains used are JM103 and JM105 (New England Biolabs), which are also mutant for *sbcBC*. Two methods of cloning transgenes are presented, both of which use the modular UAS transformation system. The advantages in using the GAL4/UAS system were outlined in the introduction to this chapter. Recombinant plasmids are injected with helper plasmid into *Drosophila* embryos and transformant flies are generated by standard P-element transformation (Spradling and Rubin 1982).

## Procedures

### Method 1: The pUAST Two-Step

> **MATERIALS**
>
> **REAGENTS**
>
> Oligonucleotides
>
> pUAST
>
> **EQUIPMENT**
>
> PCR machine

1. Two pairs of oligonucleotides are used as primers to amplify a portion of the target gene by PCR. One pair of primers results in *Eco*RI and *Bgl*II sites at either end of the fragment. Insert this fragment between the corresponding restriction sites in pUAST, which has a multicloning site downstream from the UAS promoter (*Eco*RI *Bgl*II *Not*I *Xho*I *Kpn*I *Xba*I).

2. The second pair of primers amplify the same portion of the target gene but place *Xba*I and *Kpn*I sites at either end of the fragment. Insert this fragment between the corresponding restriction sites in the plasmid construct that was made in Step 1.

   The resulting plasmid has an inverted repeat of the gene fragment with a 28-bp spacer sequence (AGATCTGCGGCCGCGGCTCGAGGGTAAC), which is nonpalindromic.

### Method 2: The pUAST One-Step

Two pairs of oligonucleotides are used as primers to amplify a portion of the target gene by PCR. One pair of primers should produce an *Sfi*I site at one end of the fragment and a restriction site compatible with pUAST (e.g., *Eco*RI) at the other end of the fragment. *Sfi*I recognizes the sequence GGCCXXXX|XGGCC where X is any base and the vertical line is the cutting site. The *Sfi*I site in the first PCR product is (top strand) GGCCATCTTGGCC.

The second pair of primers amplifies the same portion of the target gene by PCR. They should produce an *Sfi*I site at the same end of the fragment and a different site compatible with pUAST (e.g., *Xba*I) at the other end. The *Sfi*I site in the second product is (top strand) GGCCTAGATGGCC.

---

**MATERIALS**

**REAGENTS**

  Oligonucleotides

  pUAST

**EQUIPMENT**

  Gel electrophoresis equipment

  PCR machine

---

1. Ligate the two PCR products together at their *Sfi*I sites. The only productive ligation products are formed between the two different products since they cannot ligate to themselves. Gel-purify the heterodimer ligation product.

2. Insert into corresponding restriction sites in pUAST (e.g., *Eco*RI and *Xba*I).

3. The resulting plasmid has an inverted repeat of the gene fragment with a 5-bp non-palindromic spacer sequence that is in the center of the *Sfi*I site, GGCC<u>ATCTA</u>GGCC.

---

**Problems with Transgenic RNAi**

- ***Deriving stable plasmid constructs.*** This is the most problematic feature of transgenic RNAi. Some DNA is extremely difficult to clone as inverted repeats, or, if they are obtained, the resulting plasmids lose repeat sequence as they are propagated in bacteria. There are no features of the DNA sequence that can serve as flags to help guide choosing which region of a gene to target. Rather, a region must be empirically tested by the procedure itself to determine whether it is fit for cloning. Several suggestions can be given when cloning these constructs: (1) Screen as many antibiotic-resistant bacterial colonies as possible for recombinant plasmids. Sometimes 200 colonies are screened until a recombinant is found. (2) Be prepared to isolate DNA from minipreparations of plasmid grown in small volume since plasmid propagation can lead to deletion variants of the plasmid. (3) For the same reason, it is prudent to immediately store a positive clone as a frozen glycerol stock.

- ***The silencing effect is frequently variable, with only a fraction of treated animals exhibiting complete silencing.*** This partial effect is also observed at the level of target mRNA abundance where a pooled population of treated animals might exhibit at most a 90% reduction in mRNA levels. Thus, treated individuals have a spectrum of RNAi-induced phenotypes, which makes interpretation of gene function somewhat difficult. Moreover, there is frequently a variation in the strength of RNAi effects between different transformant lines carrying the same snapback transgene. This is likely due to the influence of nearby chromosomal modulation of transgene expression that depends on the point of trangene insertion. Because RNAi silencing is not complete, weak or strong snapback RNA expression translates to a corresponding weak or strong silencing effect.

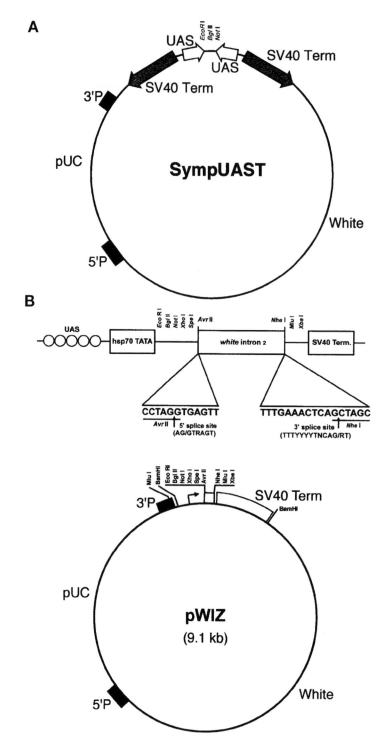

**FIGURE 17.4.** (*A*) Schematic representation of the SympUAST vector (Giordano et al. 2002). The symmetric UAS promoters drive bidirectional transcription toward the multicloning site. Distal SV40 polyadenylation signal sequences provide termination signals for the transcripts. (*B*) Schematic representation of the pWIZ vector (Lee and Carthew 2002). The pWIZ vector carries the 74-bp second intron of the *white* gene, flanked by unique *Eco*RI, *Bgl*II, *Not*I, *Xho*I, *Spe*I, and *Avr*II sites on the 5′ side, and *Nhe*I and *Xba*I sites on the 3′ side. The sequences at the junction of the 5′ and 3′ splice sites in the vector are highlighted, and arrows indicate the 5′ and 3′ splice sites. The consensus sequences for 5′ and 3′ splicing are shown in parentheses: (*slash*) The splice site; (R) purine; (Y) pyrimidine; (N) any base.

## PROTOCOL 8: SECOND-GENERATION TRANSGENIC RNAi

The problems with the first-generation vectors have inspired new approaches to produce dsRNA in vivo.

## Procedures

### Method 1: Symmetrically Transcribed Transgenes

Furia and co-workers (Giordano et al. 2002) have recently designed a construct to generate GAL4-dependent simultaneous transcription of both sense and antisense strands. According to the expression pattern of the GAL4 driver line, these transgenes are capable of repressing target gene activity. The vector, SympUAST, is a modification of pUAST in which its multicloning site is flanked on both sides by two identical but oppositely oriented regulatory regions, each composed of a UAS promoter coupled with an inversely oriented polyadenylation site (Figure 17.4A). A given DNA fragment is inserted into the multicloning site, such that the transcription of sense and antisense strands of the DNA insert can be driven by the convergent UAS promoters within *Drosophila*. The options for cloning a gene fragment within SympUAST are to use the unique sites *Eco*RI, *Bgl*II, and *Not*I. The DNA insert sizes successfully used to date have been 1.4–1.7 kb. Insufficient data are available to know yet whether there are size constraints for producing RNAi with this vector.

The obvious advantage of this system is that one does not need to clone inserts as inverted repeats, and thus can avoid associated plasmid stability problems. The strength of RNAi silencing with SympUAST is comparable to the IR method (Giordano et al. 2002). Surprisingly, it has long been known that elongating RNA polymerase II molecules interfere with transcription initiation and elongation of other RNA polymerases operating in *cis* (Proudfoot 1986; Eggermont and Proudfoot 1993; Eszterhas et al. 2002). Thus, it is interesting that two opposing polymerase II promoters are still active enough to elicit an RNAi effect. Perhaps a little dsRNA produced with this method is sufficient to achieve silencing.

### Method 2: Intron-spliced Snapback RNA

In plants, using a functional intron as a spacer between inverted repeats strongly enhanced silencing potency of snapback RNA (Smith et al. 2000). A similar observation has been recently noted in *Drosophila* when inverted repeats composed of cDNA–genomic DNA hybrids are separated by functional introns (Kalidas and Smith 2002). This motivated the construction of a modular RNAi transgene vector that produces spliced snapback RNA (Lee and Carthew 2003). The vector is derived from the pUAST transformation plasmid, which offers the advantages of the GAL4/UAS modular expression system, as outlined earlier. The vector, called pWIZ for *White Intron Zipper*, is designed so that gene fragments can be subcloned upstream and downstream from a 74-nucleotide intron of the *white* gene (Figure 17.4B). The 74-nucleotide intron of the *white* gene bears all the features of a consensus *Drosophila* intron, and it can be spliced in heterologous tissues (Mount et al. 1992; Guo et al. 1993). The intron is flanked by an *Avr*II site on the 5′ side and by an *Nhe*I site on the 3′ side. The *Avr*II and *Nhe*I sites in pWIZ conform to the consensus sequences for 5′ and 3′ splice sites, respectively. Thus, any DNA fragment inserted into the *Avr*II or *Nhe*I site is potentially competent to be treated as an exon. Moreover, the sites are unique in pWIZ, providing convenient cloning sites for gene fragments. This method is likely to be very useful for analyzing the function of the many *Drosophila* genes in a manner conditional for particular tissues and developmental times. An RNAi transformation vector that generates spliced snapback RNA has been also reported by Reichhart, Gubb, and colleagues (Reichhart et al. 2002).

The DNA fragment that is inserted into pWIZ or related vector could theoretically correspond to a complete exon from the target gene, a partial exon, or several contiguous exons. Insertion of a single complete exon from the target gene is preferred, since exons often contain sequences that facilitate the processing of transcripts. It is clear that exonic splicing enhancers (ESEs) are prevalent in many genes where they have an important role in splice site identification, particularly through recruitment of U2AF to 3´ splice sites (Cartegni et al. 2002). For this reason, placing the inverted repeats in a 3´-5´/5´-3´ orientation (coding strand) is preferable since any ESEs in the second repeat would be competent to stimulate 3´ splice site utilization. Exons that are known to be alternatively spliced should be avoided since these can contain silencing sequences that repress or restrict splicing. Finally, the exon should not have sequences in the antisense (3´-5´) orientation that match a 5´ consensus splice site. This is important to prevent splicing from cryptic sites within the first repeat if in a 3´-5´/5´-3´ orientation.

## Method 3: Construction of a pWIZ Snapback Repeat

The simplest means to insert DNA is as a PCR fragment. The system is designed so that a single PCR fragment can be inserted on each side of the intron. This is because the *Spe*I and *Avr*II sites on the 5´ side are ligation-compatible with the *Nhe*I and *Xba*I sites on the 3´ side. A *Spe*I, *Avr*II, *Nhe*I, or *Xba*I restriction site should be present at the 5´ end of the forward and reverse PCR primers. To allow efficient restriction cleavage, we add an extra four nucleotides to the 5´ side of each primer restriction site.

The region of the target gene to amplify and insert should be 500 to 700 bp in length, and it should not have internal restriction sites corresponding to the PCR primer sites.

---

**MATERIALS**

**REAGENTS**

*E. coli* SURE or JM103 strain

Oligonucleotides

pWIZ (Lee and Carthew 2003)

**EQUIPMENT**

PCR machine

---

1. Amplify the DNA fragment from genomic DNA or cDNA by PCR. Digest the PCR product with restriction enzymes that cleave sites in the forward and reverse primers.

2. Insert the DNA fragment into the *Avr*II site of pWIZ. Transform the SURE or JM103 strain of *E. coli* and select ampicillin-resistant colonies. Screen for a recombinant clone with the desired insert orientation (3´-5´) by restriction mapping or sequencing.

3. Insert the DNA fragment into the *Nhe*I site of the pWIZ recombinant made in Step 2. Transform SURE or JM103 bacteria and select ampicillin-resistant colonies. Identify recombinant clones with the insert in opposite orientation (5´-3´) to the first by mapping or sequencing.

   A feature of pWIZ is that almost all repeat constructs made thus far in our laboratory are stable in *E. coli* strains such as SURE cells.

4. Carry out transformation to generate stable transgenic lines carrying the WIZ gene.

   Upon mating transgenic animals harboring the WIZ gene with animals carrying tissue- or cell-specific GAL4 drivers, the $F_1$ progeny produce loopless snapback RNA. This induces RNAi against target genes in tissue- and cell-specific patterns.

## ACKNOWLEDGMENTS

I am thankful to several people who generously provided their time and help during the preparation of this chapter. Young Sik Lee contributed information and illustrations. Jason Kennerdell and Shinji Yamaguchi provided extensive input into the design of many of these protocols and provided unpublished results. Mike Nonet provided information and advice for construction and operation of his gene gun. I thank members of my laboratory for helpful comments on the manuscript.

## REFERENCES

Adams M.D., Celnicken S.E., Holt R.A., Evans C.A., Gocayne J.D., Amanatides P.G., et al. 2000. The genome sequence of *Drosophila melanogaster*. *Science* **287:** 2185–2195.

Alder M.N., Dames S., Gaudet J., and Mango S.E. 2003. Gene silencing in *Caenorhabditis elegans* by transitive RNA interference. *RNA* **9:** 25–32.

Ashburner M. 1989. Drosophila: *A laboratory handbook*. Cold Spring Harbor Laboratory Press, Cold Spring Harbor, New York.

Biyasheva A., Do T.V., Lu Y., Vaskova M., and Andres A.J. 2001. Glue secretion in the *Drosophila* salivary gland: A model for steroid-regulated exocytosis. *Dev. Biol.* **231:** 234–251.

Brand S. and Bourbon H.M. 1993. The developmentally-regulated *Drosophila* gene *rox8* encodes an RRM-type RNA binding protein structurally related to human TIA-1-type nucleolysins. *Nucleic Acids Res.* **21:** 3699–3704.

Caplen N.J., Fleenor J., Fire A., and Morgan R.A. 2000. dsRNA-mediated gene silencing in cultured *Drosophila* cells: A tissue culture model for the analysis of RNA interference. *Gene* **252:** 95–105.

Cartegni L., Chew S.L., and Krainer A.R. 2002. Listening to silence and understanding nonsense: Exonic mutations that affect splicing. *Nat. Rev. Genet.* **3:** 285–298.

Clemens J.C., Worby C.A., Simonson-Leff N., Muda M., Maehama T., Hemmings B.A., and Dixon J.E. 2000. Use of double-stranded RNA interference in *Drosophila* cell lines to dissect signal transduction pathways. *Proc. Natl. Acad. Sci.* **97:** 6499–6503.

Davison A. and Leach D.R. 1994. The effects of nucleotide sequence changes on DNA secondary structure formation in *Escherichia coli* are consistent with cruciform extrusion in vivo. *Genetics* **137:** 361–368.

Eggermont J. and Proudfoot N.J. 1993. Poly(A) signals and transcriptional pause sites combine to prevent interference between RNA polymerase II promoters. *EMBO J.* **12:** 2539–2548.

Elbashir S.M., Lendeckel W., and Tuschl T. 2001a. RNA interference is mediated by 21- and 22-nucleotide RNAs. *Genes Dev.* **15:** 188–200.

Elbashir S.M., Harborth J., Lendeckel W., Yalcin A., Weber K., and Tuschl T. 2001b. Duplexes of 21-nucleotide RNAs mediate RNA interference in cultured mammalian cells. *Nature* **411:** 494–498.

Eszterhas S.K., Bouhassira E.E., Martin D.I., and Fiering S. 2002. Transcriptional interference by independently regulated genes occurs in any relative arrangement of the genes and is influenced by chromosomal integration position. *Mol. Cell. Biol.* **22:** 469–479.

Fire A., Xu S., Montgomery M.K., Kostas S.A., Driver S.E., and Mello C.C. 1998. Potent and specific genetic interference by double-stranded RNA in *Caenorhabditis elegans*. *Nature* **391:** 806–811.

Fortier E. and Belote J.M. 2000. Temperature-dependent gene silencing by an expressed inverted repeat in *Drosophila*. *Genesis* **26:** 240–244.

Giordano E., Rendina R., Peluso I., and Furia M. 2002. RNAi triggered by symmetrically transcribed transgenes in *Drosophila melanogaster*. *Genetics* **160:** 637–648.

Guo M., Lo P.C., and Mount S.M. 1993. Species-specific signals for the splicing of a short *Drosophila* intron in vitro. *Mol. Cell. Biol.* **13:** 1104–1118.

Hama C., Ali Z., and Kornberg T.B. 1990. Region-specific recombination and expression are directed by portions of the *Drosophila engrailed* promoter. *Genes Dev.* **4:** 1079–1093.

Hammond S.M., Bernstein E., Beach D., and Hannon G.J. 2000. An RNA-directed nuclease mediates post-transcriptional gene silencing in *Drosophila* cells. *Nature* **404:** 293–296.

Kalidas S. and Smith D.P. 2002. Novel genomic cDNA hybrids produce effective RNA interference in adult *Drosophila*. *Neuron* **33:** 177–184.

Kennerdell J.R. and Carthew R.W. 2000. Heritable gene silencing in *Drosophila* using double-stranded RNA. *Nat. Biotechnol.* **18:** 896–898.

Kennerdell J.R., Yamaguchi S., and Carthew R.W. 2002. RNAi is activated during *Drosophila* oocyte maturation in a manner dependent upon Aubergine and Spindle-E. *Genes Dev.* **16:** 1884–1889.

Kiger A., Baum B., Armknecht S., Chang M., Jones S., Jones M., Coulson A., Sonnichsen B., Echeverri C., and Perrimon N. 2002. Functional genomic analysis of cell morphology using high-throughput RNAi screens. In *Annual* Drosophila *research conference,* pp. a328. Genetics Society of America, San Diego.

Klein R.M., Wolf E.D., Wu R., and Sanford J.C. 1992. High-velocity microprojectiles for delivering nucleic acids into living cells. 1987. *Bio/Technology* **24:** 384–386.

Lam G. and Thummel C.S. 2000. Inducible expression of double-stranded RNA directs specific genetic interference in *Drosophila. Curr. Biol.* **10:** 957–963.

Lee Y.-S. and Carthew R.W. 2003. Making a better RNAi vector for *Drosophila:* Use of intron spacers. *Methods* (in press).

Lipardi C., Wei Q., and Paterson B.M. 2001. RNAi as random degradative PCR: siRNA primers convert mRNA into dsRNAs that are degraded to generate new siRNAs. *Cell* **107:** 297–307.

Martinek S. and Young M.W. 2000. Specific genetic interference with behavioral rhythms in *Drosophila* by expression of inverted repeats. *Genetics* **156:** 1717–1725.

Misquitta L. and Paterson B.M. 1999. Targeted disruption of gene function in *Drosophila* by RNA interference (RNA-i): A role for nautilus in embryonic somatic muscle formation. *Proc. Natl. Acad. Sci.* **96:** 1451–1456.

Mount S.M., Burks C., Hertz G., Stormo G.D., White O., and Fields C. 1992. Splicing signals in *Drosophila:* Intron size, information content, and consensus sequences. *Nucleic Acids Res.* **20:** 4255–4262.

Okada C.Y. and Rechsteiner M. 1982. Introduction of macromolecules into cultured mammalian cells by osmotic lysis of pinocytic vesicles. *Cell* **29:** 33–41.

Perrimon N. 1998. New advances in *Drosophila* provide opportunities to study gene functions. *Proc. Natl. Acad. Sci.* **95:** 9716–9717.

Proudfoot N.J. 1986. Transcriptional interference and termination between duplicated alpha-globin gene constructs suggests a novel mechanism for gene regulation. *Nature* **322:** 562–565.

Reichhart J.M., Ligoxygakis P., Naitza S., Woerfel G., Imler J.L., and Gubb D. 2002. Splice-activated UAS hairpin vector gives complete RNAi knockout of single or double target transcripts in *Drosophila melanogaster. Genesis* **34:** 160–164.

Schiedlmeier B., Schmitt R., Muller W., Kirk M.M., Gruber H., Mages W., and Kirk D.L. 1994. Nuclear transformation of *Volvox carteri. Proc. Natl. Acad. Sci.* **91:** 5080–5084.

Schwarz D.S., Hutvagner G., Haley B., and Zamore P.D. 2002. Evidence that siRNAs function as guides, not primers, in the *Drosophila* and human RNAi pathways. *Mol. Cell* **10:** 537–548.

Sijen T. and Kooter J.M. 2000. Post-transcriptional gene-silencing: RNAs on the attack or on the defense? *BioEssays* **22:** 520–531.

Smith N.A., Singh S.P., Wang M.B., Stoutjesdijk P.A., Green A.G., and Waterhouse P.M. 2000. Total silencing by intron-spliced hairpin RNAs. *Nature* **407:** 319–320.

Spradling A.C. and Rubin G.M. 1982. Transposition of cloned P elements into *Drosophila* germ line chromosomes. *Science* **218:** 341–347.

Sullivan W., Ashburner M., and Hawley R.S. 2000. Drosophila *protocols.* Cold Spring Harbor Laboratory Press, Cold Spring Harbor, New York.

Tautz D. and Pfeifle C. 1989. A non-radioactive in situ hybridization method for the localization of specific RNAs in *Drosophila* embryos reveals translational control of the segmentation gene *hunchback. Chromosoma* **98:** 81–85.

Waterhouse P.M., Graham M.W., and Wang M.B. 1998. Virus resistance and gene silencing in plants can be induced by simultaneous expression of sense and antisense RNA. *Proc. Natl. Acad. Sci.* **95:** 13959–13964.

Wilm T., Demel P., Koop H.U., Schnabel H., and Schnabel R. 1999. Ballistic transformation of *Caenorhabditis elegans. Gene* **229:** 31–35.

Worby C.A., Simonson-Leff N., and Dixon J.E. 2001. RNA interference of gene expression (RNAi) in cultured *Drosophila* cells. *Sci STKE* **2001:** PL1.

# CHAPTER 18

# RNA Interference in *Trypanosoma bruce*i and Other Nonclassical Model Organisms

Elisabetta Ullu*[†] and Christian Tschudi*[‡]

*Departments of Internal Medicine, [†]Cell Biology and [‡]Epidemiology and Public Health, Yale University Medical School, New Haven, Connecticut 06520-8022

ABLATION OF GENE EXPRESSION IS CENTRAL to the understanding of biological mechanisms in all organisms. Classical genetic approaches are available for a few model systems, but are missing for many single-cell and multicellular organisms, which occupy key positions for analyzing eukaryotic evolution at the levels of both development and gene function. Genetic interference by double-stranded RNA (dsRNA) (RNA interference or RNAi), originally described in a few organisms, has now been proven to work in a remarkable variety of eukaryotes, including protists, fungi, animals, and plants. One of the most attractive aspects of RNAi is its application to functional studies in organisms not amenable to classical genetic studies. Furthermore, given the widespread occurrence of RNAi, it would be of great interest to trace the evolutionary history of RNAi for understanding its biological significance and its role in the regulation of gene expression. This chapter reviews RNAi studies in nonclassical model organisms with special emphasis on the ancient protozoan parasite *Trypanosoma brucei*.

## RNAi AS A TOOL TO UNDERSTAND EVOLUTIONARY DIVERSITY

In this first section, we summarize the application of RNAi to systems, which include highly divergent eukaryotic microorganisms, such as *Paramecium*, and classical models for developmental studies, such as *Hydra* and the spider *Cupiennius salei*.

### *Paramecium*

*Paramecium* is a free-living unicellular ciliate and is well known for nuclear dimorphism: A diploid germ-line micronucleus and a polyploid somatic macronucleus are present in each organism. The first evidence for gene silencing in *Paramecium* was reported in 1998 using microinjection of plasmids into the macronucleus (Ruiz et al. 1998). The injected DNA was maintained as extrachromosomal DNA at about 40–200 haploid genome equivalents. The observed silencing appeared to be dose-dependent, as well as homology-dependent, with the caveat that very few examples have been examined. Furthermore, transgenes containing promoter and coding sequences induced silencing, whereas coding regions including 3′-untranslated regions (3′UTRs) did not elicit silencing (Galvani and Sperling 2001). Nuclear run-on and northern blot experiments were consistent with the notion that this mode of gene silencing occurred at the posttranscriptional level (Ruiz et al. 1998; Galvani and Sperling 2001). In *Paramecium*, the link to RNAi was established by inducing silencing via feeding *Escherichia coli* expressing dsRNA from a double T7 promoter construct (Galvani and Sperling 2002), an approach pioneered in *Caenorhabditis elegans* (Timmons et al. 2001). This advance is most promising, as pointed out by the authors, because it potentially can extend the application of the RNAi technology to other organisms that feed by phagocytosis of bacteria, like *Tetrahymena*, *Acanthamoeba*, and *Entamoeba*.

## Hydra

The cnidarian *Hydra* is a freshwater polyp, and its development is being studied because it represents one of the most basal metazoan animals. To test whether RNAi can be used in these organisms, dsRNA corresponding to the head-specific gene *ks1* was introduced into whole polyps by electroporation (Lohmann et al. 1999). The analysis of mRNA levels after the polyps fully recovered (6 days) revealed specific reduction of *ks1* mRNA to about 10% of the control. Concomitantly, polyps had defects in head formation, but developmental processes in general were not affected.

## Cupiennius

Similar to the reasoning for studying *Hydra*, the spider *Cupiennius salei* serves as a model organism to understand the evolution of developmental processes in arthropods. *C. salei* belongs to the chelicerates, which occupy a basal position in the arthopode clade. Here, the Distal-less (*Dll*) gene was targeted by injecting dsRNA into the perivitelline space of spider embryos (Schoppmeier and Damen 2001). A phenotype was apparent in slightly more than half of the surviving embryos, which is distinctly lower than similar experiments in *Drosophila*, where a phenotype was obtained in up to 90% of the embryos, but quite comparable to studies in zebrafish and *Tribolium*. Furthermore, a high number of embryos showed a mosaic phenotype, in that not all appendages were affected. The authors point out that this is most likely due to the injection of dsRNA into the perivitelline space and not directly into the blastula, which unfortunately reduced the survival rate of embryos quite dramatically. Although more systematic experiments need to be performed, it appears that the success rate of RNAi varies in different insects depending on the site of injection of the dsRNA.

## Dictyostelium

In the systems described so far, RNAi was solely used as a tool to down-regulate endogenous gene expression, and no attempts have been made to explore the underlying mechanism. The road to RNAi in *Dictyostelium discoideum* was quite different, in that a search for a dsRNA activity was initiated to explain the specific degradation of highly expressed mRNA in the presence of the corresponding antisense RNA (Novotny et al. 2001). This activity, mainly present in the cytosol, processed dsRNA to approximately 23-nucleotide RNAs in vitro, fractionated as an approximately 450-kD complex, and thus is a likely candidate for the *Dictyostelium* Dicer, an RNase-III-like nuclease first described in *Drosophila* (Bernstein et al. 2001), which converts dsRNA into small 21–25-nucleotide-long fragments, known as short interfering RNAs (siRNAs) (Elbashir et al. 2001). Subsequently, it was shown that endogenous genes, as well as transgenes, can be silenced in *Dictyostelium* by expressing hairpin constructs (Martens et al. 2002). In addition, feeding cells with *E. coli* harboring plasmids containing the target sequence in-between two opposing T7 RNA polymerase promoters was also successful in eliciting RNAi (Martens et al. 2002). Database searches revealed three RNA-dependent RNA polymerase (RdRP)-related genes in *Dictyostelium* (RrpA, RrpB, and DosA), and disruption by homologous recombination of each gene showed that silencing was absent in the RrpA knockout strain, whereas both RrPB⁻ and DosA⁻ strains were comparable to wild-type strains. As expected, induction of RNAi in *Dictyostelium* resulted in the production of siRNAs of approximately 23 nucleotides. However, these siRNAs were not seen in a strain containing the RNAi construct, but lacking the target gene. In the RrpA⁻ mutant, no siRNAs were detected, although in vitro extracts from this strain generated approximately 23-nucleotide siRNAs (Martens et al. 2002).

The examples described above unquestionably illustrate that RNAi already has made an impact in organisms that are not amenable to classical genetic approaches. Given the gene-specific feature of RNAi and the ease of inducing RNAi by a number of different methods, including microinjection, electroporation, or feeding *E. coli*-expressing dsRNA, it is conceivable that we will see more and more systems taking advantage of the power of the RNAi technology.

## GENERAL FEATURES OF RNAi IN TRYPANOSOMES

Among the genetically intractable organisms are the majority of parasites, both protozoa and helminths, and it is self-evident that RNAi would be a major step forward for functional genomics and for the validation of drug targets. At the time of writing this review, African trypanosomes were the only parasitic protozoa in which the RNAi response has been shown to be functional. In the following sections, we will recapitulate our present knowledge of RNAi in *Trypanosoma brucei*, from both a practical and mechanistic perspective.

The interest in *T. brucei* as a model system stems not only from its parasitic life style, but also from its evolutionary position and its biology. Trypanosomatid protozoa represent one of the deepest branches of the eukaryotic lineage and are therefore considered descendants of "ancient" eukaryotes. Thus, understanding the mechanism of RNAi in trypanosomes might provide clues to the evolution of the RNAi mechanism and to its biological significance. Furthermore, given that in trypanosomes most gene regulation takes place at the posttranscriptional level (Clayton 2002), there is the possibility that RNAi has a central role in mRNA metabolism.

### The Early History of RNAi in *T. brucei*

The discovery of RNAi in *T. brucei* occurred in a rather serendipitous fashion and independently of the findings in *C. elegans* (Ngo et al. 1998). Briefly, while constructing an expression vector for transfection of trypanosome cells, the curious observation was made that electroporation of one of these plasmids resulted in cells with an altered morphology and multiple nuclei and kinetoplasts (mitochondrial DNA). Dissection of the sequence elements responsible for this phenotype led to the conclusion that the active molecule was a hairpin RNA homologous to the α-tubulin 5′UTR. Indeed, the phenotype could be reproduced by electroporation of dsRNA generated in vitro with T7 RNA polymerase.

These initial experiments found that dsRNA transfection specifically targeted the degradation of the corresponding mature mRNA, but had no effect on pre-mRNA. Furthermore, expression of dsRNA by plasmid transfection and the concomitant degradation of the target mRNA was short-lived and was lost after 24 hours. Similarly, electroporation of synthetic dsRNA, which was extremely efficient (nearly 100% transformants), gave an RNAi effect that also vanished after one cell generation. In a third approach to elicit RNAi, trypanosomes expressing T7 RNA polymerase were transfected with a plasmid construct expressing dsRNA from opposing T7 RNA polymerase promoters (Shi et al. 2000). Although this method had the added bonus of a transfection efficiency of almost 100%, as assayed by the RNAi response, RNAi was again transient.

## Inducible and Heritable Methodology to Trigger RNAi in *T. brucei*

Having realized that the RNAi approaches described above were transient, vectors were constructed to achieve long-lasting RNAi responses by expressing dsRNAs under the control of tetracycline-inducible promoters (Bastin et al. 2000; LaCount et al. 2000; Shi et al. 2000; Wang et al. 2000; Inoue et al. 2002). This improved methodology opened up the power of RNAi for the genetic manipulation of *T. brucei*. At present, two types of vectors are available that afford regulated expression of either a hairpin RNA or dsRNA transcribed from opposing T7 RNA polymerase promoters. Both vector types are stably integrated in the genome by homologous recombination at the nontranscribed spacer of the rDNA locus and require a recipient trypanosome cell expressing T7 RNA polymerase and the *tet* repressor (Wirtz et al. 1999). In the hairpin-type vector, dsRNA is produced from a sense and antisense fragment of the gene of interest cloned upstream and downstream from a stuffer fragment, respectively. The purpose of the stuffer fragment is to stabilize the plasmid for replication in bacteria. Expression of the dsRNA hairpin is under the control of the tetracycline-inducible promoter of the procyclin acidic repetitive protein (PARP) gene, which is repressed by the presence of two *tet* operator sequences (Wirtz et al. 1999). The PARP sequences also provide a 3′ splice acceptor for *trans*-splicing. There is a 3′UTR and a poly(A)-addition signal downstream from the antisense gene sequence. Although experimental evidence is missing, the presence of these processing signals might result in a *trans*-spliced and polyadenylated hairpin RNA, and thus facilitate exit of the RNA from the nucleus.

The double T7 vector contains two opposing T7 polymerase promoters, each of them followed by two T7 terminators arranged in tandem (LaCount et al. 2000; Wang et al. 2000). A portion of the gene of interest is cloned between the two promoters. In contrast to the hairpin-type vector, there are no pre-mRNA processing signals in the double T7 vector. The T7 promoters are also tetracycline-inducible, but repression by the *tet* repressor in the absence of tetracycline is not as tight as with the hairpin-type vector, probably because only one *tet* operator sequence was inserted immediately downstream from the T7 polymerase initiation site. As a consequence, there is production of detectable levels of dsRNA, even in the absence of tetracycline (Wang et al. 2000). Although in most cases this does not present a problem, one should keep in mind that persistent generation of low amounts of dsRNA homologous to an essential cellular gene might provoke unforeseen changes in the cell metabolism.

A direct comparison of the efficiency of the two vectors described above at triggering degradation of α-tubulin mRNA revealed similar performances (A. Djikeng et al., in prep.). However, given the ease with which dsRNA-producing plasmids can be engineered, the double-T7 promoter vector is much better suited for genome-wide analysis of gene function.

Finally, addition of either short or long dsRNA to the growth medium of insect form trypanosomes has so far not given an RNAi response (E. Ullu, unpubl.), possibly because the rate of endocytosis in these cells is relatively low as compared to that of higher eukaryotes.

## dsRNA Requirements for Eliciting RNAi

What regions of the mRNA are most accessible for targeting by dsRNA? Using the α- and β-tubulin mRNAs as a model system, by far the most efficient triggers were dsRNAs homologous to the α- and β-tubulin 5´UTRs, which are 113 and 50 nucleotides in length, respectively; dsRNAs representing the corresponding 3´UTR were the least competent in activating degradation of the cognate mRNA. Targeting the tubulin protein-coding region appeared to correlate with the size of the dsRNA: The longer the dsRNA, the more efficient degradation of the target mRNA was observed (A. Djikeng et al., in prep.).

To investigate whether the two strands of the dsRNA are equally important in RNAi, dsRNA homologous to the α-tubulin 5´UTR was synthesized containing 5-Br-uridine either in the sense or antisense strand or in both strands. This resulted in a significant reduction in the potency of RNAi, when the antisense or both strands were modified, but no effect was seen with 5-Br-uridine in the sense strand. Since the bromine atom at the 5´ position of uridine does not affect the ability of uridine to base pair, the reduced RNAi activity of 5´-Br-uridine-substituted dsRNA is likely to be due to recognition of the dsRNA by a protein component(s) of the RNAi machinery. Indeed, 5´-Br-uridine-substituted dsRNA is inefficiently processed to siRNAs (A. Djikeng et al., in prep.). Thus, as has been shown in other systems (Parrish et al. 2000; Yang et al. 2000), the structure of the antisense strand of dsRNA is more important than the sense strand for eliciting an RNAi response in trypanosomes.

## RNAi Triggers Degradation of Cytoplasmic mRNA and the Response Is Very Rapid

The observation that mature α-tubulin mRNA, but not the corresponding pre-mRNA, was targeted for degradation by electroporation of synthetic α-tubulin dsRNA suggested that RNAi takes place predominantly in the cytoplasm (Ngo et al. 1998). So far, targeted degradation of mostly nuclear RNAs, like the U-snRNAs and the spliced leader RNA, has not been achieved (C. Tschudi, unpubl.). It is possible that RNAi does not operate efficiently in the nucleus and/or the structure of pre-mRNA and small nuclear RNA (snRNA) ribonucleoprotein particles prevents access to the RNAi degradation machinery.

Kinetic experiments to assess the rate of RNAi-mediated mRNA degradation revealed that upon electroporation of α-tubulin dsRNA, about 50% of the target mRNA vanishes within the first 10 minutes; however, at later time points, the rate of mRNA disappearance decreases, and 5–10% of the tubulin mRNA remains in steady-state conditions (A. Djikeng et al., in prep.). The simplest explanation of the initial burst of mRNA degradation is that at the early time points, the rate of RNAi-mediated mRNA degradation is higher than the rate of α-tubulin mRNA synthesis, whereas at later time points, when the bulk of mRNA has already been degraded, the rate of mRNA degradation is only slightly higher than the rate of synthesis. This implies that at a first approximation, the balance between synthesis and RNAi of a given mRNA determines the extent of remaining mRNA. Although this is intuitive, the rules governing efficient targeting of mRNA in *T. brucei* are not well understood. For instance, tubulin mRNA can be efficiently targeted by dsRNAs as short as 50 nucleotides (Ngo et al. 1998), but the same approach for other less abundant mRNAs does not produce a robust RNAi response; instead, longer dsRNAs are required in these instances (Ngo et al. 1998).

In summary, it is clear that RNAi can act very rapidly, and this supports the view that trypanosomes are poised for RNAi. Along the same line of reasoning, blocking pre-mRNA *trans*-splicing (Ngo et al. 1998) or transcription (E. Ullu, unpubl.) has no effect on the extent of α-tubulin mRNA degradation triggered by electroporation of synthetic dsRNA, thus indicating that no de novo gene product must be synthesized to initiate the RNAi response.

## Functional Studies Using RNAi

RNAi has become an invaluable tool in trypanosomes for the analysis of gene function. Insights into the mechanism of flagellar ontogeny (Bastin et al. 2000), morphogenesis (Moreira-Leite et al. 2001), RNA editing (Aphasizhev et al. 2002; Drozdz et al. 2002; Huang et al. 2002), kinetoplast DNA replication (Wang and Englund 2001), and RNA capping (S. Shen et al., in prep.), just to mention a few examples, have emerged from silencing various components by RNAi. The size of the dsRNA trigger, generated by either of the two vectors described above, was in the range of 500–1000 bp. So far, the experience in many laboratories, including ours, has been that mRNA degradation always occurred; however, in about 50% of the cases, a phenotype was not observed, even when the target was an essential gene. At present, there is no clear correlation between the extent of degradation and the appearance of a phenotype. Because the mRNA of interest is never completely abolished, it thus appears that in many cases, trypanosomes can survive even when little mRNA is there. Increasing the size of the dsRNA or targeting two different regions of the mRNA has not resulted in further degradation of the mRNA.

## CHARACTERIZATION OF siRNAs

In the current model for RNAi, the first step is recognition and cleavage of dsRNA by the Dicer RNase-III-like nuclease (Bass 2000; Bernstein et al. 2001), which converts dsRNA into 21–25-nucleotide-long siRNAs (Elbashir et al. 2001). Next, siRNAs join a multicomponent nuclease complex, termed RNA-induced silencing complex (RISC) that is competent for triggering degradation of target mRNA (Bernstein et al. 2001). Whereas we know a great deal about the fate of dsRNA both in vitro and in vivo, we have little knowledge about how the RISC complex interacts with the target mRNA and about the biochemical mechanism of target mRNA destruction in vivo. In the following sections, we summarize what we have learned so far about the RNAi mechanism in *T. brucei* and discuss the role of mRNA translation in RNAi.

## Analysis of siRNAs Derived from Exogenous dsRNA

The fate of dsRNA in trypanosomes was analyzed by either electroporation of synthetic dsRNA or expression of transgenes producing hairpin RNAs (Djikeng et al. 2001). In both instances, the dsRNA was processed to small RNAs of 24–26 nucleotides, which had all the characteristics of siRNAs: Their sequence information was restricted to the dsRNA, both sense and antisense strands were present, and about 10% of the siRNAs were pelleted at $200,000g$ together with polyribosomes (see below).

Introduction of exogenous dsRNA into trypanosomes by electroporation elicits only a transient RNAi response, and thus, as expected, siRNAs derived from these dsRNAs are degraded within a few hours after electroporation (Djikeng et al. 2001). Similarly, when the expression of a hairpin RNA was halted by the removal of tetracycline from the medium, the abundance of siRNAs returned to the level found in control cells over a period of 2–3 days. At the same time, full expression of the target mRNA was restored (A. Djikeng, in prep.). Thus, it appears that in trypanosomes, siRNAs are turned over by a yet to be identified nuclease, and targeting of endogenous mRNAs by RNAi is completely reversible.

The structure of siRNAs generated from an actin transgene producing dsRNA was analyzed by cloning and sequencing 20–30-nucleotide small RNAs isolated from soluble and polyribosome-associated pools (Djikeng et al. 2001). A total of 147 and 146 actin siRNAs for the soluble and pellet fraction, respectively, were analyzed. Of these, only two represented sequences outside the dsRNA. The size of the siRNAs varied from 20 to 30 nucleotides, with 72% being between 24 and 26 nucleotides long (20% 24 nucleotides, 29% 25 nucleotides, 23% 26 nucleotides, 11% 27 nucleotides). Thus, in trypanosomes, siRNAs are a few nucleotides longer than those produced in vitro in a *Drosophila* extract, which are 20–22 nucleotides long (Elbashir et al. 2001), but similar in size to those described in plants (Hamilton and Baulcombe 1999). The difference in size between the trypanosome and *Drosophila* siRNAs might reflect differences in the cleavage specificity of the nuclease that degrades dsRNA.

The distribution of actin siRNAs relative to the sequence of the corresponding dsRNA was asymmetric: siRNAs corresponding to the 3′ end of the dsRNA were more abundant than those corresponding to the 5′ end. In particular, only a handful of siRNAs originated from the most 5′ 100 nucleotides of the dsRNA (Djikeng et al. 2001). A similar asymmetric distribution of siRNA sequences has also been observed by hybridization analysis of siRNAs derived from a transgene in a plant system (Hutvagner et al. 2000). This might suggest that degradation of dsRNA proceeds preferentially from one end of the duplex, as has been recently determined to occur in a *Drosophila* in vitro extract competent for RNAi (Elbashir et al. 2001). Finally, there was no significant difference between the actin siRNAs derived from the soluble fraction and those derived from the polyribosomes, indicating that these two populations of siRNAs are probably in flux (Djikeng et al. 2001).

## Analysis of siRNAs Derived from Endogenous Transcripts: A Role for RNAi in Silencing Retroposon Transcripts

The cloning of actin siRNAs led to an analysis of the pool of small RNAs present in *T. brucei* with the goal of identifying endogenous siRNAs that might be involved in silencing naturally occurring transcripts (Djikeng et al. 2001). By extensive sequencing of small RNAs isolated from polyribosomes, abundant sequences derived from two endogenous non-long terminal repeat (LTR)-type retroposon elements, termed INGI and SLACS, were identified. These retroposon-specific small RNAs had all the attributes of siRNAs: They represent both the sense and antisense strands of retroposon transcripts, had the same size as trypanosome actin siRNAs, and partition between the soluble and polysome fraction. Importantly, these retroposon-specific siRNAs are constitutively present in trypanosomes, regardless of whether RNAi has been activated by expressing high levels of dsRNA from a transgene. These observations suggest that in trypanosomes, RNAi is a constitutive mechanism that functions as a quality control device to down-regulate expression of deleterious gene products. Accordingly, in RNAi-deficient trypanosomes or TbAGO$^{-/-}$ cells (see below), the abundance of retroposon transcripts was increased three- to fivefold, as compared to wild-type cells (H. Shi et al., in prep.). This provided genetic evidence that one of the biological functions of RNAi in trypanosomes is to silence retroposon transcripts to prevent or reduce retroposition and its deleterious effects on genome integrity. Because siRNAs are derived from the processing of dsRNA, one question that remains is the precise structure of retroposon transcripts giving rise to dsRNA.

The extensive sequencing of polyribosome-derived small RNAs also revealed the presence of sequences derived from known mRNAs, from unannotated sequences of the *T. brucei* database, and from unsequenced regions of the genome (Djikeng et al. 2001). Although further analysis is required to determine whether these small RNA fragments are of functional significance, it is tempting to speculate that in trypanosomes, in addition to retroposon transcripts, there exist other mRNAs that are silenced via the RNAi pathway.

## A ROLE FOR mRNA TRANSLATION IN THE RNAi MECHANISM

### Association of siRNAs with Polyribosomes

Initial cell fractionation experiments in *Drosophila* S2 cells (Hammond et al. 2000; Bernstein et al. 2001), and subsequently in trypanosomes (Djikeng et al. 2001), revealed that about 20% of siRNAs are present in large complexes, sedimenting at 200,000$g$ together with polyribosomes, ribosomes, and other large ribonucleoprotein particles. This observation was further explored by fractionating trypanosome cytoplasmic extracts on sucrose density gradients (A. Djikeng et al., in prep.). By this analysis, heavy-sedimenting siRNAs precisely followed the sedimentation profile of polyribosomes. When polyribosomes were disassembled by inhibiting translation initiation, either by using the translation initiation inhibitor pactamycin or by inducing ribosome run-off, siRNAs were released from polyribosomes and were found at the top of the gradient, where small ribonucleoproteins and soluble material sediment. Furthermore, limited digestion of polyribosomes with micrococcal nuclease showed that siRNAs were associated with translating 80S particles. Thus, these studies provided supporting evidence that under steady-state conditions, there is a close interaction between siRNAs and the translation machinery.

Most siRNAs associated with polyribosomes do not appear to be hydrogen-bonded with the target mRNA, as both sense and antisense siRNAs cosedimented with polyribosomes and could be dissociated from polyribosomes by salt extraction with a moderate monovalent ion concentration of 400 mM. siRNAs derived from a transgene dsRNA, as well as from endogenous retroposon transcripts, were released with these ionic conditions as ribonucleoprotein particles of about 70 kD. This is so far the smallest siRNA-containing complex that has been identified, but whether it is only a portion of a larger salt-sensitive complex remains to be determined.

### Coupling RNAi and mRNA Translation: A Model

In thinking about the potential mechanism of interaction between the siRNA ribonucleoprotein particles and the ribosomes, it is important to take into consideration the relative number of ribosomes and siRNPs present under steady-state conditions. Using a quantitative northern blot hybridization approach, it was estimated that a few thousand siRNAs exist in a single trypanosome (A. Djikeng et al., in prep.). Considering that there are about 100,000 ribosomes/cell, the ratio between siRNAs and ribosomes is about 1:10 to 1:20. Thus, the number of siRNPs is substoichiometric relative to the number of ribosomes. How can siRNAs trigger degradation of the target mRNA? What we have learned in the past decade about the functioning of the signal recognition particle (SRP) in targeting secreted and membrane protein to the endoplasmic reticulum (ER) membrane might be used as a paradigm in thinking about how the siRNPs can target mRNAs for degradation. SRPs, like the siRNPs, are present in substoichiometric amounts relative to translating ribosomes with an estimated ratio of about 1:10 to 1:100. Evidence provided by Peter Walter's laboratory indicates that the SRP samples the nascent polypeptide chains for the presence of the appropriate signal peptide by associating with the ribosomes at specific stages during mRNA translation (Ogg and Walter 1995). Once the SRP 54-kD polypeptide binds to a signal peptide emerging from the ribosomes, the SRP causes a translational arrest or pausing. A similar mode of functioning could be envisaged for the siRNPs. It is conceivable that the protein moiety of the siRNP has affinity for some component of the ribosome and that the siRNPs are in a dynamic equilibrium between a ribosome bound and unbound form. A stable interaction could be established, once the antisense siRNAs grabs the target mRNA by hydrogen bonding with complementary sequences. This would most likely block translation and somehow signal the initiation of target mRNA degradation by a RISC-like complex. It should be pointed out that this is at present only a working model that must be analyzed both genetically and biochemically. Interestingly, however, ribosomes have a long history of being targets for mRNA quality-control devices, like, for instance, some of the factors involved in nonsense-mediated decay (Maquat 2002) or the dsRNA protein kinase PKR, which binds to dsRNA and blocks protein synthesis by phosphorylating eIF2α (Raine et al. 1998).

## Translation-dependent and Translation-independent RNAi Mechanisms?

Targeting of mRNA via the RNAi pathway can also occur in a translation-independent manner. In fact, early in vivo experiments by Montgomery and Fire showed that the maternally inherited *mex-3* mRNA is targeted by RNAi following injection of homologous dsRNA in *C. elegans* (Montgomery et al. 1998). Furthermore, in mouse oocytes, RNAi can trigger degradation of maternal mRNAs (Svoboda et al. 2000). Because maternal mRNAs are not translated until later in development, this result suggests that RNAi can function independently of mRNA translation. In addition, in vitro experiments have established that in a *Drosophila* embryo extract competent for RNAi, degradation of target mRNA is not inhibited by translation inhibitors, such as cycloheximide, anisomycin, or puromycin (Zamore et al. 2000). Thus, these observations corroborate the existence under in vitro conditions of a translation-independent RNAi pathway. However, in vitro systems, as useful as they are in dissecting the biochemistry of biological reactions, often do not fully reflect the complex interactions taking place in vivo.

Taken together, the available evidence supports the existence of translation-dependent and -independent pathways for RNAi. This would ensure that both translatable mRNAs and "aberrant" transcripts that might not be recognized by the translational apparatus are targeted for degradation by the RNAi machinery.

## GENETIC ANALYSIS OF RNAi IN TRYPANOSOMES

### The RNAi Pathway Is Not Essential for Viability

Two approaches have been exploited to identify gene products required for RNAi in trypanosomes. A "classical" tactic was to select for RNAi-resistant cells by repeated challenge with α-tubulin dsRNA (H. Shi et al., in prep.). Down-regulation of α-tubulin synthesis results in the block of cytokinesis and eventual cell death, but a small proportion of cells survive this procedure. Thus, cloned cell lines, without prior mutagenesis, were subjected to four cycles of electroporation followed by growth of the survivors, resulting in cells that were resistant to the challenge with α-tubulin dsRNA. In addition, targeting of other mRNAs by RNAi was blocked, establishing that these cells were deficient in RNAi. The selected trypanosomes displayed no major growth defect, suggesting that the RNAi pathway is dispensable for viability, as has been shown by genetic analysis in *C. elegans* (Ketting et al. 1999; Tabara et al. 1999). Analysis of several independent RNAi-deficient cell lines revealed that siRNAs from a transgene-expressing *gfp* hairpin RNA were produced in similar, if not higher, amounts than in wild-type cells. In addition, siRNAs from endogenous retroposons were unchanged in abundance. Thus, the barrier in the RNAi pathway was downstream from the production of siRNAs.

The genetic basis for the establishment of RNAi deficiency in trypanosomes by the above selection method is at present unknown. Very likely, there is an epigenetic component involved, because 10–20% of RNAi-deficient cells regain RNAi competency after about 100 generations. Perhaps RNAi deficiency is brought about by expression of an RNAi repressor or by transcriptional silencing of some of the genes involved in the RNAi pathway. In both instances, the phenotype would be reversible.

## A Gene Belonging to the Argonaute Family Is Essential for RNAi in Trypanosomes

Database mining has become a favorite shortcut over labor-intensive biochemical and genetic approaches. Although the *T. brucei* genome is not completely sequenced, computer-assisted searches have identified a gene with the signature motifs of the Argonaute family of proteins that are involved in RNAi/posttranscriptional gene silencing (PTGS) in other organisms (H. Shi et al., in prep.). The domain structure of the *T. brucei* Argonaute gene (named *TbAGO*) consists of the recognized signature motifs of members of the Argonaute family, namely, the PAZ and Piwi domains. In addition, at the very amino terminus, *TbAGO* has a domain rich in RGG repeats, a motif characteristic of RNA-binding proteins. This is different when compared to the *Drosophila* AGO-2 and the *Arabidopsis* AGO1, where the amino terminus consists of a polyglutamine domain of unknown function. The *TbAGO* gene product is essential for the RNAi response, because genetic ablation of both *TbAGO* alleles leads to complete inactivation of the RNAi response. In AGO$^{-/-}$ trypanosomes, there is a clear defect in the accumulation of siRNAs. Thus, *TbAGO* might function to stabilize siRNAs or might be part of a complex that processes dsRNA to siRNAs. This is different from what has been determined for the *C. elegans RDE-1* gene (Parrish and Fire 2001) and the *Neurospora crassa qde2* gene (Catalanotto et al. 2002), two Argonaute family members involved in RNAi and "quelling," respectively, where in the absence of the *RDE-1* or *qde2* gene product, the abundance of siRNAs does not change relative to controls. It is quite evident from these studies that although members of the Argonaute family of proteins share signature motifs, the function of individual members might be distinct depending on the presence of additional domains.

## SUMMARY

Since its discovery in 1998 (Fire et al. 1998), RNAi has been shown to be functional in a remarkable variety of organisms representing all of the eukaryotic phyla, thus demonstrating the ancient origin of the RNAi mechanism. Moreover, given the ease with which RNAi can be elicited in all organisms, dsRNA-mediated gene silencing has become an extremely valuable tool to ablate gene function in nonclassical model organisms, where classical genetic approaches are not available or poorly developed. The analysis of the RNAi mechanism in the ancient eukaryote *T. brucei* has revealed that as shown in plants and animals, RNAi is dispensable for cell viability and functions as a genome defense mechanism to down-regulate expression of retroposon transcripts. Intriguingly, in these organisms, a proportion of retroposon-derived as well as transgene-derived siRNAs are found associated with translating ribosomes, suggesting a potential role for mRNA translation in the RNAi mechanism.

## ACKNOWLEDGMENTS

Work in our laboratory was supported by National Institutes of Health grant AI28798 to E.U. C.T. is the recipient of a Burroughs Wellcome Fund New Investigator Award in Molecular Parasitology.

# REFERENCES

Aphasizhev R., Sbicego S., Peris M., Jang S.H., Aphasizheva I., Simpson A.M., Rivlin A., and Simpson L. 2002. Trypanosome mitochondrial 3′ terminal uridylyl transferase (TUTase): The key enzyme in U-insertion/deletion RNA editing. *Cell* **108:** 637–648.

Bass B.L. 2000. Double-stranded RNA as a template for gene silencing. *Cell* **101:** 235–238.

Bastin P., Ellis K., Kohl L., and Gull K. 2000. Flagellum ontogeny in trypanosomes studied via an inherited and regulated RNA interference system. *J. Cell Sci.* **113:** 3321–3328.

Bernstein E., Caudy A.A., Hammond S.M., and Hannon G.J. 2001. Role for a bidentate ribonuclease in the initiation step of RNA interference. *Nature* **409:** 363–366.

Catalanotto C., Azzalin G., Macino G., and Cogoni C. 2002. Involvement of small RNAs and role of the *qde* genes in the gene silencing pathway in *Neurospora. Genes Dev.* **16:** 790–795.

Clayton C.E. 2002. Life without transcriptional control? From fly to man and back again. *EMBO J.* **21:** 1881–1888.

Djikeng A., Shi H., Tschudi C., and Ullu E. 2001. RNA interference in *Trypanosoma brucei:* Cloning of small interfering RNAs provides evidence for retroposon-derived 24–26-nucleotide RNAs. *RNA* **7:** 1522–1530.

Drozdz M., Palazzo S.S., Salavati R., O'Rear J., Clayton C., and Stuart K. 2002. TbMP81 is required for RNA editing in *Trypanosoma brucei. EMBO J.* **21:** 1791–1799.

Elbashir S.M., Lendeckel W., and Tuschl T. 2001. RNA interference is mediated by 21- and 22-nucleotide RNAs. *Genes Dev.* **15:** 188–200.

Fire A., Xu S., Montgomery M.K., Kostas S.A., Driver S.E., and Mello C.C. 1998. Potent and specific genetic interference by double-stranded RNA in *Caenorhabditis elegans. Nature* **391:** 806–811.

Galvani A. and Sperling L. 2001. Transgene-mediated post-transcriptional gene silencing is inhibited by 3′ non-coding sequences in *Paramecium. Nucleic Acids Res.* **29:** 4387–4394.

———. 2002. RNA interference by feeding in *Paramecium. Trends Genet.* **18:** 11–12.

Hamilton A.J. and Baulcombe D.C. 1999. A species of small antisense RNA in posttranscriptional gene silencing in plants. *Science* **286:** 950–952.

Hammond S.M., Bernstein E., Beach D., and Hannon G.J. 2000. An RNA-directed nuclease mediates post-transcriptional gene silencing in *Drosophila* cells. *Nature* **404:** 293–296.

Huang C.E., O'Hearn S.F., and Sollner-Webb B. 2002. Assembly and function of the RNA editing complex in *Trypanosoma brucei* requires band III protein. *Mol. Cell. Biol.* **22:** 3194–3203.

Hutvagner G., Mlynarova L., and Nap J. P. 2000. Detailed characterization of the posttranscriptional gene-silencing-related small RNA in a GUS gene-silenced tobacco. *RNA* **6:** 1445–1454.

Inoue N., Otsu K., Ferraro D.M., and Donelson J.E. 2002. Tetracycline-regulated RNA interference in *Trypanosoma congolense. Mol. Biochem. Parasitol.* **120:** 309–313.

Ketting R.F., Haverkamp T.H., van Luenen H.G., and Plasterk R.H. 1999. Mut-7 of *C. elegans,* required for transposon silencing and RNA interference, is a homolog of Werner syndrome helicase and RNaseD. *Cell* **99:** 133–141.

LaCount D.J., Bruse S., Hill K.L., and Donelson J.E. 2000. Double-stranded RNA interference in *Trypanosoma brucei* using head-to-head promoters. *Mol. Biochem. Parasitol.* **111:** 67–76.

Lohmann J.U., Endl I., and Bosch T.C. 1999. Silencing of developmental genes in *Hydra. Dev. Biol.* **214:** 211–214.

Maquat L.E. 2002. Nonsense-mediated mRNA decay. *Curr. Biol.* **12:** R196–197.

Martens H., Novotny J., Oberstrass J., Steck T.L., Postlethwait P., and Nellen W. 2002. RNAi in *Dictyostelium:* The role of RNA-directed RNA polymerases and double-stranded RNase. *Mol. Biol. Cell* **13:** 445–453.

Montgomery M.K., Xu S., and Fire A. 1998. RNA as a target of double-stranded RNA-mediated genetic interference in *Caenorhabditis elegans. Proc. Natl. Acad. Sci.* **95:** 15502–15507.

Moreira-Leite F.F., Sherwin T., Kohl L., and Gull K. 2001. A trypanosome structure involved in transmitting cytoplasmic information during cell division. *Science* **294:** 610–612.

Ngo H., Tschudi C., Gull K., and Ullu E. 1998. Double-stranded RNA induces mRNA degradation in *Trypanosoma brucei. Proc. Natl. Acad. Sci.* **95:** 14687–14692.

Novotny J., Diegel S., Schirmacher H., Mohrle A., Hildebrandt M., Oberstrass J., and Nellen W. 2001. *Dictyostelium* double-stranded ribonuclease. *Methods Enzymol.* **342:** 193–212.

Ogg S.C. and Walter P. 1995. SRP samples nascent chains for the presence of signal sequences by interacting with ribosomes at a discrete step during translation elongation. *Cell* **81:** 1075–1084.

Parrish S. and Fire A. 2001. Distinct roles for RDE-1 and RDE-4 during RNA interference in *Caenorhabditis elegans*. *RNA* **7:** 1397–1402.

Parrish S., Fleenor J., Xu S., Mello C., and Fire A. 2000. Functional anatomy of a dsRNA trigger. Differential requirement for the two trigger strands in RNA interference. *Mol. Cell* **6:** 1077–1087.

Raine D.A., Jeffrey I.W., and Clemens M.J. 1998. Inhibition of the double-stranded RNA-dependent protein kinase PKR by mammalian ribosomes. *FEBS Lett.* **436:** 343–348.

Ruiz F., Vayssie L., Klotz C., Sperling L., and Madeddu L. 1998. Homology-dependent gene silencing in *Paramecium*. *Mol. Biol. Cell* **9:** 931–943.

Schoppmeier M. and Damen W.G. 2001. Double-stranded RNA interference in the spider *Cupiennius salei:* The role of Distal-less is evolutionarily conserved in arthropod appendage formation. *Dev. Genes Evol.* **211:** 76–82.

Shi H., Djikeng A., Mark T., Wirtz E., Tschudi C., and Ullu E. 2000. Genetic interference in *Trypanosoma brucei* by heritable and inducible double-stranded RNA. *RNA* **6:** 1069–1076.

Svoboda P., Stein P., Hayashi H., and Schultz R.M. 2000. Selective reduction of dormant maternal mRNAs in mouse oocytes by RNA interference. *Development* **127:** 4147–4156.

Tabara H., Sarkissian M., Kelly W.G., Fleenor J., Grishok A., Timmons L., Fire A., and Mello C.C. 1999. The *rde-1* gene, RNA interference, and transposon silencing in *C. elegans*. *Cell* **99:** 123–132.

Timmons L., Court D.L., and Fire A. 2001. Ingestion of bacterially expressed dsRNAs can produce specific and potent genetic interference in *Caenorhabditis elegans*. *Gene* **263:** 103–112.

Wang Z. and Englund P.T. 2001. RNA interference of a trypanosome topoisomerase II causes progressive loss of mitochondrial DNA. *EMBO J.* **20:** 4674–4683.

Wang Z., Morris J.C., Drew M.E., and Englund P.T. 2000. Inhibition of *Trypanosoma brucei* gene expression by RNA interference using an integratable vector with opposing T7 promoters. *J. Biol. Chem.* **275:** 40174–40179.

Wirtz E., Leal S., Ochatt C., and Cross G.A. 1999. A tightly regulated inducible expression system for conditional gene knock-outs and dominant-negative genetics in *Trypanosoma brucei*. *Mol. Biochem. Parasitol.* **99:** 89–101.

Yang D., Lu H., and Erickson J.W. 2000. Evidence that processed small dsRNAs may mediate sequence-specific mRNA degradation during RNAi in *Drosophila* embryos. *Curr. Biol.* **10:** 1191–1200.

Zamore P.D., Tuschl T., Sharp P.A., and Bartel D.P. 2000. RNAi: Double-stranded RNA directs the ATP-dependent cleavage of mRNA at 21 to 23 nucleotide intervals. *Cell* **101:** 25–33.

# Cautions

## GENERAL CAUTIONS

The following general cautions should always be observed.

- **Become completely familiar with the properties of substances** used before beginning the procedure.

- **The absence of a warning** does not necessarily mean that the material is safe, since information may not always be complete or available.

- **If exposed to toxic substances,** contact your local safety office immediately for instructions.

- **Use proper disposal procedures** for all chemical, biological, and radioactive waste.

- **For specific guidelines on appropriate gloves**, consult your local safety office.

- **Handle concentrated acids and bases with great care.** Wear goggles and appropriate gloves. A face shield should be worn when handling large quantities.

  **Do not mix strong acids** with organic solvents as they may react. Sulfuric acid and nitric acid especially may react highly exothermically and cause fires and explosions.

  **Do not mix strong bases** with halogenated solvent as they may form reactive carbenes which can lead to explosions.

- **Handle and store pressurized gas containers with caution** as they may contain flammable, toxic, or corrosive gases; asphyxiants; or oxidizers. For proper procedures, consult the Material Safety Data Sheet that must be provided by your vendor.

- **Never pipette solutions using mouth suction.** This method is not sterile and can be dangerous. Always use a pipette aid or bulb.

- **Keep halogenated and nonhalogenated solvents separately** (e.g., mixing chloroform and acetone can cause unexpected reactions in the presence of bases). Halogenated solvents are organic solvents such as chloroform, dichloromethane, trichlorotrifluoroethane, and dichloroethane. Some nonhalogenated solvents are pentane, heptane, ethanol, methanol, benzene, toluene, *N,N*-dimethylformamide (DMF), dimethylsulfoxide (DMSO), and acetonitrile.

- **Laser radiation, visible or invisible, can cause severe damage to the eyes and skin.** Take proper precautions to prevent exposure to direct and reflected beams. Always follow manufacturers safety guidelines and consult your local safety office. See caution below for more detailed information.

- **Flash lamps, due to their light intensity, can be harmful to the eyes.** They also may explode on occasion. Wear appropriate eye protection and follow the manufacturer's guidelines.

- **Photographic fixatives and developers also contain chemicals that can be harmful.** Handle them with care and follow manufacturer's directions.

- **Power supplies and electrophoresis equipment pose serious fire hazard** and electrical shock hazards if not used properly.

- **Microwave ovens and autoclaves in the lab require certain precautions.** Accidents have occurred involving their use (e.g., to melt agar or bacto-agar stored in bottles or to sterilize). If the screw top is not completely removed and there is not enough space for the steam to vent, the bottles can explode and cause severe injury when the containers are removed from the microwave or autoclave. Always completely remove bottle caps before microwaving or autoclaving. An alternative method for routine agarose gels that do not require sterile agar is to weigh out the agar and place the solution in a flask.

- **Ultrasonicators use high-frequency sound waves** (16–100 kHz) for cell disruption and other purposes. This "ultrasound," conducted through air, does not pose a direct hazard to humans, but the associated high volumes of audible sound can cause a variety of effects, including headache, nausea, and tinnitus. Avoid direct contact of the body with high-intensity ultrasound (not medical imaging equipment). Use appropriate ear protection and display signs on the door(s) of laboratories where the units are used.

- **Use extreme caution when handling cutting devices** such as microtome blades scalpels, razor blades, or needles. Microtome blades are extremely sharp! Use care when sectioning. If unfamiliar with their use, have someone demonstrate proper procedures. For proper disposal, use the "sharps" disposal container in your lab. Discard used needles *unshielded*, with the syringe still attached. This prevents injuries (and possible infections; see Biological Safety) while manipulating used needles since many accidents occur while trying to replace the needle shield. Injuries may also be caused by broken Pasteur pipettes, coverslips, or slides.

## GENERAL PROPERTIES OF COMMON CHEMICALS

The hazardous materials list can be summarized in the following categories:

- Inorganic acids, such as hydrochloric, sulfuric, nitric, or phosphoric, are colorless liquids with stinging vapors. Avoid spills on skin or clothing. Dilute spills with large amounts of water. The concentrated forms of these acids can destroy paper, textiles, and skin as well as cause serious injury to the eyes.

- Inorganic bases such as sodium hydroxide are white solids which dissolve in water and under heat development. Concentrated solutions will slowly dissolve skin and even fingernails.

- Salts of heavy metals are usually colored powdered solids which dissolve in water. Many of them are potent enzyme inhibitors and therefore toxic to humans and to the environment (e.g., fish and algae).

- Most organic solvents are flammable volatile liquids. Avoid breathing the vapors which can cause nausea or dizziness. Also avoid skin contact.

- Other organic compounds, including organosulphur compounds such as mercaptoethanol or organic amines, can have very unpleasant odors. Others are highly reactive and should be handled with appropriate care.

- If improperly handled, dyes and their solutions can stain not only the sample, but also skin and clothing. Some of them are also mutagenic (e.g., ethidium bromide), carcinogenic, and toxic.

- All names ending with "ase" (e.g., catalase, β-glucuronidase, or zymolase) refer to enzymes. There are also other enzymes with nonsystematic names like pepsin. Many of them are provided by manufacturers in preparations containing buffering substances, etc. Be aware of the individual properties of materials contained in these substances.

- Toxic compounds are often used to manipulate cells. They can be dangerous and should be handled appropriately.

- Be aware that several of the compounds listed have not been thoroughly studied with respect to their toxicological properties. Handle each chemical with the appropriate respect. Although the toxic effects of a compound can be quantified (e.g., $LD_{50}$ values), this is not possible for carcinogens or mutagens where one single exposure can have an effect. Also realize that dangers related to a given compound may also depend on its physical state (fine powder vs. large crystals/diethylether vs. glycerol/dry ice vs. carbon dioxide under pressure in a gas bomb). Anticipate under which circumstances during an experiment exposure is most likely to occur and how best to protect yourself and your environment.

## HAZARDOUS MATERIALS

In general, proprietary materials are not listed here. Follow the manufacturer's safety guidelines that accompany the product.

**[α-$^{32}$P]dCTP, *see* Radioactive substances**

**[α-$^{32}$P]ATP, *see* Radioactive substances**

**[α-$^{32}$P]UTP, *see* Radioactive substances**

**[$^{32}$P]UTP, *see* Radioactive substances**

**[γ-$^{32}$P]ATP, *see* Radioactive substances**

**[5,6-$^{3}$H]UTP, *see* Radioactive substances**

**Acetic acid (glacial)** is highly corrosive and must be handled with great care. Liquid and mist cause severe burns to all body tissues. It may be harmful by inhalation, ingestion, or skin absorption. Wear appropriate gloves and goggles and use in a chemical fume hood. Keep away from heat, sparks, and open flame.

**Acetone** causes eye and skin irritation and is irritating to mucous membranes and upper respiratory tract. Do not breathe the vapors. It is also extremely flammable. Wear appropriate gloves and safety glasses.

**Acrylamide** (unpolymerized) is a potent neurotoxin and is absorbed through the skin (the effects are cumulative). Avoid breathing the dust. Wear appropriate gloves and a face mask when weighing powdered acrylamide and methylene-bisacrylamide. Use in a chemical fume hood. Polyacrylamide is considered to be nontoxic, but it should be handled with care because it might contain small quantities of unpolymerized acrylamide.

**α-Amanitin** is highly toxic and may be fatal by inhalation, ingestion, or skin absorption. Symptoms may be delayed for as long as 6–24 hours. Wear appropriate gloves and safety glasses and always use in a chemical fume hood.

**Ammonium persulfate, $(NH_4)_2S_2O_8$,** is extremely destructive to tissue of the mucous membranes and upper respiratory tract, eyes, and skin. Inhalation may be fatal. Wear appropriate gloves, safety glasses, and protective clothing and use only in a chemical fume hood. Wash thoroughly after handling.

**Benzyl alcohol** is an irritant and may be harmful by inhalation, ingestion, or skin absorption. Wear appropriate gloves and safety glasses. Keep away from heat, sparks, and open flame.

**Benzyl benzoate** is an irritant and may be harmful by inhalation, ingestion, or skin absorption. Avoid contact with the eyes. Wear appropriate gloves and safety glasses.

**Bisbenzimide** may be harmful by inhalation, ingestion, or skin absorption. Wear appropriate gloves and safety glasses and use in a chemical fume hood. Do not breathe the dust.

**Bleach (Sodium hypochlorite), NaOCl,** is poisonous, can be explosive, and may react with organic solvents. It may be fatal by inhalation and is also harmful by ingestion and destructive to the skin. Wear appropriate gloves and safety glasses and use in a chemical fume hood to minimize exposure and odor.

**Boric acid, $H_3BO_3$,** may be harmful by inhalation, ingestion, or skin absorption. Wear appropriate gloves and goggles.

**Bromophenol blue** may be harmful by inhalation, ingestion, or skin absorption. Wear appropriate gloves and safety glasses and use in a chemical fume hood.

**$CaCl_2$,** *see* **Calcium chloride**

**Calcium chloride, $CaCl_2$,** is hygroscopic and may cause cardiac disturbances. It may be harmful by inhalation, ingestion, or skin absorption. Do not breathe the dust. Wear appropriate gloves and safety goggles.

**$CHCl_3$,** *see* **Chloroform**

**Chloroform, $CHCl_3$,** is irritating to the skin, eyes, mucous membranes, and respiratory tract. It is a carcinogen and may damage the liver and kidneys. It is also volatile. Avoid breathing the vapors. Wear appropriate gloves and safety glasses and always use in a chemical fume hood.

**$N,N$-Dimethylformamide (DMF), $HCON(CH_3)_2$,** is a possible carcinogen and is irritating to the eyes, skin, and mucous membranes. It can exert its toxic effects through inhalation, ingestion, or skin absorption. Chronic inhalation can cause liver and kidney damage. Wear appropriate gloves and safety glasses and use in a chemical fume hood.

**Dithiothreitol (DTT)** is a strong reducing agent that emits a foul odor. It may be harmful by inhalation, ingestion, or skin absorption. When working with the solid form or highly concentrated stocks, wear appropriate gloves and safety glasses and use in a chemical fume hood.

**DMF,** *see* **$N,N$-Dimethylformamide**

**DTT,** *see* **Dithiothreitol**

**Ethanol (EtOH), $CH_3CH_2OH$,** may be harmful by inhalation, ingestion, or skin absorption. Wear appropriate gloves and safety glasses.

**Ethidium bromide** is a powerful mutagen and is toxic. Consult the local institutional safety officer for specific handling and disposal procedures. Avoid breathing the dust. Wear appropriate gloves when working with solutions that contain this dye.

**EtOH,** *see* **Ethanol**

**Formaldehyde, HCHO,** is highly toxic and volatile. It is also a possible carcinogen. It is readily absorbed through the skin and is irritating or destructive to the skin, eyes, mucous membranes, and upper respiratory tract. Avoid breathing the vapors. Wear appropriate gloves and safety glasses and always use in a chemical fume hood. Keep away from heat, sparks, and open flame.

**Formamide** is teratogenic. The vapor is irritating to the eyes, skin, mucous membranes, and upper respiratory tract. It may be harmful by inhalation, ingestion, or skin absorption. Wear appropriate gloves and safety glasses and always use a chemical fume hood when working with concentrated solutions of formamide. Keep working solutions covered as much as possible.

### Glacial acetic acid, *see* Acetic acid (glacial)

**Glycine** may be harmful by inhalation, ingestion, or skin absorption. Wear gloves and safety glasses. Avoid breathing the dust.

**Guanidine thiocyanate** may be harmful by inhalation, ingestion, or skin absorption. Wear appropriate gloves and safety glasses.

### $H_3BO_3$, *see* Boric acid

### HCHO, *see* Formaldehyde

### HCl, *see* Hydrochloric acid

**Heptane** may be harmful by inhalation, ingestion, or skin absorption. Wear appropriate gloves and safety glasses. It is extremely flammable. Keep away from heat, sparks, and open flame.

### Hoechst No. 33342, *see* Bisbenzimide

**Hydrochloric acid, HCl,** is volatile and may be fatal if inhaled, ingested, or absorbed through the skin. It is extremely destructive to mucous membranes, upper respiratory tract, eyes, and skin. Wear appropriate gloves and safety glasses and use with great care in a chemical fume hood. Wear goggles when handling large quantities.

**Isoamyl alcohol (IAA)** may be harmful by inhalation, ingestion, or skin absorption and presents a risk of serious damage to the eyes. Wear appropriate gloves and safety goggles. Keep away from heat, sparks, and open flame.

**Isopropanol** is flammable and irritating. It may be harmful by inhalation, ingestion, or skin absorption. Wear appropriate gloves and safety glasses. Do not breathe the vapor. Keep away from heat, sparks, and open flame.

### KCl, *see* Potassium chloride

### $KH_2PO_4/K_2HPO_4/K_3PO_4$, *see* Potassium phosphate

### KOH, *see* Potassium hydroxide

**Lactic acid** is corrosive and causes severe irritation and burns to any area of contact. It may be harmful by inhalation, ingestion, or skin absorption. Wear appropriate gloves and safety goggles. Do not breathe vapor or mist.

**Magnesium acetate** may be harmful by inhalation, ingestion, or skin absorption. Wear appropriate gloves and safety glasses.

**Magnesium chloride, $MgCl_2$,** may be harmful by inhalation, ingestion, or skin absorption. Wear appropriate gloves and safety glasses and use in a chemical fume hood.

**Magnesium sulfate, $MgSO_4$,** may be harmful by inhalation, ingestion, or skin absorption. Wear appropriate gloves and safety glasses and use in a chemical fume hood.

**MeOH or H₃COH,** *see* **Methanol**

**MES,** *see* **2-(N-Morpholino)ethanesulfonic acid**

**β-Mercaptoethanol (2-Mercaptoethanol), HOCH₂CH₂SH,** may be fatal if inhaled or absorbed through the skin and is harmful if ingested. High concentrations are extremely destructive to the mucous membranes, upper respiratory tract, skin, and eyes. β-Mercaptoethanol has a very foul odor. Wear appropriate gloves and safety glasses and always use in a chemical fume hood.

**Methanol, MeOH** or **H₃COH,** is poisonous and can cause blindness. It may be harmful by inhalation, ingestion, or skin absorption. Adequate ventilation is necessary to limit exposure to vapors. Avoid inhaling these vapors. Wear appropriate gloves and goggles and use only in a chemical fume hood.

**MgCl₂,** *see* **Magnesium chloride**

**MgSO₄,** *see* **Magnesium sulfate**

**2-(N-Morpholino)ethanesulfonic acid (MES)** may be harmful by inhalation, ingestion, or skin absorption. Wear appropriate gloves and safety glasses.

**NaH₂PO₄/Na₂HPO₄/Na₃PO₄,** *see* **Sodium phosphate**

**NaH₂PO₄,** *see* **Sodium dihydrogen phosphate**

**NaN₃,** *see* **Sodium azide**

**NaOAc,** *see* **Sodium acetate**

**NaOCl,** *see* **Bleach**

**NaOH,** *see* **Sodium hydroxide**

**(NH₄)₂S₂O₈,** *see* **Ammonium persulfate**

**Nitrogen (gaseous or liquid)** may be harmful by inhalation, ingestion, or skin absorption. Wear appropriate gloves and safety glasses. Consult your local safety office for proper precautions.

**Paraformaldehyde** is highly toxic. It is readily absorbed through the skin and is extremely destructive to the skin, eyes, mucous membranes, and upper respiratory tract. Avoid breathing the dust. Wear appropriate gloves and safety glasses and use in a chemical fume hood. Paraformaldehyde is the undissolved form of formaldehyde.

**Phenol** is extremely toxic, highly corrosive, and can cause severe burns. It may be harmful by inhalation, ingestion, or skin absorption. Wear appropriate gloves, goggles, protective clothing, and always use in a chemical fume hood. Rinse any areas of skin that come in contact with phenol with a large volume of water and wash with soap and water; do not use ethanol!

**Phenol red** may be harmful by inhalation, ingestion, or skin absorption. Wear appropriate gloves and safety glasses and use in a chemical fume hood.

**Polyvinylpyrrolidone** may be harmful by inhalation, ingestion, or skin absorption. Wear appropriate gloves and safety glasses and use in a chemical fume hood.

**Ponceau S** is an irritant and may be harmful by inhalation, ingestion, or skin absorption. Wear appropriate gloves and safety glasses.

**Potassium chloride, KCl,** may be harmful by inhalation, ingestion, or skin absorption. Wear appropriate gloves and safety glasses.

**Potassium hydroxide, KOH and KOH/methanol,** is highly toxic and may be fatal if swallowed. It may be harmful by inhalation, ingestion, or skin absorption. Solutions are corrosive and can cause severe burns. It should be handled with great care. Wear appropriate gloves and safety goggles.

**Potassium phosphate, $KH_2PO_4$/$K_2HPO_4$/$K_3PO_4$,** may be harmful by inhalation, ingestion, or skin absorption. Wear appropriate gloves and safety glasses. Do not breathe the dust. *$K_2HPO_4$ • $3H_2O$ is dibasic and $KH_2PO_4$ is monobasic.*

**Proteinase K** is an irritant and may be harmful by inhalation, ingestion, or skin absorption. Wear appropriate gloves and safety glasses.

**Radioactive substances:** When planning an experiment that involves the use of radioactivity, include the physicochemical properties of the isotope (half-life, emission type and energy), the chemical form of the radioactivity, its radioactive concentration (specific activity), total amount, and its chemical concentration. Order and use only as much as really needed. Always wear appropriate gloves, lab coat, and safety goggles when handling radioactive material. X-rays and gamma rays are electromagnetic waves of very short wavelengths either generated by technical devices or emitted by radioactive materials. They may be emitted isotropically from the source or may be focused into a beam. Their potential dangers depend on the time period of exposure, the intensity experienced, and the wavelengths used. Be aware that appropriate shielding is usually of lead or other similar material. The thickness of the shielding is determined by the energy(s) of the X-rays or gamma rays. Consult the local safety office for further guidance in the appropriate use and disposal of radioactive materials. Always monitor thoroughly after using radioisotopes.

**SDS,** *see* **Sodium dodecyl sulfate**

**Sodium acetate (NaOAc),** *see* **Acetic acid**

**Sodium azide, $NaN_3$,** is highly poisonous. It blocks the cytochrome electron transport system. Solutions containing sodium azide should be clearly marked. It may be harmful by inhalation, ingestion, or skin absorption. Wear appropriate gloves and safety goggles and handle it with great care. Sodium azide is an oxidizing agent and should not be stored near flammable chemicals.

**Sodium dihydrogen phosphate, $NaH_2PO_4$, (sodium phosphate, monobasic)** may be harmful by inhalation, ingestion, or skin absorption. Wear appropriate gloves and safety glasses and use in a chemical fume hood.

**Sodium dodecyl sulfate (SDS)** is toxic, an irritant, and poses a risk of severe damage to the eyes. It may be harmful by inhalation, ingestion, or skin absorption. Wear appropriate gloves and safety goggles. Do not breathe the dust.

**Sodium phosphate, $NaH_2PO_4$/$Na_2HPO_4$/$Na_3PO_4$,** is an irritant to the eyes and skin. It may be harmful by inhalation, ingestion, or skin absorption. Wear appropriate gloves and safety goggles. Do not breathe the dust.

**Streptomycin** is toxic and a suspected carcinogen and mutagen. It may cause allergic reactions. It may be harmful by inhalation, ingestion, or skin absorption. Wear appropriate gloves and safety glasses.

**TEMED,** *see N,N,N´,N´*-**Tetramethylethylenediamine**

*N,N,N´,N´*-**Tetramethylethylenediamine (TEMED)** is highly caustic to the eyes and mucus membranes and may be harmful by inhalation, ingestion, or skin absorption. Wear appropriate gloves and tightly sealed safety goggles.

**Tris** may be harmful by inhalation, ingestion, or skin absorption. Wear appropriate gloves and safety glasses.

**Trypsin** may cause an allergic respiratory reaction. It may be harmful by inhalation, ingestion, or skin absorption. Do not breathe the dust. Wear appropriate gloves and safety goggles. Use with adequate ventilation.

**Xylene** is flammable and may be narcotic at high concentrations. It may be harmful by inhalation, ingestion, or skin absorption. Wear appropriate gloves and safety glasses and use only in a chemical fume hood. Keep away from heat, sparks, and open flame.

**Xylene cyanol,** *see* **Xylene**

# Suppliers

With the exception of those suppliers listed in the text with their addresses, all suppliers mentioned in this manual can be found in the BioSupplyNet Source Book and on the Web Site at:

http://www.biosupplynet.com

If a copy of the BioSupplyNet Source Book was not included with this manual, a free copy can be ordered by any of the following methods:

- Complete the Free Source Book Request Form found at the Web Site at:

http://www.biosupplynet.com

- E-mail a request to info@biosupplynet.com
- Fax a request to 1-919-659-2199

# Index